Chemical Engineering Methods and Technology

nova
science publishers
www.novapublishers.com

Chemical Engineering Methods and Technology

Properties and Applications of Nanocomposite Membranes
Greg C. Clements (Editor)
2023. ISBN: 979-8-89113-200-9 Softcover)
2023. ISBN: 979-8-89113-279-5 (eBook)

Triazoles and Their Uses
Karl Roland (Editor)
2023. ISBN: 979-8-88697-800-1 (Hardcover)
2023. ISBN: 979-8-88697-830-8 (eBook)

Chromophores with Nonlinear Optical Properties and Their Applications
Maria Marinescu, PhD (Editor)
2023. ISBN: 979-8-88697-525-3 (Softcover)
2023. ISBN: 979-8-88697-726-4 (eBook)

Properties and Uses of Calcium Silicate
Amanda G. Carlton (Editor)
2022. ISBN: 979-8-88697-128-6 (Hardcover)
2022. ISBN: 979-8-88697-184-2 (eBook)

Chemical Vapor Deposition (CVD): Methods and Technologies
Levi Karlsson (Editor)
2021. ISBN: 978-1-53619-949-9 (Hardcover)
2021. ISBN: 978-1-53619-990-1 (eBook)

More information about this series can be found at https://novapublishers.com/product-category/series/chemical-engineering-methods-and-technology/

Susarla V. A. R. Sastry
Shravan Kumar
and Ashish Kapoor
Editor

Sustainability in Chemical Processes through Digitalization and Green Chemistry Approaches

nova
science publishers
www.novapublishers.com

Library of Congress Cataloging-in-Publication Data

ISBN: 979-8-89113-257-3

Published by Nova Science Publishers, Inc. † New York

Contents

Preface

The sustainability of chemical processes extends beyond the physical and chemical sciences to include the biological sciences – reflected by bio and green modifiers, and other resources like alternate energy, alternate feedstock, green processes with minimal footprint and minimal environmental impact, process intensification, recycling etc. Digitalization has transformed the chemical industry rapidly across its entire value chain through predictive maintenance and process automation. There are applications of artificial intelligence in the chemical engineering field, such as modeling, optimization, process control, reaction engineering, fault detection and diagnosis. Artificial Intelligence (AI) could provide a significant impact in problem-solving, mathematical analysis and information technology by optimizing processes in chemical industries. AI has promising applications in new areas that include materials designs such as catalysts, nanostructures, pharmaceuticals, additives, polymeric composites, rubber compounds and alloys. Also, machine learning is an area of AI that has been used to integrate physical models and computational techniques for autonomous molecular design in the chemical and pharmaceutical industries, especially in the manufacturing of soft bio-inspired materials. AI can make industrial processes safer and cleaner, and help invent new molecules for specific customer needs, including the increased circularity of materials. Over the next few years it will be critical for the chemical sector to find the right applications of AI technologies that can bring benefits in terms of reducing energy consumption, the reduction of environmental pollution, the realization of a full circular economy, cost reductions and product quality improvements. Sustainability chemistry focuses on the design of products and processes that eliminate the use and generation of hazardous substances and make the global environment clean and safe. Increased drive for sustainable products and technologies, enhanced customer specifications, new performance requirements and the ground-breaking integration of mass data technologies is creating a vibrant and evolving landscape of opportunities and challenges. This book will provide an excellent opportunity to all the stake holders in the industries, consumer interest organizations, regulatory agencies, policy implementers and research organizations by providing future directions in the areas of soaps and detergents; vegetable oils; food and beverages; paints and coatings; bimolecular and bioprocesses technologies; packaging; leather products; petrochemicals and polymers; essential oils and cosmetics, and biofuels and lubricants.

Chapter 1 - Chemical engineers rely on mathematical models for design, research, and daily decision-making, often with potentially large financial and safety implications. The fields of artificial intelligence (AI), machine learning, deep learning, convolutional networks, and recurrent networks are comprised of data-driven techniques that have proven powerful in many applications where system modeling, future prediction based on past data, and optimized input for desired output are needed. These techniques are even more powerful when perfect mathematical modeling is not known or possible. In the last few years, the increasing availability of data and computational resources has led to a greater focus on machine learning-based research. The transformation of the chemical industry to renewable energy and feedstock supply requires new paradigms for the design of flexible plants, biocatalysts, and functional materials. Recent breakthroughs in artificial intelligence (AI) and machine learning provide unique opportunities to perform interdisciplinary research between AI and chemical engineering. This interdisciplinary research can open a new way of looking at chemical engineering that has never been seen before. In this paper, we discuss perspectives for future interdisciplinary research along with existing research and challenges that will enable the transformation of chemical engineering with the help of artificial intelligence (AI).

Chapter 2 - Chromium (VI) metal present in aqueous solution at low concentrations is also considered a toxic pollutant. By adsorption, the ppm to ppb chromium content can be separated from the aqueous mass. The available adsorbent is mallet flower leaf powder (MFLP), which can be used as an adsorbent to perform the adsorption operation. By batch studies, the interaction of the solid and liquid phases of known quantities led to the generation of variables correlated to the removal percentage of chromium (VI), which is the capability of the proposed biosorbent. The pH value, dosage quantity, and starting concentration obtained from each experiment run are used as condition variables that are related to the specific percentage removal calculated by the balancing formula. Using MATLAB software, an ANN model was built to train, validate, and test the data obtained from 48 experiments as input variables. After the training, the model could predict the sample data set with good accuracy.

Chapter 3 - In our work, we use the NFFA-EUROPE-SEM dataset, which has a total of 18,577 SEM images produced at CMR-IOM. SEM images are a type of image that uses the focal beam of an electron to produce a high-resolution image. This dataset comprises 10 different categories from the process industries. We are building a deep neural network model to classify the different categories of surfaces. Most of the classification models in this work are based on convolutional neural networks. Convolutional neural networks are a class of artificial neural networks that are used to visualize images for different deep learning tasks. We propose a model that uses attention networks with weighted pooling in the neural network along with squeeze excitation to achieve state-of-the-art accuracy. Additionally, to address the imbalance in the dataset, we use focal cross-entropy loss.

Chapter 4 - The leather industries are a notable sector of the economic development of a country. However, it is one of the most polluting sectors because it produces enormous amounts of chemical waste in effluent water. Chromium is one of the pollutants in effluents from tannery waste. Membrane separation technology is widely used in various sectors, such as the biomedical, water treatment, and pharmaceutical industries. Forward osmosis (FO) is a promising technology that can sustainably treat and produce high-quality water. In this process, a semipermeable membrane is used, and water passes through that membrane to draw solution from the feed due to an osmotic pressure gradient. The advantages of the FO process are the low hydraulic pressure required for separation processes and the low fouling tendency. Experiments were performed on a laboratory scale using a commercial thin FO membrane with monovalent ions Na2SO4 for the draw solution and K2Cr2O7 for the feed solution. Different parameters, such as pH and flow rates, were varied during the filtration process. The performance of the membrane was determined at various pH levels. It has been observed that membrane performance depends on pH. High flux and low fouling of the membrane were observed at different pH levels of the feed solution during the filtration process. It was also noted that water flux is lower in acidic conditions and higher in alkaline conditions. The main objective of the process was to enhance the filtration rate with low concentration polarization and a high rate of rejection (> 95%).

Chapter 5 - One of the most common industrial devices in chemical industries, pharmaceutical firms, and petrochemical refineries are heat exchangers. It can either be used in the form of a plate heat exchanger or a shell and tube heat exchanger. Mild steel, on the other hand, is used as the basic structural material for manufacturing heat exchangers. The reason behind this is its easy availability and vast range of usage. However, atmospheric corrosion, industrial fouling, and microbial attacks can degrade the material drastically. The disadvantages stated above can be effectively minimized using the technique of electroless coating. It not only improves the surface texture but also increases the material's long-term value. The present study focuses on the comparative analysis of heat transfer using electroless Ni-P and Ni-P-Cu-coated mild steel. The technique of spray cooling has been utilized for analyzing the heat transfer study. The surface morphology and phase structure analysis were carried forward from the SEM, EDX, and XRD results. The microhardness study of the sample has been analyzed and reported. The heat transfer data were collected using an in-house spray cooling setup. The experiment is executed with INTEMP software. The results using both Ni-P and Ni-P-Cu were recorded and compared in the given study. It forecasts and optimizes the sample to give the most suitable result under the spray cooling technique. Scaling up the analysis for industrial heat exchangers is a potential future scope for the current analysis.

Chapter 6 - Researchers have developed pipeline leak detection and localization technologies in response to the enormous environmental and economic damage caused by pipeline spills. These systems use various measures to guarantee the security and dependability of pipes. Recent research on pipeline leak detection systems has been compiled, and their results and conclusions are summarized and synthesized in this paper. The findings will be helpful for pipeline integrity experts who are investigating the projected risk to the sector. In addition, the paper covers the research needs and challenges ahead for effective pipeline leak detection systems (LDSs).

Chapter 7- The co-pyrolysis of Erythrina indica and Azadirachta indica was performed in a tubular batch reactor at seven different ratios. The reactions were performed at a temperature of 600°C and pressure of 1 bar under inert atmospheric conditions. The heating rates of the reactions were considered to be 10°C/min. The liquid product obtained from the co-pyrolysis reactions consisted of a fuel and non-fuel phase, with the former separated from the latter using n-hexane. The yield of the fuel phase varied in the range of 4.917-7.907 wt.%, whereas the non-fuel phase varied in the range of 25.326-29.826 wt.%. The density of the fuel phase and non-fuel phase obtained varied in the range of 0.819-1.012 g/ml and 1.005-1.013 g/ml, respectively. A maximum calorific value of 36.798 MJ/kg was attained for the fuel phase obtained from the feed with a ratio of 1:1, which also had a minimum density of 0.819 g/ml.

Chapter 8 - Now, the continued existence of all life forms, including humans, is in danger due to climate change and global warming brought on by massive anthropogenic CO2 emissions. Therefore, reducing the global carbon dioxide burden is crucial. Researchers worldwide are interested in low-cost bio-capture and CO2 sequestration through photosynthesis to produce various value-added products, including biofuels, biofertilizers, and biopolymers. Considering flue gas CO2 sequestration with microalgae as one of the viable solutions for energy and environmental sustainability, several countries have done so in light of the critical state of the environment. This review considered all factors affecting CO2 sequestration in microalgal flue gases, such as culture conditions, growth nutrients, microalgae strains, and flue gas composition. Additionally, several bioreactor design strategies for improving CO2 fixation, as well as the benefits and drawbacks of outdoor and indoor growth systems, are comprehensively discussed.

Chapter 9 - A method is proposed for obtaining the outlet gas composition of a primary reformer operating with natural gas containing higher hydrocarbons as feed stock. For a specific case, where the feed contained CH_4 = 84.7%, C_2H_6 = 8.59%, C_3H_8 = 3.39%, i-C_4H_{10} = 0.33%, n-C_4H_{10} = 0.37%, i-C_5H_{12} = 0.06%, n-C_5H_{12} = 0.05, C_6H_{14} = 0.03%, CO_2 = 2.14%, N_2 = 0.28%, H_2O = 0.05%, inlet temperature = 830°K, steam/carbon ratio =3.91, and outlet pressure =3137.28 kPa, the complete outlet gas composition was CH_4 = 16.098%, H_2 =70.63%, CO = 6.20%, CO_2 = 6.99%, and N_2 = 0.069% (dry basis). Under the operating conditions, coke formation was found to be minimal. The proposed method of solution should be of particular interest to design engineers who are confronted with the estimation of the outlet gas composition of the primary reformer at the design stage and plant personnel for performance evaluation and who do not have access to proprietary software.

Chapter 10 - Core-annular flow (CAF) is one of the liquid-liquid flow patterns observed in microchannels when used for solvent extraction. The liquid that preferentially wets the wall of the microchannel forms the annulus, while the other liquid forms the core. Thus, depending on wettability, different liquid–liquid systems may have different directions of mass transfer (either from core to annulus or annulus to core). The direction of mass transfer will also be different for extraction and stripping modes. The diameter of the microchannel also affects the rate of mass transfer in the CAF. In this study, CFD modeling of liquid-liquid mass transfer in CAF is performed to study the effect of the direction of mass transfer and channel diameter. Despite the change in the concentration profile in the core and annulus due to a change in the direction of mass transfer, mass transfer variables such as the concentration boundary layer thickness (δ), local mass transfer coefficient (KL) and Sherwood number (Sh) of the core and annulus do not change significnatly. However, on interchanging the properties of the liquid in the core and liquid in the annulus in addition to the change in the direction of mass transfer, the change in mass transfer variables is significant. A decrease in microchannel diameter increases the mass transfer coefficient for the core (KLc) and the annulus (KLa), but the Sherwood number for the core (Shc) and the annulus (Sha) remain the same. This means that the convective flux and diffusive flux increase in the same proportion when the diameter of the microchannel is increased.

Chapter 11 - This chapter presents the results of numerical simulations aimed at understanding the drop formation phenomena at nozzles or hole in a plate submerged in a quiescient continuous phase. Three different inlet geometries, i.e., a flat tip nozzle, a sharp tip nozzle and a sieve plate hole, are used in the studies. The phase-field algorithm implemented in COMSOL Multiphysics 5.3 is used to track the interface of the drop being formed. The impacts of flow velocity and dispersed phase inlet diameter on the drop formation phenomenon are predicted using the developed model.

Chapter 12 - Bamboo fiber comes into the category of natural fiber and has more advantages over synthetic fibers. Lower cost, renewability and eco-friendliness make natural fibers more suitable than synthetic fibers. When it is mixed with matrix, it forms a reinforced composite. Composites have many applications in the automotive industry, construction applications, furniture applications, etc. Fibers are water-loving, so the absorption of water is high, resulting in lower compatibility between the fiber and matrix. The surface treatment method of fiber results in an increase in the strength of composites. To understand the water absorption behavior of composites, the kinetic behavior is established, and the diffusivity is calculated.

Chapter 13 - This research paper presents the data from experiments conducted in the laboratory on CTAB and mixed surfactants. Surfactants find a wide range of applications in domestic and industrial areas. These include soaps, detergents, dispersions, paints, emulsions, foams and anti-foams. Recently, there has been an interest in biosurfactants due to the adverse effects of conventional surfactants on the environment. More recently, there has been considerable research on certain dimeric surfactants, which have two hydrophilic tails and two hydrophobic tails attached together. These are called Gemini surfactants.

Chapter 14 - A novel class of surfactants derived from maleic anhydride has incredible properties. Maleic anhydride is an acid anhydride of maleic acid that is a colorless organic compound that is mostly used as an additive in the paint and coating industries. These surfactants provide excellent solubilization, dispersion, low foaming, and synergic properties. In addition, maleic anhydride-based surfactants are less toxic, biodegradable and can be produced from green synthesis. These surfactants are produced by reacting maleic anhydride with triglycerides in the presence of acid catalysts, which impart certain enhanced characteristics to the surfactant. These properties can be further demonstrated with the help of NMR, IR and elemental analysis. Additionally, maleic anhydride-based surfactants are widely used in the manufacturing of laundry detergents, hard surface cleaners, fabric softeners, cement setters, unsaturated polyester resins, coatings, pharmaceuticals and pesticides. The present paper addresses the routes of synthesis, properties and applications of maleic anhydride-based surfactants.

Chapter 15 - With the diminishing range of fossil fuel resources, there is an urgent need for sustainable alternative resources, so researchers are focused on developing sustainable energy resources from a renewable source; developing energy from waste is more advantageous. This work focuses on using biomass waste resources to make porous carbon electrodes with efficient ion-conducting electrolytes for supercapacitor applications. The synthesis methodology involves simple steps: the Artocarpus hirsutus dried leaves were activated using phosphoric acid with a 1:1 impregnation ratio, followed by carbonization at 650°C and then pyrolysis at 850°C in a tubular furnace in a nitrogen atmosphere; the porous carbon was finally mixed in melamine and heated for nitrogen doping in this process. The polymer gel electrolyte is made of a mixture of neutral salt and polymer, i.e., KOH and cationic polyacrylamide. The electrochemical performance was analyzed using a three-electrode system; this gave a high specific capacitance value of 353.3 F g-1 at a current density of 0.5 A g-1, an energy density of 24.54 Wh kg-1 and a power density of 36 kW kg-1. Thus, a dual benefit of waste being used as feedstock for energy and waste management is possible.

Chapter 16 - Energy storage systems have been of great interest since the advent of the industrial revolution. Energy storage systems are defined by the amount of power and energy they can store. It basically stores energy and uses it later. The presence of an energy storage device in the system reduces the imbalance between power demand and production. A large variety of energy storage systems are available in the industrial sector, which includes energy discharge on both short-term and long-term scales.

Research regarding a novel molten battery has been in progress for the past few years. Research regarding state-of-the-art lithium-ion batteries has been ongoing for even longer. The replacement of the metal lithium with a more abundant metal such as Sodium has been under discussion for the past two decades. In comparison to its counterpart, the lithium-Sulphur battery, the Sodium-Sulphur battery has seen very little exposure to the world. The construction of this battery involves the use of Sodium and Sulphur as the electrodes and a solid electrolyte.

The electrical circuit model and electrochemical model are considered for modeling. The parameters analyzed using the model are the state of charge, voltage and current generated by the battery, which are then compared to its more successful counterpart, the lithium-Sulphur battery. The purpose of this research is to model a Sodium Sulphur battery that can be used to store large amounts of energy by combining multiple cells and to reduce the operating temperature of the original model.

Chapter 17 - Roxarsone (ROX) and P-arsanilic acid (PA) are persistent, toxic, bio accumulative Organo arsenic contaminants emerging from feeding additives that pose high potential risks to the environment. In the present study, a novel biodegradable NH2-functionalized nano cellulosic fibrous mat was synthesized using the waste aquatic weed Eichhornia crassipes (NH2/CA-EC) and employed to adsorb ROX and PA. The chemical and morphological structures of as-synthesized NH2/CA-EC were investigated by scanning electron microscopy (SEM) and Fourier transform infrared spectroscopy (FTIR). The results of NH2/CA-EC exhibit ample nanofiber structures and abundant functional groups, such as -NH2, C=O and -OH. Various parameters, such as adsorbent loading (0–5 g/L), solution pH (1–12), initial concentration (ROX, PA) (10–100 mg/L), and contact time (0–120 min), on the adsorption behavior of the fabricated NH2/CA-EC cellulosic fibrous mat was investigated.

High removal efficiencies of 97.62% (ROX) and 95.23% (PA) and adsorption capacities of 96.23 mg/g (ROX) and 103.25 mg/g (PA) were achieved under the optimum conditions.

Chapter 18 - Microwave radiation is a clean alternative to conventional energy and hence is used in green chemistry. Blanching is an effective pretreatment technique used to inactivate browning enzymes in vegetables. The effect of microwave blanching at 360 W, 540 W and 900 W was studied in comparison to hot water blanching at 90°C and 50°C and steam blanching on lotus rhizome slices. This vegetable crop was selected because it is an amalgamation of good taste, rich nutrient content and crisp texture. The time of all the blanching methods was optimized to minimize nutritional and quality losses. Blanching efficacy was checked by the inactivation of peroxidase and catalase enzymes. Several quality parameters, such as ascorbic acid content, hardness, residual activity of the enzymes peroxidase and polyphenol oxidase, browning index and microstructural analysis, were determined. An increase in ascorbic acid content was observed after microwave blanching. Similarly, texture and microstructure were also found to be intact in microwave-blanched lotus rhizome slices compared to steam- and hot-water-blanched samples. Higher nutritional loss was observed in the case of hot water blanching due to leaching, making it an inefficient method of blanching. Browning decreased under all three treatments equally. Microwave blanching can therefore be regarded as the best and most ecologically sustainable pretreatment method.

Chapter 19 - A proton accelerator is being developed at BARC as the front end of an accelerator-driven subcritical reactor system. Heat generated within the proton accelerator structure during its operation is removed by a low conductivity water (LCW) system with stringent control of the supply water temperature. One such typical accelerator structure in the Low Energy High Intensity Proton Accelerator (LEHIPA) is the Drift Tube Linac (DTL), which requires the LCW supply temperature to be controlled within ±0.2°C at 20°C. The LCW system for DTL consists of different components, such as a heat exchanger, control valves, holding tank, heater, pump, piping and instrumentation. In the present paper, the mathematical modeling of each of these components of the LCW system has been discussed. Most of these models are derived from first principles. For some components, test data obtained during inspection or testing at the manufacturer's facility were used to derive the model. Finally, all the mathematical models of the individual components were integrated to form a complete system model for the LCW system, and simulation was carried out to predict the system response against various possible load disturbances. The mathematical model will be quite useful in the process control design of cooling water systems with high-precision control requirements.

Chapter 20 - Biogas mainly contains methane, carbon dioxide, water and nitrogen. Methane and carbon dioxide are the sources of carbon and hydrogen generated during syngas processing that can be used to convert them to Dimethyl Ether (DME). It has low particulate emissions, sulfur-free emissions and a high cetane number (55–60). The production of DME can be a two-step process or a one-step process. One-step DME synthesis is a highly exothermic process because it carries out the water gas shift reaction, methanol formation and methanol dehydration. In the present study, a single-step process is chosen because of its inherent capacity to replace several process units, namely, the dehydration reactor, preheater and separator, by a single reactor. One-step process technology is adopted in the present study for the processing of 5 kmol/hr or 125 kg/hr of biogas from four different sources as input to the process. The process was developed with the Aspen Plus simulator. It is also envisaged to know the sensitivity of the process for different operating parameters, namely, temperature, pressure, stream flow rates, stream composition, stream enthalpies, etc., and to identify optimum operating conditions for the lowest possible equipment sizes. A sensitivity analysis is carried out to evaluate the influence of different possible changes on the process. Therefore, the model can facilitate the development of small-scale DME production units at community poultry farms, dairy farms, community biogas units and even industrial and municipal waste treatment plants.

Chapter 21 - The problem of efficient disposal of municipal solid waste has puzzled humans since the industrial revolution. Moisture-laden solid waste of approximately 60-70% (by wt.) has become one of the main hurdles in the field of waste-to-energy conversion technologies. It is necessary to eliminate excess moisture present in the solid waste using hybrid technological pathways, namely, biodrying, in which the heat generated by inherent microorganisms in the solid waste is used for drying. The performance of the biodrying process is estimated in terms of moisture removal by adjusting the air flow rate as the primary process control parameter. In the present simulation study, the aeration rate is varied from 115.2-172.8 m3/d for the developed mass balance model, which integrates 18 first-order ordinary differential equations for a single layer by applying first principles based on process kinetics. The simulation is carried out using Simulink software with the MATLAB R2022a interface. The predicted final moisture content in the waste is found to change from 28.62% to 16.99% for the selected input range of aeration rates, which is fixed based on the experimental studies. From the predicted results, the optimum value of aeration flow rate is 172.8 m3/d, which provides a final moisture content of 16.99%, since this is the minimum value required for the survival of microorganisms in municipal solid waste. The predicted output was compared with literature data and found to fit well.

Chapter 22 - Progressively, in day-to-day life, the amount of carbon dioxide emitted into the atmosphere is increasing, thereby increasing the carbon footprint. It has become a serious issue that requires our attention. Carbon dioxide capture and conversion can be a solution to this issue. This is a process where large amounts of carbon dioxide are captured from various sources using different capture techniques and later converted into useful products. This work focuses on analyzing and comparing different carbon dioxide capture techniques and finding the most effective capture technique. This work also focuses on converting the captured CO2 into methanol, analyzing the methanol production plant by making changes in the physical, thermodynamic, and kinetic properties of the process, and finding the most efficient process conditions. Sensitivity analysis was carried out on different CO2 capture techniques, such as post-combustion capture, precombustion capture, oxy-fuel combustion technique, direct air capture method, chemical looping method, and downstream to the conversion of methanol from carbon dioxide capture, to find the effect of different parameters, such as pressure, temperature, energy requirement, and other properties, on the amount of CO2 captured and methanol conversion. The most effective CO2 capture and the methanol plant were then fitted into a thermal power generation plant, which is run by coal, in an attempt to achieve zero discharge. ASPEN PLUS software was used to simulate the problem.

Chapter 23 - The world is moving toward the utilization of gaseous fuel for controlling CO2 emissions. Currently, natural gas is the primary gaseous fuel that is being used. This natural gas is transported mainly as liquefied natural gas (LNG) and compressed natural gas (CNG). LNG has a vast network across the globe for transportation but faces challenges while distributing on a small scale. At the same time, CNG has serious safety concerns, which can be observed from the accidents in the last two decades. The solidified natural gas (SNG) formed when natural gas is solidified in an ice-like crystalline structure in the presence of water (also known as natural gas hydrate) may be a better alternative for this bottleneck.

This work investigates thermodynamic and kinetic studies on SNG formation with pure water and chemical promoters. In these experiments, synthetic natural gas containing the required proportion of methane, ethane, propane, butane, carbon dioxide, and hydrogen, similar to natural gas, replicates the behavior of commercial natural gas. The modeling study was performed using fundamental and artificial neural network (ANN)-based models, which were developed using MATLAB R2022. Both models took specific input parameters, such as temperature, pressure, and time, to predict the natural gas mole consumption as an output parameter. To find an efficient combination of parameters for the scale-up of the study, an optimization study was performed, and preliminary results are reported here.

Chapter 24 - Continuous stirred tank reactors (CSTRs) are the most important element in the process industry. Investigations into continuous hydrogen production using stirred tank reactors (CSTRs) are common. Since chemical processes are either exothermic (release energy) or endothermic (need energy input), energy must either be withdrawn from or added to the reactor to maintain a steady temperature. In this paper, the control problem of an ideal CSTR system is considered. Harris Hawks optimization (HHO) is implemented for temperature control. The controller gains are optimally tuned by minimizing the integral square error as an objective function. The validity of the proposed algorithm is verified using MATLAB simulations.

Chapter 25 - Poorly water-soluble drugs have limited therapeutic effects, low bioavailability, and incomplete absorption by the gastrointestinal tract. Hence, these drugs become incapable of treating the affected organ or tissue. Many studies have been conducted to enhance solubility through nanoformulations. Particle size reduction is one of the tools of modern nanotechnology to overcome poor aqueous solubility. Recent studies have found that cavitation is a greener method for the production of nanoparticles. Cavitation is a phenomenon that involves the nucleation, growth, and collapse of bubbles in a very short time. The collapse of bubbles occurs at millions of places with the release of high temperatures and pressure, known as "hot spots." This released energy causes the formation of nanoparticles. In this review paper, an effort has been made to investigate the various reactors and the effects of varying the conditions for nanoparticle formation. This chapter also discusses the literature on the factors that contribute to the higher yield efficiency and small size of nanoparticles. There are several methods for the formation of nanoparticles, such as thermal decomposition, polyol method, microwave-assisted method, laser ablation, sputtering deposition, lithography, pulsed electrochemical etching, vapor deposition, and ol-gel. Of them, the use of the ultrasound-assisted method is considered good due to its high surface-to-volume ratio, high yield, environmental friendliness, no toxicity, etc. Therefore, the cavitation-assisted method is a step toward the development of green energy sources.

Chapter 26 - The recovery of oil in the existing reservoirs has to be increased to meet the rising energy demand, which is possible using enhanced oil recovery processes. The incorporation of nanoparticles in polymers has become trending in the development of nanotechnology for enhanced oil recovery processes. The viscosity of polymers and flooding process efficiency decrease because of polymer deformation. To prevent such deformation, nanoparticles are added to the injection fluid during the polymer flooding process. In this study, graphene oxide (GO) nanosheets were used, which are best known for their unique properties. They are first synthesized and then characterized. This study investigates the rheology, interfacial tension (IFT), and contact angle of GO sheets in the oil recovery process by enhancing the performance of the polymer hydrolyzed polyacrylamide (HPAM) and its synergistic effect with the presence of the cationic surfactant cetyltrimethylammonium bromide (CTAB). The performance of polymer flooding along with the surfactant using sand packs was investigated. The results showed significant improvements in the nanofluid viscosity, interfacial interactions, and wettability. The GO/HPAM/CTAB system exhibited excellent effects in decreasing IFT to a significant level and thereby attained an additional 10.62% of oil recovery attained using polymers in the core flooding experiment.

Chapter 27 - Spirulina platensis sp. is used for the extraction of many value-added compounds, such as proteins, lipids, and omega-3 fatty acids. After extraction, this species is still found to be rich in flavonoids, terpenoids, phenolics, and other biological compounds, which act as reducing agents for metal ions. These residues are used in this study as a potential source for nanostructures, resulting in nanoparticles with optical and magnetic properties. Synthesized nanoparticles have fluorescent properties, which are suitable for many biosensing applications. Biotiol groups are nonessential amino acids that are found at higher levels in the blood and can cause serious health issues, so they must be detected in human blood. Therefore, we have developed a simple technique for the detection of biothiol groups in blood by using a fluorescent spectrophotometer. The antimicrobial properties of the synthesized nanoparticles were also studied.

Chapter 28 - Nanotechnology is a forefront technology with wide applications in the areas of biomedical fields, optics, electronics, chemical industries, defense, etc. NPs have an extremely small size and a large S:V ratio, which is why their physical and chemical properties are altered. This makes nanotechnology the center of attention for the scientific community. There are many methods for the synthesis of NPs, but several nanoparticles possess toxicity, which leads to the development of clean, nontoxic, biocompatible, and eco-friendly methods that bring the GS of nanoparticles into the picture. GS-zed NPs are eco-friendly, nontoxic, and cost-effective. Silver NPs are one of the most essential and captivating nanomaterials among many metallic NPs, comprising unique properties (e.g., size and shape depending on optical, electrical, and magnetic properties). The properties of silver NPs can be incorporated into wide applications in biomedical fields, such as their bactericidal ability, wound and burn healing properties, disinfection properties, and cleaning properties. The characterization techniques used for Ag NPs are FTIR, UV–vis, SEM, AFM, etc. In this paper, specific applications of silver NPs in self-cleaning and as an antimicrobial agent will be focused on.

Chapter 29 - Growing populations, industrial development, and improperly managed domestic and agricultural practices contribute to water pollution and drinking shortages. Hence, efficient water purification technologies are needed. Nanocellulose-based nanomaterials contribute a key function in the purification of effluents from different sources. The nontoxic, biodegradable, biocompatible, and renewable nature of nanocellulose (NC) drives more utilisation of it. Lower hydrophilicity and poor physical and chemical stability of cellulose, which reduce its absorption capacity, have prompted the researchers to look at ways of synthesising nano-sized cellulose using plants, bacteria, and agro-waste as they upgrade the adsorption capabilities. Nanocellulose consists of hydroxyl groups that make it a potent alternative for the scavenging of different contaminants such as pharmaceuticals, heavy metals, and dyes. That is why synthesis techniques become crucial for the effective removal of pollutants. Several surface-modified nanocelluloses are utilised in water purification. In this work, we examine the potential of nanocellulose in diverse water purification technologies.

Chapter 30 - Municipal solid waste disposal is an enormous challenge for developing countries such as India due to the increase in population, rapid urbanization, and waste generation. The practice of open dumping of municipal solid waste (MSW) in India leads to major global warming issues due to the presence of methane and carbon dioxide in landfill gases. The present study assesses the potential of methane generation from MSW determined in Kanpur city by using the LandGEM 3.02 version model developed by USEPA for the duration of 2015–2020. It was calculated that the average and total methane generation for the study period were 228.89 x 104 m3/year and 1373.35 x 104 m3, respectively. The default input parameters for the values of k and Lo were modified to be more representative of the study conditions. The k and Lo values used for the study were 0.05 and 38 m3/Mg, respectively. It was further observed that the methane generation potential from MSW increased by a factor of 7.72 for 2020 compared to the base year of 2016. This suggests that enormous volumes of methane gases are emitted from the open dumpsite in Kanpur, which should be collected and harnessed for fuel and electricity production in Kanpur city.

Chapter 31 - Mechanical ventilation is a technique in which air and oxygen mixtures are pumped into the lungs of patients who have low levels of oxygen and are unable to breathe on their own. During emergencies such as the outbreak of a pandemic such as COVID-19, hospitals run short of ventilator devices to meet the growing number of patients who require ventilator support. A single ventilator device should be used to ventilate multiple patients using a flow splitter as an emergency resort. This study aims to analyze the effect of different junction geometries and split angles on the hydrodynamics of flow through a typical ventilator splitter used for two patients having the same compliance and requiring equal flow rates. Three-dimensional spherical junctions and Y-shaped flow splitters were constructed, having an inlet diameter of 21.8 mm and splitting into two exit arms of 19 mm each. For the analysis, the Y-splitter was compared with the spherical junction splitter, which has a split angle over a range of 30° to 180°. Additionally, spherical junction splitters with diameters varying from 40 to 60 mm were analysed. A transient 3D model was developed in ANSYS 2019 R1 to simulate the flow through the ventilator splitter for a range of inlet mass flow rates. A splitter with an equal outlet mass flow rate was identified and recommended.

Chapter 32 - Dye effluents are non-biodegradable wastes that are produced by a multitude of industries, including pharmaceuticals, leather, paper, textiles, etc. Typically, dyes scatter or absorb sunlight that penetrates a water system, preventing light from absorption in the water. This, in turn, significantly slows photosynthesis and has a negative impact on the growth of microorganisms. Moreover, it reduces the amount of dissolved oxygen that aquatic organisms may absorb. Acute and chronic toxicity are caused by the presence of dye in water bodies. Therefore, the treatment of dyes is an important aspect. Among various techniques, the adsorption process is one of the simplest and most effective. The advantage of this process is its operational simplicity. Treatment of dye waste-water using bio-adsorbent made from different leaves is ecological and eco-friendly. Leaf-based bio-adsorbents are prepared by the following steps: washing, drying, crushing, acid treatment, and pyrolysis. These were found to be very effective in dye removal. In this work, an overview is given for comparison of various leaf-based bio-adsorbents used for the treatment of dyes.

Chapter 33 - For a long time, a conventional rotating biological contactor (RBC) has been used as a secondary treatment for domestic and industrial wastewater using microbial activity. The prime objective of the present study was to treat synthetic wastewater containing Auramine-O dye and enhance the biological oxygen demand (BOD) by modifying conventional RBCs. To achieve this objective, conventional RBCs were modified by replacing the bank of discs with a rotating drum with a capacity of 10 L, which was divided into three sections. The middle section of the drum consisted of six stationary paddles, which were covered with scrubber material on both sides. The remaining two end sections of the drum were filled with Pall rings covered with porous media. The use of a rotating drum in RBCs enhanced the biofilm attachment surface area up to 2.29 m2 ($\approx$ 108.2% higher than conventional RBCs), which enhances the performance of RBCs. To prevent dissolved oxygen (DO) from being a limiting factor, four small aerators were placed as a supplemental aeration system in the wastewater collection tank of the modified RBC. To prevent the influence of atmospheric temperature on the biological process, a blast-proof aquarium glass heater was placed in the tank. The wastewater collection tank and the rotating drum of the modified RBCs were fabricated from lightweight high-density polyethylene (HDPE) sheets. Compared to the conventional RBCs, an enhancement of 6.7% in the BOD removal efficiency from the auramine-o-dye was accomplished using the modified RBCs.

Chapter 34 - The photocatalytic activity of a carbonized MOF-5/g-C3N4 heterostructured photocatalyst was investigated in this study. Herein, a metal–organic framework (MOF-5) was carbonized at 800 °C and utilized as a co-catalyst of graphitic carbon nitride (g-C3N4). A metal–organic framework (MOF)-templated strategy has been developed to prepare heterostructured nanocatalysts with superior photocatalytic activity. After carbonization, metal organic framework (MOF)-derived ZnO forms heterostructures with g-C3N4, significantly narrowing the band gap. All of the nanomaterials generated were effectively analyzed using Fourier transform infrared (FTIR) spectroscopy, scanning electron microscopy (SEM), and X-ray diffraction (XRD), which showed that CMOF800-g-C3N4 was better in the presence of visible light than all other synthesized composites. The carbonized MOF's proper conduction band location allows photogenerated electrons to be transferred from g-C3N4 to the carbonized MOF, subsequently increasing the charge carrier separation, which shows considerable potential in environmental remediation. This work might open up new possibilities for creating environmentally stable, highly effective photocatalysts for the treatment of wastewater.

Chapter 35 - The global water shortage is a problem that has yet to be solved. Although water covers 71% of the surface of the world, the bulk of it is not deemed suitable for drinking or other uses. Numerous methods for purifying water have been developed, including RO (reverse osmosis), ED (electrodialysis), MSF (multistage flash distillation), and others.

In addition to the methods already stated, capacitive deionization (CDI) is becoming more popular for purifying water. In a CDI cell, which has two porous electrodes (either static or flow electrodes) between which our feed travels, an open channel or porous dielectric material acts as the separator. The ions in the wastewater will travel to their corresponding oppositely charged electrode as a result of electrosorption, and they will be trapped in the electrical double layer that is created due to the ions adhering to the electrode. A CDI model was simulated using COMSOL6 software, wherein the variation in concentration with time for different values of V (electric potential) and U (flow velocity) by deploying standard carbon electrodes and carbon (95% impregnated with silver) for water desalination was studied. The salt adsorption capacity and salt adsorption rate for AC (activated carbon) and silver-doped activated carbon electrodes were calculated. COMSOL Multiphysics software interfaced with MATLAB applications was used to model a novel electrode in different rigorous conditions for CDI processes and to check its performance capacity and efficiency.

Chapter 36 - The toxic metals discharged into the environment from industrial effluents accumulate at an exponential rate. Even with an increase in PPM levels, these metals show adverse effects on human health and all naive animals that depend on water for survival. In the age of modernization, technologies have come up to effectively remove pollutants from industrial effluent, but when pollutants are at a lower concentration, they may face an economic lag in operation. The objective of this paper is to exploit the use of available Mallet flower leaf powder (MFLP) for the elimination of lead from wastewater. Experiments are performed to determine the effects of stirring time, pH of the solution, initial concentration of the adsorbate, temperature of the system, solid adsorbent dosage, and particle size. While keeping the rpm constant, the optimum pH of the solution was 5 at 100 ppm and 1.0 gm of biosorbent with a metal uptake of 8.9 mg/g and a lead metal removal of 91.1% at room temperature. The optimum concentration of the solution was 80 ppm of 1.0 g of biosorbent at room temperature, with the removal of lead metal at 87.37%. The lead metal adsorption isotherms for MFLP fit well with the Freundlich isotherm and Temkin isotherm equations. Atomic absorption spectroscopy was used for the analysis of lead. The final concentrations of the metal ions after adsorption were analyzed through an AAS analysis instrument. The surface of the MFLP was examined by scanning electron microscopy to understand the various functional groups present at the surface sites by FTIR analysis before and after adsorption.

Chapter 37 - India is one of the many countries where God is worshipped via sculptures and idols. Plaster of Paris, gypsum, clay, textiles, iron fillings, bamboo, and paints for embellishment are the most commonly used raw materials in the creation of these sculptures and idols. Broken or damaged idols should not be dumped in the open but rather submerged in water bodies, according to Hindu ceremonies and customs. This immersion includes not only damaged or shattered idols but also several festivals, such as Durga Pooja and Ganesh Chaturthi. Immersion of these plaster of Paris idols is capable of causing significant alterations in the water quality at the immersion site, including eutrophication, acidification, and heavy metal toxicity (including mercury, cadmium, arsenic, zinc, and lead). Although the government is attempting to prohibit the production of POP idols, this is impossible because idol production is a source of income for thousands of individuals.

Chapter 38 - Municipal solid waste disposal is an enormous challenge for a developing country such as India due to its growing population, rapid urbanization, and increase in waste generation. The practice of open dumping of municipal solid waste in India leads to major global warming issues due to the presence of methane and carbon dioxide in landfill gases. This chapter assesses the potential of methane generation from MSW determined in Kanpur city by using the LandGEM 3.02 version model developed by USEPA for the duration of 2015–2020. It was calculated that the average and total methane generation for the study period were 228.89 x 104 m3/year and 1373.35x104 m3, respectively. The default input parameters for the values of k and Lo were modified to be more representative of the study conditions. The k and Lo values used for the study were 0.05 and 38 m3/Mg, respectively. It was further observed that the methane generation potential from MSW increased by a factor of 7.72 for 2020 compared to the base year of 2016. This suggests that enormous volumes of methane gases are emitted from the open dumpsite in Kanpur, which should be collected and harnessed for fuel and electricity production in Kanpur city.

Chapter 1

The Transformation of Chemical Engineering with Artificial Intelligence (AI) and Machine Learning (ML): A Review of Future Opportunities and Challenges

Vivaksha Patel*
Haresh Dave
and Kaushik Nath
Department of Chemical Engineering, GCET (CVMU), Vallabh Vidyanagar, Gujarat, India

Abstract

Chemical engineers rely on mathematical models for design, research, and daily decision-making, often with potentially large financial and safety implications. The fields of artificial intelligence (AI), machine learning, deep learning, convolutional networks, and recurrent networks are comprised of data-driven techniques that have proven powerful in many applications where system modeling, future prediction based on past data, and optimized input for desired output are needed. These techniques are even more powerful when perfect mathematical modeling is not known or possible. In the last few years, the increasing availability of data and computational resources has led to a greater focus on machine learning-based research. The transformation of the chemical industry to renewable energy and feedstock supply requires new paradigms for the design of flexible plants, biocatalysts, and functional materials. Recent breakthroughs in artificial intelligence (AI) and machine learning provide unique opportunities to perform interdisciplinary research between AI and chemical engineering. This interdisciplinary research can open a new way of looking at chemical engineering that has never been seen before. In this paper, we discuss perspectives for future interdisciplinary research along with existing research and challenges that will enable the transformation of chemical engineering with the help of artificial intelligence (AI).

Keywords: artificial intelligence (AI), chemical engineering, deep learning, system modeling, machine learning

* Corresponding Author's Email: vivakshaFpatel@gcet.ac.in.

In: Sustainability in Chemical Processes through Digitalization and Green Chemistry Approaches
Editors: Susarla V. A. R. Sastry, Shravan Kumar and Ashish Kapoor
ISBN: 979-8-89113-257-3

1. Introduction

Artificial intelligence (AI) is now a trending topic in both the research and professional industrial communities to enable advanced growth and solve complex problems in almost every sector and field. The latest advancements in hardware research and data storage capacity that are essential to running complex algorithms in machine learning have enabled numerous opportunities for communities from different sectors to apply and solve complex data-driven problems. Researchers and industrial communities in the chemical engineering field have mostly relied on complex mathematical modeling to solve complex problems such as system modeling, simulation, maximizing desired output with constrained input, chemical reaction design and optimization, and chemical process monitoring and control. Artificial intelligence (AI)-based algorithms that use data-driven techniques have tremendous potential to solve different chemical engineering problems. In this paper, we have mentioned different artificial intelligence (AI)-based methods that can be applied to solve complex chemical engineering problems, along with future opportunities and challenges in this area.

2. Machine Learning Algorithms

Machine learning is part of artificial intelligence in which a machine or system is trained with already available data to learn patterns and predict future states based on past data. An artificial neural network is used to learn based on already available data, where weight and bias matrices are adjusted during training time, and the same trained network can be used at inference time for prediction with new data.

Below are common types of machine learning algorithms [1]:

(1) Supervised learning algorithms
(2) Unsupervised learning algorithms
(3) Semi-supervised learning algorithms
(4) Reinforcement Learning Algorithms

3. Supervised Learning Algorithms

In supervised learning algorithms, an artificial neural network is trained with labeled data. Labeled data contain a dataset with the desired output for each known input [2]. Labeled data can be either data collected from previous experiments or manually labeled data from past collected data. During supervised learning, a neural network learns to predict patterns out of previously labeled data or previous input–output mapping so that when new data are provided to the network, it can make predictions based on learning.

Below are categories that come under the supervised learning algorithm:

(1) Classification
(2) Regression
(3) Forecasting

Classification algorithms predict output in terms of specified categories or classes based on certain inputs. These types of algorithms are trained from data that have input and certain classes for each fixed input.

Regression algorithms predict output values, not a fixed number of classes based on previously trained data. In this type of task, labeled training data contain input and output in terms of values, and the system is trained or modeled to predict the output value for a certain problem based on previous training data.

Forecasting algorithms are used to predict the future based on past and present data to predict the trend of any complex problem where the future trend is dependent on current and past inputs. Supervised learning algorithms include the Naive Bayes algorithm, the Decision Tree algorithm, the nearest neighbor algorithm, linear and logistic regression, support vector machines, neural networks (artificial neural networks, convolutional networks, recurrent networks), etc.

4. Unsupervised Learning Algorithms

Unlabeled data are used by unsupervised machine learning algorithms for training machines or systems. Unlabeled data are data that do not contain input-to-output mapping and can be used for training. In this type of algorithm, the system tries to find patterns, groups, rules, or clusters from unlabeled data [3]. Certain widely used unsupervised learning algorithms are clustering algorithms, association rules, etc.

5. Semi-Supervised Learning Algorithms

Semi-supervised machine learning algorithms are somewhat in between supervised machine learning and unsupervised machine learning. It uses both labeled and unlabeled data for system modeling.

6. Reinforcement Learning Algorithms

Reinforcement machine learning algorithms use observations gathered from environmental interactions to make decisions about actions so that reward can be maximized and risk can be minimized [4]. In reinforcement learning, the agent continuously learns from the environment iteratively until the software agent explores all possible states. Based on reward feedback, these algorithms make software agents behave optimally within a specific context so that performance can be maximized. In these algorithms, the agent needs to select the best possible action based on its current state.

As shown in Figure 1, the following steps are performed:

(1) The agent observes the input state.
(2) Agents perform actions based on their decision-making functions.

(3) After action from the agent, the environment provides reward or reinforcement to the agent based on the performed action.
(4) Information about states and actions is stored.

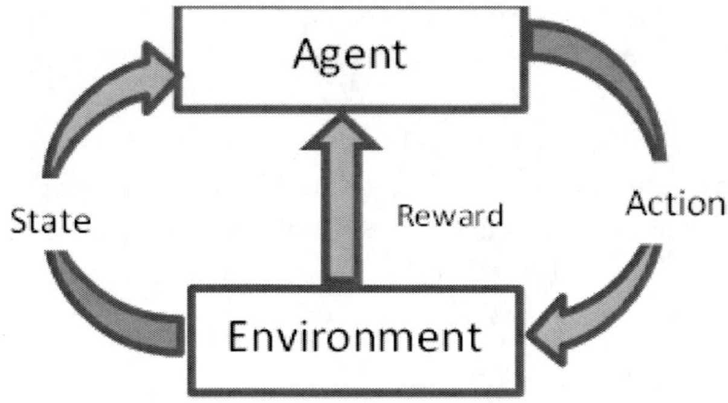

Figure 1. Reinforcement Learning Algorithm.

7. Artificial Intelligence (AI) in Chemical Engineering

Researchers and industry professionals in chemical engineering must deal with a large amount of data. They collect data from experiments and analyze it to design chemical reactions, optimize chemical reactions, develop models, control different chemical processes, understand patterns of flow, control different chemical systems and subsystems, and perform many more similar applications [5].

Computer-aided synthesis planning (CASP) focuses on accelerating the process by deciding to synthesize small-molecule compounds. It takes the structure of molecules as input, and sorted reactions will be output. The nearest neighbor model can be used to generate high-quality reaction data. An artificial neural network-based approach that is pretrained on previous experimental training data is useful in the task of forward prediction [6].

Artificial neural networks are a tool that is used to develop catalysts. Neural networks are applied to estimate catalytic activity performance, for example, lanthanide oxide catalytic activity during butane oxidation and product selectivity in the dehydrogenation of ethyl benzene on $SnO2$ catalysts [7].

Pressurized natural gas is used by gas-lift oil wells for improved production. If not enough gas is available, then the oil well becomes unstable when the slugging phenomenon occurs. Unknown internal states, unmeasurable disturbances, and other well parameters are other causes of this problem. Parameter estimation is performed with gathered data using the nonlinear model predictive controller (NMPC). Machine learning algorithms are used for the inference of sensor internal states using internal states generated during parameter estimation [8]. Random forests and neural networks are useful for understanding the flux of oil-in-water emulsions in cross-flow microfiltration [9]. Machine learning algorithms are also effective for predicting pure compound properties from the structures of molecules [10].

The erosion effects due to particle impingement are always concerning for various energy industries. In steam distribution headers, long short-term memory (LSTM) and convolutional neural networks (CNNs) are effective for the prediction of surface erosion profiles [11].

In a bidisperse solid–liquid fluidized bed (SLFB), the combination of convolutional networks (CNNs) and long-short-term memory (LSTM) is effective in predicting segregation behaviors and mixing [12]. In the application of scheduling energy systems that are closed loops, the deep reinforcement learning (DRL) method is useful [13].

Supervised machine learning algorithms are effective in developing property relations with quantitative structures [14].

Recognition of electro hydrodynamic (EHD) patterns was developed using machine learning models for guidance of process operation so that cone-jet mode stability can be achieved. Pre trained neural networks are effective in predicting spray patterns in electro hydrodynamics [15].

Proper reaction conditions are necessary to validate experiments, and they have an impact on the success or failure of transformation. An artificial neural network is effective in suggesting catalysts, regents, solvents, and temperatures that are most suitable for any reaction [16].

Flow and transport modeling of porous media is very important in chemical engineering, which includes catalytic reactors and batteries. Convolutional neural networks (CNNs) are used to predict the rate of filtration and permeability in porous media systems [17].

A deep learning-based approach is useful for optimal battery management and frequency regulation. It can be achieved using several supervised learning methods and reinforcement learning [18].

Electrochemical reduction of CO2 is currently important due to climate change awareness and the high demand for clean energy supply. The CO2 electrocatalytic reactor is still in its infancy because of the complexity of unknown electron transfer pathways and reaction mechanisms. However, an artificial neural network-based approach is useful to predict electrochemical reactor output for given inputs, and this network is trained with training data obtained from sensors [19].

Machine learning is used in the prediction of the diffusion coefficient of supercritical water (SCW) mixtures [20].

8. Challenges

The greatest challenge in artificial intelligence (AI) and machine learning is the availability of quality data because the accuracy and precision of any machine learning model depend on how accurate the collected data is and how accurately it has been labeled along with the source of the data [21].

Conclusion

As discussed in this paper, chemical engineering deals with a lot of data in different fields and subfields. It has been proven that artificial intelligence (AI), artificial neural networks, deep neural networks, convolutional networks, reinforcement learning algorithms, machine learning, etc., have enormous potential to solve complex chemical engineering problems.

References

[1] Sarker, I. H. (2021). Machine learning: Algorithms, real-world applications and research directions. *SN computer science*, 2(3), 160.

[2] Singh, A., Thakur, N., & Sharma, A. (2016, March). A review of supervised machine learning algorithms. In *2016 3rd International Conference on Computing for Sustainable Global Development* (INDIACom) (pp. 1310-1315). Ieee.

[3] Alloghani, M., Al-Jumeily, D., Mustafina, J., Hussain, A., & Aljaaf, A. J. (2020). A systematic review on supervised and unsupervised machine learning algorithms for data science. *Supervised and unsupervised learning for data science,* 3-21.

[4] Wiering, M. A., & Van Otterlo, M. (2012). Reinforcement learning. *Adaptation, learning, and optimization*, 12(3), 729.

[5] Dobbelaere, M. R., Plehiers, P. P., Van de Vijver, R., Stevens, C. V., & Van Geem, K. M. (2021). Machine learning in chemical engineering: strengths, weaknesses, opportunities, and threats. *Engineering*, 7(9), 1201-1211.

[6] Coley, C. W., Green, W. H., & Jensen, K. F. (2018). Machine learning in computer-aided synthesis planning. Accounts of chemical research, 51(5), 1281-1289.

[7] Hattori, T., & Kito, S. (1995). Neural network as a tool for catalyst development. *Catalysis Today*, 23(4), 347-355.

[8] Rojas Soares, F. D., Secchi, A. R., & Bezerra de Souza Jr, M. (2022). Development of a Nonlinear Model Predictive Control for Stabilization of a Gas-Lift Oil Well. *Industrial & Engineering Chemistry Research*, 61(24), 8411-8421.

[9] Tanudjaja, H. J., & Chew, J. W. (2022). Application of machine learning-based models to understand and predict critical flux of oil-in-water emulsion in crossflow microfiltration. *Industrial & Engineering Chemistry Research*, 61(24), 8470-8477.

[10] Dobbelaere, M. R., Ureel, Y., Vermeire, F. H., Tomme, L., Stevens, C. V., & Van Geem, K. M. (2022). Machine learning for physicochemical property prediction of complex hydrocarbon mixtures. *Industrial & Engineering Chemistry Research*, 61(24), 8581-8594.

[11] Yang, S. D., Ali, Z. A., Kwon, H., & Wong, B. M. (2022). Predicting Complex Erosion Profiles in Steam Distribution Headers with Convolutional and Recurrent Neural Networks. *Industrial & Engineering Chemistry Research*, 61(24), 8520-8529.

[12] Xie, Z., Gu, X., & Shen, Y. (2022). A machine learning study of predicting mixing and segregation behaviors in a bidisperse solid–liquid fluidized bed. *Industrial & Engineering Chemistry Research*, 61(24), 8551-8565.

[13] Campos, G., El-Farra, N. H., & Palazoglu, A. (2022). Soft Actor-Critic Deep Reinforcement Learning with Hybrid Mixed-Integer Actions for Demand Responsive Scheduling of Energy Systems. *Industrial & Engineering Chemistry Research*, 61(24), 8443-8461.

[14] Shi, Y., Wang, J., Wang, Q., Jia, Q., Yan, F., Luo, Z. H., & Zhou, Y. N. (2022). Supervised machine learning algorithms for predicting rate constants of ozone reaction with micropollutants. *Industrial & Engineering Chemistry Research*, 61(24), 8359-8367.

[15] Wang, J., Dong, T., Cheng, Y., & Yan, W. C. (2022). Machine learning assisted spraying pattern recognition for electrohydrodynamic atomization system. *Industrial & Engineering Chemistry Research*, 61(24), 8495-8503.

[16] Gao, H., Struble, T. J., Coley, C. W., Wang, Y., Green, W. H., & Jensen, K. F. (2018). Using machine learning to predict suitable conditions for organic reactions. ACS central science, 4(11), 1465-1476.

[17] Marcato, A., Boccardo, G., & Marchisio, D. (2022). From computational fluid dynamics to structure interpretation via neural networks: an application to flow and transport in porous media. *Industrial & Engineering Chemistry Research*, 61(24), 8530-8541.

[18] Li, Y., Wang, Y., Chen, Y., Lu, Y., Hua, K., Ren, J., & Cao, Y. (2022). Deep Learning Based Predictive Control of Battery Management for Frequency Regulation. *Industrial & Engineering Chemistry Research*, 61(24), 8432-8442.

[19] Luo, J., Canuso, V., Jang, J. B., Wu, Z., Morales-Guio, C. G., & Christofides, P. D. (2022). Machine learning-based operational modeling of an electrochemical reactor: Handling data variability and improving empirical models. *Industrial & Engineering Chemistry Research*, 61(24), 8399-8410.

[20] Zhao, X., Luo, T., & Jin, H. (2022). Predicting diffusion coefficients of binary and ternary supercritical water mixtures via machine and transfer learning with deep neural network. *Industrial & Engineering Chemistry Research*, 61(24), 8542-8550.

[21] Venkatasubramanian, V. (2019). The promise of artificial intelligence in chemical engineering: Is it here, finally? *AIChE Journal*, 65(2), 466-478.

Chapter 2

Modeling the Biosorption of Heavy Metal Ions on MFLP as a Biosorbent Using an Artificial Neural Network Approach

D. Venkata Padma[1,*]
S. V. A. R. Sastry[2]
S. Kishore Kumar[3]
R. Hema Latha[1]
and Ch. Vinodh Kumar[4]

[1] Faculty, Department of Chemical Engineering, MVGRCE(A), Vizianagaram, A. P., India
[2]Faculty, School of Chemical Technology, HBTU, Kanpur, U. P., India
[3]PG student, Department of Chemical Engineering, AUCE(A), Vizag, A. P., India
[4]Faculty, Government Institute of Chemical Engineering, Vizag, A. P., India

Abstract

Chromium (VI) metal present in aqueous solution at low concentrations is also considered a toxic pollutant. By adsorption, the ppm to ppb chromium content can be separated from the aqueous mass. The available adsorbent is mallet flower leaf powder (MFLP), which can be used as an adsorbent to perform the adsorption operation. By batch studies, the interaction of the solid and liquid phases of known quantities led to the generation of variables correlated to the removal percentage of chromium (VI), which is the capability of the proposed biosorbent. The pH value, dosage quantity, and starting concentration obtained from each experiment run are used as condition variables that are related to the specific percentage removal calculated by the balancing formula. Using MATLAB software, an ANN model was built to train, validate, and test the data obtained from 48 experiments as input variables. After the training, the model could predict the sample data set with good accuracy.

Keywords: MFLP, biosorption, heavy metal ions, artificial neural networks (ANNs)

---

[*] Corresponding Author's Email: padmakishore9@gmail.com.

In: Sustainability in Chemical Processes through Digitalization and Green Chemistry Approaches
Editors: Susarla V. A. R. Sastry, Shravan Kumar and Ashish Kapoor
ISBN: 979-8-89113-257-3

1. Introduction

Chromium (VI) causes physiological damage to all living organisms [1–3]. It leads to the growth of cancer in the human body [4, 5]. However, pollution-control technologies were developed to treat industrial effluents. PPM to PPB levels of chromium (VI) are mostly left out. Eventually, with time, the accumulation of metal ions in the environment increases. Thus, economically sustainable operations are needed to purify treated wastewater. Available technologies that perform separation and purification operations are ion exchange, evaporation, membrane separation, electrochemical precipitation, solvent extraction, and foam separation. These technologies are difficult to implement due to their demand for power and machinery operation expenses. If the complexity of the operation increases, then there is the possibility of dropping out of the process.

Industrial zones that included all levels of production capacity of metals, refineries, and treatment units were built with wastewater treatment plants where contaminants disperse into the groundwater and soil. Modern industrialization is considered a significant source of contamination of the environment [6–8] with heavy metals. At the cellular level, the organism tends to accumulate this metal and sluggishly develop toxic compounds that become lethal to life sustainability, even at relatively minor levels of exposure.

In this view, natural biosorbents are the ones that could take up the challenge; studies of adsorption with metal are needed and should be optimized for real-life applications [9-11]. The literature is building up with biosorbent and metal ion adsorption studies. They are abundant in gathering much data on the desorption and regeneration of adsorbents [12].

The data obtained from these studies cannot be easily programmed to perform simulations of adsorption operations to obtain optimal conditions. However, the neural network method is dependent on the training of data, which has advantages over programming to simulate the mathematical model [13, 14]. Each variable obtained from the research studies on the specific operation becomes a weight, and to it, a bias is added. Not only slow learning methods such as the Sig modal function but also Re-LU are used to train the model to recognize the pattern of the adsorption operation under particular conditions. By testing the data, the network is made to generate error until it minimizes to considerable levels that find the appropriate weights and bias for the precise condition of the adsorption process.

2. Materials and Methods

2.1. Adsorbent Preparation and Characterization

Mallet leaves were collected from the MVGR College and used as absorbents. The leaves are washed and reduced to a size of approximately 74 microns to 149 [15–17] microns for future study. On the other hand, chromium stock solutions are prepared at different concentrations using potassium dichromate as a source for hexavalent chromium. Morphological study attempts on the surface of the absorbent were performed by FTIR.

2.2. Experimental Procedure

Laboratory batch adsorption experiments were performed to determine the effect of the initial concentration of metal ions, pH of the solution, size of the adsorbent dosage, temperature of the solution, and running time on the removal of chromium metal ions at different sets. The final concentration was analyzed by the [18–20] AAS analysis instrument. Finally, we obtained the results from various parameters by changing one parameter while keeping others constant. The optimized values were noted for the future scope. For chromium, the optimum values obtained are as follows: pH value at 2, 80 PPM of initial concentration of the metal ion, one gram of dosage, and 74 micron particle size at 30 degrees [21, 22] Celsius. For these conditions, a percentage removal of 71% is achieved. Various studies on kinetics, equilibrium, and thermodynamics calculated and plotted the outcomes. The experimental results were used to predict the test data using the ANN tool.

3. Results and Discussion

The present study was aimed at an ANN-based model to describe the adsorption of chromium. The experimental data from a real situation were used by ANN to determine and optimize the values. The actual data set, containing 48 experimental data points, was divided into 3 subgroups: training, validation, and [23] test sets. Finally, these tests were used to estimate the performance of adsorption studies. The value of the regression coefficient gives the exact proof for artificial neural network model fit, and minimizing the MSE (mean square error) values gives a satisfactory approach.

Optimization of adsorption studies using artificial neural networks by introducing mathematical functions that include both linear and nonlinear solvers. This approach is helpful for us to solve the requirement of varieties, which leads to a higher number of experiments and the usage of chemicals in the real world. In the ANN language, there are three main components: input, output, and a hidden layer. The present studies optimize the parameters and adsorption phenomena. For this, MATLAB software was used to predict the data using the ANN toolbox.

3.1. Characterization of the MFLP Surface

Figures 1 and 2 of SEM express the various functional groups present on the surface before and after adsorption. SEM images reveal that the morphological studies are on the surface of the MFLP.

The adsorption of metal ions using an adsorbent (MFLP) is achieved. The effects of the parameters studied in the present investigation are adsorbent size, adsorbent dosage, initial metal concentration, pH of the solution, and solution temperature.

Figure 3 shows the effect of adsorbent dosage on the percentage removal of Cr(VI) and metal uptake. Figure 4 shows the effect of the initial batch concentration in ppm on the percentage recovery of Cr(VI). Figure 5 shows the effect of the size of the biosorbent on the percentage removal of Cr(VI). Figure 6 shows the effect of pH on the percentage recovery of Cr(VI).

Figure 1. SEM image of the MFLP adsorbent before chromium adsorption.

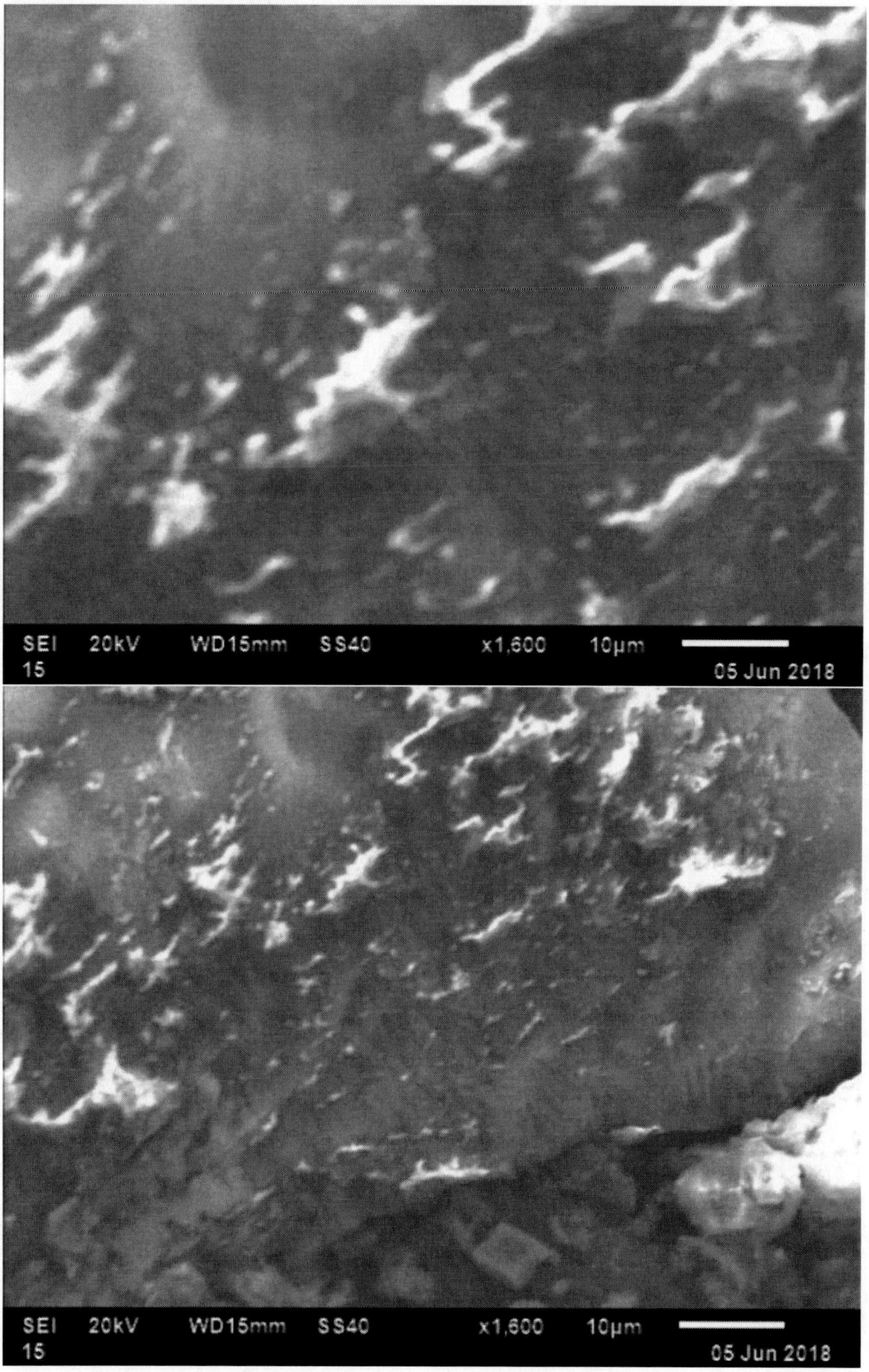

Figure 2. SEM image of the MFLP adsorbent after chromium adsorption.

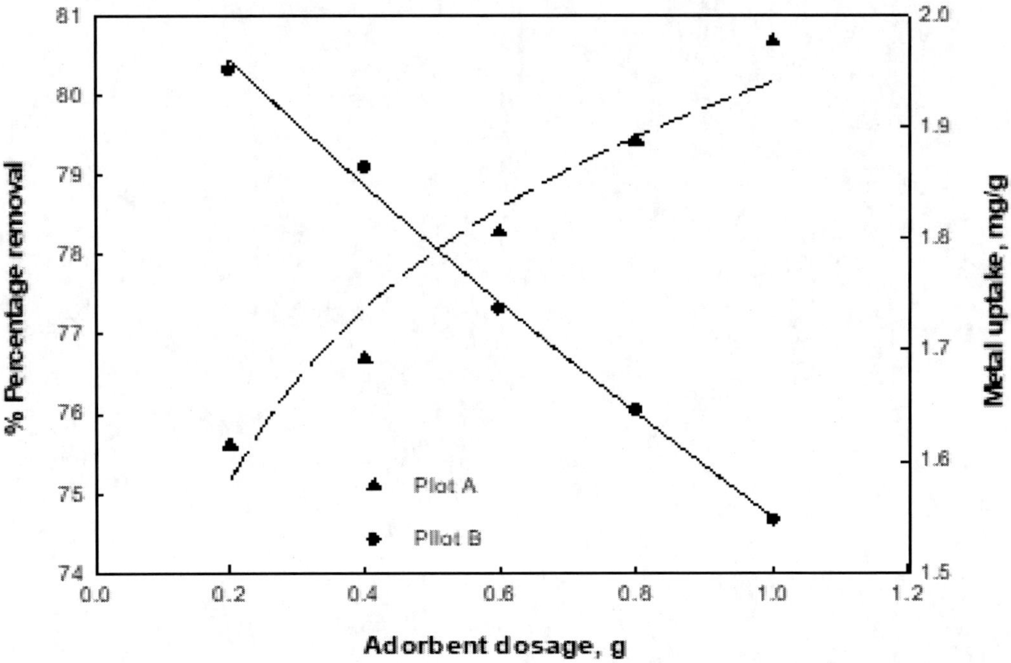

Figure 3. Effect of adsorbent dosage on percentage removal of Cr(VI) and metal uptake.

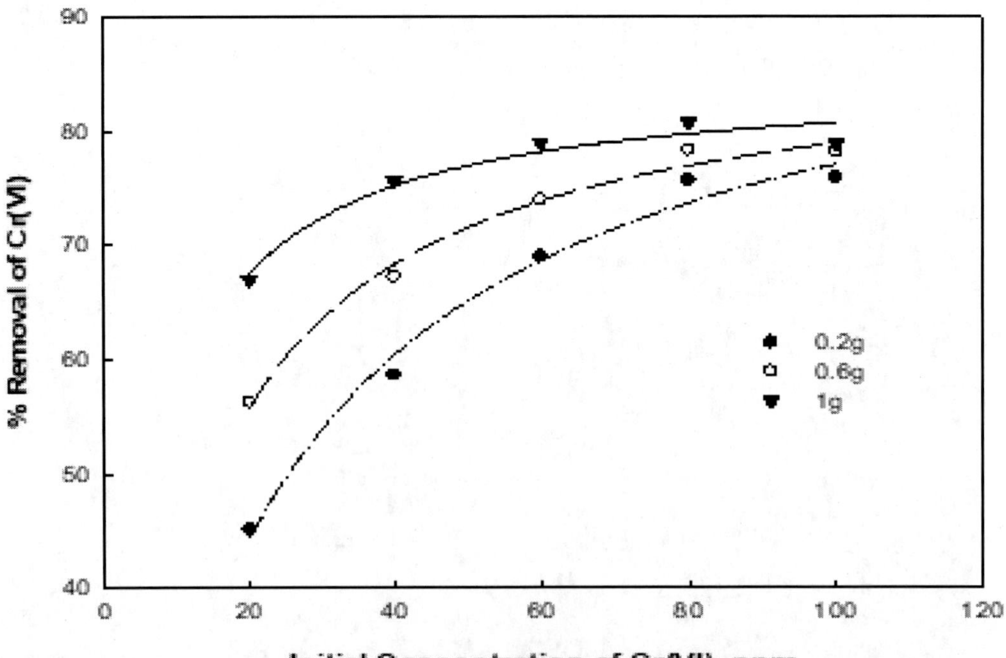

Figure 4. Effect of initial batch concentration in ppm on percentage recovery of Cr(VI).

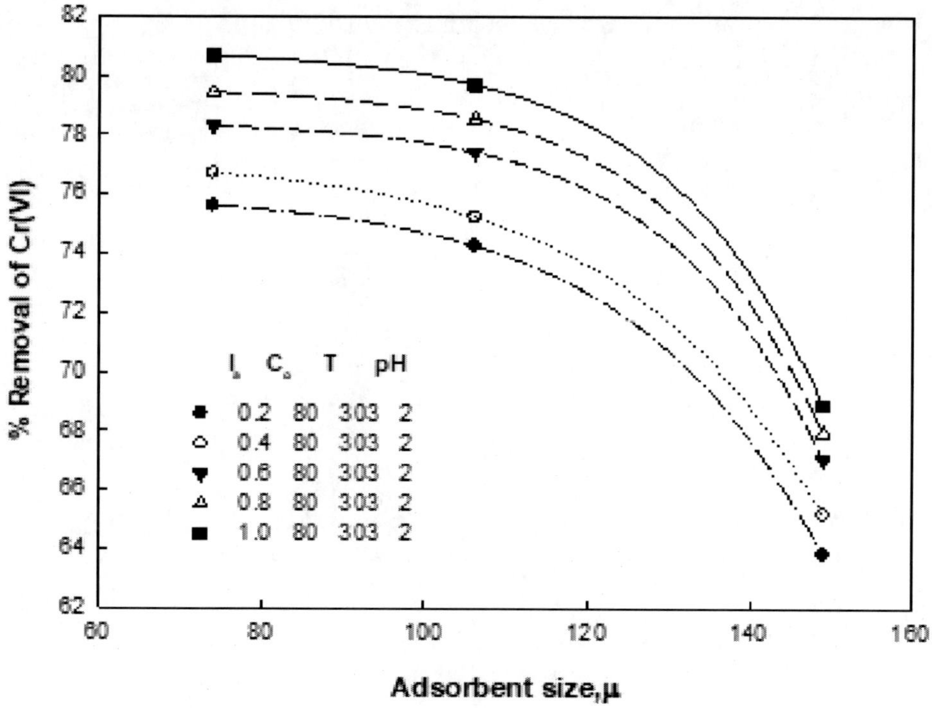

Figure 5. Effect of the size of the biosorbent on the percentage removal of Cr(VI).

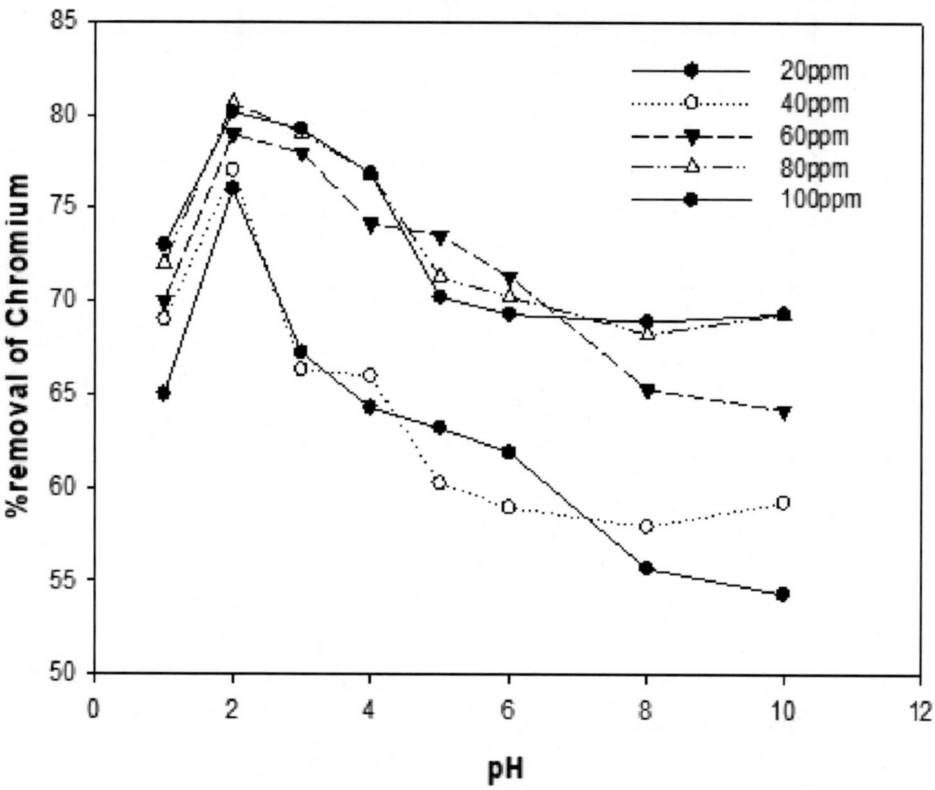

Figure 6. Effect of pH on the percentage recovery of Cr(VI).

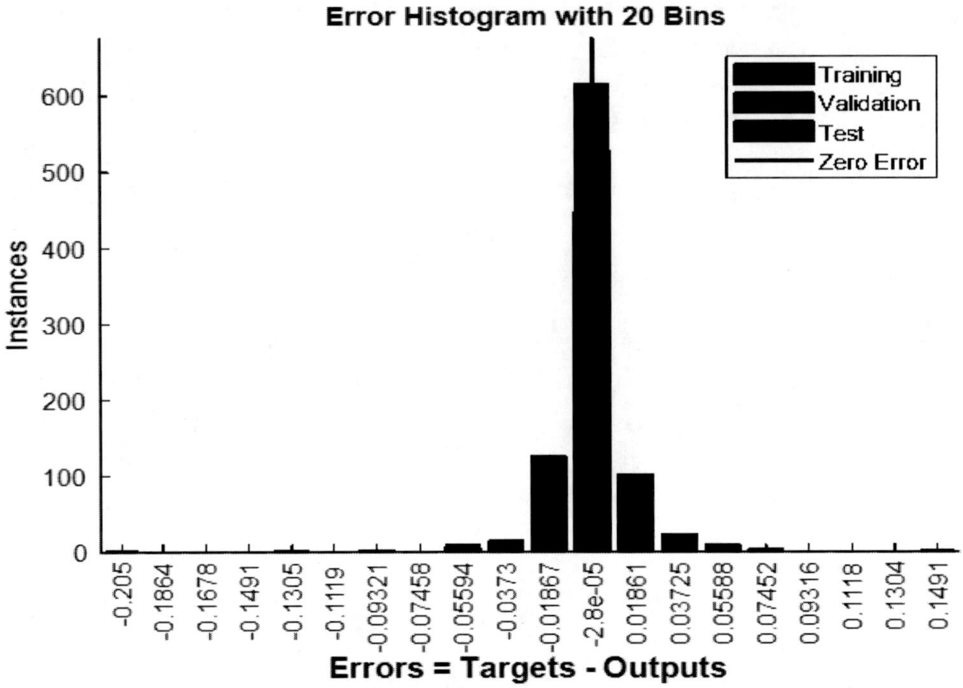

Figure 7. MFLP ANN model Error Histogram plot.

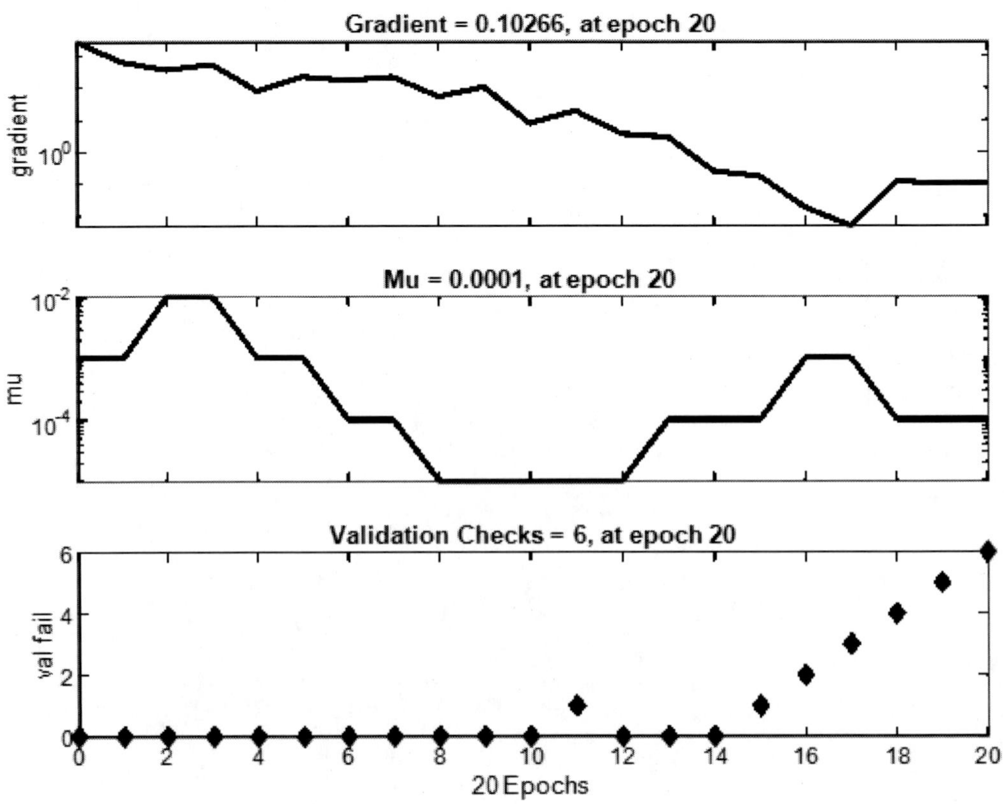

Figure 8. MFLP ANN model Training state plot.

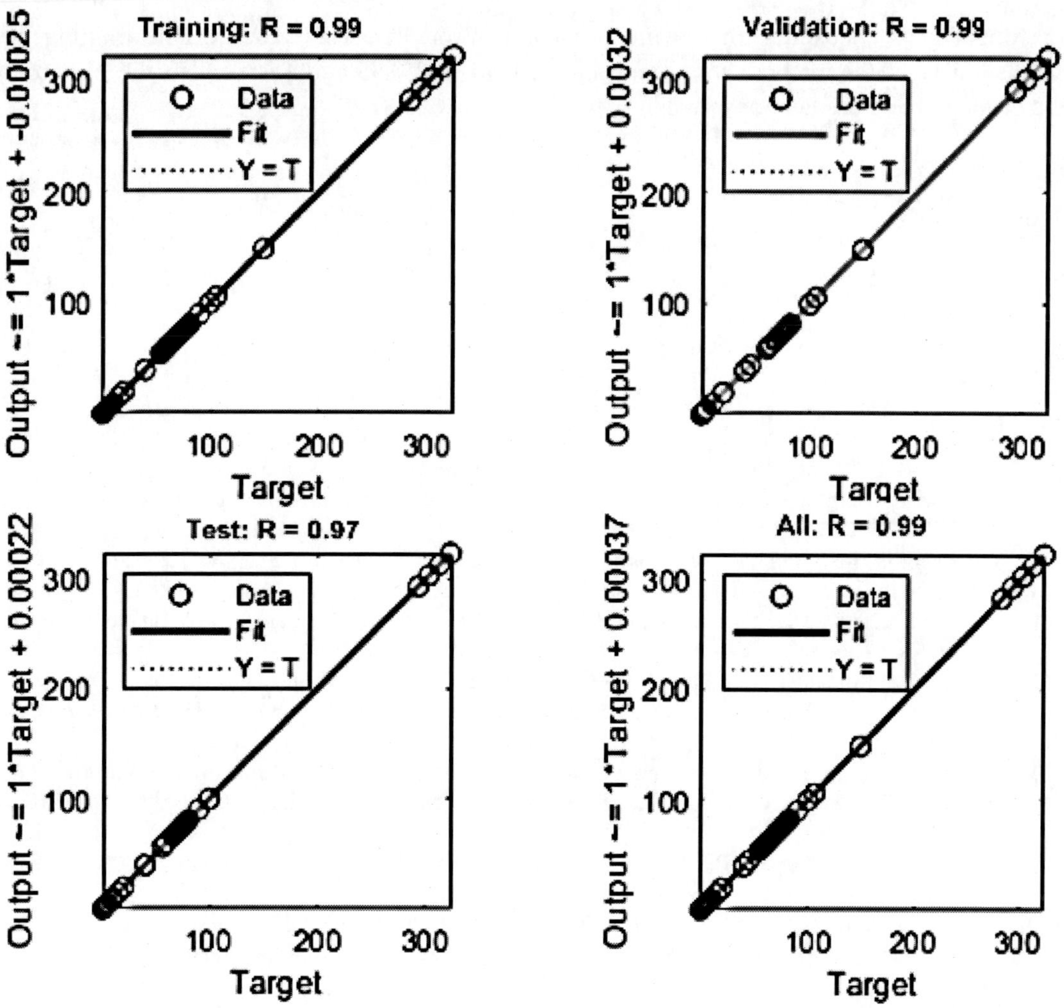

Figure 9. MFLP ANN model regression plot.

The error histogram in Figure 7 shows the instances with an error zero bar. The training state plots in Figures 8 and 9 show the gradient and Mu with validation checks, which show a good validation check of 6 at epoch 20 in the second run.

Conclusion

The removal of chromium from aqueous media by using the adsorption process was established. Using a neural network to see these results gives an idea about how the experimental results and prediction data exactly match the future expectations of the removal of chromium using various biosorbents. Here are the different types of transfer functions used for the predicted value. The purlin transfer function is used at the output layer in the neural network tool. a two-layer feed-forward network with sigmoid hidden neurons and a linear output neuron, suitable for regression tasks. With layer sizes of 10 and validation data and test data of 15 each. The Levenberg–Marquardt training algorithm is used in the network, and the performance is checked with the mean squared error. The R2 values of the training and validation are 0.99 for both tests, and all are 0.97 and 0.99 with respect to the chromium experimental and predication values.

References

[1] Volesky, B. and Holan, Z. R. (1995). Biosorption of heavy metals. *Biotechnology Progress,11*(3): 235-250.
[2] Kowalshi, Z. (1994). Treatment of chromic tannery wastes. *Journal of Hazardous Materials,37*: 137-144.
[3] Veglio, F. and Beolchini, F. (1997). Removal of metals by biosorption: a review. *Hydrometallurgy, 44*(3): 301-316.
[4] Sikaily, A. E., Nemr, A. E., Khaled, A., and Abdelwahab, O. (2007). Removal of toxic chromium from wastewater using green alga Ulva lactuca and its activated carbon. *Journal of Hazardous Materials, 148*(1-2): 216-228.
[5] Tiravanti, G. Petruzzelli, D. and Passino, R. (1997). Pretreatment of tannery wastewaters byan ion-exchange process for Cr (III) removal and recovery. *Water Science and Technology, 36*(2-3): 197-207.
[6] Das, A. K. (2004). Micellar effect on the kinetics and mechanism of chromium (VI) oxidation of organic substrates. *Coordination Chemistry Reviews, 248*: 81-99.
[7] Zhou, X., Korenaga, T., Takahashi, T., Moriwake, T., and Shinoda, S. (1993). A process monitoring/controlling system for the treatment of waste water containing (VI). *Water Research, 27:* 1049-1054.
[8] Li, H., Li, Z., Liu, T., Xiao, X., Peng, Z., and Deng, L. (2008). A novel technology for biosorption and recovery hexavalent chromium in wastewater by biofunctional magnetic beads. *Bioresource Technology, 99*: 6271-6279.
[9] Gomez, V. and Callo, M. P. (2006). Chromium determination and speciation since 2000. *Trends in Analytical Chemistry, 25*: 1006-1015.
[10] Seaman, J. C., Bertsch, P. M., and Schwallie, L. (1999). In-Situ Cr (VI) reduction within coarse-textured oxide-coated soil and aquifer systems using Fe (II) solutions. *Environmental Science and Technology, 33*: 938-944.
[11] Pagilla, K. R. and Canter, L. W. (1999). Laboratory studies on remediation ofChromium-contaminated soils. *Journal of Environmental Engineering, 125*: 243-248.
[12] Chakravathi, A. K., Chowadary, S. B., Chakrabarty, S., Chakrabarty, T., and Mukherjee, D. C. (1995). Liquid membrane multiple emulsion process of chromium (VI) separation from waste waters. *Colloids and Surfaces A: Physicochemical and Engineering Aspects, 103*(1-2): 59-71.
[13] Kongsricharoern, N. and Polprasert, C. (1996). Chromium removal by a bipolar electrochemical precipitation process. *Water Science and Technology, 34*: 109-116.
[14] World Health Organization. (2004). "Guidelines for drinking water quality," 3rd ed., Genrva,1: 334.
[15] Huang, S. D. Fann, C. F. and Hsiech, H. S. (1982). Foam separation of chromium (VI) from aqueous solution. *Journal of Colloid and Interface Science, 89*: 504-513.
[16] Park, D. Yun, Y. S. and Park, J. M. (2010). The past, present and, and future trends of biosorption. *Biotechnology and Bioprocess Engineering, 15*: 86-102.
[17] Aksu, Z. Ozer, D. Ekiz, H. I. Kutsal, T. and Calar, A. (1996). Investigation of Biosorption of Chromium (VI) on Cladophora Crispata in Two-Staged Batch Reactor. *Environmental Technology, 17*: 215-220.
[18] Padma, D. V. and Venkata Ananta Rama Sastry, S. (2020). Kinetic studies on simultaneous biosorption of divalent copper and hexavalent chromium using mallet flower leaf powder. *Journal of Water and Environmental Nanotechnology, 5*(3): 204-217.
[19] Padma, D. V. and Sastry, S. V. A. R. (2022). Biosorption of divalent copper from aqueous solution using mallet flower leaves powder as adsorbent. *International Journal of Environment and Waste Management 29*(4): 444-458.

[20] Padma, D. V. and Sastry, S. V. A. R. (2021). Kinetics, Isotherms and Thermodynamic Studies on Removal of Divalent Copper using Mallet Flower Leaf Powder as Bio-Adsorbent. *Trends in Sciences 18*(24): 1431-1431.

[21] Sastry, S. V. A. R. Rao, B. S. Padma, D. V. (2020). Studies on Selective Batch Adsorption of Cu(II) and Cr(VI) from aqueous solution. *TEST Engineering and Management 83*: 6794 – 6797.

[22] Padma, D. V. and Sastry, S. V. A. R. (2020). Biosorption of Hexavalent Chromium using Mallet Flower Leaves Powder as adsorbent, *TEST Engineering and Management 83*:15714- 15729.

[23] Kapoor, A. and Vijayraghavan, T. (1998). Biosorption of heavy metals on Aspergillus Niger: effect of pretreatment. *Bioresource Technology 63*(2): 109-113.

Chapter 3

The Identification of Various Materials in Process Industries Using Advanced Intelligent Techniques

Pawar Ajay Abhaysing*
Nilesh Shriram Vishwakarma
Patil Sharvari Narendra
Piyush Bhatee
Raju Yerolla
and Chandra Shekar Besta
Department of Chemical Engineering, National Institute of Technology, Calicut, India

Abstract

In our work, we use the NFFA-EUROPE-SEM dataset, which has a total of 18,577 SEM images produced at CMR-IOM. SEM images are a type of image that uses the focal beam of an electron to produce a high-resolution image. This dataset comprises 10 different categories from the process industries. We are building a deep neural network model to classify the different categories of surfaces. Most of the classification models in this work are based on convolutional neural networks. Convolutional neural networks are a class of artificial neural networks that are used to visualize images for different deep learning tasks. We propose a model that uses attention networks with weighted pooling in the neural network along with squeeze excitation to achieve state-of-the-art accuracy. Additionally, to address the imbalance in the dataset, we use focal cross-entropy loss.

Keywords: attention networks, CNN, focal cross entropy loss, squeeze excitation

1. Introduction

Deep learning (DL) already has myriad applications in a variety of fields. Recently, its development has been expeditious in different fields, which comprise information retrieval,

* Corresponding Author's Email: ajay_b190702ch@nitc.ac.in.

In: Sustainability in Chemical Processes through Digitalization and Green Chemistry Approaches
Editors: Susarla V. A. R. Sastry, Shravan Kumar and Ashish Kapoor
ISBN: 979-8-89113-257-3

visual data processing, natural language processing (NLP), and many more. To perform the task of classification given the dataset, which is large enough, deep learning models are the most efficient [1]. The field of materials science also observed the utilization of deep learning techniques. These techniques provide the internal structure of the different materials and structure-property linkage analysis. Currently, the most interesting research areas are the identification of elements from the images with the help of microscopy methods.

To carry out the task of image classification, input images are given a particular label from the available categories. It is an emerging and highly dynamic research area. To examine substances at the nanoscale, scanning image microscopy (SEM) has proven to be highly useful [2]. To extrapolate the capacity of SEM to reach subnanometer spatial resolution, it is photographed with electrons instead of photons. It provides details of obscure compositional and topological characteristics that are not possible with standard light microscopy. Due to this property, SEM finds its application in a plethora of domains, such as nanofabrication, physics, chemistry, biology, and forensics.

A convolutional neural network (CNN) is one of the most often used DL networks [3]. CNN provides an advantage in comparison to its predecessor in that it voluntarily focuses on the substantial features without assistance from humans, which makes it highly preferable. CNN, popularly known as ConvNet, has the magnitude to generalize more efficiently compared to other networks that are highly connected. It consists of a feed-forward architecture. The primary idea behind it takes inspiration from biologically hierarchical feature detectors. It is extremely quick to identify things and recognize very abstract features.

CNN and other deep learning techniques are capable of understanding context information; for this particular property, it found its exercise in image recognition. However, the main caveat of this technique is that it relies on the large depth and large width of the image, which may result in high computational costs.

2. Related Works

Image classification is a major subdomain of artificial intelligence. The development of image classification is presented using machine learning techniques from a high-level perspective. The most widely used convolutional neural network-based approach can use ResNet [4] and VGG (Visual Geometry Group) Net [5] architectures as backbones. This provides better accuracy. On the other hand, SEM images can also be reconstructed using GANSCs (generative adversarial networks) [6]. Another 3D reconstruction can be performed using FIB-SEM (focused ion beam scanning electron microscopy) images. The microstructural analysis of images has recently been performed using U-Net CNN [7, 8]. For the comparison of models based on defect patterns, RetinaNet models can be trained using different ResNet and VGGNet architectures. The model can thus be presented as a comparison between accuracies and performance analysis on SEM images. It is necessary to segment images for multiphase materials. The most commonly used model for this segmentation is the GMM (Gaussian mixture model) [9]. To train numerous SEM images on a small dataset, the image classification can be optimized using geometric similarities.

2.1. Focal Loss

During the training of a focal loss function [10], a class imbalance is rectified. To emphasize learning in difficult misclassified cases, a modifying term to the cross-entropy loss is added up by focal loss. As confidence in the right class rises, the scaling factor decreases to zero along with cross-entropy loss, which is dynamically scaled. Subjectively, the scaling factor can hastily concentrate the model on complex cases.

Focal loss requires the inclusion of a component $-(1-pt)\gamma$ in the standard cross-entropy criterion. In the case of well-categorized examples, $(pt > 0.5)$, fixing $\gamma > 0$ decreases its relative loss by placing extra emphasis on difficult misclassified examples. Here, there is a tunable focusing parameter$\gamma \geq 0$.

The conventional cross-entropy criteria are formulated as the focal loss in addition to a factor$(1-pt)\gamma$. Assuming $\gamma > 0$ lessens the proportionate loss for correctly categorized cases $(pt > 0.5)$, focusing greater attention on difficult, incorrectly classified examples. There is a variable focusing parameter of $\gamma \geq 0$ here.

$$FL(pt) = -(1-pt)\gamma \, log(pt) \quad (1)$$

3. Methodology

Resnet: Resnet has a residual learning framework that helps in better understanding features extracted from images. Residual connections are the type of skip connections that help pass the data information to the later part of the neural network. The residual connection was able to handle overfitting, but it was not able to adapt to new data, and there was a high failure rate.

Vgg19: The VGG19 model is a very deep convolutional neural network. This model involves increasing the depth of architecture with small 3 x 3 filters, which results in an improvement in the accuracy of the model as it eventually reduces bias in the model.

AlexNet: AlexNet is an eight-layer-based model that has three fully connected layers and a total of five convolutional layers. The tanh activation function was the standard activation function used at that time, but it introduced the rectified linear unit (ReLU) activation function, whose values start at zero. AlexNet has a total of 60 million parameters but has a significant overfitting problem.

The vision transformer model is used for the classification task [11–13]. The vision transformer architecture is further developed based on the original transformer paper. In the paper including transformers, each word is taken as a token and then converted into word embeddings, which are taken as input in the transformer encoder model. Similarly, in the vision transformer, images are divided into N patches, where each patch is of size P and P is the hyperparameter. The value of P used in this study is 16. Now, each patch behaves as an embedding similar to that of the words and is taken as input in the model. With these input patches, adding positional embeddings gives positional information about the pixels that are present in the RGB (red, green, and blue) image that helps to obtain the spatial information. Adding the CLS (classification) token with embeddings gives global information about each and every token. At the top of this are the outputs generated by the vision transformer model, the introduction of the multilayer perceptron (MLP) layer for the classification task, and softmax at the last layer, which gives the probability of 10 different SEM images.

The model was trained using different image sizes, including 224, 256, and 512. During the training of the model, the following methods were used. The proposed architecture is shown in Figure 1.

Augmentations: Using augmentations [14] to convert the images and adding noise to the dataset to reduce the problem of variance led to a more reliable model.

Stochastic weighted averaging (SWA): SWA saves the weights of the particular models at the end of each learning rate. Therefore, it reduces obtaining different weights for the model that is closer to the region that has relatively lower losses and takes the average of these weights. This stochastic weighted averaging helps to obtain an ensemble model by using a single model.

Pretrained weight decay: Excessive learning of the model creates the problem of overfitting, i.e., high variance and low bias. Therefore, while fine-tuning the decay, some of the weights from the model help stabilize the transformer model and learn the new dataset.

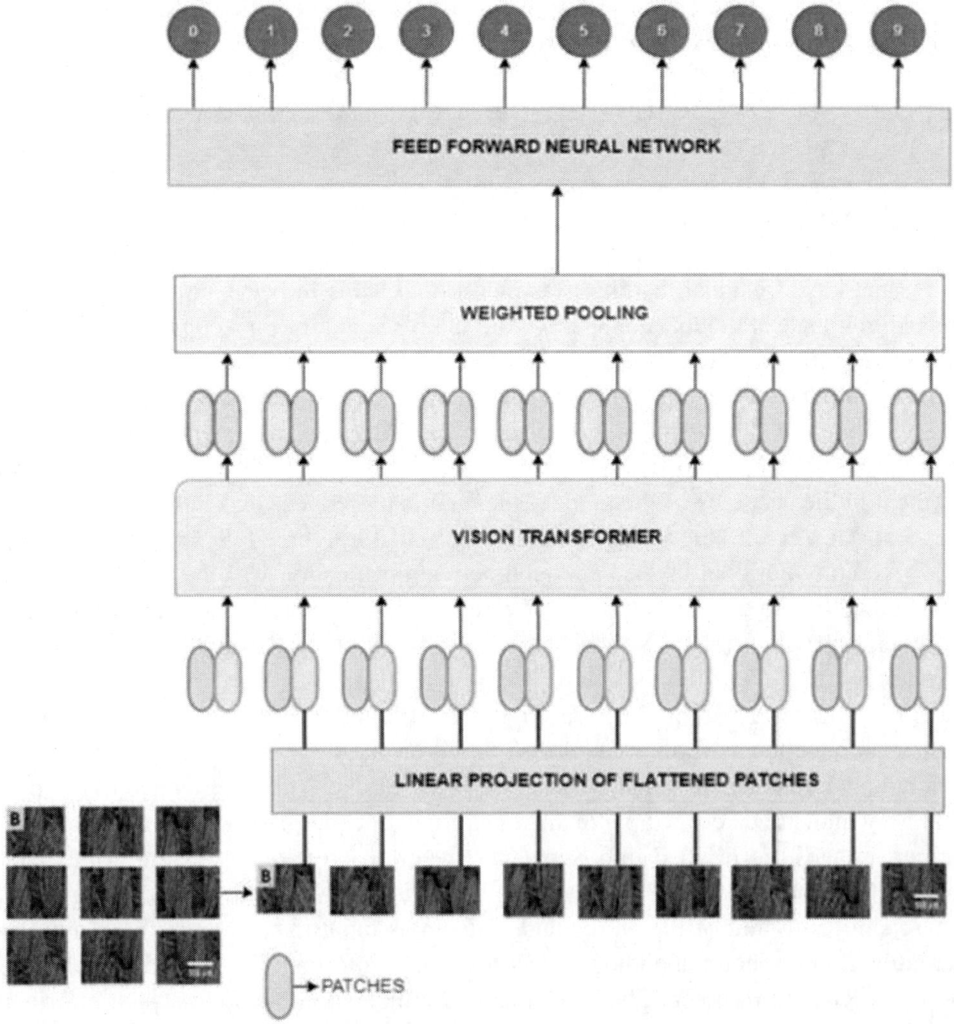

Figure 1. Proposed Architecture.

4. Results and Discussions

Dataset

We use the NFFA-EUROPE SEM (Scanning Electron Microscopy) dataset obtained under the NFFA-EUROPE project. This dataset contains 25,430 SEM images. Images are categorized into 10 different categories in folder form. Images comprise different classes, such as biological, fibers, film-coated surfaces, tips, meme devices, electrodes, nanowires, particles, patterned surfaces, porous images, and powder.

Hyperparameter Settings

To train the model, we set several hyperparameters, which are summarized in the table given below.

Table 1. Various parameters and their values for training of the model

Parameters	Value
Batch size	16
Iterations	15
Folds (cross-validation)	5
Learning Rate	3e-5
Dropout	0.25

Evaluation Metrics

There are different types of matrices based on necessity and requirements. In classification problems, when we have an equal number of positive and negative samples in a binary metric, we generally use accuracy, precision, recall, and f1. Specificity quantifies the percentage of genuine negatives that the model successfully detects. A few terms related to evaluation metrics.

True Positive (TP): It is a true positive if the class predicted by the model and the actual class are positive.

True Negative (TN): It is a true negative if the class predicted by the model and the actual class are both negative.

False Positive (FP): It is a false positive if the class predicted by the model is positive and the actual class is negative.

False Negative (FN): It is a true negative if the class predicted by the model is negative and the actual class is positive.

Types of Evaluation Metrics:

$$\text{Precision} = \frac{\textit{True Positive}}{\textit{True Positive}+\textit{False Positive}}$$

$$\text{Recall} = \frac{\textit{True Positive}}{\textit{True Positives+False Negative}}$$

$$\text{F1 Score} = \frac{2*(\textit{Recall} * \textit{Precision})}{(\textit{Recall} + \textit{Precision})}$$

$$\text{Accuracy} = \frac{\textit{True Positive+True Negative}}{\textit{True Positve+False Positive+False Negative+True Negative}}$$

5. Results and Analysis

We used weighted pooling to fine-tune the Vision Transformer model, VGG-19, ResNet, Efficient-Net, and AlexNet. To address the imbalance in the dataset, we used focus loss, which assigns weights to the class with fewer samples. We used the SEM image dataset to train our own in-house Vision Transformer model and the other models mentioned above. With image resolution sizes of 224, 256, and 512, we trained various models. Using the Adam optimizer for backpropagation, its momentum leads to a smooth convergence point and an optimal loss. Table 2 provides a summary of the outcomes based on focal loss for each of the models provided. It tells about the precision, recall, and F1 score that lead the metrics in the classification task.

Table 2. Results of different evaluation metrics

Models	*Precision*	*Recall*	*F1 Score*
Our Model	0.96	0.99	0.97
Efficient Net	0.92	0.90	0.91
ResNet	0.91	0.89	0.92
VGG19	0.59	0.86	0.70

Conclusion

We employed the use of a vision transformer model to address the problem of classification on the SEM dataset. In parallel, we also trained state-of-the-art VGG19, AlexNet, and ResNet models and compared the accuracy of these models with the vision transformer model. We modified the vision transformer model by adding weighted pooling to obtain a good understanding of the pixels in the images. We introduced focal cross-entropy loss to address the problem of dataset imbalance. During training, we applied different optimization and regularization techniques to avoid overfitting and underfitting and ensure a robust model. The results obtained from the proposed model outperformed other models on evaluation metrics such as the f1 score.

References

[1] Sarker, I. H. (2021). Deep Learning: A Comprehensive Overview on Techniques, Taxonomy, Applications and Research Directions. *SN Computer Science*, *2*(6), 420. https://doi.org/10.1007/s42979-021-00815-1.

[2] Wang, H., Li, S., Ding, J., Li, S., Dong, L., & Lu, Z. (2022). SEM Image Quality Assessment Based on Intuitive Morphology and Deep Semantic Features. *IEEE Access*, *10*, 111377–111388. https://doi.org/10.1109/ACCESS.2022.3216388.

[3] O'Shea, K., & Nash, R. (2015). *An Introduction to Convolutional Neural Networks* (arXiv:1511.08458). arXiv. https://doi.org/10.48550/arXiv.1511.08458.

[4] He, K., Zhang, X., Ren, S., & Sun, J. (2016). *Deep Residual Learning for Image Recognition*. 770–778. https://openaccess.thecvf.com/content_cvpr_2016/html/He_Deep_Residual_Learning_CVPR_2016_paper.html.

[5] Simonyan, K., & Zisserman, A. (2015). *Very Deep Convolutional Networks for Large-Scale Image Recognition* (arXiv:1409.1556). arXiv. https://doi.org/10.48550/arXiv.1409.1556.

[6] Goodfellow, I., Pouget-Abadie, J., Mirza, M., Xu, B., Warde-Farley, D., Ozair, S., Courville, A., & Bengio, Y. (2020). Generative adversarial networks. *Communications of the ACM*, *63*(11), 139–144. https://doi.org/10.1145/3422622.

[7] Ronneberger, O., Fischer, P., & Brox, T. (2015). U-Net: Convolutional Networks for Biomedical Image Segmentation. In N. Navab, J. Hornegger, W. M. Wells, & A. F. Frangi (Eds.), *Medical Image Computing and Computer-Assisted Intervention – MICCAI 2015* (pp. 234–241). Springer International Publishing. https://doi.org/10.1007/978-3-319-24574-4_28.

[8] Bangaru, S. S., Wang, C., Zhou, X., & Hassan, M. (2022). Scanning electron microscopy (SEM) image segmentation for microstructure analysis of concrete using U-net convolutional neural network. *Automation in Construction*, *144*, 104602. https://doi.org/10.1016/j.autcon.2022.104602.

[9] Chatzigeorgiou, M., Vrigkas, M., Beazi-Katsioti, M., Katsiotis, M., Boukos, N., & Constantoudis, V. (2023). Segmentation of SEM images of multiphase materials: When Gaussian mixture models are accurate? *Journal of Microscopy*, *289*(1), 58–70. https://doi.org/10.1111/jmi.13150.

[10] Lin, T.-Y., Goyal, P., Girshick, R., He, K., & Dollar, P. (2017). *Focal Loss for Dense Object Detection*. 2980–2988. https://openaccess.thecvf.com/content_iccv_2017/html/Lin_Focal_Loss_for_ICCV_2017_paper.html.

[11] Dosovitskiy, A., Beyer, L., Kolesnikov, A., Weissenborn, D., Zhai, X., Unterthiner, T., Dehghani, M., Minderer, M., Heigold, G., Gelly, S., Uszkoreit, J., & Houlsby, N. (2021). *An Image is Worth 16x16 Words: Transformers for Image Recognition at Scale* (arXiv:2010.11929). arXiv. https://doi.org/10.48550/arXiv.2010.11929.

[12] Dey, B., Goswami, D., Halder, S., Khalil, K., Leray, P., & Bayoumi, M. A. (2022). Deep learning-based defect classification and detection in SEM images. *Metrology, Inspection, and Process Control XXXVI*, *PC12053*, PC120530Y. https://doi.org/10.1117/12.2622550.

[13] Shorten, C., & Khoshgoftaar, T. M. (2019). A survey on Image Data Augmentation for Deep Learning. *Journal of Big Data*, *6*(1), 60. https://doi.org/10.1186/s40537-019-0197-0.

[14] Wong, C. H., Ng, S. M., Leung, C. W., & Zatsepin, A. F. (2022). The Effectiveness of Data Augmentation of SEM Images on a Small Database Based on Deep-Learning Intelligence. *Brazilian Journal of Physics*, *52*(2), 59. https://doi.org/10.1007/s13538-021-01008-0.

Chapter 4

Chromium Removal from Tannery Effluent Waste by a Forward Osmosis System

Chhanda Mondal Roy[1,2,*]
Urjoshi Chatterjee[3]
Dipankar Pal[4]
and Jaya Sikder[1]

[1]Department of Chemical Engineering, National Institute of Technology, Durgapur, India
[2]Department of Civil Engineering, Dr. B. C. Roy Polytechnic College, Durgapur, India
[3]Heritage Institute of Technology, Kolkata, India
[4]Department of Chemical Engineering, Indian Institute of Petroleum and Energy, Visakhapatnam, India

Abstract

The leather industries are a notable sector of the economic development of a country. However, it is one of the most polluting sectors because it produces enormous amounts of chemical waste in effluent water. Chromium is one of the pollutants in effluents from tannery waste. Membrane separation technology is widely used in various sectors, such as the biomedical, water treatment, and pharmaceutical industries. Forward osmosis (FO) is a promising technology that can sustainably treat and produce high-quality water. In this process, a semipermeable membrane is used, and water passes through that membrane to draw solution from the feed due to an osmotic pressure gradient. The advantages of the FO process are the low hydraulic pressure required for separation processes and the low fouling tendency. Experiments were performed on a laboratory scale using a commercial thin FO membrane with monovalent ions Na_2SO_4 for the draw solution and $K_2Cr_2O_7$ for the feed solution. Different parameters, such as pH and flow rates, were varied during the filtration process. The performance of the membrane was determined at various pH levels. It has been observed that membrane performance depends on pH. High flux and low fouling of the membrane were observed at different pH levels of the feed solution during the filtration process. It was also noted that water flux is lower in acidic conditions and higher in alkaline conditions. The main objective of the process was to enhance the filtration rate with low concentration polarization and a high rate of rejection (> 95%).

Keywords: concentration polarization, forward osmosis, draw solution, water flux, rejection rate

[*] Corresponding Author's Email: cmr.20ch1502@phd.nitdgp.ac.in.

In: Sustainability in Chemical Processes through Digitalization and Green Chemistry Approaches
Editors: Susarla V. A. R. Sastry, Shravan Kumar and Ashish Kapoor
ISBN: 979-8-89113-257-3

1. Introduction

The tannery industry generates a high quantity of solid, liquid, and gaseous waste [1]. Tannery industries produce significant waste from pre-tanning, tanning, and post tanning units. Currently, tannery waste is treated by physio-chemical treatment with catalytic oxidation of sulfate to sulfide [2].

Different types of chemicals come from tannery wastewater effluent. The tanning process has three major phases: the collection of raw hides, the pretreatment of raw animal hides, and the tanning of hides with a tanning agent. The tannery sector is a feasible sector of the global economy and one of the oldest industries in the world. Tanning has two types, i.e., mineral tanning and vegetable tanning. In various tanning industries, mineral tanning and chrome tanning are available. High levels of inorganic and organic pollutants are present, such as biological oxygen demand (BOD), chemical oxygen demand, salinity, ammonia, total Kjeldahl nitrogen (TKN), dissolved solids (DS), suspended solids (SS), chloride, chromium, sulphide, sodium, heavy metals, and other salt residues. A large amount of water is used, and 90% of the effluent is discharged, with a few amounts of solid and gaseous waste discharged in the tanning process. Approximately 40% of chromium salts are discharged during final effluents, which severely affects the environment.

As a tanning producer and exporter of leather, the second-largest country in the world, Brazil, is achieving excellence in the industry with investments, qualifications, and technology. Leather comes from hide and skin, which are obtained from fallen animals. Leather is made from smaller animals, such as goats, sheep, lambs, and pigs, called skin, and larger animals, such as cows and buffalo, are called hide. The three main sections of hide treatment are performed in the beam house, tanning, and finishing operations. Cleaning the hides and preparing for the consequent tanning process in the beam house unit is responsible for 70% of the total pollution from leather processing [3]. For tanning preparation, washed skins or hides are treated with alkali to remove hair and natural fat and then acidify that material. The second stage is tan-yard operation, i.e., actual tanning and other chemical treatment. The third stage is post-tanning, which applies to finishing the surface, dyeing, fat-liquoring, and retanning under wet conditions. The final stage's objective is to determine the color and gloss of the finished leather. The final finishing operation is applied to the surface of the leather. In the pretanning areas, water consumption is the highest, but a substantial amount of water is also required in the post-tanning processes. The most polluting stage is the soaking stage, which accounts for approximately 50–55% of the total polluting load [4]. The posttanning process generates polluting agents such as chrome salts, fat liquoring agents, dyestuff residues, organic ammonia, nitrogen, dissolved and suspended solids, and other organic matter with high pH values [5]. All types of effluents with heavy metals cause serious health hazards, and it is unsafe to dispose of untreated effluents in affected soil and water [6].

Membrane processes are one of the possible operations for water shortage mitigation and increasing the water supply. Microfiltration (MF) and ultrafiltration (UF) are used for pretreatments in seawater desalination treatment. Reverse osmosis and nanofiltration are used for brackish water [7]. However, forward osmosis processes have been introduced for seawater desalination and industrial wastewater treatment purposes. The FO process is preferable to the RO process for membrane fouling or cleaning. The FO membrane has three parameters: the solute permeability coefficient, the pure water permeability coefficient, and mass transport across the support layer of the membrane. These parameters are used to define the solute fluxes and permeate water flux of the FO membrane, which follow the standard criteria for determining FO membrane characteristics and comparing their performance. The objective of this study is to evaluate the performance of commercial thin FO membranes that use monovalent ions Na_2SO_4 for the draw solution and $K_2Cr_2O_7$ for the feed solution [8]. Different parameters, such as pH and flow rate, were varied during the filtration process and determined at various pH levels. High flux and low fouling of the membrane have been observed during the filtration process at different pH levels of the feed solution [9, 10].

2. Materials and Methods

2.1. Membrane and Membrane Characterization Methods

The chemicals were used as raw wastewater materials from the tanning process and as feed solutions with different concentrations of $K_2Cr_2O_7$, i.e., 50, 100, 150, 200, and 300 mg/l. The draw solution was prepared using various concentrations of Na_2SO_4 solution, such as 10,000 to 20,000 mg/l. The flat sheet is composed of the channel, and the membrane is a commercial flat sheet membrane of a thin semipermeable active nonporous skin layer used in the flat membrane module [11, 12]. All solutions were prepared in a 200 ml Erlenmeyer flask after the prepared solution was kept in 5-liters containers and poured into the volume chamber of 5 liters.

2.2. FO Process Setup

In the separation method, a semipermeable water solution could be separated between FS and DS water solutions. Water passes through the membrane from the FS side to the DS side due to the osmotic pressure difference. This dispersing water dilutes to the Draw solution; concurrently, the FS is concentrated. Usually, there is no need for any physical pressure; therefore, the pumping of FS and DS through the flow channels next to the membrane is only for the energy demand. The FO process has a few advantages: low energy consumption and two-stream treatment, but it only has one treatment step due to the absence of compression that is easily removable from the fouling layer. The FO instrumental setup is shown in Figure 1.

The flat sheet is composed of two channels, and a flat membrane has been taken as one layer and is hydrophilic in nature. It has an active area of 74x10-4 m2. The dimension of the membrane is 5.5 cm, and the channel length is 13.5 cm. At ambient temperature, flow rates were variable between 20 and 45 Lh-1 for DS and FS solutions; both solutions were circulated toward the membrane using a peristaltic pump. The water flux is measured by decreasing the volume of the feed solution. The rate of volume change is divided by the effective area. Therefore, the flux is

$$J_{W} = \frac{\Delta V}{At} \quad (1)$$

The change in volume of FS is ΔV over a fixed time Δt (h), and the effective membrane area of FO is A (m^2).

The solute rejection, R

$$R = (1 - \frac{C_P}{C_F})x100\% \quad (2)$$

where R = solute rejection, C_P (mg/l) = permeate concentration, and C_F (mg/l) = solute concentration.

2.3. Forward Osmosis Experiments

The FO membrane of its active layer is in contact with the feed solution during the experiment. Then, 0.5 ml/liter of H_2SO_4 was added to the feed solution to determine the effect of pH on water flux. After that, it gradually increased, and its pH gradually decreased. Similarly, initially, 0.5 mg/l NaOH was added to the feed solution, gradually increasing the pH value. For every set of experiments, the flow time was controlled at 2 hrs. The draw solutions and feed were at ambient temperature (30°C) for the experiment.

Figure 1. In-laboratory FO instrumental setup.

3. Results and Discussion

Forward osmosis laboratory experiments were carried out using sodium sulfate (Na2SO4) and potassium dichromate as draw solutions and feed solutions, respectively. The effect of different parameters, such as draw and feed concentration and pH, on water flux and rejection parameters was evaluated. FO performance is evaluated by water flux performance. The effect of pH, DS, and FS concentrations on the water flux was studied by the FO process [13, 14].

3.1. The Effect of Variable Feed Concentration to Determine Water Flux

At ambient temperature, to determine the water flux, we used varying feed concentrations with variable draw solutions. The water flux was determined by different feed concentrations, such as 50, 100, 150, 200, and 300 mg/l. Figure 2 represents the water flux vs. the feed concentration. It shows a linearly decreasing water flux as a function of feed concentration. At low feed concentration levels, the water flux is quite high because, at high concentration levels, the osmotic driving force S ($\Delta\pi = \pi D - \pi F$) decreases and may reduce the water flux level. Zhao et al., (2016) studied the effect of FS concentration on the water flux and organic micro pollutants. It was observed that water fluxes decreased from 5.6 to 3.5 m/lit.

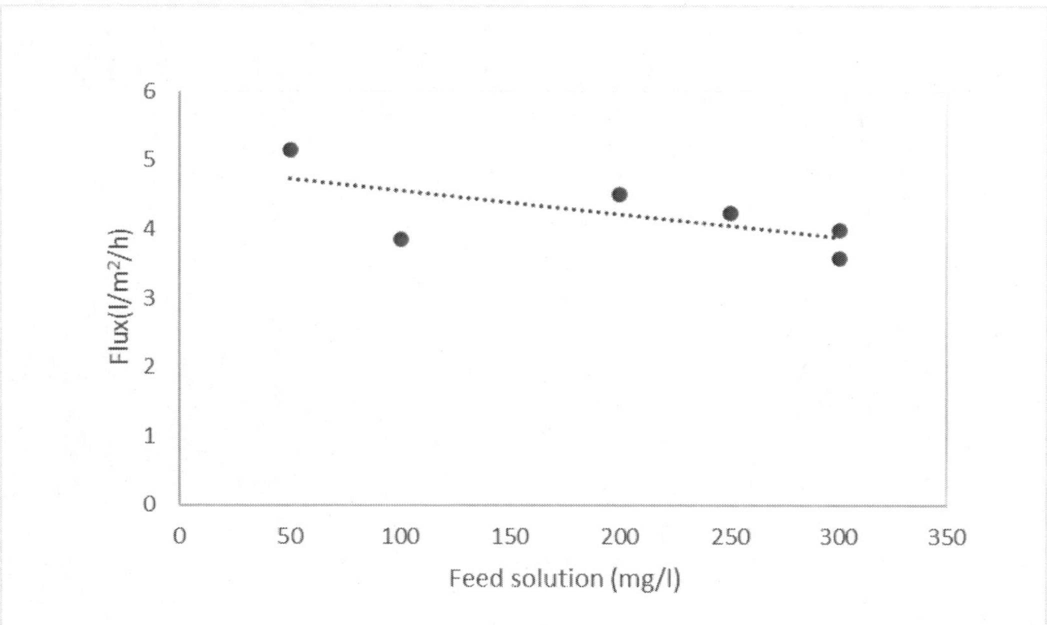

Figure 2. Water flux vs feed solution concentration at ambient temperature.

3.2. Effect of Water Flux at Variable DS Concentrations

At ambient temperature, the variation in water flux vs. DS concentration, represented by Figure 3, shows that the water flux increases with increasing DS concentration. For high concentrations, we studied whether an increase in deviation may be characterized between the model and experimental results. Concentration polarization may affect high water flux, creating higher concentration differences [15].

An increasing draw concentration may increase the driving force and, as a result, increase the water flux. The FO membrane performance depends directly on the DS concentration because it has a higher driving force and higher osmotic pressure than the FS concentration.

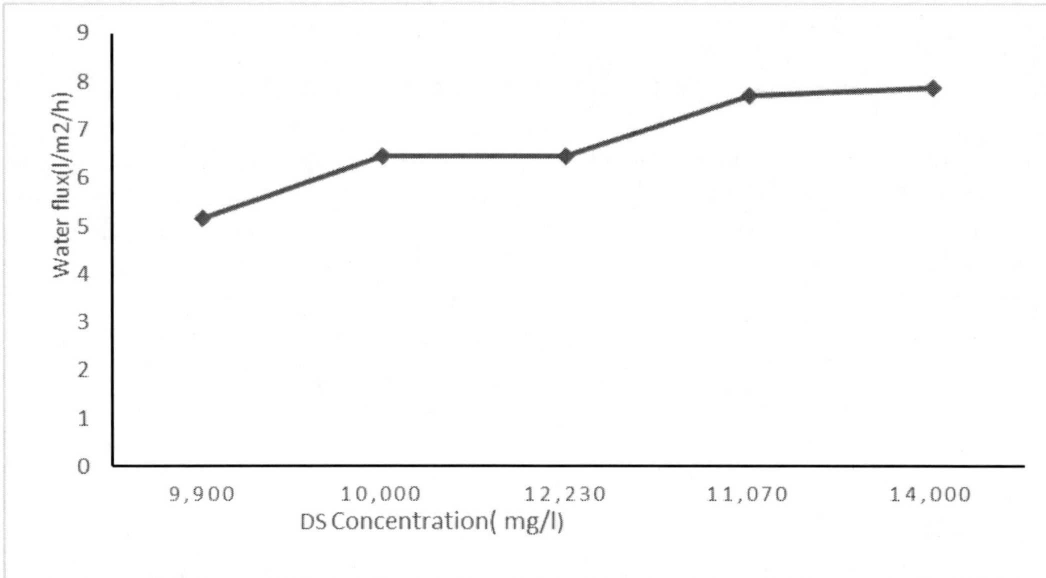

Figure 3. DS concentration vs water flux.

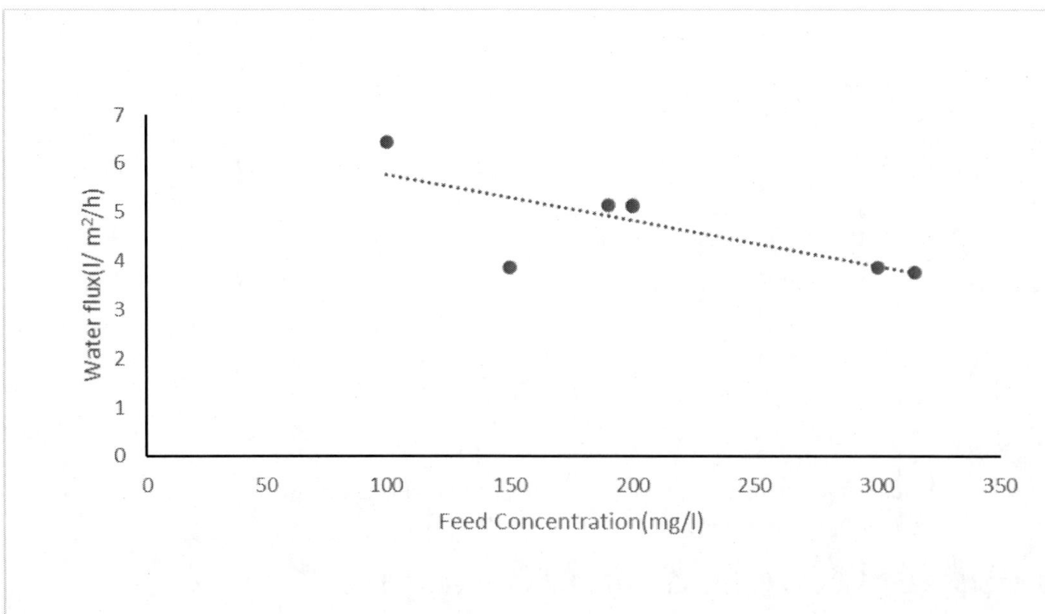

Figure 4. Feed concentration vs water flux under acidic conditions.

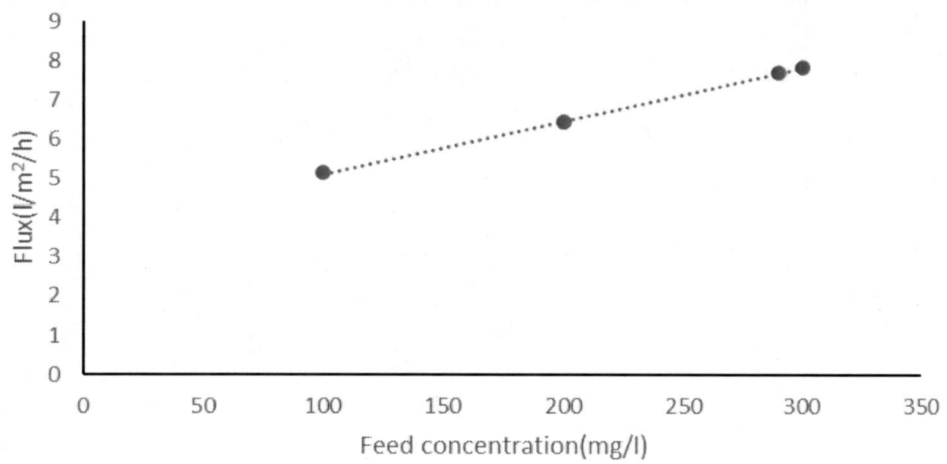

Figure 5. Feed concentration vs water flux under alkaline conditions.

3.3. The Effect of Water Flux at Different Levels of Feed Concentration with Variable pH

NaOH (0.5 mg/l) and H_2SO_4 (0.5 mg/l) were mixed separately to draw solutions to inspect the effects of pH on the membrane flux. The ambient temperature was 30°C. The experimental results showed that the variations in pH values of acidic and alkaline conditions in the feed solution ranged from 2-3 and 8-10. Due to changes in pH, membrane flux has changed; it depends on the osmotic pressure. From Figures 4 and 5, we observe that the alkaline condition of the water flux is much better than the acidic condition. The osmotic potential of FS water has the most significant influence on the membrane flux.

3.5. Determination of the Water Flux at Different DS Concentrations

To determine the water flux of the membrane, forward osmosis was used at ambient temperature. The changes in pH in both the feed and draw solutions may affect the water flux. The osmotic pressure of the FO membrane mainly caused the increase in water flux. According to data analysis under acidic conditions with a high DS concentration (20,000 mg/l), the permeate flux is 6.8 l/m^2.hr. Figure 6 shows the DS concentration vs. water flux.

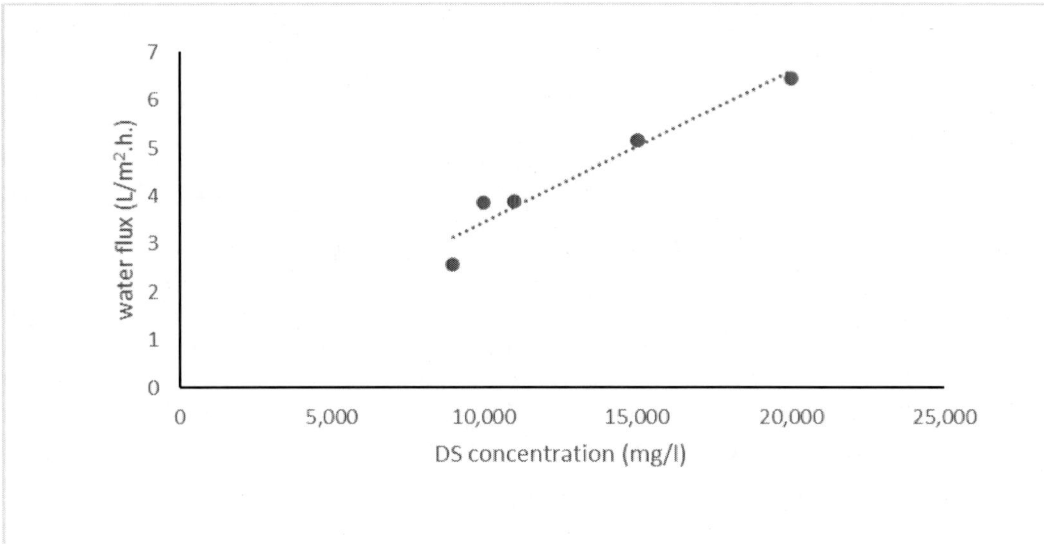

Figure 6. DS concentration vs water flux.

Conclusion

The FO membrane separation process is excellent for wastewater and drinking water treatment due to its high rejection performance. This study investigates the FO membrane performance by measuring flux and rejection under various conditions, e.g., DS and FS solution concentrations, flow rate, and acidic and alkaline conditions of feed solutions. It has been experimentally proven that the test performance of the FO process indicates an increase in chromium removal depending on alkaline conditions with high DS concentrations. In this study, chromium was successfully removed using the FO process. The disadvantage is that the fouling of FO membranes reduces the water flux, so regular membrane washing is necessary.

References

[1] Chowdhury M., M. G. Mostafa, T. K. Biswas, Abul Mondal, A. K. Saha" Characterization of the effluents from leather processing Industries" *Environ. Process.* (2015)2, 173-187. DOI 10.1007/s40710-015-0065-7.

[2] Cooman K., Gajardo M., Nieto J., Bornhardt C., Vidal G., "Tannery waste water characterization and toxicity effects on Daphnia" Spp, *Environ Toxicol.* (2003).

[3] Durai G. and M. Rajasimman. "Biological Treatment of Tannery wastewater- A Review, *Journal of Environmental Science and Technology* 4(1): 1-17,2011. ISSN 1994-7887/DOI:10.3923jest.2011.1.17.

[4] Gadiula S., L. Ndloka Ndlovu, L. K. Ncube. "Membrane technology in tannery wastewater management: A review" Vol.14[2019], e-ISSN 2409-0360. *Zimbabwe Journal Science & Technology*, pp. 57- 72.

[5] Judd S. & Jefferson B. *"Membranes for industrial wastewater recovery and reuse."* Oxford, UK, Elsevier Advanced Technology. (2003).

[6] Kim B., Sangyoup Lee, Seungkwan Hong. "A novel analysis of reverse draw and feed solute fluxes in forward osmosis membrane process" *Desalination*, Volume 352, 3 (2014), Pages 128-135.

[7] Korenak, SubhankarBasu, MaliniBalakrishnan, Claus Hélix-Nielsen and Irena Petrinic, ActaChim. "Forward Osmosis in Wastewater Treatment Processes Jasmina" *Slov*. 83 2017, 64, 83–94.

[8] Kruith J. C., Schippers J. C., Kamp P. C., Folmer H. C. & Hofman J. A. M. H. (1998). "Integrated multiobjective membrane systems for surface water treatment: pretreatment of reverse osmosis by conventional treatment and ultrafiltration." *Desalination* 117(1–3): 37-48.

[9] Maouiaa D. B., Ali Boubakria, Raja Bouchrita, Amor Hafianea, Salah Bouguecha "Chromium removal by forward osmosis: a flux modeling and experimental validation" March 2019, *Desalination and Water Treatment* 154:30-38.

[10] Meenachi S., Dr. S. Kandasamy, "Review on wastewater treatment methods in tannery wastewater," *International Journal of Advanced Science and Engineering Research*, Volume 2, Issue1, June 2017.

[11] Wendong W., Yinting Z., MariemEsparra-Alvarado, Xiaomao Wang, Hongwei Yang, Yuefeng Xie. "Effects of pH and temperature on forward osmosis membrane flux using rainwater as the makeup for cooling water dilution," *Desalination*. Volume 351, (2014), Pages 70-76.

[12] Wu Minjie, Panel Lijun Meng, Haisheng Chen, Yu Xi, Manhong Huang, Xubiao Luo. "Rejection of antimony in dyeing and printing wastewater by forward osmosis" *Science of The Total Environment* Volume 745, 25 November 2020.

[13] Zakmout A., Fatma Sadi, Carla A. M. Portugal, Joao G. Crespo, "Tannery Effluent treatment by nanofiltration, reverse osmosis and Chitosan Modified Membranes," *Membrane*, November 2020, 10, 378.

[14] Zhao J., Qinglian Wu, Yuling Tang, Jianfei Zhou and Hongguang Guo. "Tannery wastewater treatment: conventional and promising processes, an updated 20-year review" *Journal of Leather Science and Engineering*, Vol.4.10, (2022).

[15] Zhao P., B. Gao, Q. Yue, S. Liu, H. K. Shon," Effect of high salinity on the performance of forward osmosis: water flux, membrane scaling and removal efficiency," *Desalination*, 378(2016) 67-73.

Chapter 5

The Utilization of the Spray Cooling Mechanism for the Comparison of Heat Transfer Analysis between Ni-P and Ni-P-Cu-Coated Mild Steel

Suparna Bhattacharyyaa*
Deepak Choudharya
Surya K. Palb
Prashant Kumar
and Sudipto Chakrabortya
Department of Chemical Engineering, Indian Institute of Technology Kharagpur, India
Department of Mechanical Engineering, Indian Institute of Technology Kharagpur, India
Department of Chemical Engineering, IIT Kharagpur

Abstract

One of the most common industrial devices in chemical industries, pharmaceutical firms, and petrochemical refineries are heat exchangers. It can either be used in the form of a plate heat exchanger or a shell and tube heat exchanger. Mild steel, on the other hand, is used as the basic structural material for manufacturing heat exchangers. The reason behind this is its easy availability and vast range of usage. However, atmospheric corrosion, industrial fouling, and microbial attacks can degrade the material drastically. The disadvantages stated above can be effectively minimized using the technique of electroless coating. It not only improves the surface texture but also increases the material's long-term value. The present study focuses on the comparative analysis of heat transfer using electroless Ni-P and Ni-P-Cu-coated mild steel. The technique of spray cooling has been utilized for analyzing the heat transfer study. The surface morphology and phase structure analysis were carried forward from the SEM, EDX, and XRD results. The microhardness study of the sample has been analyzed and reported. The heat transfer data were collected using an in-house spray cooling setup. The experiment is executed with INTEMP software. The results using both Ni-P and Ni-P-Cu were recorded and compared in the given study. It forecasts and optimizes the sample to give the most suitable result under the spray cooling technique. Scaling up the analysis for industrial heat exchangers is a potential future scope for the current analysis.

* Corresponding Author's Email: suparna.ju2015@gmail.com.

In: Sustainability in Chemical Processes through Digitalization and Green Chemistry Approaches
Editors: Susarla V. A. R. Sastry, Shravan Kumar and Ashish Kapoor
ISBN: 979-8-89113-257-3

Keywords: electroless coating, INTEMP, mild steel, microhardness, spray cooling

Nomenclature and Abbreviations

q	surface heat flux, W/m2
Ts	surface temperature, K
Tc	coolant temperature, K
TC	Thermocouple
MS	Mild steel sample
Ni-P	Nickel phosphorous composite-coated mild steel sample
Ni-P-Cu	Nickel Phosphorous Copper tertiary composite-coated mild steel sample
Wt. %	weight percent
Cu	Copper
INTEMP	Inverse heat conduction problem
HTC	Heat transfer coefficient
DAQ	Data Acquisition Unit

1. Introduction

Thermal operations are quite common in core industries such as chemical, mechanical firms, refineries and pharmaceutical industries [1]. The usage of heat exchangers is a forefront priority among them [2]. The main function of heat exchangers is transferring thermal energy from one fluid to another at a low range of temperatures. The operation of heat exchangers is quite easy and effective. Its budget-friendly operation also suits the purpose quite well [3]. On the other hand, mild steel is the most common constructional unit used for manufacturing industrial heat exchangers. However, along with its large range of advantages, there are various disadvantages in terms of count [4]. To overcome these problems, different techniques are used by industries, such as adsorption [5] and organic procedures [6, 7, 1]]. One such effective method is coating. It not only improves the surface texture but also enhances different mechanical and thermal properties [8]. Electroless nickel phosphorous coatings are quite famous for such purposes [9]. However, as the study is related to the enhancement of heat transfer analysis, a novel approach of utilizing copper composite coatings along with nickel and phosphorous has been analyzed [10]. The literature proves that different types of copper-based composite coatings are currently utilized to enhance the heat transfer properties of different surfaces [11]. Hence, the current study focuses on utilizing different nickel-based composite coatings to enhance the heat transfer phenomenon for shell and tube heat exchangers (the focus of this study).

The shell and tube heat exchanger mainly comprises two parts: (i) the shell side and (ii) the tube side. In a typical heat exchanger, the cylindrical shell is the housing of mounted tubes placed parallel to the shell axis [12]. The direction of fluid flow inside the shell and tube heat exchanger can be parallel and cross-flow. In addition, the fluid flow direction is guided by several insertion-like vertical plates, baffles, etc., which play an important role in improving the heat transfer mechanism [12]. An illustration of a typical shell and tube heat exchanger is provided in Figure 1, with cold fluid on the tube side and hot fluid on the shell side.

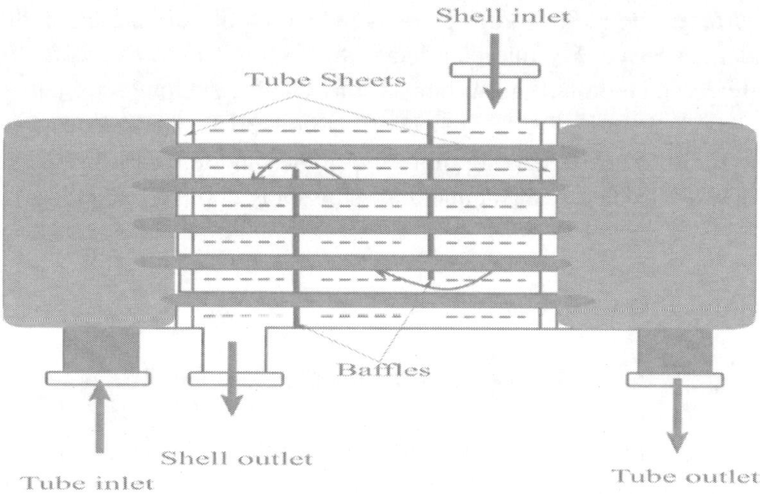

Figure 1. An illustration of a shell and tube heat exchanger.

The literature shows different studies on the utilization of nickel-based composite coatings for enhancing several engineering purposes, such as corrosion [13], cooling rate enhancement and heat transfer studies [14], and antifouling phenomena [15]. A study illustrating nickel-based chemical surface modification of nickel foil was used to develop various kinds of nickel-based electrochemical coatings that are effective for surface pool boiling and efficiently enhance the heat transfer coefficient (HTC) [16], indicating that modified steel surfaces using nickel composite coatings are greatly effective for the purpose of cooling rate enhancement and heat transfer capacity. Another piece of literature experiments with the electroless deposition of Ni-Cu-PTFE on mild steel surfaces to alter the surface morphology of the base metal, which results in great efficiency for antifouling characteristics and eventually an improved heat transfer coefficient [15]. The process of coating is among those basic processes that serve as a fine protective layer on different surfaces [17], along with various multifunctional activities [18]. Similar to Biswas et al. [19], in his experiment, he correlated the tribological behavior of a nickel-phosphorous (Ni-P)-based coating with the heat treatment temperature and its duration. He inferred that both heat treatment time and temperature play a key role in affecting the microstructure formation of coatings, thus having a profound effect on hardness, wear, and friction performance that are applied to the synthesized material. Furthermore, there are many such studies that have modified the base fluid with surfactants [20], additives [21], double-layered hydrocarbons [22], and nanomaterials [23], producing a new group of nanofluids for improved heat transfer analysis. However, the surface modification of the base metal (mild steel for the current case), especially utilizing nickel-phosphorous-based electroless coatings for cooling rate improvement and a better heat transfer analysis, is truly scant.

Thus, the current study utilizes the concept of electroless coating to modify the surface of mild steel, which is considered the structural unit of shell and tube heat exchangers, to enhance the heat transfer properties and improve the cooling rate and heat transfer coefficient. The study gives a comparison between the heat transfer analysis of electroless Ni-P and Ni-P-Cu coated mild steel with an optimization of the best suit between them. The heat transfer analysis is carried forward using a spray cooling setup. SEM, EDX, and XRD analyses of both samples showcase the surface morphology and phase structure analysis. A brief overview of viscosity analysis and surface roughness studies predicts the suitability of surface modification for the required research objective. The microhardness of the sample before and after the heat transfer study was analyzed and reported. An in-house spray cooling setup was used to analyze the heat transfer using INTEMP software. The results using both Ni-P and Ni-P-Cu were recorded and compared. The study forecasted and optimized that the sample Ni-P-Cu mild steel is far more suitable than the Ni-P-coated steel sample under the spray cooling technique.

2. Experimental Procedure

2.1. Synthesis of Ni-P-Cu Coatings at Various Cu Concentrations

In the current study, the substrate chosen is AISI 1040 mild steel for the deposition of electroless coating. According to the requirements of various tests, the sample measurements are (i) a flat sample (with dimensions of 15 mm x 15 mm x 1 mm) and (ii) a cylindrical sample (dimensions of Ø6 x 30 mm). The samples are supplied by the local vendor. Both types of substrate are introduced to a grinding process following the roughness scale grade N_5 (i.e., Ra ~ 0.4 ± 1% μm as per the standardized scale of BS EN ISO 1302:2002. The maintenance of surface roughness is treated as an important step in the coating procedure, as the coating deposition is highly dependent on the surface roughness according to a previous study [24]. The sample must be pretreated before starting the experiment. Thus, in the following, Figure 2 illustrates the different sample preparation techniques in detail.

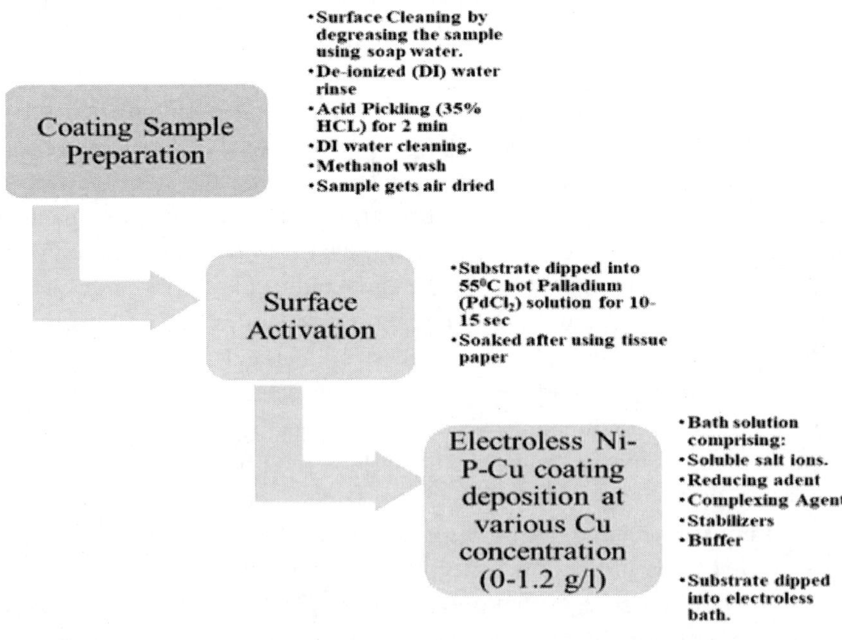

Figure 2. Preparatory steps before starting the Ni-P-Cu coating at different concentrations.

The coating substrate needs to be activated by dipping the sample in a solution of warm palladium chloride with 0.1 g/l of $PdCl_2$ and 0.1 ml of dilute HCL in 100 ml of DI water. The surface activator lowers the initial energy barrier of the reaction kinetics. The chemical composition and operating conditions of the bath solution at different concentrations are shown in Table 1. The experimental setup is shown in Figure 3. Thus, the introduction of this step ensures great adhesion of the coating deposit on the mild steel substrate. The treated samples are hung using clips and submerged into the Ni-P-Cu plating bath using thin wires. The experiment was repeated at different copper concentrations varying from 0 g/l to 1.2 g/l. The cylindrical and flat samples are coated separately, each using 200 ml of bath solution at a time. The operating conditions and chemicals used are tabulated as follows:

Table 1. Chemical composition and operating conditions of bath solutions with different concentrations of Ni-P-Cu

S. No.	Composition of bath solution and operating condition	Ni-P-Cu-0 (with copper conc approximately 0 g/l)	Ni-P-Cu (copper conc varying from (0.2-1.2) g/l
1	Nickel sulphate (g/l)	5	5
2	Nickel chloride (g/l)	5	-
3	Sodium citrate (g/l)	-	3
4	Sodium succinate (g/l)	2.4	-
5	Sodium hypophosphate (g/l)	3	3
6	Copper sulphate (g/l)	-	0.04 (0.2 wt%), 0.08, 0.12, 0.16, 0.2 (1 wt%)
7	pH	4.5-5.5	4.5-5.5
8	Temperature	85-90	85-90
9	Bath Temperature	200	200
10	Time given (min)	45	45

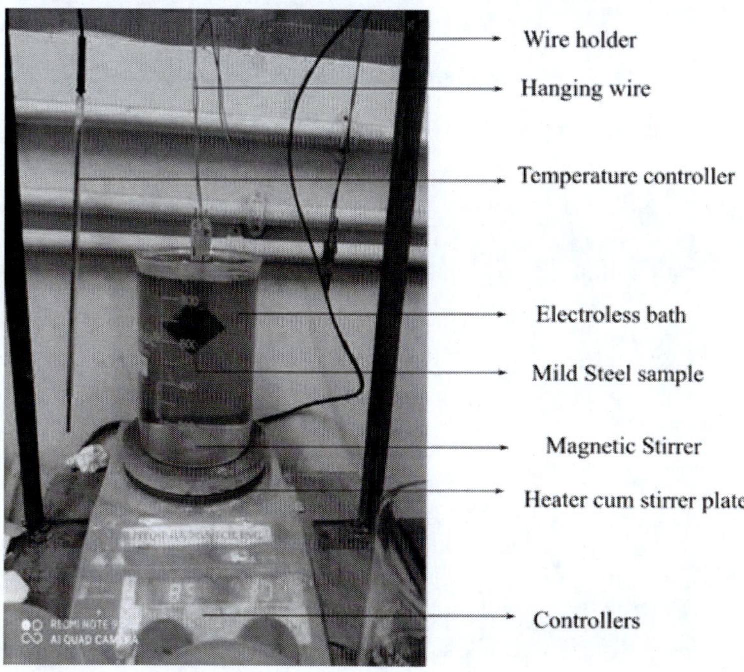

Figure 3. Experimental setup.

2.2. Surface Morphology Characterization and Analysis Tests

To observe the surface morphology and coating microstructure, the rectangular samples are analyzed using SEM on a Zeiss FESEM-MERLIN associated with EDAX-AMETEK software for EDS and elemental mapping. The sample thickness is measured using FESEM images (JEOL JSM 6410F Instrument) of polished and ground edges of the specimen associated with INCA Analyzer B Oxford Instruments for the corresponding EDS study. An angular grinder (Bosch GWS 6-125 Professional) was used to cut the specimen for the FESEM study. The coating phases are analyzed with XRD analysis under Cu Kα radiation with a wavelength of 1.54059 Å. The varying range of analysis is kept at approximately 2Θ = 10-90^0 with a step size of 0.05^{0,} and the time per step is maintained at 0.4 s.

Figure 4 (a). Experimental heat transfer analysis.

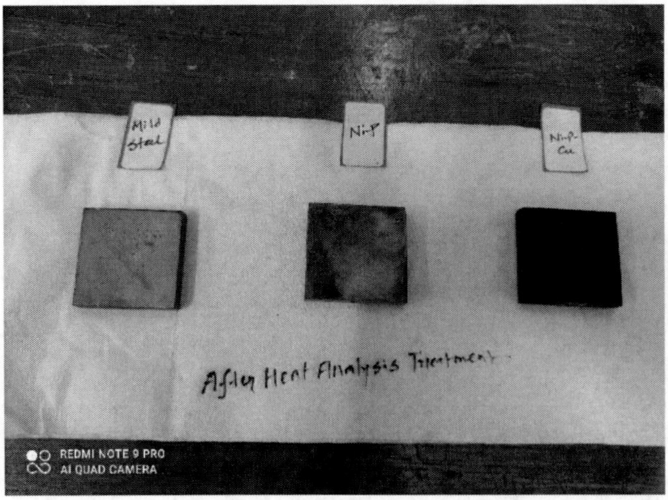

Figure 4 (b). Uncoated and coated samples after experimental analysis.

To study the changes in the sample coating thickness, a sample was prepared following the procedure in the literature [25]. A thickness study after heat treatment is presented in a later discussion section. The data of the most suitable and optimized sample, Ni-P-Cu-coated mild steel, have been presented, which have been analyzed and studied in detail in the following. The viscosity measurement was completed using a Cannon-Fenske viscometer, which works on the principles of ASTM D445 and ISO 3104. The roughness analysis simply underwent atomic force microscopy (AFM) analysis with a large scanner measurement of 90 μm x 90 μm and a small scanner specification of 10 μm x 10 μm.

For the analysis, a flat rectangular sample was used to study the microhardness variation. The microhardness measurements on rectangular samples were performed with the UHL-VMHT 001 microhardness tester at a 100 gf load. An average of five readings was calculated and reported for the current study. Additionally, the typical characteristic indentation is presented below in the microhardness study, as shown in Figure 4.

For executing heat transfer analysis, an in-house setup consisting of four components was used [26, 27].

1. Test Specimen comprising:
   - Mild steel (AISI 1040 with dimensions of 30 mm x 30 mm x 6 mm with a hole in the middle having a depth of 15 mm)
   - K-Type Thermocouple
2. The heating unit consists of:
   - Furnace (Up to 500^{0}C),
   - Temperature Controller
3. Cooling Unit
   - Tank, stirrer with motor, temperature sensor, heating element, pump, rotameter, solenoid valve, pressure gauges, valve, nozzle, bypass outlet, and drainage outlet
4. Data acquisition unit
   - Personal computer (PC)
   - Data Acquisition Unit (DAQ)

3. Results and Discussion

3.1. Surface Morphology Analysis Using SEM and XRD

The SEM images presented below show and discuss the detailed surface morphology of the uncoated and coated samples. Researchers have already discussed the surface morphology and its analytics in previous works [27]. Here, mainly the morphological changes of the samples after the heat treatment analysis are provided and discussed properly.

For experimentation with heat transfer analysis, samples are heated up to 400°C. Before this heat treatment, a sphere-shaped nodular structure is generally visible on Ni-P-coated steel samples, which changes to a coarse grain nodular structure for Ni-P-Cu-coated steel. After the heat treatment, this nodular surface morphology tends toward a flat round-shaped structure with much denser accumulation, as visible in the FESEM images of the after-treatment samples provided above in Figure 5. The reason behind such a change is the coarsening of the granular particle. Additionally, the porous structure that remains effective in plated Ni-P-coated samples due to the presence of absorbed gases decreases in size and availability [28], mainly due to the formation of oxide bonds at the coating surface, which can be further validated following elemental composition using the EDAX study.

The EDAX analysis shown in Figure 6 provides the weight percentage of oxygen to be ~5% along with the presence of phosphorous (~3 wt%) and nickel (~80%) for Ni-P and approximately 12% oxygen together with the presence of copper (~50%), phosphorous (~0.5 wt%) and nickel (~33%) for sample Ni-P-Cu. Different iron and oxide phases are identified at higher temperatures, which may be due to the interdiffusion of the base metal with the coating components along with the occurrence of oxidation at the same time.

Figure 5. FESEM images of (a) MS (b) Ni-P & (c) Ni-P-Cu coated sample after the heat treatment analysis.

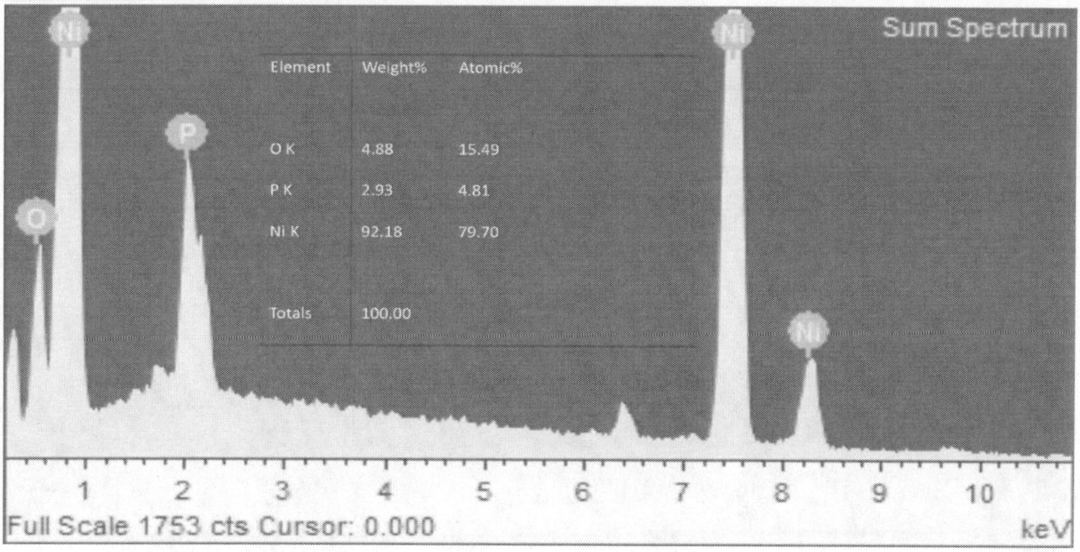

(a)

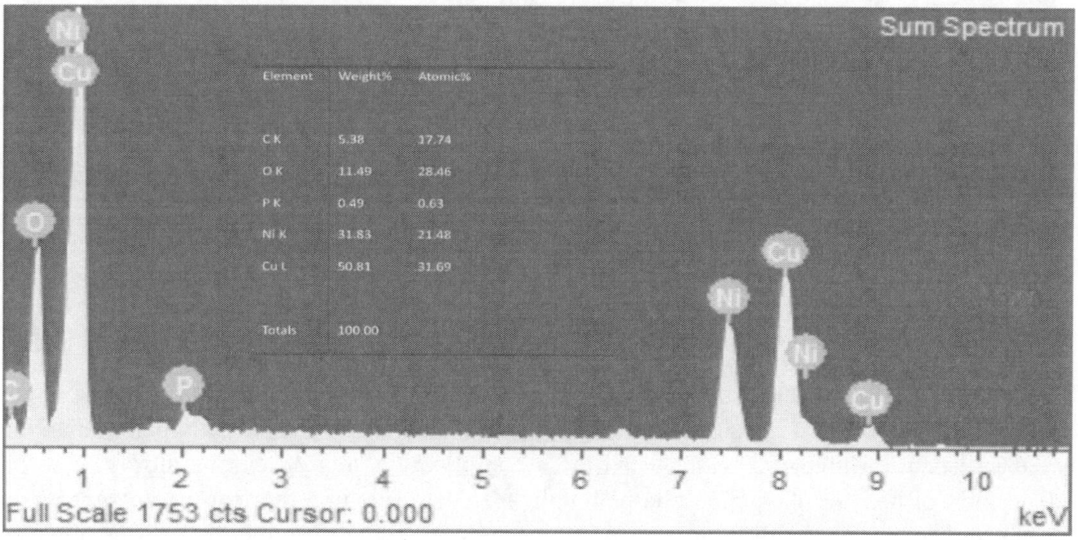

(b)

Figure 6. (a) Elemental composition of Ni-P and (b) Ni-P-Cu.

3.2. Thickness and Viscosity Study

Figure 7 shows the thickness changes of the most suitable and optimized sample for the study among the three, which is Ni-P-Cu-coated mild steel. The coating thickness of the sample has already been studied previously [29]. The currently presented picture clearly depicts that the heat treatment of the samples due to heat transfer analysis definitely increases the coating thickness by quite a bit. The thickness of the coated Ni-P-Cu samples increases from approximately 25 μm to ~69 μm as per the study evaluated. This increased coating thickness not only increases the mechanical stability of the coated material but also protects it from erosional depletion caused by fluid flow on the sample surfaces.

Table 2. Viscosity analysis of different samples

Serial Number	Sample	Viscosity(in mPa-sec)	Remarks
1	Mild steel (without coating)	1.05	
2	Ni-P (as plated)	1.57	
3	Ni-P-Cu-1 (as plated) (with Copper concentration 1 g/l)	1.35	Best Optimized study

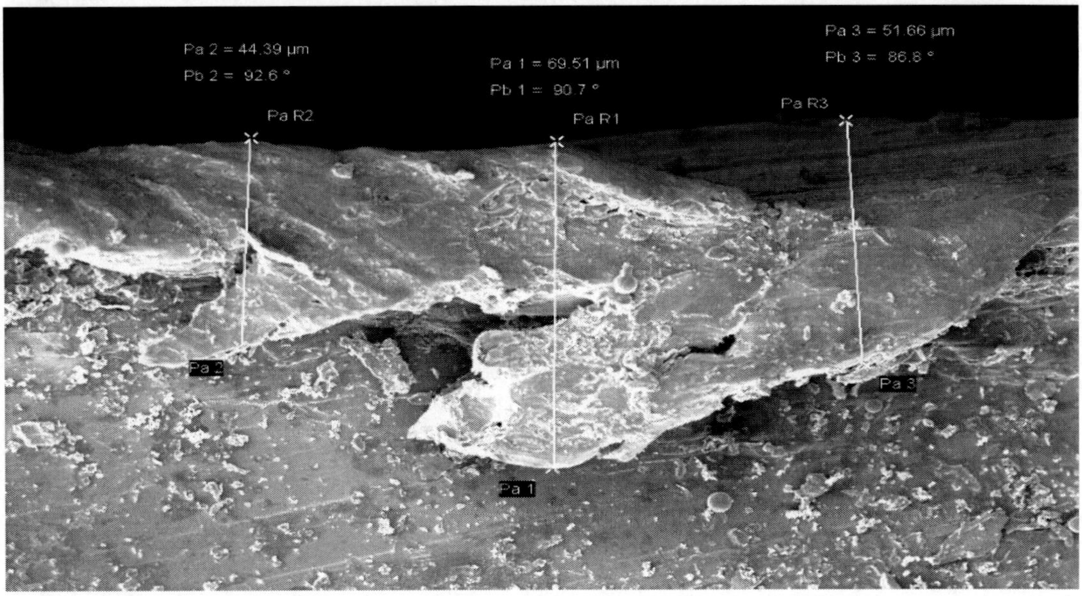

Figure 7. SEM image of the cross-cut section of the Ni-P-Cu-coated sample.

The application of heat transfer is highly related to the viscosity of the coating solution. Table 2 shows that the viscosity of the coating suspension for Ni-P is slightly higher than that of the suspension prepared with the Ni-P-Cu complex. The reason can be attributed to the deposition of heavier phosphide phases within the coating region that can be analyzed from XRD studies, as well as the action of the Van der Waal force of attraction and the adhesive force between suspension molecules and the substrate particles [30]. Higher viscosity can be beneficial for a good residence time of the coating suspension, but lower viscosity proves to be an optimum parameter for a uniform coating surface, mainly because of its ease of spreadability [31]. With a lower viscous force, it is easier for the suspension to spread on the base material, giving a uniform finish to the coated particles, which results in the formation of more nucleation sites and accelerates the heat transfer capacity of the experimental suspension [31]. Henceforth, the viscosity study indicates the favorability of the Ni-P-Cu composite coating for the selected study.

3.3. Sample Roughness Analysis

Table 3 shows the surface roughness of different samples, starting with uncoated mild steel, which reports a roughness value of 28.6 nm, followed by Ni-P-coated mild steel with a roughness value of 40.1 nm, and a Ni-P-Cu-coated sample with a roughness reading of 392 nm. The analysis clearly points toward an increase in roughness with the formation of composites. The roughness of Ni-P-coated samples has increased from that of the uncoated ones, which further rises with quite higher values in the case of Ni-P-Cu-coated samples. This phenomenon directly affects the heat transfer analysis of the synthesized samples [32], which can be vividly depicted by the pattern of cooling curves formed during the analysis. The roughness increases with the addition of coating composites, which in turn increases the number of additional nucleation sites [33]. The rise in nucleation sites improves the cooling rate, which is further discussed in detail in Section 3.5.1.

Table 3. Surface roughness of uncoated and different composite coating samples

Serial Number	Sample	Root mean square roughness value (in nm)	Surface Roughness Illustration (by AFM study)
1	Mild steel (without coating)	28.6	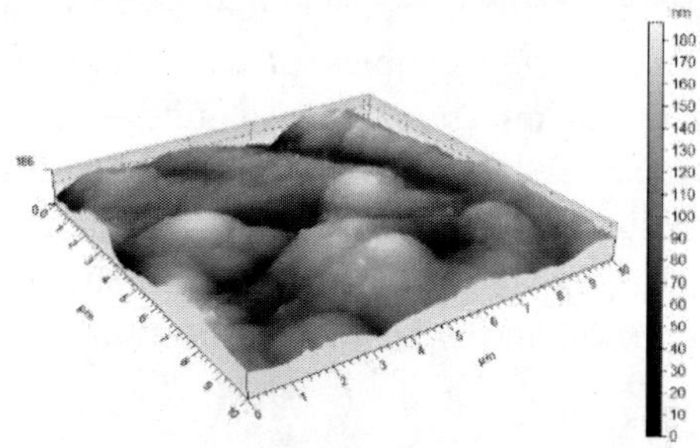
2	Ni-P	40.1	
3	Ni-P-Cu	392	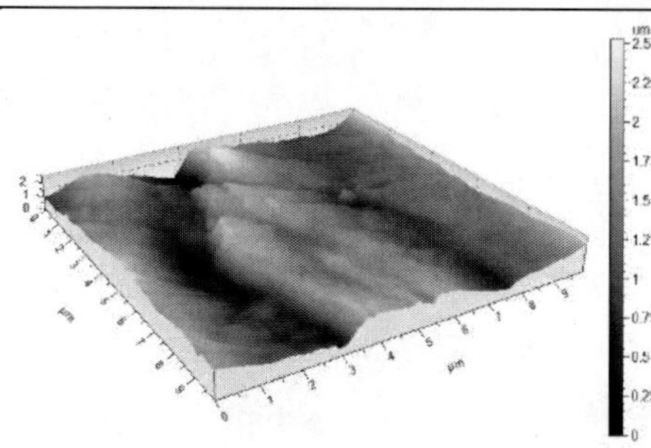

3.4. Changes in the Microhardness Behavior of the Coated Sample

The comparison of coating microhardness of uncoated and composite-coated mild steel as-prepared is presented and tabulated in detail in previous literature [33]. The current study focuses more on the microhardness analysis of both coated and uncoated materials after sample heat treatment and heat transfer analysis using synthesized lab-scale wastewater. The major application of the present study is to analyze the cooling effect and heat transfer of the synthesized materials. Thus, the change in sample microhardness alongside the heat treatment following the wastewater treatment plays a crucial role in sample preparation and selecting the appropriate one for the role.

Figure 8 clearly depicts that the hardness of the coated samples clearly increases in quite a large range after the treatment because of the formation of stronger Ni3P precipitates [25, 26] and the solid solution strengthening technique due to the application of heat [19]. The microhardness of the as-plated and treated coated samples is recorded to be 515 HV for Ni-P-Cu-coated samples before heat treatment, which is approximately 72% higher than that of the uncoated samples, proving the inclusion of harder phases Ni3P and Ni8P3 within the coating realm. This hardness quite efficiently increases upon application of heat, raising the microhardness to 626 HV for the above-stated samples, which records a 23% rise compared to the as-prepared Ni-P-Cu composite samples. The reason can be the obstruction of dislocated material movement within the crystal matrix, along with the inclusion of virgin phosphorous with nickel during heating application.

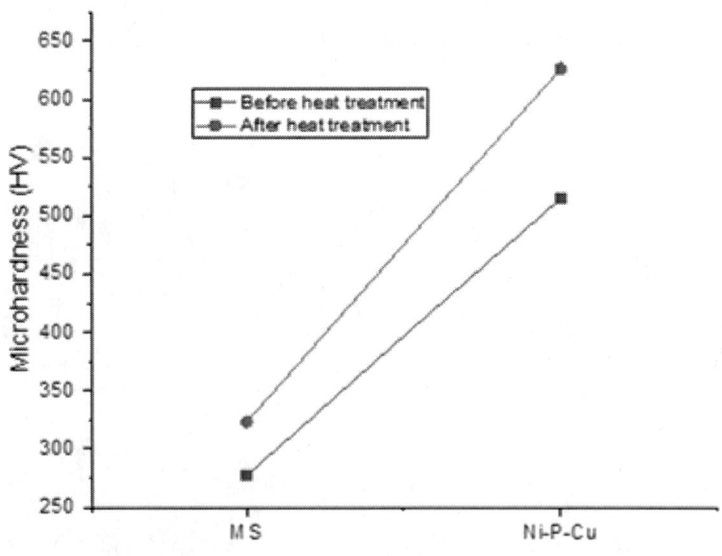

Figure 8. Variation in the sample microhardness before and after the heat transfer analysis.

3.5. Measurement of the Heat Transfer Coefficient

The base liquid used for the experiment is water itself, following the techniques and methods used in some of our previous works [28].

3.5.1. Determination of the Heat Transfer Coefficient

For the process of spray cooling, the heat transfer coefficient (HTC) determines the convective heat transfer between the hot plate and impinging droplets, which is again proportional to the available surface flux for extraction and the temperature difference between the surface and coolant [27]. The following equation properly derives the expression of the heat transfer coefficient.

$$h = \frac{q}{Ts - Tc}$$

where q = surface heat flux, W/m^2

Ts = surface temperature, K

Tc = coolant temperature, K

Table 4. The heat transfer coefficient

S. No.	Sample Name	Heat transfer coefficient (HTC) (in KW/m^2. K)
1	Uncoated sample (only Mild steel)	2.329116629
3	Ni-P-Cu coated sample	3.686047361

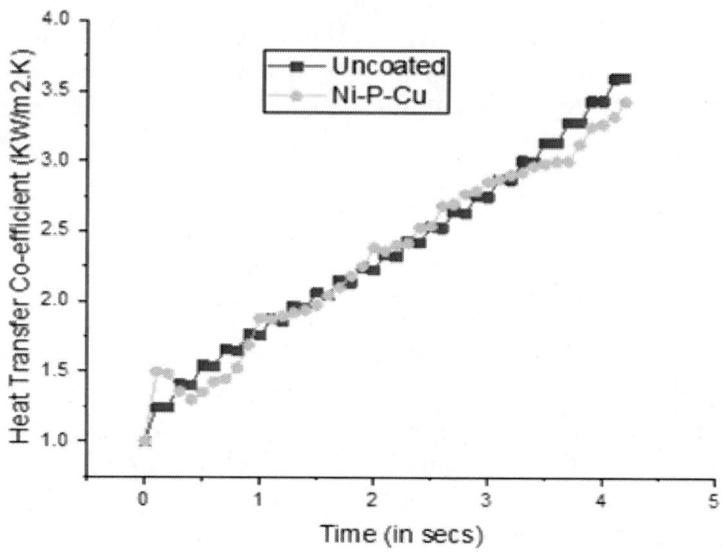

Figure 9. Demonstration of the heat transfer coefficient.

The heat transfer coefficients derived from the following equation are tabulated in Table 4.

Figure 9 shows the change in the heat transfer coefficient in kW/m2. K with time measured in seconds. The graphical plot points toward the clear analysis that the changes in HTC are much faster and dominant for the coated sample coded Ni-P-Cu than those of the uncoated mild steel, validating the cooling curve demonstrated above in a justified way. The work clearly follows the pattern of some previous works carried forward following a near-similar setup [27, 28].

3.5.2. Measurement Uncertainty

The uncertainties of the current work have been measured using the PTC 19.8–1983 methodology according to the ASME test code [29]. The reported errors followed the terms of bias and precision error values. The origin of the measuring instrument faults can definitely be minimized by calibrating the instruments before the start of the measurement. Flow was determined using a calibrated rotameter. The calibration of the thermocouples (TC) used for the experiment is performed using normal surrounding temperature water at approximately 1000°C. The furnace is also properly calibrated before use. The TC is used for measuring the temperature of water and is calibrated at normal room temperature and the boiling point of the fluid used. The shortcoming in the measured quantities explains the precision errors of the study.

The temperature measurement for the current study was completed using INTEMP software. Most of the prediction error has been successfully minimized using L-curve analysis. It makes sure to eliminate the unintended data to which inverse problems are quite sensitive. The DAQ NI-9211 (National Instruments Corporation) and extension wires of the TCs are also maintained properly. An offset of 0.7°C and a sensitivity value $\leq$0.07°C in the zone of temperature measurement were noticed in the DAQ. The wire extension used alongside the thermocouple is also calibrated to minimize the error. Additionally, an airtight thermocouple tip is maintained by filling the air gaps using conducting paste before inserting the thermocouples [27]. All experiments are repeated three times to ensure accuracy. Their average values are taken and reported. Furthermore, the maximum error consideration of 10% was maintained between the calculated and measured parameters.

Conclusion

The conclusion of the present study can be derived from the comparison study between the uncoated samples and the coated samples of Ni-P-Cu, which concludes with the fact that the coated samples are far superior to the uncoated ones for all the analyses put forward. The study analyses various samples with respect to their surface morphology, phase transformation microhardness, and novel use of wastewater for heat transfer analysis, optimizingg Ni-P-Cu as the best suit. The important findings from the study are stated as follows:

1. The nodular structure of the samples is seen and analysed, and the coarse-grained structure is accustomed to Ni-P-Cu-coated steel that changes to a flat round-shaped structure with much denser accumulation due to coarsening of the granular particle. The porous structure decreases in size and availability due to the formation of an oxide bond at the coating surface, as validated by the EDAX study.
2. The coating microhardness improved for the electroless-coated samples. The microhardness of the as-plated coated samples is recorded to be 515 HV for Ni-P-Cu-coated samples before heat treatment, which is approximately 72% higher than that of the uncoated samples, proving the inclusion of harder phases Ni3P and Ni8P3 within the coating realm. This hardness rises to 626 HV for both the above-stated samples, which records a 23% rise compared to the as-prepared Ni-P-Cu

composite samples due to the obstructed material movement and coarsening nature due to the inclusion of copper and oxide composite.

3. The heat transfer analysis concludes that the sample coated with Ni-P-Cu takes the least amount of time for cooling, mainly due to the addition of copper (a good conductor of heat) in the form of various composites. The heat transfer coefficient analysis further added value to the theory by calculating the maximum heat transfer coefficient for Ni-P-Cu at approximately 3.69 KW/m^2. K. The study can be readily scaled up for the application of pilot plant shell and tube heat exchangers.

References

[1] S. Bhattacharyya, Waste Management and Resource Efficiency, Springer Singapore, 2019. https://doi.org/10.1007/978-981-10-7290-1.

[2] Rezaei, M., Mahidashti, Z., Eftekhari, S., & Abdi, E. (2021). A corrosion failure analysis of heat exchanger tubes operating in petrochemical refinery. *Engineering Failure Analysis, 119*(October 2020), 105011. https://doi.org/10.1016/j.engfailanal.2020.105011.

[3] Faes, W., Lecompte, S., Ahmed, Z. Y., Van Bael, J., Salenbien, R., Verbeken, K., & De Paepe, M. (2019). Corrosion and corrosion prevention in heat exchangers. *Corrosion Reviews*, *37*(2), 131–155. https://doi.org/10.1515/corrrev-2018-0054.

[4] Kakac, S., Liu, H., & Pramuanjaroenkij, A. (2002). *Heat exchangers: selection, rating, and thermal design*. CRC press.

[5] Nguyen, T. K., Sheikholeslami, M., Jafaryar, M., Shafee, A., Li, Z., Chandra Mouli, K. V. V., & Tlili, I. (2020). Design of heat exchanger with combined turbulator. *Journal of Thermal Analysis and Calorimetry*, *139*(1), 649–659. https://doi.org/10.1007/s10973-019-08401-7.

[6] Bhattacharyya, S., Banerjee, P., Bhattacharya, S., Rathour, R. K. S., Majumder, S. K., Das, P., & Datta, S. (2018). Comparative assessment on the removal of ranitidine and prednisolone present in solution using graphene oxide (GO) nanoplatelets. *Desalination and Water Treatment*, *132*, 287–296. https://doi.org/10.5004/dwt.2018.23170.

[7] Bhattacharyya, S., Das, P., & Datta, S. (2019). Removal of Ranitidine from Pharmaceutical Waste Water Using Activated Carbon (AC) Prepared from Waste Lemon Peel. *Waste Water Recycling and Management*, 123–141. https://doi.org/10.1007/978-981-13-2619-6_11.

[8] Bhattacharyya, S., & Ghosh, S. K. (2019). Waste Valorization and Recycling. In *Waste Valorization and Recycling*. Springer Singapore. https://doi.org/10.1007/978-981-13-2784-1.

[9] Pungaiya, S. (2018). *A Review of Surface Coating Technology to Increase the Heat Transfer*. 7(5), 458–465. https://doi.org/10.18178/ijmerr.7.5.458-465.

[10] Kundu, S., Das, S. K., & Sahoo, P. (2018). *Tribological Behavior of Electroless Ni-P Deposits Under Elevated Temperature*. https://doi.org/10.1007/s12633-016-9450-8.

[11] Biswas, A., Das, S. K., & Sahoo, P. (2021b). Oxidation issues during heat treatment and effect on the tribo-mechanical performance of electroless Ni-P–Cu deposits. *Proceedings of the Institution of Mechanical Engineers, Part L: Journal of Materials: Design and Applications*, *235*(7), 1665–1685. https://doi.org/10.1177/1464420721999823.

[12] Bai, D., Zhang, C., Chen, Z., Zhang, Y., Li, G., & Lu, X. (2022). Thermal conductivity characterization of Ni60A/WC composites with different copper dilutions. *Materials Letters*, *311*(November 2021), 131546. https://doi.org/10.1016/j.matlet.2021.131546.

[13] Padhee, S. (2014). Controller design for temperature control of heat exchanger system: Simulation studies. *WSEAS Transactions on Systems and Control*, *9*(1), 485–491.

[14] Zhao, G. L., Zou, Y., Hao, Y. L., & Zou, Z. D. (2015). Corrosion resistance of electroless Ni-P/Cu/Ni-P multilayer coatings. *Archives of Metallurgy and Materials*, *60*(2A), 1003–1008. https://doi.org/10.1515/amm-2015-0250.

[15] Ren, L., Cheng, Y., Yang, J., & Wang, Q. (2020). *Study on Heat Transfer Performance and Anti - Fouling Mechanism of Ternary Ni-W-P Coating*. 1–20. https://doi.org/10.3390/app10113905.

[16] Cheng, Y. H., Chen, S. S., Jen, T. C., & Zhu, Z. C. (2014). *Effect of copper addition on the properties of electroless Ni-Cu-P coating on heat transfer surface. Dlc.* https://doi.org/10.1007/s00170-014-6437-8.

[17] Li, Y., Liu, Z., & Zheng, B. (2015). International Journal of Heat and Mass Transfer Experimental study on the saturated pool boiling heat transfer on nanoscale modification surface. *International Journal of Heat and Mass Transfer, 84*, 550–561. https://doi.org/10.1016/j.ijheatmasstransfer.2014.12.064.

[18] Bhattacharyya, S., Manna, S., & Medda, S. K. (2019). ZrO_2 incorporated TiO_2 based solar reflective nanocomposite coatings on glass to be used as energy saving building components. *SN Applied Sciences*, *1*(11), 1–8.

[19] Bhattacharyya, S., Medda, S. K., & Naskar, M. K. (2021). A Preparative Approach of TiO_2-ZrO_2 Coating Using Aquo-Based TiO_2 Precursor Useful for Light Reflective Application. *Transactions of the Indian Ceramic Society*, *80*(4), 227–233.

[20] Biswas, A., Das, S. K., & Sahoo, P. (2017). Correlating tribological performance with phase transformation behavior for electroless Ni-(high)P coating. *Surface and Coatings Technology, 328*, 102–114. https://doi.org/10.1016/j.surfcoat.2017.08.043.

[21] Ravikumar, S. V, Haldar, K., Jha, J. M., Chakraborty, S., Sarkar, I., Pal, S. K., & Chakraborty, S. (2015). International Journal of Thermal Sciences Heat transfer enhancement using air-atomized spray cooling with water - Al_2O_3 nanofluid. *International Journal of Thermal Sciences, 96*, 85–93. https://doi.org/10.1016/j.ijthermalsci.2015.04.012.

[22] Ravikumar, S. V, Jha, J. M., Haldar, K., & Pal, S. K., https://doi.org/10.1115/1.4029815.

[23] Chakraborty, S., Sarkar, I., Ashok, A., Sengupta, I., Pal, S. K., & Chakraborty, S. (2018). Synthesis of Cu-Al LDH nanofluid and its application in spray cooling heat transfer of a hot steel plate. *Powder Technology, 335*, 285–300. https://doi.org/10.1016/j.powtec.2018.05.004.

[24] Sarkar, I., Chakraborty, S., Roshan, A., Behera, D. K., Pal, S. K., & Chakraborty, S. (2018). Application of TiO_2 nanofluid-based coolant for jet impingement quenching of a hot steel plate. *Experimental Heat Transfer*, *00*(00), 1–15. https://doi.org/10.1080/08916152.2018.1517835.

[25] Biswas, A., Das, S. K., & Sahoo, P. (2021a). Effect of copper incorporation on phase transformation behavior of electroless nickel–phosphorous coating and its effect on the tribological behavior. *Proceedings of the Institution of Mechanical Engineers, Part L: Journal of Materials: Design and Applications, 235*(4), 898–916. https://doi.org/10.1177/1464420720981602.

[26] Bhattacharyya, S., Mohanty, D., Kumar, P., Das, S. K., Sahoo, P., Pal, S. K., & Chakraborty, S. (2023). A corrosion and tribo-failure analysis of Ni-P-Cu coated mild steel (AISI-1040) at varied copper concentration Indian Institute of Technology Kharagpur. *Engineering Failure Analysis*, 107063. https://doi.org/10.1016/j.engfailanal.2023.107063.

[27] Biswas, A., Das, S. K., & Sahoo, P. (2019). A comparative study in microstructural and tribological aspects of phosphorus enriched electroless Ni-P and Ni-P-Cu coating. *Materials Today: Proceedings*, *19*(xxxx), 403–408. https://doi.org/10.1016/j.matpr.2019.07.625.

[28] Kumar, P., Bhattacharyya, S., & Chakraborty, S. (2022). Investigation of spray cooling in an inclined nozzle-plate configuration with varying coolant temperature. *Experimental Heat Transfer*, *00*(00), 1–18. https://doi.org/10.1080/08916152.2022.2088895.

[29] Kumar, P., Chaurasia, C. K., Das, S., Bhattacharyya, S., & Chakraborty, S. (2023). Synthesis, characterization and application of SiO_2 and CuO nanofluid in spray cooling of hot steel plate. *Heat and Mass Transfer*. https://doi.org/10.1007/s00231-023-03345-z.

[30] Abernethy, R. B., Benedict, R. P., & Dowdell, R. B. (1983). Asme Measurement Uncertainty. *American Society of Mechanical Engineers (Paper), 107*(83)..

[31] Chakraborty, S., Chakraborty, A., Sumitesh, D. A. S., Mukherjee, T., Bhattacharjee, D., & Ray, R. K. (2010). Application of water based-tiO_2 nanofluid for cooling of hot steel plate. *ISIJ International*, *50*(1), 124–127. https://doi.org/10.2355/isijinternational.50.124.

[32] Danaci, S., Protasova, L., Lefevere, J., Bedel, L., Guilet, R., & Marty, P. (2016). Efficient CO_2 methanation over Ni/Al_2O_3 coated structured catalysts. *Catalysis Today*, *273*, 234–243. https://doi.org/10.1016/j.cattod.2016.04.019.

[33] Elmardi, O. M. (n.d.). *Fundamentals of Heat Exchangers.* Associate Professor Dr. Osama Mohammed Elmardi Suleiman Khayal Department of Mechanical Engineering, Faculty of Engineering and Technology, Nile Valley University, Atbara.

[34] Jabbar, A., Yasin, G., Khan, W. Q., Anwar, M. Y., Korai, R. M., Nizam, M. N., & Muhyodin, G. (2017). Electrochemical deposition of nickel graphene composite coatings effect of deposition temperature on its surface morphology and corrosion resistance. *RSC Advances*, *7*(49), 31100–31109. https://doi.org/10.1039/c6ra28755g.

Chapter 6

Recent Developments in Leak Detection Techniques for Pipelines: A Critical Evaluation

Adarsh Kumar Arya[1,*]
S. V. A. R. Sastry[1]
Sneha[1]
Bhoomi Gupta[1]
and Murli Pujari[2]
[1]Department of Chemical Engineering, Harcourt Butler Technical University, Kanpur, Uttar Pradesh, India
[2]Department of Chemical Engineering, College of Engineering Studies,
University of Petroleum and Energy Studies, Energy Acres, Bidholi, Dehradun, Uttarakhand, India

Abstract

Researchers have developed pipeline leak detection and localization technologies in response to the enormous environmental and economic damage caused by pipeline spills. These systems use various measures to guarantee the security and dependability of pipes. Recent research on pipeline leak detection systems has been compiled, and their results and conclusions are summarized and synthesized in this paper. The findings will be helpful for pipeline integrity experts who are investigating the projected risk to the sector. In addition, the paper covers the research needs and challenges ahead for effective pipeline leak detection systems (LDSs).

Keywords: pipeline leaks, hardware-based LDS, software-based LDS, pipeline integrity, sensors

1. Introduction

Pipelines are the most reliable, safest, and cost-effective means of transporting oil and gas [1–4]. Pipelines transport vast amounts of oil and gas; hence, maintaining pipeline integrity is essential for the safe delivery of products. Pipelines are designed and operated according to strict codes. However, due to the aging and corrosion of pipe material, leaks may still occur [5–

* Corresponding Author's Email: aarya@hbtu.ac.in.

In: Sustainability in Chemical Processes through Digitalization and Green Chemistry Approaches
Editors: Susarla V. A. R. Sastry, Shravan Kumar and Ashish Kapoor
ISBN: 979-8-89113-257-3

9]. A pipeline leak results in several casualties, injuries, and environmental pollution [10]. Earth movement, pipeline aging, damage from adjacent excavation, punching and theft, subversion, terrorism, and corrosion contribute to pipeline leaks [11]. Leaks may also occur due to the poor functioning of isolating or bypassing valves in supply systems, creating a water hammer at high flow rates and velocity, resulting in a pipeline leak [12].

In the event of a pipeline leak, shutdowns increase and maintenance expenses increase. Thus, developing effective leak detection technology is vital [13]. A pipeline leak induces changes in the gas flow rate and pressure and generates ultrasonic waves. Consequently, by tracking the linked metrics, it may be possible to detect a pipeline leak.

For decades, techniques for identifying pipeline leaks have been proposed, including real-time transient modeling, tracer gases, and acoustic techniques [14]. The leak detection system's primary function is to determine the leak volume and adequately identify the pipeline when it leaks. Pressure, flow, and other indications from the pipeline must be used to their maximum potential to identify leaks, prolonged leaks, and multiple leaks that account for less than one percent of the pipe's traffic. Real-time detection of gas pipeline leaks is required to avert the worst-case scenario.

2. Major Accidents due to Pipeline Leakage

Among the most prevalent accidents, leakage from petroleum pipelines is also one of the most significant contributors to large financial losses [5].

Leaks in a pipeline disrupt the pipeline's regular operation and threaten the ecosystem and human safety. Pipeline catastrophes in the United States have cost approximately $7 billion in property damage, more than 500 human deaths, and countless wounded people during the last three decades [14].

For example, on September 6, 2010, a pipeline explosion in San Bruno, California, USA, killed eight people and wounded more than fifty. Another event occurred on July 26, 2010, when over 840,000 gallons of crude oil flowed into the Kalamazoo River in Michigan, generating an estimated $800 million in losses—pipeline leaks caused massive damage to property [15]. For example, on June 28, 2012, a pipeline transporting natural gas in British Columbia, Canada, began leaking and burning. The overall quantity of leakage was m3, and the total amount of burned area was m2. Similarly, an intercontinental pipeline carrying natural gas in West Virginia, United States, burst and caught fire on December 11, 2012 [14]. Consequently, it resulted in an economic loss of $8 million and 680 thousand in its dispersion region.

Pipelines delivering crude oil have also taken a hit. An oil leak was discovered in China's Donghuang oil pipeline on November 22, 2013. Upon entering the drainage culvert, crude oil and natural gas began to collect in the pipes, culminating in an explosion when sparks were ignited. Seventy-two people died, and another one hundred thirty-six were wounded, with an estimated direct economic loss of approximately 752 million Chinese yuan [14].

Thus, it is evident that pipelines, although the most economical mode of transportation, leaks lead to a catastrophic impact on the environmental and financial sectors. For the past several decades, researchers have developed techniques to find pipeline leaks. The paper focuses on the recent methods developed by researchers and implemented by industries to identify pipeline leaks.

3. Literature Survey

An investigation was done by [16] to detect pipeline leaks. Pressure sensors are housed in a shielded cable in an optical fiber sensor array with two FBG sensors. A voice-coil pressure wave generator, which provides sustained excitation, is used to quantify the pressure responses caused by the probe. The simulated leak may be readily seen in the coupled IRF, which proves its presence.

Al-Tofan et al., 2020, proposed a transfer matrix technique for locating liquid pipeline breaches. The method simplifies and solves the frequency domain governing continuity and momentum equations. The amplitude of the pressure oscillations is proportional to the leak's location and the pipe system's characteristics. This link is used to create leak detection curves.

Bohorquez et al., 2020, found that ANNs, or artificial neural networks, can detect pipeline leaks. Automated learning systems allow computers to tackle complex problems without being explicitly programmed. This research introduces a novel approach for predicting features in a pipeline by using neural networks. Ninety-five percent of the samples evaluated were able to find the leak with an inaccuracy of approximately 3.0 m (out of a 1,000 m pipe length).

The following paper [17] utilizes the optical fiber detection technique to identify multiple pipeline leaks. The Lyapunov function and Barbalat lemma demonstrate the stability of the fault diagnostic observer. Tests reveal that the proposed strategy is successful.

Liu et al., 2019a, developed an 'acoustic leak localization' system for detecting leaks of natural gas, a more potent greenhouse gas than CO_2. In the event of a natural gas accident, the average economic loss is substantial. New approaches are provided in the research paper for various application scenarios in distinct ways.

Liu et al., 2019b With Markov characteristics and a two-stage decision mechanism, this study provides an innovative leak detection method. A transformation may be used to convert pressure data into a Markov chain. It is essential to utilize short- and long-term detection algorithms to promptly and adequately monitor pipeline conditions. Measurements of pipeline pressure at an industrial plant and an adjacent field support the proposed method.

Xiao et al. (2019) presented an integrated leak detection approach for acoustic signals with wavelet transform and supported vector machine methods. An accuracy of 99.4 percent was achieved in determining the leak state, and 95.6 percent was achieved in classifying the typical and various leak severity situations.

A study by Rai and Kim in 2021 indicated that pipeline condition monitoring is essential for recognizing fluid leaks and the associated costs and risks. Data from pipeline failures are often used to train contemporary supervised AI models. This is a severe drawback for artificial intelligence systems. The research uses multiscale analysis, the KS test, and the Gaussian mixture model to produce a health index-based approach.

A review paper [18-20] demonstrates that major pipeline accidents occur due to leaks. Oil and gas pipeline leaks may lead to serious safety incidents and significant monetary losses. The study presents the current detection techniques and assesses their benefits, limits, application scenarios, and performance.

The subject of this research is the detection of several leaks in a water-filled conduit utilizing transient waves. According to numerical simulations, five leaks can be reliably recognized, and the total number of leaks can be precisely calculated using the technique [21-23].

A novel convolutional neural network (SE-CNN) is being developed to detect gas pipeline leaks. In 6 different categories, the SE-CNN can achieve an average accuracy of 94.3% with a detection time of 1.04 s. Real-time leak detection in industrial settings is made possible because of its accuracy and efficiency [24–26].

An aspherical detector that can go very near a leak location and record the sound of the leak from the pipeline's inside is being developed. During forward motion, collisions and friction produce acoustic sounds that obscure leak acoustic signals. The pipe is filled with water, pressured, and tiltable, and the SD may stay stationary or roll to gather a range of leak sound samples from the system [27–29].

Pipelines in remote locations might be difficult to monitor for damage. Despite several attempts, only a few leak detection systems can determine the leak's location and volume simultaneously. The researchers suggest dynamic modeling as a way to circumvent this restriction [30–32].

Pressure testing has been used for decades to detect pipeline leaks. Although this technique is uncomfortable and time-consuming, it is challenging to locate the specific site. This study focuses on developing model control systems that may be used on and off the job to find oil leaks. A prototype test rig was used to accomplish this project [33–35].

Pipeline status monitoring is severely hindered by partial data issues caused by equipment failures or network interruptions. An adversarial network (tnGAN) based on the tri-network form is proposed in this article to address leak detection issues using inadequate sensor data. Network architectures are suggested to identify pipeline status by calculating the similarity of latent properties among samples. It may also combine the output and characteristics of discriminative networks to obtain a pipeline leak detection result [36–38].

Nonlinear first-order hyperbolic partial differential equations are utilized to model a long pipeline leak detection and location issue. Models for leak detection, amount estimation, and location estimation are constructed using data collected in typical and leaky cases [39–41].

This study aims to find and locate leaks in a complex pipeline manifold structure. The researchers rebuilt the robot's position trajectory using information from the IMU's gyroscope and accelerometer. The wavelet transforms the sound input into feature vectors, and the Bayesian classifier categorizes each feature to detect the leak [42–44].

Pipeline leaks will cause energy waste, environmental harm, and safety risks. Methods for detecting and locating pipeline leaks are provided in this paper using a model-free approach to anomaly detection. An acoustic signal is broken down into "subsignals" using the zero-crossing line. Identifying the leak's location is possible by employing this information and cross-correlation analysis.

Transient-based leakage detection based on multiple signal classification (MUSIC) approaches was used to locate leaks. A one-dimensional search for the leak location and the pipe results in a low computational cost. It can detect leaks if the distance between them is less than half the length of the shortest probing wavelength [45–46].

To ensure the safe functioning of pipelines, leak detection is critical. External techniques (hardware) and internal methods are the two primary categories of leak detection methods (software). A mathematical model of transient gas flow is proposed as a unique leak detection method [21].

An improved spline-local mean decomposition (ISLMD) was used to remove noise interference. Using a convolutional neural network for leak detection in image identification is offered as a technique for picture recognition. Upstream and downstream transmitter signal time delays must be taken into account. In this way, the leak's location may be pinpointed with accuracy [45].

An acoustic emission (AE) approach for identifying and locating a pipeline leak and its causes is discussed. Sound sensors positioned on the pipeline's exterior surface detect acoustic signals generated by a leak and are used to detect the breach. When an auditory signal arrives at two nearby sensors simultaneously, the time difference between them determines the leak's location [20].

The study examines leak detection systems (LDSs) and information fusion techniques for pipeline monitoring. There has been a detailed examination and comparison of all the most current publications [10].

In the Raman distributed fiber sensor (RDFS) system, researchers propose a dynamic threshold identification method (DTIM) for identifying pipeline leaks that are both theoretically and experimentally sound. Dispersed temperature readings from the RDFS may reveal the location of pipeline leaks. Pipeline leak detection and monitoring are made easier using this technology's distributed monitoring solution [42].

Regarding efficiency and cost-effectiveness, pipeline transportation of fluids such as oil, natural gas, and water is the best option. Pipelines are susceptible to leaks due to corrosion, aging, weld faults, or damage from a third party. With an intelligent hybrid technique, this work employs KPCA to locate pipeline leaks and Cas-SVDD to describe cascade support vector data (Cas-SVDD) [46].

The current study seeks to detect gas distribution leaks with a 0% false alarm rate under pseudorandom boundary situations (FAR). Under leak-free circumstances, the intake pressure and mass flow rate data were utilized to predict the output mass flow rate using the Hammerstein model. Several leak sites, variable leak sizes, and different signal-to-noise ratios were used to test the suggested approach for leak identification (SNR). Effective detection of 0.15 kg/s (3%) leakage under transient boundary conditions yielded an F score of 99.7 percent. A high probability of avoiding false alarms and finding tiny leaks under transitory settings may be derived from the suggested technique [26].

The release of oil, gas and radioactive fluids from pipelines has put people and marine life at risk. Solutions to identify pipeline leaks, particularly software-based methods, are examined in this research. Research on wireless sensor networks for leak detection is also included in this overview [30].

This study reviews pipeline leak-detecting methods and summarizes current developments. The pros and limitations of several pipeline leak detection and localization methods are discussed. Comparative performance analysis may assist in deciding which leak detection method is optimal for a specific operating setting. Also highlighted are research gaps and ongoing concerns about developing reliable pipeline leak detection systems [24].

k-nearest neighbor (k-NN) classifiers and hybrid features from acoustic emission data, a novel approach is proposed for locating leaks in a gas pipeline. In the study, k-NN classifiers are adapted to the unique context in which the system is deployed [28].

An innovative pigging methodology for pipeline leak detection is investigated in this research as an active leak detection tool. Detailed mathematical descriptions of fluid–solid and solid–solid interactions surrounding the pig and network modeling to estimate pressure and rate throughout the pipeline are two essential components of intelligent pigging. This work proposes a novel way to predict a leak's location and amount by combining these several causes [35].

Machine learning, artificial intelligence, big data, and other cutting-edge technologies are being examined by the industry to see whether they can enhance present procedures. This study used five intelligent models (gradient boosting, decision trees, random forest, support vector machine, and artificial neural network). Researchers found that the two-hour detection of a 0.1% leak using random forest and decision tree models is the most sensitive [2].

People and the environment might be in danger if industrial gas leaks occur. Most importantly, anomalous concentration signals are specified by a preset concentration value, such as 25 percent of the lower explosive limit. When the leak is modest, it is not easy to accumulate to a fixed point rapidly enough. Researchers proposed the autocorrelation function of the standard concentration segment for detecting leaks. A wireless sensor network is set up to test the suggested strategy in the pipeline leak scenario [15].

To enhance pipeline leak detection, denoising pipeline signals is critical. Particle swarm optimization may be used to improve the VMD parameters, according to this work (PSO-VMD). The suggested approach is more effective in removing noise and detecting leaks. On average, the SNR (signal-to-noise ratio) of the denoised pipeline signal is enhanced by 2.1127, the MSE and MAE are lowered by 0.00035 and 0.0043, and the recognition accuracy of the SVM (support vector machine) is improved by using the approach provided in [36].

4. Literature Summary (Table 1)

Table 1. Literature review matrix

S. No.	Author (year)	Key observations
1.	[8]	The authors suggest **a hydraulic transient paired-IRF (impulse response function)** technology for leak detection. A protective wire surrounds two fiber Bragg grating pressure sensors. It is connected to a laboratory pipeline to monitor pressure responses produced by a voice-coil pressure generator.
2.	[11]	The transfer matrix approach simplifies and solves the continuity and momentum equations in the frequency domain (TMM). The amplitude of the pressure oscillations is linked to the relative leak site. LDCs are used to create leak detection curves. The effects of constant and varying friction are investigated.
3.	[18]	In various domains, **artificial neural networks** (ANNs) have shown to be a beneficial tool. Automated learning systems allow computers to tackle complex problems without having to be explicitly trained to do so. In this study, an ANN-based approach to predicting the existence of leaks in a pipeline is introduced.
4.	[19]	**Optical fiber** detection techniques may be used to identify multiple leaks in pipelines. The fault diagnostic observer produces the residual generator when optical fiber sensors receive a pressure signal. By using this method, leak coefficients may be detected and estimated. This approach's efficacy has been shown by testing.
5.	[20]	Environmental concerns have been raised after pipeline leaks. **Leak detection characteristics** such as mass flow rate, temperature and pressure are examined in this article. It was shown that the intake and outflow parameters were affected by leaks of various sizes.
6.	[21]	Part of the methane released into the atmosphere comes from natural gas leaks, which are more alarming than CO_2. In the event of a natural gas accident, the average economic loss is substantial. Natural gas pipelines may be safeguarded and monitored using an **acoustic leak-localization device**.
7.	[22]	Pipeline leaks are detected using a unique approach based on Markov feature extraction and a two-stage decision system. Using a transformation function, pressure data may be transformed into a Markov chain. The proposed method is very accurate and has a low false-positive rate.

S. No.	Author (year)	Key observations
8.	[16]	In recent years, leak detection in gas pipelines has received much attention. This research uses **wavelet transform and Support Vector Machines (SVM)** to provide an integrated leak detection approach (SVM). An accuracy rating of 99.4% was achieved for the leak state determination and a 95.66% accuracy rate for classifying various leak severity levels.
9.	[23]	Preventing financial losses and accidents due to fluid leaks may be controlled by inspecting pipes regularly. A significant drawback of existing supervised AI systems is that they largely depend on adequate pipeline failure history data for their training. **Multiscale analysis, Kolmogorov–Smirnov (KS) test, and Gaussian mixture model (GMM)** are used in this research to develop a health index-oriented strategy (GMM).
10.	[14]	Pipeline leaks may result in serious safety and economic consequences. Leak detection in pipelines is an effective method for reducing the risk of pipeline failure. **Existing detection techniques** are introduced, and their benefits, limits, applications, and performance are discussed in detail in this work.
11.	[25]	Detecting several leaks in a water-filled pipe using transient waves is the focus of this article. Leak detection may be improved by using a **maximum likelihood (ML)** technique proposed by the authors in an earlier study. The current study focuses on the identification of many leaks when the number of leaks is significant.
12.	[26]	Gas pipeline leaks may be detected using a new convolutional neural network **(SE-CNN)**. As a result of the SE, the leak signals are amplified, and background noise is reduced. The SE-average CNN's accuracy may rise to 94.3 percent in the presence of heavy background noise.
13.	[27]	Leak noises may be picked up from within a long pipeline by using an **SD** that can get near enough to the source of a leak. The pipe is filled with water, pressured, and may be tilted to generate a leak. With a specificity of 89.6 percent, the identification accuracy may reach 93 percent.
14.	[28]	Leak prevention and early leak detection are critical to the pipeline's integrity management. According to a literature assessment, leak detection systems may be found in various technical and application areas.
15.	[29]	Only a few leak detection systems are capable of simultaneously determining the leak's location and magnitude. **Dynamic modeling** may be an excellent solution to this issue.
16.	[30]	This research investigates an approach for detecting oil leaks to create a model control system that can detect leaks without human involvement. During this project, a prototype test rig was built and tested. The rig contains four separate flow lines, each with a pressure sensor and a gauge attached to it.
17.	[31]	When a device fails, or the network is disrupted, pipeline status monitoring cannot be implemented. This paper introduces a **generative adversarial network (tnGAN)** based on the tri-networks form to deal with leak detection issues. The suggested technique may handle a variety of partial data recovery scenarios, including personal loss and random missing.
18.	[32]	Nonlinear first-order hyperbolic partial differential equations are used to simulate a lengthy pipeline leak detection and localization problem. An **SVM model** for leak detection, quantity estimate and position measurement is trained using data obtained under varied normal and leaking situations.
19.	[33]	Leak detection and localization in a nontrivial pipeline manifold layout is the goal of this research. The authors utilized data obtained from the gyroscope and accelerometer sensors to recreate the robot's position trajectory. Each feature is classified using a **Bayesian classifier** to locate the leak.
20.	[34]	The authors offer a **model-free** pipeline leak detection and locating approach based on anomalous (leak and operation) signals. The experimental findings show that computational efficiency and a decreased false/missing alarm rate are achieved.
21.	[35]	Transient leakage detection utilizing **multiple signal classification** (MUSIC) is used in this work. As a result of this one-dimensional search, the location of the leak is pinpointed. Detection of all leaks at a distance of less than half the shortest probing wavelength is possible with this technique.
22.	[36]	The authors have presented a transient state mathematical model that can be used for leak detection in pipeline networks.
23.	[37]	To reduce noise interference, and improved spline-local mean decomposition (**ISLMD**) was devised. A convolutional neural network-based leak detection technique is proposed. The signal delay between the pressure transmitters upstream and downstream is computed. As a result, the location of the leak is known

Table 1. (Continued)

S. No.	Author (year)	Key observations
24.	[38]	An Acoustic Emission (AE) approach is presented to identify and locate a leak in a pipeline. Sensors that pick up acoustic signals caused by a leak are positioned on the outside of the pipeline. Acoustic signals that arrive at nearby sensors may be used to pinpoint the location of a leak by deciphering the difference in arrival times.
25.	[39]	**LDS and data fusion methods** that may be used for pipeline monitoring are discussed in this research. Based on established criteria, LDS are compared. Classification and critical evaluation of various methods is performed. The most recent works have been thoroughly analyzed and compared in the review paper
26.	[40]	A **DTIM** based on an **RDFS** system is proposed and experimentally demonstrated to detect pipeline leaks. The RDFS may be used to determine the location of a pipeline leak based on the surrounding area's temperature. It is an excellent way to monitor pipeline leaks in a distributed manner.
27.	[41]	Pipe leaks may be caused by various circumstances such as corrosion, age, poor weld quality, or even deliberate injury. The authors suggest a **hybrid intelligent approach** that combines kernel principal component analysis with cascade support vector data representation.
28.	[42]	The study aims to identify leaks in gas mixture pipelines with a 0% false alarm rate (FAR). The **Hammerstein model** employed pressure and mass flow rate data at the pipeline intake to predict mass flow rate at the pipeline exit.
29.	[43]	Researchers and the general public should focus on pipeline leak detection and monitoring. Pipeline leaks transporting oil, gas, and radioactive substances have put humans and marine life at risk. An examination of **software-based strategies** for detecting pipeline leaks is presented in this research.
30.	[44]	Hydrocarbon fluids may be transported across millions of kilometers via pipelines. Thousands of pipeline operators and the environment have been harmed by pipeline leaks. Pipeline leak detection systems are discussed in this study and the **current state of the art**.
31.	[45]	This work offers a method for identifying leaks in a gas pipeline using a **k-nearest neighbor** (k-NN) classifier and **acoustic emission** (AE) signal hybrid features. Real-time leak detection is possible thanks to an integrated microcontroller unit (MCU). ALEOR (accumulative leak occurrence rate) is monitored in combination with a predetermined threshold.
32.	[22]	An active leak-detection technology known as "smart pigging" is the focus of this investigation.
33.	[19]	This research looks at intelligent models' capacity to identify tiny pipeline breaches. The study used five intelligent models to identify pipeline leaks: gradient boosting, decision trees, random forest, SVM, and ANN.
34.	[26]	Industrial gas leaks result in accidents and put the environment and human lives at risk. This article provides a detection strategy based on the regular concentration segment's **autocorrelation function** (ACF). It is established to detect threats at a rate of up to 96.7 percent and with an average detection time of fewer than 30 seconds.
35.	[29]	Denoising pipeline signals are critical for improving pipeline leak detection accuracy. This research aims to offer a technique for enhancing **VMD parameters** using **particle swarm optimization** (PSO-VMD). The suggested approach provides improved denoising performance and a greater rate of leak identification.

5. Categories of Leak Detection Systems

The most popular categorization approach is based on the features of the detection technology and may be separated into two categories: hardware-based methods and software-based methods [35]. A short description of these methodologies is presented.

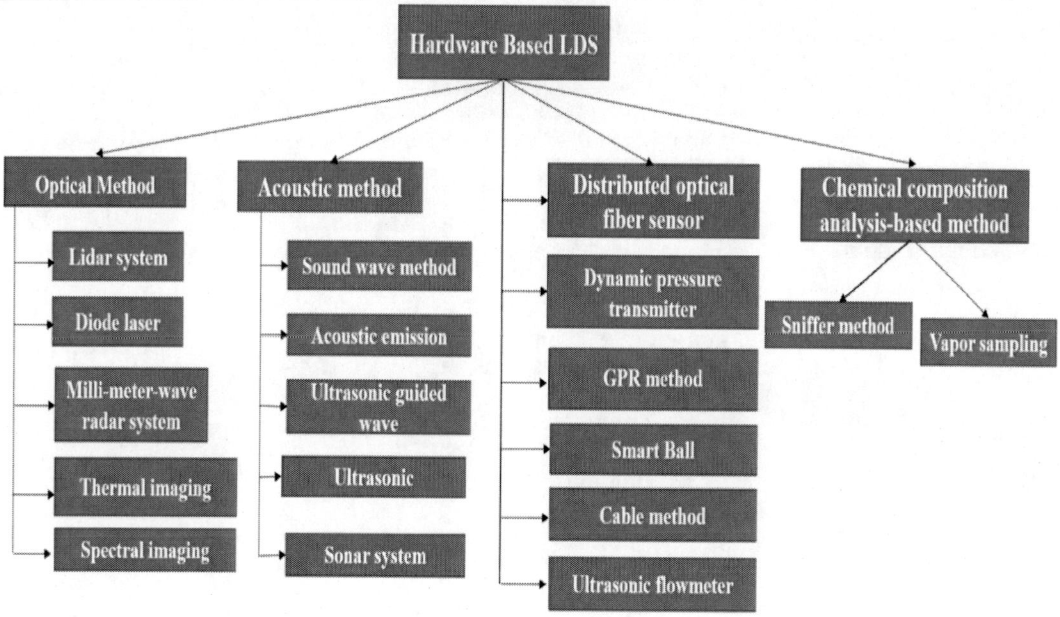

Figure 1. Hardware-based leak detection systems.

5.1. Hardware-Based LDS

In hardware-based leak detection techniques, hardware sensors are used to directly detect a leak and aid in pinpointing the leak's location. As a general rule, hardware techniques are susceptible to leaks since they have instrumentation throughout the whole pipeline and can discover both significant and minor leaks at a rapid pace. In the case of a leak, the position of the leak may also be approximated by the instrumentation placed on the pipeline's surface, which saves time and material waste. For these reasons, hardware approaches are widely utilized on pipelines traveling through high-risk zones where the loss of life in the event of a leak is quite likely. Despite these advantages, a high degree of instrumentation adds to the method's shortcomings. The cost of installation and maintenance is also increased. There is also a high level of installation difficulty since most hardware approaches need much work below the surface to position the sensors along pipes accurately. **Figure 1** shows the different types of hardware-based LDS.

5.2. Software-Based LDS

Instead of relying on physical sensors to monitor pressure, temperature, and flow rate, software-based systems use computer software to detect pipeline leaks. The complexity and dependability of these programs vary greatly. The more parameters a software technique employs, the more exact the results. Figure 2 shows the different types of software-based LDSs.

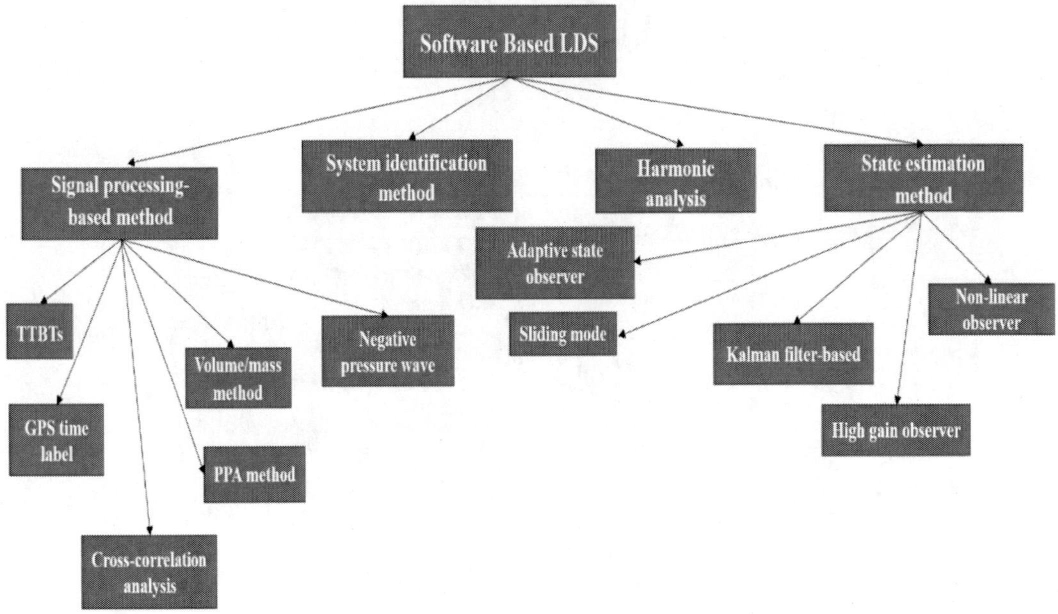

Figure 2. Software-based leak detection systems.

6. Challenges and Future Scope

The need for pipeline leak detection and locating technologies increases as more and more pipelines reach the end of their useful lives. As a result, future leak detection technology development is predicted to follow the patterns shown below.

(a) With the advancement of detection technology in recent years, LDS are becoming increasingly diverse. Integrating hardware and software-based LDS represents the future of progress.
(b) Machine learning has evolved in recent years. The technique facilitates the data to be monitored in real-time, thus reducing the failure occurrence of automated judgment and early warning.
(c) Detecting minor leaks is still a challenge for existing detection technologies; hence, the focus should be on increasing the accuracy and stability of detection.
(d) Methods for detecting the distribution of optical fibers have demonstrated promising results. There is further work to be done in multipoint detection, environmental resilience, improved positioning accuracy, and automated identification of vibration source types [46].

7. Conclusion

To ensure the safety of pipelines and limit environmental pollution, there is a constant need to build effective pipeline integrity management systems. Hardware-based systems are more expensive than software-based solutions. Volume-balance and pressure-point analyses are part of the software-based range of approaches. Pipelines with frequent operational changes are not suited for these technologies. Leak detection systems that are cost-effective and highly sensitive to detect minor leaks and estimate their location require functional organizations. As a result, if the detection system's sensitivity is set too high, data uncertainties will trigger alarms even if there are no leaks to be detected at all. It is usual for pipeline operating modifications to result in false alarms, resulting in decreased efficiency and a loss of faith in the leak-detection system. There is a lack of trust in the leak detection system, which results in all alerts being disregarded or the system being shut down. Safety and environmental concerns necessitate the development of dependable pipeline integrity management systems. In every effort to increase the quality of real-time pipeline leak-detection systems, reducing complexity is essential. This article covers leak detection systems for pipeline inspection due to a thorough examination of the literature and industry technology. There is currently no technology that has a distinct edge over the others.

References

[1] Adegboye, M. A., Fung, W. K., and Karnik, A. (2019). Recent advances in pipeline monitoring and oil leakage detection technologies: Principles and approaches. *Sensors (Switzerland)*, 19(11). https://doi.org/10.3390/s19112548.

[2] Akinsete, O., and Oshingbesan, A. (2019). Leak detection in natural gas pipelines using intelligent models. *Society of Petroleum Engineers - SPE Nigeria Annual International Conference and Exhibition 2019*, NAIC 2019, August, 5–7. https://doi.org/10.2118/198738-MS.

[3] Al-Tofan, M., Elkholy, M., Khilqa, S., and Chaudhry, M. H. (2020). Leak Detection in Liquid Pipelines Using Lower Harmonics of Pressure Oscillations. *Journal of Pipeline Systems Engineering and Practice*, 11(4), 04020033. https://doi.org/10.1061/(asce)ps.1949-1204.0000479.

[4] Antaki, G. A. (2003). *Piping and pipeline engineering: design, construction, maintenance, integrity, and repair*. CRC Press.

[5] Arya, A. K. (2021). Optimal operation of a multidistribution natural gas pipeline grid: an ant colony approach. *Journal of Petroleum Exploration and Production Technology*, 11(10), 3859-3878.

[6] Arya, A. K. (2022a). A critical review on optimization parameters and techniques for gas pipeline operation profitability. *Journal of Petroleum Exploration and Production Technology*, 12(11), 3033-3057.

[7] Arya, A. K., Jain, R., and Bisht, S. (2022b). Corrosion Inhibitors in Oil and Gas Industry—A Critical Review. *Tailored Functional Materials: Select Proceedings of MMETFP 2021*, 355-369.

[8] Arya, A. K., Gautam, S., and Yadav, S. (2022c). Impact of Hydrogen Embrittlement in Pipeline Structures—A Critical Review. *Tailored Functional Materials: Select Proceedings of MMETFP 2021*, 409-428.

[9] Arya, A. K., Jain, R., Yadav, S., Bisht, S., and Gautam, S. (2022). Recent trends in gas pipeline optimization. *Materials Today: Proceedings*, 57, 1455-1461.

[10] Baroudi, U., Al-Roubaiey, A. A., and Devendiran, A. (2019). Pipeline Leak Detection Systems and Data Fusion: A Survey. *IEEE Access*, 7, 97426–97439. https://doi.org/10.1109/ACCESS.2019.2928487.

[11] Bie, Q., ZHENG, Y. P., Fu, M., and Song, X. (2007). Leakage Detection Technology of Oil and Gas Transmission Pipelines and Its Development Trend. *Petroleum Engineering Construction*, 03.

[12] Boaz, L., Kaijage, S., and Sinde, R. (2014, July). An overview of pipeline leak detection and location systems. In *Proceedings of the 2nd Pan African International Conference on Science, Computing and Telecommunications (PACT 2014)* (pp. 133-137). IEEE.

[13] Bohorquez, J., Alexander, B., Simpson, A. R., and Lambert, M. F. (2020). Leak Detection and Topology Identification in Pipelines Using Fluid Transients and Artificial Neural Networks. *Journal of Water*

Resources Planning and Management, 146(6), 04020040. https://doi.org/10.1061/(asce)wr.1943-5452.0001187.

[14] Daniyan, I. A., Dahunsi, O. A., Oguntuase, O. B., Daniyan, O. L., and Mpofu, K. (2019). Development of a prototype test rig for leak detection in pipelines. *Procedia CIRP*, 80, 524–529. https://doi.org/10.1016/j.procir.2019.01.016.

[15] Dong, L., Qiao, Z., Wang, H., Yang, W., Zhao, W., Xu, K., Wang, G., Zhao, L., and Yan, H. (2019). The Gas Leak Detection Based on a Wireless Monitoring System. *IEEE Transactions on Industrial Informatics*, 15(12), 6240–6251. https://doi.org/10.1109/TII.2019.2891521.

[16] Ge, Z., Zhang, S., Xie, J., Tang, Z., and Dubljevic, S. (2020). Development of a swimming robot for pipeline leak detection. *Proceedings of the American Control Conference*, *2020-July*, 5231–5236. https://doi.org/10.23919/ACC45564.2020.9148038.

[17] Gupta, S. S., and Arya, A. K. (2019). An approach to determine probability of third-party damage to cross-country pipelines in India.

[18] Hu, X., Zhang, H., Ma, D., and Wang, R. (2021). A tnGAN-Based Leak Detection Method for Pipeline Network Considering Incomplete Sensor Data. *IEEE Transactions on Instrumentation and Measurement*, 70. https://doi.org/10.1109/TIM.2020.3045843.

[19] Kim, J., Chae, M., Han, J., Park, S., and Lee, Y. (2021). The development of leak detection model in subsea gas pipeline using machine learning. *Journal of Natural Gas Science and Engineering*, 94(March), 104134. https://doi.org/10.1016/j.jngse.2021.104134.

[20] Kousiopoulos, G. P., Papastavrou, G. N., Karagiorgos, N., Nikolaidis, S., and Porlidas, D. (2019). Pipeline Leak Detection in Noisy Environment. *2019 8th International Conference on Modern Circuits and Systems Technologies, MOCAST 2019*, 2(4), 1–5. https://doi.org/10.1109/MOCAST.2019.8741673.

[21] Kwestarz, M. A., Osiadacz, A. J., and Kotynski, L. (2019). Method for leak detection and location for gas networks. *Archives of Mining Sciences*, 64(1), 131–150. https://doi.org/10.24425/ams.2019.126276.

[22] Li, J., Wu, Y., and Lu, C. (2020). Pipeline leak detection using the multiple signal classification-like method. *Journal of Hydroinformatics*, 22(5), 1321–1337. https://doi.org/10.2166/hydro.2020.194.

[23] Liu, C., Li, Y., Fang, L., and Xu, M. (2019a) New leak-localization approaches for gas pipelines using acoustic waves. *Measurement: Journal of the International Measurement Confederation*, 134, 54–65. https://doi.org/10.1016/j.measurement.2018.10.052.

[24] Liu, J., Zang, D., Liu, C., Ma, Y., and Fu, M. (2019b). A leak detection method for oil pipeline based on markov feature and two-stage decision scheme. *Measurement: Journal of the International Measurement Confederation*, 138, 433–445. https://doi.org/10.1016/j.measurement.2019.01.029.

[25] Lu, H., Iseley, T., Behbahani, S., and Fu, L. (2020). Leakage detection techniques for oil and gas pipelines: State-of-the-art. *Tunnelling and Underground Space Technology*, 98(September 2019). https://doi.org/10.1016/j.tust.2019.103249.

[26] Mujtaba, S. M., Lemma, T. A., Ammar Taqvi, S. A., Ofei, T. N., and Vandrangi, S. K. (2020). Leak detection in gas mixture pipelines under transient conditions using hammerstein model and adaptive thresholds. *Processes*, 8(4). https://doi.org/10.3390/PR8040474.

[27] Ning, F., Cheng, Z., Meng, D., Duan, S., and Wei, J. (2021). Enhanced spectrum convolutional neural architecture: An intelligent leak detection method for gas pipeline. *Process Safety and Environmental Protection*, 146, 726–735. https://doi.org/10.1016/j.psep.2020.12.011.

[28] Quy, T. B., and Kim, J. M. (2021). Real-time leak detection for a gas pipeline using a k-nn classifier and hybrid ae features. *Sensors (Switzerland)*, 21(2), 1–14. https://doi.org/10.3390/s21020367.

[29] Rai, A., and Kim, J. M. (2021). A novel pipeline leak detection approach independent of prior failure information. *Measurement: Journal of the International Measurement Confederation*, 167(December 2019), 108284. https://doi.org/10.1016/j.measurement.2020.108284.

[30] Sheltami, T. R., Bala, A., and Shakshuki, E. M. (2016). Wireless sensor networks for leak detection in pipelines: a survey. *Journal of Ambient Intelligence and Humanized Computing*, 7(3), 347–356. https://doi.org/10.1007/s12652-016-0362-7.

[31] Thakur, A. K., Arya, A. K., and Sharma, P. (2022a). Prediction and mitigation of AC interference on the pipeline system. *Corrosion Reviews*, 40(2), 149-157.

[32] Thakur, A. K., Arya, A. K., and Sharma, P. (2022b). Analysis of cathodically protected steel pipeline corrosion under the influence of alternating current. *Materials Today: Proceedings*, 50, 789-796.

[33] Thakur, A. K., Arya, A. K., and Sharma, P. (2021). Corrosion of pipe steels under alternating currents. *Int J Electrochem Sci*, 16(11).

[34] Thakur, A. K., Arya, A. K., and Sharma, P. (2020). The science of alternating current-induced corrosion: a review of literature on pipeline corrosion induced due to high-voltage alternating current transmission pipelines. *Corrosion Reviews*, 38(6), 463-472.

[35] Thiberville, C., Wang, Y., Waltrich, P., Williams, W., and Kam, S. I. (2020). Modeling of smart pigging for pipeline leak detection. *SPE Production and Operations*, 35(3), 610–627. https://doi.org/10.2118/198648-PA.

[36] Wang, D., Zhu, L., Yue, J., Lu, J., Li, D., and Li, G. (2020). Application of variational mode decomposition based on particle swarm optimization in pipeline leak detection. *Engineering Research Express*, 2(4). https://doi.org/10.1088/2631-8695/abcc47.

[37] Wang, F., Lin, W., Liu, Z., and Qiu, X. (2019). Pipeline leak detection and location based on model-free isolation of abnormal acoustic signals. *Energies*, 12(16). https://doi.org/10.3390/en12163172.

[38] Wang, X., and Ghidaoui, M. S. (2019). Identification of multiple leaks in pipeline II: Iterative beamforming and leak number estimation. *Mechanical Systems and Signal Processing*, 119, 346–362. https://doi.org/10.1016/j.ymssp.2018.09.020.

[39] Xiao, R., Hu, Q., and Li, J. (2019). Leak detection of gas pipelines using acoustic signals based on wavelet transform and Support Vector Machine. *Measurement: Journal of the International Measurement Confederation*, 146, 479–489. https://doi.org/10.1016/j.measurement.2019.06.050.

[40] Xie, J., Xu, X., and Dubljevic, S. (2019). Long range pipeline leak detection and localization using discrete observer and support vector machine. *AIChE Journal*, 65(7). https://doi.org/10.1002/aic.16532.

[41] Xu, T., Zeng, Z., Huang, X., Li, J., and Feng, H. (2021). Pipeline leak detection based on variational mode decomposition and support vector machine using an interior spherical detector. *Process Safety and Environmental Protection*, 153, 167–177. https://doi.org/10.1016/j.psep.2021.07.024.

[42] Xu, Y., Li, J., Zhang, M., Yu, T., Yan, B., Zhou, X., Yu, F., Zhang, J., Qiao, L., Wang, T., and Gao, S. (2020). Pipeline Leak Detection Using Raman Distributed Fiber Sensor with Dynamic Threshold Identification Method. *IEEE Sensors Journal*, 20(14), 7870–7877. https://doi.org/10.1109/JSEN.2020.2980366.

[43] Yang, L., Guo, Y., and Gao, S. (2020). Multileak detection in pipeline based on optical fiber detection. *Optik*, 220(June), 164996. https://doi.org/10.1016/j.ijleo.2020.164996.

[44] Zeng, W., Gong, J., Cook, P. R., Arkwright, J. W., Simpson, A. R., Cazzolato, B. S., Zecchin, A. C., and Lambert, M. F. (2020). Leak Detection for Pipelines Using In-Pipe Optical Fiber Pressure Sensors and a Paired-IRF Technique. *Journal of Hydraulic Engineering*, 146(10), 06020013. https://doi.org/10.1061/(asce)hy.1943-7900.0001812.

[45] Zhou, M., Pan, Z., Liu, Y., Zhang, Q., Cai, Y., and Pan, H. (2019). Leak Detection and Location Based on ISLMD and CNN in a Pipeline. *IEEE Access*, 7, 30457–30464. https://doi.org/10.1109/ACCESS.2019.2902711.

[46] Zhou, M., Zhang, Q., Liu, Y., Sun, X., Cai, Y., and Pan, H. (2019). An integration method using Kernel Principal component analysis and cascade support vector data description for pipeline leak detection with multiple operating modes. *Processes*, 7(10). https://doi.org/10.3390/pr7100648.

Chapter 7

The Co-Pyrolysis of *Erythrina Indica* and *Azadirachta Indica* in a Tubular Reactor

Gaffer Ahmed*
and Nanda Kishore

Department of Chemical Engineering, Indian Institute of Technology Guwahati, Assam, India

Abstract

The co-pyrolysis of ***Erythrina indica*** and ***Azadirachta indica*** was performed in a tubular batch reactor at seven different ratios. The reactions were performed at a temperature of 600°C and pressure of 1 bar under inert atmospheric conditions. The heating rates of the reactions were considered to be 10°C/min. The liquid product obtained from the co-pyrolysis reactions consisted of a fuel and non-fuel phase, with the former separated from the latter using n-hexane. The yield of the fuel phase varied in the range of 4.917-7.907 wt.%, whereas the non-fuel phase varied in the range of 25.326-29.826 wt.%. The density of the fuel phase and non-fuel phase obtained varied in the range of 0.819-1.012 g/ml and 1.005-1.013 g/ml, respectively. A maximum calorific value of 36.798 MJ/kg was attained for the fuel phase obtained from the feed with a ratio of 1:1, which also had a minimum density of 0.819 g/ml.

Keywords: co-pyrolysis, fuel phase, non-fuel phase, calorific value, density

Nomenclature

EI	***Erythrina indica***
AI	***Azadirachta indica***
TGA	Thermogravimetric analysis
DTG	Differential thermogravimetry
FT-IR	Fourier **t**ransform **i**nfrared **s**pectroscopy
XRD	X-ray **d**iffractogram

* Corresponding Author's Email: gaffer@iitg.ac.in.

In: Sustainability in Chemical Processes through Digitalization and Green Chemistry Approaches
Editors: Susarla V. A. R. Sastry, Shravan Kumar and Ashish Kapoor
ISBN: 979-8-89113-257-3

1. Introduction

The ever-increasing global energy demand, as well as its negative environmental consequences, has pushed the world toward renewable energy and its development. Because of its abundant availability, biomass to energy has grown in popularity among renewable energy sources in recent years. As a clean and sustainable feedstock, lignocellulosic biomass could be an excellent source for the production of biofuels and other value-added chemicals. There are various biomass conversion pathways to convert biomass to liquid, solid and gaseous fuels [1]. However, when the severity and residence time of the conversion method were considered, pyrolysis was discovered to be advantageous due to its moderate operating conditions [2]. Regrettably, its commercial application has been found to be limited due to lower conversion, low calorific value, the presence of highly oxygenated compounds, water content and other factors.

There are several ways to improve the process, including co-pyrolysis, catalytic pyrolysis and catalytic co-pyrolysis. However, co-pyrolysis is found to be more advantageous because it is a less expensive technique than catalytic pyrolysis. ***Delonix regia*** and pine wood sawdust were co-pyrolyzed by Kawale and Kishore [3] at 625°C and 1 bar pressure in 25:75, 50:50 and 75:25 ratios. The resulting pyrolytic bio-oil had the highest heating value at a co-feed ratio of 50:50. Hua and Li co-pyrolyzed green algae with sugarcane bagasse at ratios of 100:0, 70:30, 60:40, 50:50, 40:60, 30:70 and 0:100 in a fixed bed reactor [4]. With a calorific value of 26.4 MJ/kg and a ratio of 50:50, the maximum bio-oil yield of 56.12 wt.% was accomplished. In a fixed bed reactor operating at 550°C, Wang et al., studied the co-pyrolysis of ***Enteromorpha clathrata*** and rice husk [5]. The following feed ratios were used in the study: 1:1, 1:2, 1:3, 2:1 and 3:1. For a 1:1 ratio, a maximum bio-oil output of 28 wt.% was noted. Thus, this study focuses on the co-pyrolysis of two widely available lignocellulosic biomasses, "***Erythrina indica***" and "***Azadirachta indica***," to produce higher grade biofuels.

2. Materials and Procedures

2.1. Collection and Characterization of Materials

The trees **or plant wastes** used in this study are the widely distributed evergreen *Azadirachta indica* (AI) and the deciduous *Erythrina indica* (EI), both of which are native to the Indian subcontinent. EI and AI are referred to as Modar and Mahaneem in India. These trees are also widely distributed on the IIT Guwahati campus, where they are planted purely for aesthetic reasons. Because of these factors, waste is constantly produced, and managing it is a major problem. Therefore, this research project's objectives are to manage this rubbish and create useful things for society.

Therefore, dry branches from these trees are gathered and manually cut into small pieces. The excess moisture in these parts is then removed by drying them outside in the sun. After that, the dried branches are crushed in a wood cutter that operates similarly to a hammer mill. The size-reduced biomass was next ground in a mixer grinder and then sieved through a 400-micron mesh. After being ground into a powder, the biomass was subsequently dried in a hot air furnace to remove any remaining moisture. However, it is crucial to ascertain the biomass's physiochemical and fuel qualities before moving on to the pyrolysis trials. As a result, the elemental analysis, the gross calorific value, and the contents of moisture, volatile matter, ash and fixed carbon were all calculated. The temperature for pyrolysis is derived using the feed's thermogravimetric analysis report. Approximately 10 mg of the materials are analyzed for this purpose at temperatures ranging from 25 to 900°C using a TGA analyzer type TG 209 F1 Libra (M/S Netzsch, Germany). The study's heating rate was taken to be 10°C per minute, and nitrogen was used to make the atmosphere inert for this analysis. The assumed nitrogen flow rate was 40 ml/min. The crystallinity of the feed material was also studied using an X-ray diffractogram analyzer (Rigaku, Japan).

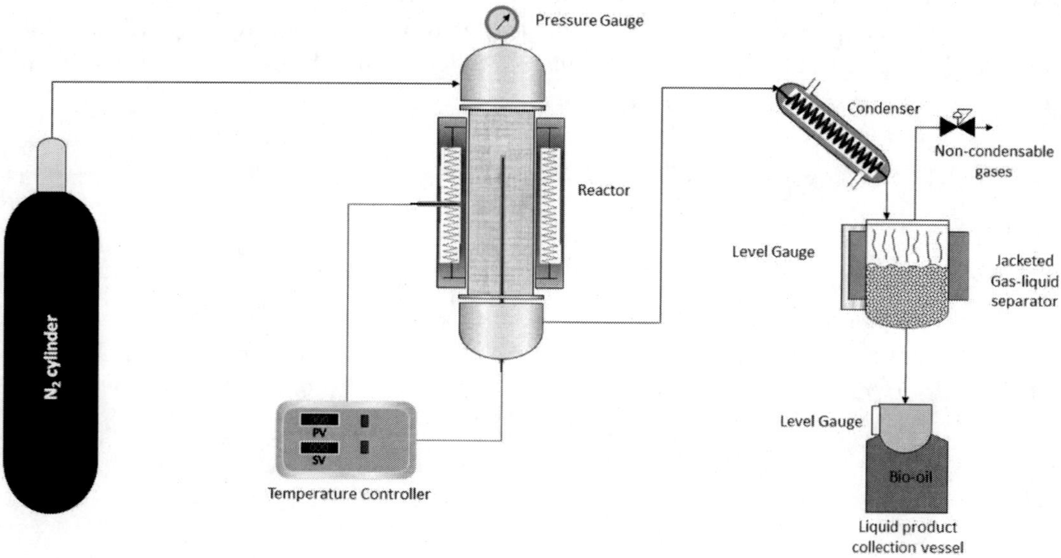

Figure 1. Schematic illustration of the experimental setup used for co-pyrolysis reactions.

2.2. Co-pyrolysis Experiments with *Erythrina Indica* and *Azadirachta Indica*

The co-pyrolysis research was performed in a batch tubular reactor. Figure 1 displays a schematic illustration of the configuration employed for this study. Each experiment's feed was made by combining the two biomasses in a variety of ratios. The reactor receives 300 g of each feed, and the reactions take place at 600°C, 1 bar of pressure and inert air conditions. The reactor's residence time was estimated to be 2 hours after attaining the target temperature in the pyrolysis tests, which used a predetermined heating rate of 10°C/min. The pyrolytic conversion of the biomass produces vapors that are fed through a condenser to condense the condensable vapors while separating the non-condensable vapors through a gas–liquid separator. The furnace is shut off after the pyrolysis experiment is finished, and the reactor is left to cool to room conditions before the products are collected. Pyrolysis reactions produce products in all three phases: liquid products known as bio-oil, solid products known as biochar and gaseous products known as non-condensable gases. The liquid bio-oil is then gathered from the liquid collecting tank following the release of the non-condensable gases. Using an organic solvent, the bio-oil is divided into two phases: fuel and non-fuel. The organic solvent used for this study was n-hexane. The raw bio-oil was mixed with 10 vol.% n-hexane, shaken vigorously for 5 minutes and then allowed to settle. The organic solvent separates the fuel and non-fuel phases due to the density difference. The solid biochar was collected from the reactor by removing it from the system.

2.3. Pyrolysis Product Distribution

Equations (1), (2) and (3) are used to calculate the pyrolysis product distribution in weight percent. The biochar output was computed by dividing the biochar weight by the feed weight. By dividing the sum of the product of volume and corresponding density of each phase by its feed weight, the yield of the bio-oil is calculated. By deducting the yields of biochar and bio-oil from 100, the gaseous yield was calculated.

$$Biochar\ yield\ wt.\% = \frac{Weight\ of\ biochar\ obtained}{Weight\ of\ biomass\ fed} \times 100 \tag{1}$$

$$Bio-oil\ yield\ wt.\% = \frac{(Weight\ of\ fuel\ phase)+(Weight\ of\ non-fuel\ phase)}{Weight\ of\ biomass\ fed} \times 100 \tag{2}$$

$$Non-condensable\ gases\ yield\ wt.\% = 100 - (\boldsymbol{wt.}\%\ Yield\ of\ bio-oil + \boldsymbol{wt.}\%\ Yield\ of biochar) \tag{3}$$

2.4. Characterization of Pyrolysis Products

For both the fuel and non-fuel phases of the bio-oil, physical and fuel parameters such as density, pH and gross calorific value are assessed. The fuel phase was examined using a Fourier transform infrared spectroscopy analyzer (M/S Shimadzu, Japan) to detect any potential functional groups that may be present. In addition to the calorific value, the biochar was examined using an X-ray diffractogram to determine the crystallinity as described for biomass.

3. Results and Discussions

3.1. Properties of the Feedstock

The physical and fuel properties, such as moisture content, volatile matter content, ash content, elemental analysis and gross calorific value, of *Erythrina indica* and *Azadirachta indica* biomass are presented in Table 1. Both EI and AI possessed high volatile matter and carbon contents. Both biomass samples were found to have low moisture and ash contents but high calorific values, which is a good sign for a feed for the pyrolytic conversion pathway.

Table 1. Physical and fuel properties of *Erythrina indica* and *Azadirachta indica* biomass

Proximate analysis (wt.%)					Ultimate analysis (wt.%)							
Biomass	Moisture content	Volatile matter content	Ash content	Fixed Carbon content[a]	C	H	N	S	O[b]	H/C ratio	O/C ratio	Calorific value (MJ/kg)
Erythrina indica	11.226	79.734	4.753	15.513	45.114	6.856	1.123	-	46.907	1.824	0.780	18.000
Azadirachta indica	14.179	82.727	3.224	14.050	50.000	7.991	5.660	-	36.349	1.917	0.533	18.020

[a] calculated by difference.
[b] calculated by difference.

The TGA and DTG analyses of the EI and AI biomass are shown in Figure 2 (A) and (B), respectively. The TGA curve showed that biomass degradation took place in three phases. Up to 180°C, smaller molecular weight hydrocarbons and moisture were removed during the first stage of degradation. The second step entails the destruction of cellulose and hemicellulose, which took place between 180 and 350°C. The final stage, which begins at 350°C and lasts until the analysis is complete, is connected to the destruction of lignin. At the conclusion of the examination, it was found that approximately 25 wt.% of the sample was still present, which may be related to the sample's inorganic matter content.

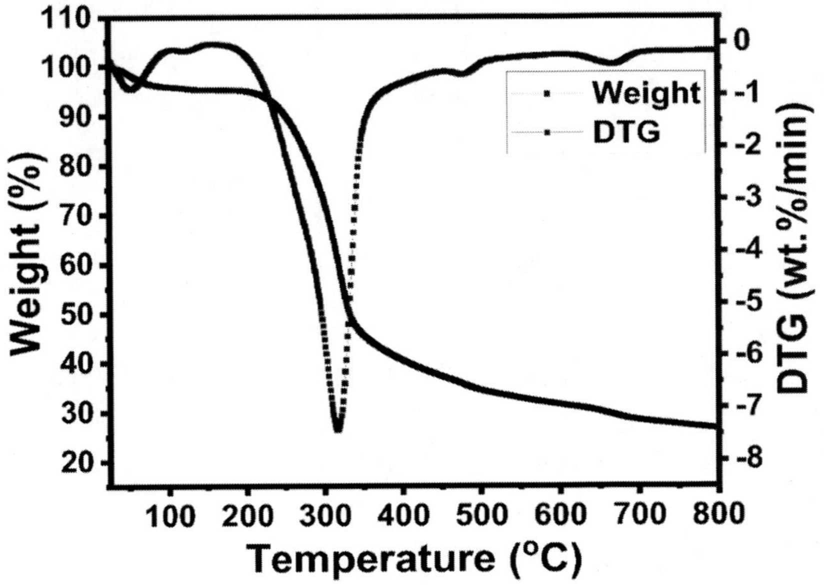

Figure 2. (A). Analysis of *Erythrina indica* biomass using TGA and DTG.

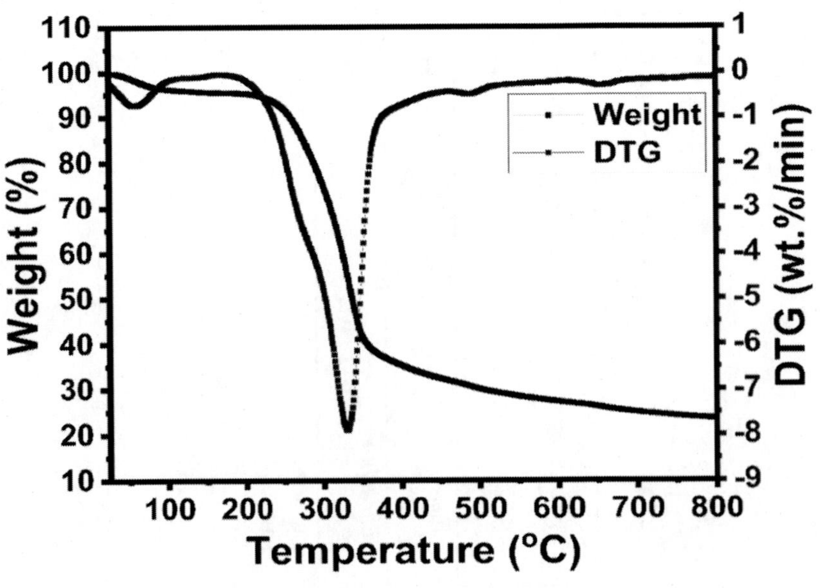

Figure 2. (B). Analysis of *Azadirachta indica* biomass using TGA and DTG.

A few pertinent studies on the thermal decomposition of lignocellulosic biomass are discussed below. In a thermogravimetric examination of *Delonix regia* between 30-750°C, Kawale & Kishore discovered that moisture was removed up to 150°C on a dry basis, while hemicellulose degradation began at approximately 270-280°C [6]. Beginning at 300°C, the degradation of cellulose and lignin proceeded until 400°C for the former and persisted until 700°C for the latter. Between 25 and 900°C, the thermal degradation of sawdust made from Acacia timber was investigated. Moisture removal and smaller molecular volatile chemicals were suggested by the TGA profile. The cellulose and hemicellulose components degraded between 200 and 500°C, while the lignin component underwent continuous degradation with a lengthy tail that lasted until 900°C [7].

EI and AI biomass are therefore determined to be well within the range of other lignocellulosic biomass as reported by other researchers based on their physical, fuel characteristics, and thermal degradation profile [8].

3.2. Pyrolysis Product Distribution

A graphical representation of the **yield of products** has been presented in Figure 3. The yield of the fuel phase was found to vary between 4.917 and 7.907 weight percent, whereas the yield of the non-fuel phase ranged between 25.326 and 29.826 weight percent. The ratio of EI:AI-3:1 yields the highest fuel phase yield of 7.907 wt.%, whereas the ratio of EI:AI-1:1 yields the lowest fuel phase yield of 4.917 wt.%. The production of biochar ranged from 38.249 to 42.480 wt.%, whereas the output of non-condensable gas varied from 23.223-26.523 wt.%.

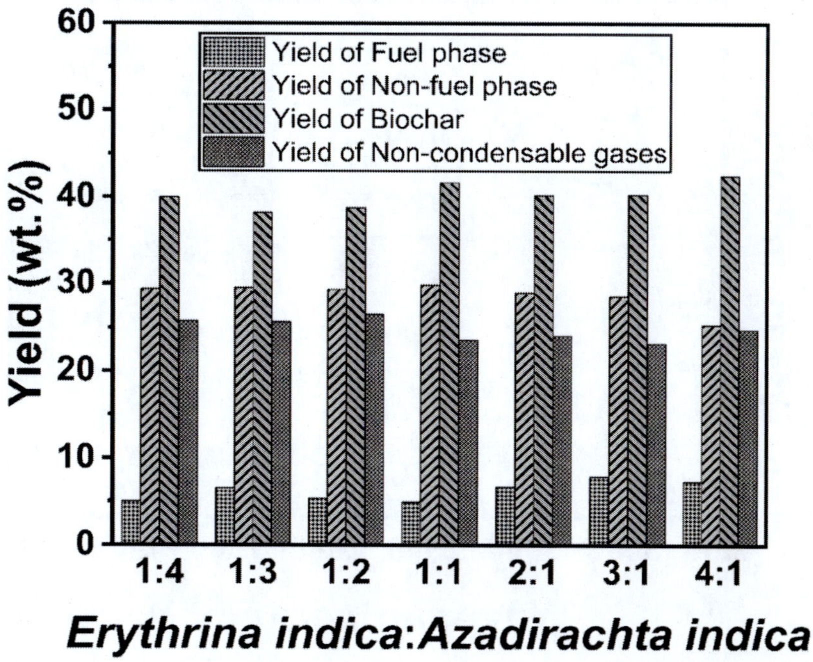

Figure 3. Graphical representation yields of products obtained from the co-pyrolysis experiments.

3.3. Characterization of the Products

The physical and fuel characteristics of the fuel and non-fuel phases are presented **below**. The fuel phase density is found to be between 0.819 and 1.012 g/ml, and the non-fuel phase density is found to be between 1.005 and 1.013 g/ml. The fuel phase pH is found to be between 3.080-3.640, and the non-fuel phase pH is found to be between 2.670 and 3.140. The fuel phase is found to have a calorific value in the range of 31.351-36.798 MJ/kg, while the non-fuel phase did not, since water makes up the majority of the non-fuel phase. The calorific value of the biochar varied between 26.687 and 29.233 MJ/kg. The maximum calorific value of the fuel phase and biochar was obtained at an EI:AI-1:1 ratio.

The FT-IR spectra of the fuel phases obtained by the co-pyrolysis of EI and AI biomass revealed intensities at cm^{-1} wavenumbers of 3400, 2933, 2858, 1707, 1609, 1509, 1458, 1370, 1220, 1222, 1110, 1027 and 745. Figure 4 shows a graphic illustration of the same. The presence of polymeric-OH and water impurities was suggested by the wide peaks seen between 3420 and 3250 cm^{-1}. The peaks between 2990 and 2850 cm^{-1} indicate the presence of methyl and methylene groups, which fall under the category of aliphatic molecules. Peaks occurring between 1720 and 1700 cm^{-1} suggested the presence of ketones, whereas the peaks at 1615-1590 and 1515-1485 cm^{-1} in aromatic compounds are indicative of the benzene ring. *t*-butyl groups were confirmed by peaks at 1400-1370 cm^{-1}. The CH_2 group of aliphatic groups was confirmed by peaks appearing at 1475-1450 cm^{-1}. Peaks at 1280-1180 cm^{-1} and 1150-1085 cm^{-1} confirmed the presence of aromatic and aliphatic ethers, while peaks at 1065-1015 cm^{-1} confirmed the presence of cyclic alcohols. Peaks at 760-740 cm^{-1} confirmed the presence of o-benzenes.

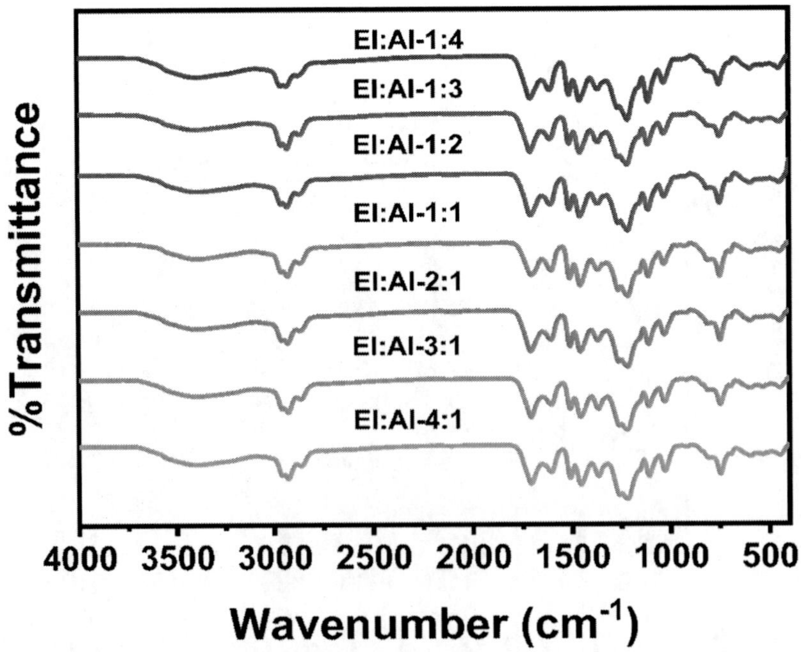

Figure 4. Fuel phase FT-IR spectrum from the co-pyrolysis experiments.

The crystallinity of the biomass feed and biochar samples obtained from the co-pyrolysis studies of EI and AI biomass was determined using XRD analysis. Figure 5 provides a graphical representation of the XRD analysis. The broad peak at 24° for the biomass samples indicates the crystalline nature of biomass owing to the spacing between the lattice of cellulosic and lignocellulosic compounds. The sharp peaks obtained at 28°, 29°, 36°, 39°, 40°, 43°, 47.5°, 48° and 50° confirm the amorphous nature of the biochar samples. All of the samples obtained from the co-pyrolysis experiments had similar diffractogram patterns. The degradation of cellulose began with increasing temperatures, which resulted in a decrease in the crystallinity of the material.

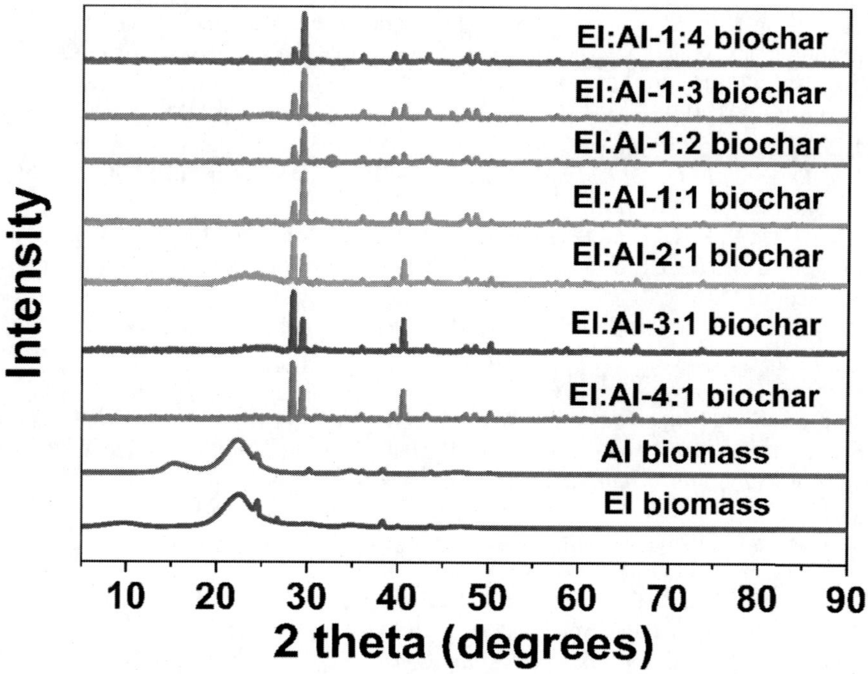

Figure 5. XRD spectra of biomass and biochar obtained from the co-pyrolysis experiments.

Conclusion

In a tubular batch reactor with an inert environment and seven different ratios, the effects of co-pyrolyzing *Erythrina indica* and *Azadirachta indica* were examined. The density, pH and calorific value showed varied trends. At a 1:1 ratio of EI and AI biomass, the fuel phase and biochar were found to have maximum calorific values of 36.798 MJ/kg and 29.233 MJ/kg, respectively. For a 3:1 ratio of EI and AI biomass, a maximum yield of 7.907 wt.% of fuel phase was attained. Aliphatic chemicals, ketones, ethers, alkyl compounds, aromatic compounds, hydrocarbons and alcohols were the major constituents of the fuel phase produced by the co-pyrolysis processes. Overall, the co-pyrolysis of EI and AI biomass was shown to be beneficial in creating biofuels with higher value.

Acknowledgments

The Central Instrument Facility and Department of Chemical Engineering at IIT Guwahati are gratefully acknowledged by the authors for providing access to advanced equipment such as XRD, TGA and FT-IR. Last but not least, acknowledgment should be given to Guwahati Biotech Park in Guwahati for carrying out the CHNS elemental analysis.

References

[1] Acharya, S., & Kishore, N. (2022). Hydro-liquefaction of Lemna minor (Duckweed) with hydrogen-donor solvent at varying hydrogen pressures. *International Journal of Sustainable Energy*. https://doi.org/10.1080/14786451.2022.2140157.

[2] Gupta, S., Kawale, H. D., Ahmed, G., Acharya, S., & Kishore, N. (2021). Effect of temperature on catalytic pyrolysis of Polyalthia Longifolia leaves solid waste and characterization of their products. *Current Research in Green and Sustainable Chemistry*, *4*(December 2020), 100062. https://doi.org/10.1016/j.crgsc.2021.100062.

[3] Kawale, H. D., & Kishore, N. (2020a). Comparative study on pyrolysis of Delonix Regia, Pinewood sawdust and their co-feed for plausible biofuels production. *Energy*, *203*, 117921. https://doi.org/10.1016/j.energy.2020.117921.

[4] Hua, M. Y., & Li, B. X. (2016). Copyrolysis characteristics of the sugarcane bagasse and Enteromorpha prolifera. *Energy Conversion and Management*, *120*, 238–246. https://doi.org/10.1016/j.enconman.2016.04.072.

[5] Wang, S., Cao, B., Feng, Y., Sun, C., Wang, Q., Abomohra, A. E. F., Xu, L. (2019). Copyrolysis and catalytic copyrolysis of Enteromorpha clathrata and rice husk: Toward high-quality products. *Journal of Thermal Analysis and Calorimetry*, *135*(4), 2613–2623. https://doi.org/10.1007/s10973-018-7334-4.

[6] Kawale, H. D., & Kishore, N. (2020b). Pyrolysis of Delonix Regia and Characterization of its Pyrolytic Products: Effect of Pyrolysis Temperature. *Journal of Energy Resources Technology*, *142*(August), 1–11. https://doi.org/10.1115/1.4046226.

[7] Ahmed, A., Abu Bakar, M. S., Sukri, R. S., Hussain, M., Farooq, A., Moogi, S., & Park, Y.-K. (2020). Sawdust pyrolysis from the furniture industry in an auger pyrolysis reactor system for biochar and bio-oil production. *Energy Conversion and Management*, *226*(August), 113502. https://doi.org/10.1016/j.enconman.2020.113502.

[8] Ahmed, G., Acharya, S., Kawale, H., Singh, A., Kishore, N., & Pal, S. (2020). Thermochemical conversion of Polyalthia longifolia leaves at different temperatures and characterization of their products. *Fuel*, *280*(April). https://doi.org/10.1016/j.fuel.2020.118574.

Chapter 8

CO_2 Sequestering and Its Utilization for the Production of Bioenergy Using Microalgae Potential

Shravan Kumar[1,*]
Prateek Mishra[1]
Anjali Awasthi[2] and
S. V. A. R. Sastry[2]
[1]Department of Biochemical Engineering, Harcourt Butler Technical University, Kanpur, Uttar Pradesh, India
[2]Department of Chemical Engineering, Harcourt Butler Technical University, Kanpur, Uttar Pradesh, India

Abstract

Now, the continued existence of all life forms, including humans, is in danger due to climate change and global warming brought on by massive anthropogenic CO_2 emissions. Therefore, reducing the global carbon dioxide burden is crucial. Researchers worldwide are interested in low-cost bio-capture and CO_2 sequestration through photosynthesis to produce various value-added products, including biofuels, biofertilizers, and biopolymers. Considering flue gas CO_2 sequestration with microalgae as one of the viable solutions for energy and environmental sustainability, several countries have done so in light of the critical state of the environment. This review considered all factors affecting CO_2 sequestration in microalgal flue gases, such as culture conditions, growth nutrients, microalgae strains, and flue gas composition. Additionally, several bioreactor design strategies for improving CO_2 fixation, as well as the benefits and drawbacks of outdoor and indoor growth systems, are comprehensively discussed.

Keywords: microalgae, flue gases, CO_2 sequestering

1. Introduction

Carbon dioxide (CO_2) emissions have increased over the past few decades due to industrialization, urbanization and increased road transport. As a result, there are now more

[*] Corresponding Author's Email: hbti.shravan@gmail.com.

In: Sustainability in Chemical Processes through Digitalization and Green Chemistry Approaches
Editors: Susarla V. A. R. Sastry, Shravan Kumar and Ashish Kapoor
ISBN: 979-8-89113-257-3

severe environmental issues, and the greenhouse effect has been aggravated [1]. An important greenhouse gas that contributes to climate change and global temperature rise is carbon dioxide. Different manmade and natural sources contribute to atmospheric CO_2 levels. In nature, it is produced by respiration, volcanic eruptions, forest fires, and the release of organic debris into the ocean. Natural emissions of carbon dioxide from soils and oceans have increased dramatically over the past few decades as a result of rising temperatures leading to microbial decomposition of organic compounds [2]. Research states that the ocean produces 42.84% of the world's carbon dioxide, followed by microbial decomposition of organic matter (28.56%), plant and animal respiration (28.56%), and volcanic eruptions (0.03%) [3].

The concentration of CO_2 in the atmosphere is greatly increased by both natural resources and anthropogenic factors related to industrialization and human activities, such as cars, power plants, fossil fuel burning, deforestation, agriculture, and crop burning [4]. When compared to other nations, many industrialized and emerging nations emit comparatively more CO_2 through various industrial processes, the burning of fossil fuels, and other industrial activities, as shown in Figure 1 [5]. Industrialization requires vast amounts of energy in most countries worldwide, mainly provided by powerful thermal power plants (TPPs). CO_2 emissions from thermal power plants, which mostly burn coal and natural gas as fuel, are substantial [6]. According to the International Coal Association, coal-fired power plants currently produce more than 41% of the world's electricity, and by 2040, increased energy demand will cause this proportion to rise by 15% [7]. The industrial revolution caused atmospheric CO_2 to rise by approximately 32% this century (from 280 ppm to 400 ppm), and it is now at a worrisome level [8]. Global CO_2 emissions from various industries, including thermal power plants, steel production, cement production and refineries, amount to approximately 6.6 gigatons per billion metric tons (GT) per year [9]. Coal, oil, and natural gas burning account for approximately 40%, 32%, and 21%, respectively, of the world's CO_2 emissions, according to a report by Global Carbon Tracker [10]. Due to the COVID pandemic, global CO_2 emissions from coal and gas in 2020 were observed to decrease by 5.4% year-on-year, from 36.7 $GTCO_2$ in 2019 to 34.8 $GTCO_2$.2021 is projected to see a return to pre-COVID levels of global CO_2 emissions from fossil fuels (36.4 GT), which is 0.8% and 5% less than the CO_2 levels of 2019 and 2020, respectively [11]. Mainly found are carbon dioxide (CO_2) and carbon monoxide (CO), as well as nitrogen oxides (NOx), sulfur oxides (SOx), particulates (PM), hydrocarbons, ammonia, volatile organic compounds, heavy metals, etc. in various industrial sources, such as power plants and boilers [11]. Flue gas has various components that are extremely harmful to both the environment and living things. When deposited on plants, the PM found in flue gas interferes with photosynthesis and causes respiratory issues in both humans and animals [6]. Due to corrosive interactions, SOX and NOX emissions from flue gases damage the structure of buildings and historical monuments, and CO_2 emissions cause global warming and climate change [6]. According to the International Energy Agency (IEA) Combustion Report of 2019, major countries such as China, the USA and India emitted 9.3, 4.8 and 2.2 GT of $CO_{2,\ \text{respectively}}$ [12]. Saudi Arabia, Australia, Canada, the United States, South Korea and India all have different levels of carbon dioxide emissions. They turned out to be 16.1, 15.6, 14.9, 14.6, 11.7 tons and 1.6 tons, respectively [12]. The IEA predicts a 63% increase in CO_2 emissions compared to current CO_2 levels by 2030 due to increased industrial activity [12]. Global temperatures rose by 0.85°C due to rising CO_2 levels between 1880 and 2012, and by the end of the 21st century, they are expected to have increased by 1.4°C to 5.8°C. This will result in a 30% increase in ocean acidity.

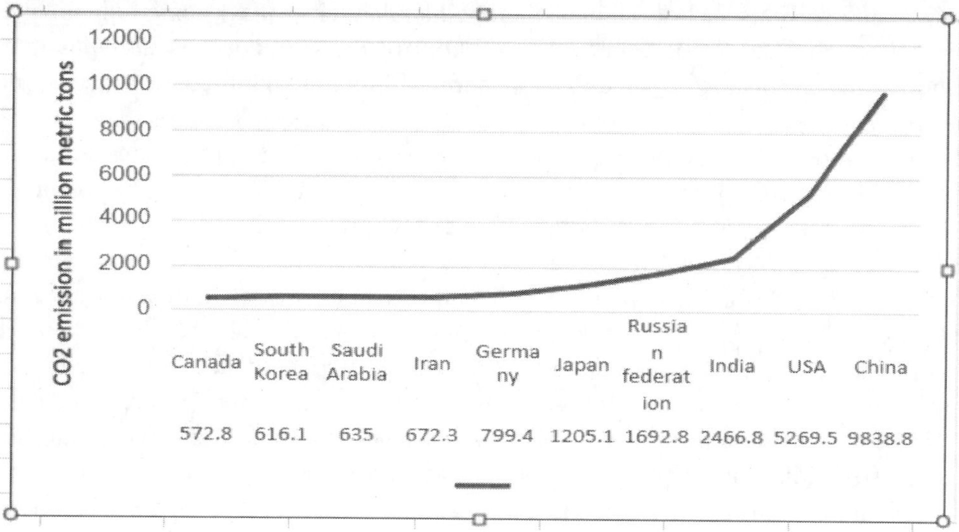

Figure 1. Ten countries with the highest CO_2 emissions in the world [5].

We have been compelled to take drastic actions due to the rapidly rising atmospheric CO_2 concentration, which has the potential to cause serious ecological and economic disruptions and unpredictable climate shifts [13].

As a result, nations have established a set of strict CO_2 management guidelines to encourage the development of clean, green alternatives that will cut CO_2 emissions while supplying energy demands and fostering a world with a cap on carbon emissions [14]. By committing to significantly decreasing carbon emissions by 2045, Britain and Ireland became the first and second governments in the world, respectively, to declare a "climate emergency" in May 2019 [13]. In a similar vein, the US has set a target for its energy efficiency program to reduce CO_2 emissions by approximately 17% by 2020 [15]. In these cases, the Kyoto Protocol and the Paris Agreement aim to protect our planet from the harmful effects of global warming. They established a number of goals and obligations for participating nations to meet to reduce greenhouse gas emissions, primarily CO_2, by lowering the use of fossil fuels and enhancing natural carbon sequestration [3]. Through pushing policies, imposing high carbon prices, and developing clean technology, the Organization for Economic Cooperation and Development (OECD) is now working to minimize CO_2 emissions through the use of solar energy and low-carbon biofuels [16]. More trees should be planted, less coal should be burned, coal-fired power plants should be run more efficiently, the use of biofuels should be increased, and renewable energy should be used more. These are only a few of the significant measures that can be employed to reduce CO_2 emissions [3, 17].

Traditional physical and chemical methods such as solid physical adsorption, chemical adsorption, membrane separation and cryogenic separation can be used to remove CO_2 from the environment [3, 10]. The ability of CO_2 to adhere through weak interactions independent of chemical bonds and its affinity for a specific surface of a material (adsorbents) are the main driving forces behind the solid physical adsorption process. Due to their high porosity, crystallinity and surface area, conventional solid adsorbents such as activated carbon, zeolites, calcium oxide and organometallic structures (MOFs) are used [18]. Chemical absorption technologies dissolve ambient CO_2 or collect it from other smoke constituents using various chemicals and solvents such as amines, mono-ethanolamine (MEA), ionic liquids, hydroxide potassium, alkalis, etc. [19]. Separation of CO_2 from gas mixtures is also possible using membrane and cryogenic separation, gas permeability and chill-condensation processes, respectively. To separate CO_2 from fuel and exhaust gases, hydrate-based gas separation (HBGS) has recently been developed [20]. However, currently used physicochemical carbon capture and storage (CCS) methods have many drawbacks, such as hazards, high cost, possible CO_2 leakage over time, scaling issues, reduced CO_2 storage efficiency and large area requirements [21]. Due to the size of the available area, marine CO_2 sequestration may appear to be a significant CO_2 sink, but it also significantly worsens ocean acidification and endangers aquatic life [12]. Meanwhile, new research has shown that biological CCS using photosynthetic organisms can replace conventional physicochemical carbon sequestration methods. Some common physical, chemical and biological methods for sequestering CO_2 are shown in Figure 1. To reduce CO_2 emissions from stationary sources, including coal-fired power plants, cement plants and steel mills, autotrophic organisms such as terrestrial plants, algae, cyanobacteria and marine phytoplankton are used in bio-CCS. This method is regarded as safe, moral, and advantageous. It is possible. And other fixed sources [3]. A total of 55-65% of all anthropogenic CO_2 emissions are absorbed by natural biosinks [22]. The biological process of CCS can use sunlight as an energy source to convert atmospheric CO_2 into organic matter (glucose), thereby achieving CO_2 biostorage (Eq. 1).

$$6CO_2 + 6H_2O \xrightarrow{\text{Sunlight}} C_6H_{12}O_6 + 6O_2$$

Algae typically photosynthesize carbon dioxide into organic molecules 10 to 15 times more efficiently than land plants [10]. Due to their ease in mass production, quick growth, increased photosynthetic efficiency, good environmental adaptation, low operational costs, and ability to utilize wastewater and flue gas as nutrition sources, microalgae are often more favorable than macroalgae for storing carbon [23]. Certain organic and inorganic pollutants contained in wastewater can be used as growth nutrients by microalgae [24]. During times of high metabolic activity, it can also accumulate harmful intracellular heavy metals such as Cr, Cd, Pb, Fe, and Co [25]. The production of diverse raw materials for biofuels, animal feed, fertilizers, lubricants, bioplastics, medicines, food additives, cosmetics, and other valuable bioactive products currently uses microalgae in combination with bioremediation processes for CO_2 sequestration. It is attracting the attention of researchers around the world, as shown in Figure 2 [5, 26]. This review focuses on flue gas enhancement, CO_2 emissions from flue gas produced by microalgae cultivation, influence of various flue gas components, characteristics affecting algae growth, various methods of culture, including binary culture, and their pros and cons. This review, which may be useful to scientific readers in the field, also reviews recent data on the use and possible impact of various species of microalgae on flue gas sequestration in various laboratory environments.

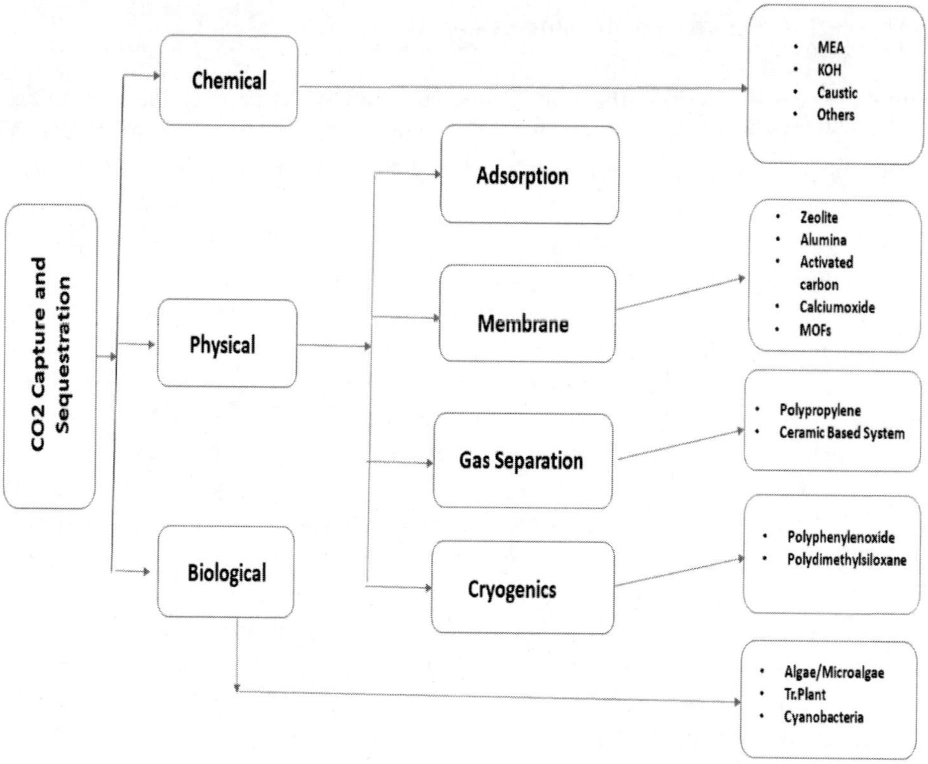

Figure 2. Various physicochemical and biological CO_2 capture techniques widely applied for flue gas CO_2 sequestration.

2. Capturing and Sequestering CO_2 by Microalgae Cultivation

The most important carbon management strategy for stabilizing atmospheric CO_2 is carbon capture and sequestration. Capture, transport, and storage are the three basic processes that can be used to manage atmospheric CO_2. Because of their better adaptability and aptitude for rapid growth in a variety of environmental circumstances, such as salinity, pH, and temperature, photosynthetic microorganisms such as microalgae can present commercial potential for achieving global sustainability [12]. The capacity of microalgae to take in CO_2 from the atmosphere is another intriguing aspect of their evolution. For every kilogram of biomass generated, 1.83 kg of CO_2 is thought to be consumed. The carbon footprint is diminished as a result [27]. As a result, the CO_2-to-microalgae pathway has attracted increased attention as a bioconversion method for CO_2 that produces a large amount of biomass. Microalgae are capable of fixing carbon dioxide from the atmosphere and industrial emissions (such as flue and flare gases) and precipitating it as soluble carbonates (e.g., $NaHCO_3$ and Na_2CO_3) [21]. In comparison to previous photosynthetic CCS techniques. Moreover, microalgae have the ability to fix and store CO_2 from air and flue gas at high quantities (i.e., 6–15% (v/v) and 0.03–0.06% (v/v), respectively) [28]. A variety of photoautotrophic microalgal species employ the process of photosynthesis to collect and store atmospheric CO_2 in the form of sugars, lipids, and proteins that can subsequently be used to produce biofuel, foods, and other high-value materials.

3. Sequestration Strategies of Flue Gas CO_2

The main sources of flue gases, also called flue gases or flue gases, are combustion equipment such as power plants and boilers. Carbon monoxide (CO), carbon dioxide (CO_2), nitrogen oxides (NOx), sulfur oxides (SOx), particulates (PM) or soot, fly ash, water vapor, any air produced and oxygen not used (O2) Yes residual composition of the main pollutants in the combustion gases [29]. The fumes produced by the combustion of raw materials containing few hydrocarbons may contain other toxic metals (Pb, Cd, Mn, Se, Cr, Zn, Hg, As, Co, Cu, Ni, Sb, Se, Sn, V, etc.), hydrogen chloride, hydrogen fluoride, hydrogen sulphide, hydrochloric acid (HCN), ammonia (NH_3) and hydrocarbons (coal and waste) [28-29]. With a share of 3–30% v/v, CO_2 is a significant component of industrial flue gas. The main factors affecting smoke composition are the type, quality and characteristics of the fuel consumed and the combustion conditions (such as the air ratio, combustion chamber volume, combustion temperature, etc.). To manage and eliminate flue gas pollutants such as CO_2 before they are discharged into the atmosphere, many nations have created stringent restrictions. Flue gas cleaning equipment, such as dust collectors and scrubbers, is placed in the incinerator to comply with government regulations [29]. Various industrial plants have successfully removed priority pollutants from flue gases (terrestrial plants and algae) using a variety of technologies, biofertilizerical adsorption, chemical, cryogenic and membrane separations, as well as geological and biological processes. As shown in Figure 2, the biological method employing microalgae has been shown to be the most effective, least harmful, and environmentally safe of these approaches, providing a blooming circular economy with useful byproducts [30].

4. Using a Coculture Approach, Increase Biomass Output and Flue Gas CO_2 Fixation

Microalgae can be employed in a number of potential fields, including environmental, bioenergy, biofertilizer, food, and nutraceuticals, in addition to sequestering carbon dioxide. Therefore, to meet the demand for microalgae feedstock, massive amounts of biomass must be produced by outdoor cultivation on a large scale. The majority of investigations on the cultivation of microalgae outdoors used a uni-algal growing technique that required an axenic environment. Axenic culture conditions can only be kept in a clean, sterile environment, much like laboratory conditions; hence, this poses a limitation [31-32]. In addition, the biomass productivity and CO_2 fixation capacity of photoautotrophic monotropic microalgal systems are reduced due to the inhibition of metabolic activity at night or in the absence of light. In contrast, heterotrophic microalgae produce significant amounts of carbon dioxide during oxidative metabolism, which can then be used in the growth medium by coexisting autotrophic microalgae. They are also less photoperiod sensitive. From this viewpoint, the notion of microalgae coculture has developed [32].

5. Limits of CO_2 Séquestration by Photosynthetic Microalgae

The ability of microalgae to effectively change CO_2 into other products with added value, such as biofuel, depends largely on their capability to sequester the gas. However, the technology for large-scale industrial CO_2 sequestration using microalgae has certain drawbacks that are outlined below [33-35].

1. The outdoor cultivation process for large-scale microalgal carbon dioxide sequestration may be impacted by variables including illumination and culture temperature. This process is difficult to control and necessitates the joint efforts of engineering technicians and scientists [36].
2. Due to pilot testing of photosynthetic algae using industrial fumes, it is difficult to concentrate large areas of land into industrial areas. A financial and practical feasibility analysis of the removal of flue gas from industrial zones is also needed. The collaboration of interdisciplinary scientific professionals in chemistry, biology, materials, manufacturing, and engineering can resolve this issue.
3. Due to the lack of theoretical studies on the design and construction of photo bioreactors, the growth process of photoautotrophic microalgae is difficult and involves many complex mechanisms of light transport, heat transport, mass transport and biological reactions. To increase the production of affordable, simple-to-use, and high-performance microalgae culture systems, careful analysis of theoretical calculations and system design is needed.
4. Currently, the selection of highly efficient carbon-fixing microalgae relies on predicting the rates of three key processes: fluid movement, mass transfer, and biotransformation. However, the study results are unstable, nonrepetitive, and subject to departure because of the unpredictability of the mass transfer process. Therefore, choosing effective algal strains and comprehending the algae development rule for CO_2 fixation may be misleading. A research approach is therefore required to improve their scientific accuracy.
5. CO_2 fixation is mainly driven by a haphazard decision or experiential method. Due to the lack of theoretical kinetic and thermodynamic studies on the CO_2 absorption process, the experimental results cannot provide an ideal and accurate design for industrial production and its quantitative management.
6. By creating algal products with added value, such as food and medicine, it is possible to make the CO_2 fixation process mediated by microalgae profitable. Therefore, it is necessary to assess and establish whether it is feasible to cultivate high-quality algae utilizing industrial waste gas.

Conclusion

Numerous researchers throughout the world have embraced microalgal-based CO_2 biofixation due to its better photosynthetic efficiency. Coculture techniques and improved indoor/outdoor reactor design are essential for increasing microalgal biomass output and flue gas CO_2 sequestration. Future efforts should focus on the large-scale growth of microalgae employing efficient local or genetically modified algal strains for the removal of industrial CO_2 flue gas, as well as the creation of biofuels and other economically significant products. To create a cost-effective, energy-efficient CO_2 treatment system based on algae, research is needed in the area of photosynthetic carbon sequestration.

References

[1] Singh, N., Batghare, A. H., Choudhury, B. J., Goyal, A., & Moholkar, V. S. (2020). Microalgae based biorefinery: Assessment of wild fresh water microalgal isolate for simultaneous biodiesel and β-carotene production. *Bioresource Technology Reports*, *11*, 100440. https://doi.org/10.1016/J.BITEB.2020.100440.

[2] Mondal, M., Ghosh, A., Gayen, K., Halder, G., & Tiwari, O. N. (2017). Carbon dioxide biofixation by Chlorella sp. BTA 9031 toward biomass and lipid production: Optimization using Central Composite Design approach. *Journal of CO_2 Utilization*, *22*, 317–329. https://doi.org/10.1016/J.JCOU.2017.10.008.

[3] Bhatia, S. K., Bhatia, R. K., Jeon, J. M., Kumar, G., & Yang, Y. H. (2019). Carbon dioxide capture and bioenergy production using biological system – A review. *Renewable and Sustainable Energy Reviews*, *110*, 143–158. https://doi.org/10.1016/J.RSER.2019.04.070.

[4] Goswami, R. K., Mehariya, S., Karthikeyan, O. P., Gupta, V. K., & Verma, P. (2022). Multifaceted application of microalgal biomass integrated with carbon dioxide reduction and wastewater remediation: A flexible concept for sustainable environment. *Journal of Cleaner Production*, *339*, 130654. https://doi.org/10.1016/J.JCLEPRO.2022.130654.

[5] Razzak, S. A., Ali, S. A. M., Hossain, M. M., & deLasa, H. (2017). Biological CO_2 fixation with production of microalgae in wastewater – A review. *Renewable and Sustainable Energy Reviews*, *76*, 379–390. https://doi.org/10.1016/J.RSER.2017.02.038.

[6] Singh, H. M., Kothari, R., Gupta, R., & Tyagi, V. V. (2019). Biofixation of flue gas from thermal power plants with algal biomass: Overview and research perspectives. *Journal of Environmental Management*, *245*, 519–539. https://doi.org/10.1016/J.JENVMAN.2019.01.043.

[7] De Silva, G. P. D., Ranjith, P. G., & Perera, M. S. A. (2015). Geochemical aspects of CO_2 sequestration in deep saline aquifers: A review. *Fuel*, *155*, 128–143. https://doi.org/10.1016/J.FUEL.2015.03.045.

[8] Bhola, V., Swalaha, F., Ranjith Kumar, R., Singh, M., & Bux, F. (2014). Overview of the potential of microalgae for CO_2 sequestration. *International Journal of Environmental Science and Technology*, *11*(7), 2103–2118. https://doi.org/10.1007/s13762-013-0487-6.

[9] Friedlingstein, P., Jones, M. W., O'Sullivan, M., Andrew, R. M., Bakker, D.C., Hauck, J., Le Qu´er´e, C., Peters, G. P., Peters, W., Pongratz, J., Stephen Sitch, Josep G. Canadell, Philippe Ciais, Rob B. Jackson, Simone R. Alin, Peter Anthoni, Nicholas R. Bates, Meike Becker, Nicolas Bellouin, Laurent Bopp, Thi Tuyet Trang Chau, Frédéric Chevallier, Louise P. Chini, Margot Cronin, Kim I. Currie, Bertrand Decharme, Laique M. Djeutchouang, Xinyu Dou, Wiley Evans, Richard A. Feely, Liang Feng, Thomas Gasser, Dennis Gilfillan, Thanos Gkritzalis, Giacomo Grassi, Luke Gregor, Nicolas Gruber, Özgür Gürses, Ian Harris, Richard A. Houghton, George C. Hurtt, Yosuke Iida, Tatiana Ilyina, Ingrid T. Luijkx, Atul Jain, Steve D. Jones, Etsushi Kato, Daniel Kennedy, Kees Klein Goldewijk, Jürgen Knauer, Jan Ivar Korsbakken, Arne Körtzinger, Peter Landschützer, Siv K. Lauvset, Nathalie Lefèvre, Sebastian Lienert, Junjie Liu, Gregg Marland, Patrick C. McGuire, Joe R. Melton, David R. Munro, Julia E. M. S. Nabel, Shin-Ichiro Nakaoka, Yosuke Niwa, Tsuneo Ono, Denis Pierrot, Benjamin Poulter, Gregor Rehder, Laure Resplandy, Eddy Robertson, Christian Rödenbeck, Thais M. Rosan, Jörg Schwinger, Clemens Schwingshackl, Roland Séférian, Adrienne J. Sutton, Colm Sweeney, Toste Tanhua, Pieter P. Tans, Hanqin Tian, Bronte Tilbrook, Francesco Tubiello, Guido R. van der Werf, Nicolas Vuichard, Chisato Wada, Rik Wanninkhof, Andrew J. Watson, David Willis, Andrew J. Wiltshire, Wenping Yuan, Chao Yue, Xu Yue, Sönke Zaehle, and Jiye Zeng. 2022. Global carbon budget 2021. *Earth Syst. Sci. Data* 14, 1917–2005. https://doi.org/10.5194/essd-14-1917-2022.

[10] Nayak, M., Karemore, A., & Sen, R. (2016). Sustainable valorization of flue gas CO_2 and wastewater for the production of microalgal biomass as a biofuel feedstock in closed and open reactor systems. *RSC Advances*, *6*(94), 9111–91120. https://doi.org/10.1039/c6ra17899e.

[11] Arun, J., Gopinath, K. P., Sundar Rajan, P. S., Felix, V., Joselyn Monica, M., & Malolan, R. (2020). A conceptual review on microalgae biorefinery through thermochemical and biological pathways: Biocircular approach on carbon capture and wastewater treatment. *Bioresource Technology Reports*, *11*, 100477. https://doi.org/10.1016/J.BITEB.2020.100477.

[12] Cheah, W. Y., Ling, T. C., Juan, J. C., Lee, D. J., Chang, J. S., & Show, P. L. (2016). Biorefineries of carbon dioxide: From carbon capture and storage (CCS) to bioenergies production. *Bioresource Technology*, *215*, 346–356. https://doi.org/10.1016/J.BIORTECH.2016.04.019.

[13] Yadav, G., Dash, S. K., & Sen, R. (2019). A biorefinery for valorization of industrial waste-water and flue gas by microalgae for waste mitigation, carbon-dioxide sequestration and algal biomass production. *Science of The Total Environment*, *688*, 129–135. https://doi.org/10.1016/J.SCITOTENV.2019.06.024.

[14] Li, Y., Zhao, R., Liu, T., & Zhao, J. (2015). Does urbanization lead to more direct and indirect household carbon dioxide emissions? Evidence from China during 1996–2012. *Journal of Cleaner Production*, *102*, 103–114. https://doi.org/10.1016/J.JCLEPRO.2015.04.037.

[15] Yadav, G., Dubey, B. K., & Sen, R. (2020). A comparative life cycle assessment of microalgae production by CO_2 sequestration from flue gas in outdoor raceway ponds under batch and semicontinuous regime. *Journal of Cleaner Production*, *258*, 120703. https://doi.org/10.1016/J.JCLEPRO.2020.120703.

[16] Zhang, L., & Mabee, W. E. (2016). Comparative study on the life-cycle greenhouse gas emissions of the utilization of potential low carbon fuels for the cement industry. *Journal of Cleaner Production*, *122*, 102–112. https://doi.org/10.1016/J.JCLEPRO.2016.02.019.

[17] Kumar, A., Ergas, S., Yuan, X., Sahu, A., Zhang, Q., Dewulf, J., Malcata, F. X., & van Langenhove, H. (2010). Enhanced CO_2 fixation and biofuel production via microalgae: recent developments and future directions. *Trends in Biotechnology*, *28*(7), 371–380. https://doi.org/10.1016/J.TIBTECH.2010.04.004.

[18] Nandasiri, M. I., Jambovane, S. R., McGrail, B. P., Schaef, H. T., & Nune, S. K. (2016). Adsorption, separation, and catalytic properties of densified metal-organic frameworks. *Coordination Chemistry Reviews*, *311*, 38–52. https://doi.org/10.1016/J.CCR.2015.12.004.

[19] Ziobrowski, Z., Krupiczka, R., & Rotkegel, A. (2016). Carbon dioxide absorption in a packed column using imidazolium based ionic liquids and MEA solution. *International Journal of Greenhouse Gas Control*, *47*, 8–16. https://doi.org/10.1016/J.IJGGC.2016.01.018.

[20] Babu, P., Linga, P., Kumar, R., & Englezos, P. (2015). A review of the hydrate based gas separation (HBGS) process for carbon dioxide precombustion capture. *Energy*, *85*, 261–279. https://doi.org/10.1016/J.ENERGY.2015.03.103.

[21] Zhang, S., & Liu, Z. (2021). Advances in the biological fixation of carbon dioxide by microalgae. *Journal of Chemical Technology and Biotechnology*, *96*(6), 1475–1495. https://doi.org/10.1002/jctb.6714.

[22] Farrelly, D. J., Everard, C. D., Fagan, C. C., & McDonnell, K. P. (2013). Carbon sequestration and the role of biological carbon mitigation: A review. *Renewable and Sustainable Energy Reviews*, *21*, 712–727. https://doi.org/10.1016/j.rser.2012.12.038.

[23] Nayak, M., Suh, W. I., Lee, B., & Chang, Y. K. (2018). Enhanced carbon utilization efficiency and FAME production of Chlorella sp. HS2 through combined supplementation of bicarbonate and carbon dioxide. *Energy Conversion and Management*, *156*, 45–52. https://doi.org/10.1016/J.ENCONMAN.2017.11.002.

[24] Sevda, S., Garlapati, V. K., Sharma, S., Bhattacharya, S., Mishra, S., Sreekrishnan, T. R., & Pant, D. (2019). Microalgae at niches of bioelectrochemical systems: A new platform for sustainable energy production coupled industrial effluent treatment. *Bioresource Technology Reports*, *7*, 100290. https://doi.org/10.1016/J.BITEB.2019.100290.

[25] Mohapatra, R. K., Parhi, P. K., Pandey, S., Bindhani, B. K., Thatoi, H., & Panda, C. R. (2019). Active and passive biosorption of Pb(II)using live and dead biomass of marine bacterium Bacillus xiamenensis

PbRPSD202: Kinetics and isotherm studies. *Journal of Environmental Management*, *247*, 121–134. https://doi.org/10.1016/J.JENVMAN.2019.06.073.

[26] Aliyu, A., Lee, J. G. M., & Harvey, A. P. (2021). Microalgae for biofuels: A review of thermochemical conversion processes and associated opportunities and challenges. *Bioresource Technology Reports*, *15*, 100694. https://doi.org/10.1016/J.BITEB.2021.100694.

[27] Vijay, A. K., Salim, S. A. M., Prabha, S., & George, B. (2021). Exogenous carbon source and phytohormone supplementation enhanced growth rate and metabolite production in freshwater microalgae Scenedesmus obtusus Meyen. *Bioresource Technology Reports*, *14*, 100669. https://doi.org/10.1016/J.BITEB.2021.100669.

[28] Cheah, W. Y., Show, P. L., Chang, J. S., Ling, T. C., & Juan, J. C. (2015). Biosequestration of atmospheric CO_2 and flue gas-containing CO_2 by microalgae. *Bioresource Technology*, *184*, 190–201. https://doi.org/10.1016/j.biortech.2014.11.026.

[29] Shmakov, V. D. (2019). Unconventional gas. *Vestnik Sankt-Peterburgskogo Universiteta, Seriya Geologiya i Geografiya*, *2010*(1), 59–98. https://doi.org/10.1016/B978-0-12-809570-6.00003-5.

[30] Yadav, G., & Sen, R. (2017). Microalgal green refinery concept for biosequestration of carbon-dioxide vis-à-vis wastewater remediation and bioenergy production: Recent technological advances in climate research. *Journal of CO_2 Utilization*, *17*, 188–206. https://doi.org/10.1016/J.JCOU.2016.12.006.

[31] Magdouli, S., Brar, S. K., & Blais, J. F. (2016). Coculture for lipid production: Advances and challenges. *Biomass and Bioenergy*, *92*, 20–30. https://doi.org/10.1016/J.BIOMBIOE.2016.06.003.

[32] Rashid, N., Park, W. K., & Selvaratnam, T. (2018). Binary culture of microalgae as an integrated approach for enhanced biomass and metabolites productivity, wastewater treatment, and bioflocculation. *Chemosphere*, *194*, 67–75. https://doi.org/10.1016/J.CHEMOSPHERE.2017.11.108.

[33] Xu, X., Gu, X., Wang, Z., Shatner, W., & Wang, Z. (2019). Progress, challenges and solutions of research on photosynthetic carbon sequestration efficiency of microalgae. *Renewable and Sustainable Energy Reviews*, *110*, 65–82. https://doi.org/10.1016/J.RSER.2019.04.050.

[34] Zhou, W., & Ruan, R. (2014). Biological mitigation of carbon dioxide via microalgae: Recent development and future direction. *SCIENTIA SINICA Chimica*, *44*(1), 63–78. https://doi.org/10.1360/032013-256.

[35] Li, F. F., Yang, Z. H., Zeng, R., Yang, G., Chang, X., Yan, J. B., & Hou, Y. L. (2011). Microalgae capture of CO_2 from actual flue gas discharged from a combustion chamber. *Industrial and Engineering Chemistry Research*, *50*(10), 6496–6502. https://doi.org/10.1021/ie200040q.

[36] Mohapatra, R. K., Padhi, D., Sen, R., & Nayak, M. (2022). Bioinspired CO_2 capture and utilization by microalgae for bioenergy feedstock production: A greener approach for environmental protection. *Bioresource Technology Reports*, *19*(June), 101116. https://doi.org/10.1016/j.biteb.2022.101116.

Chapter 9

Primary Reformer Outlet Gas Composition Estimation with Natural Gas Feed Stocks Containing Higher Hydrocarbons

Vishwas Deshpande*

(Ex). Assistant Vice President, Reliance Industries Limited, Jamnagar, Gujarat, India

Abstract

A method is proposed for obtaining the outlet gas composition of a primary reformer operating with natural gas containing higher hydrocarbons as feed stock. For a specific case, where the feed contained CH_4 = 84.7%, C_2H_6 = 8.59%, C_3H_8 = 3.39%, i-C_4H_{10} = 0.33%, n-C_4H_{10} = 0.37%, i-C_5H_{12} = 0.06%, n-C_5H_{12} = 0.05, C_6H_{14} = 0.03%, CO_2 = 2.14%, N_2 = 0.28%, H_2O = 0.05%, inlet temperature = 830°K, steam/carbon ratio =3.91, and outlet pressure =3137.28 kPa, the complete outlet gas composition was CH_4 = 16.098%, H_2 =70.63%, CO = 6.20%, CO_2 = 6.99%, and N_2 = 0.069% (dry basis). Under the operating conditions, coke formation was found to be minimal. The proposed method of solution should be of particular interest to design engineers who are confronted with the estimation of the outlet gas composition of the primary reformer at the design stage and plant personnel for performance evaluation and who do not have access to proprietary software.

Keywords: higher hydrocarbons, natural gas, primary reformer, steam reforming

Nomenclature

Symbols

a	Hydrogen in reformed gas, kmol
b	Carbon monoxide kmol
c	Carbon dioxide kmol
d	Methane in reformed gas kmol
e	Steam in reformed gas kmol

* Corresponding Author's Email: deshpandev07@gmail.com.

In: Sustainability in Chemical Processes through Digitalization and Green Chemistry Approaches
Editors: Susarla V. A. R. Sastry, Shravan Kumar and Ashish Kapoor
ISBN: 979-8-89113-257-3

f Inerts in reformed gas kmol
A Total carbon in compounds in the system, kmol
B Total hydrogen in compounds in the system, kmol
C Total oxygen in compounds in the system, kmol
K Equilibrium constants
P Reformer pressure, Atm
ΔG_f^0 Standard free energy of formation, Kcal/mol

Subscripts

1 Reforming reaction
2 Water-gas shift reaction

Greek Letters

∅ Intermediate calculation parameter
θ Intermediate calculation parameter
∑ Sum

1. Introduction

The equilibrium composition of multiple reactions is calculated either by Gibb's free energy minimization [1] or by the equilibrium constant method [2]. In previous work [3], a comprehensive review of the various methods for estimating the outlet gas composition of the primary reformer was given while reviewing the analytical solution proposed by Hampson [4, 5]. The author used the Newton Raphson method to obtain the equilibrium gas composition at the exit of the reformer.

The Newton–Raphson method requires estimation of the derivative of the function, which is somewhat complicated; therefore, there is a possibility of encountering local optima in the search region. The objective of the present work is to apply the method proposed by Belton [6] to obtain the outlet gas composition of a primary reformer. For the sake of ease, the data from previous work, symbols, and abbreviations have been retained.

The following equations represent the steam methane reforming process:

$CH_4 + H_2O \leftrightarrow CO + 3\,H_2 \quad \Delta H_{298}^0 = +206$ kJ/mol (R1)

$CH_4 + 2H_2O \leftrightarrow CO_2 + 4\,H_2 \quad \Delta H_{298}^0 = -165$ kJ/mol (R2)

$CH_4 + CO_2 \leftrightarrow 2\,CO + 2\,H_2 \quad \Delta H_{298}^0 = +247$ kJ/mol (R3)

$CO + H_2O \leftrightarrow CO_2 + H_2 \quad \Delta H_{298}^0 = -41$ kJ/mol (R4)

$C_nH_{2m} + nH_2O \leftrightarrow nCO + (n+0.5\,m)\,H_2 \quad \Delta H_{298}^0 < 0$ (R5) where equations (R1) through (R5) describe the complete reforming process. The reforming process requires solutions to the reactions described by (R1) and (R4) since these are thermodynamically independent. Other reactions describing coke deposition rendering deactivation of the catalyst are represented as follows:

$2CO \rightleftharpoons CO_2 + C$ $\Delta H^0_{298} = -172.6$ kJ/mol Boudouard coking (R6)

$CO + H_2 \rightleftharpoons C + H_2O$ $\Delta H^0_{298} = -131$ kJ/mol CO reduction (R7)

$CH_4 \rightleftharpoons C + 2H_2$ $\Delta H^0_{298} = +74.8$ kJ/mol Methane Cracking (R8)

$C_mH_{2n} \rightleftharpoons m\,C + n\,H_2$ Higher hydrocarbons (R9)

$C_mH_{2n} \rightarrow$ Polymer $\rightarrow$ Coke

2. Formulation of the Problem

Case Study

For an operating primary reformer of a urea plant, the following operating data [3] are available, which is required to estimate the outlet gas composition of the primary reformer. The composition of the natural gas feed to the primary reformer is given in Table 1. Rate constants for reforming and water gas shift reactions are given in Table 2.

Data

- Natural gas flow rate 19800 nM^3/Hr.
- Steam flow 73.2 T/Hr.
- Inlet gas temperature 830°K
- Outlet gas temperature 1031°K
- Pressure at the inlet of reformer 3137.28 kPa
- Pressure drop across the reformer 310.66 kPa
- $\left[\frac{Steam}{Carbon}\right]$ Ratio 3.9102

Table 1. Composition of natural gas feed to the primary reformer [3]

CH_4	C_2H_6	i-C_3H_8	n-C_4H_{10}	i-C_4H_{10}	n-C_5H_{12}	i-C_5H_{12}	C_6H_{14}	CO_2	*N_2	#H_2O
84.7	8.59	3.39	0.33	0.37	0.06	0.05	0.03	2.14	0.28	0.05

(*: Air as N_2; # Moisture)

Table 2. Rate constants for reforming and water gas shift reactions [3]

Temp °K	1031	1026	1021	1016	1011
K_1	60.59	53.29	46.81	41.07	35.99

$K_2 = 1.229$.

Solution

The composition of the feed entering the primary reformer is given in Table 3. To set up material balance equations across the primary reformer, the species involved include carbon, hydrogen, oxygen and inert elements present per 100 kmol of feed.

Table 3. Kilomoles of C, H, O and Inerts per 100 moles of feed

Species	Moles	C	H	O	Inert
CH_4	84.7	84.7	169.4	0	0
C_2H_6	8.59	17.18	25.77	0	0
C_3H_8	3.39	10.17	13.56	0	0
i- C_4H_{10}	0.33	1.32	1.65	0	0
n- C_4H_{10}	0.37	1.48	1.85	0	0
i-C_5H_{12}	0.06	0.3	0.36	0	0
n-C_5H_{12}	0.05	0.25	0.3	0	0
n- C_6H_{12}	0.03	0.18	0.21	0	0
CO_2	2.14	2.14	0	1.07	0
N_2	0.28	0	0	0	0.28
H_2O	0.05	0	0	0.025	0
	99.99	117.72	213.1	1.095	0.28
$\sum$ For 1 kmol NG		1.1772	2.131	0.01095	0.0028

Estimation of moles of steam added, Kmol = 3.9102 x 1.1772 = 4.603
Estimate of H_2 Kmole = 4.603
Estimate of O_2 Kmole = 2.3015

Material Balance Equations

The material balance equations for carbon, oxygen and hydrogen species across the primary reformer can be written as follows:

Carbon Balance

IN =OUT
$1.1772 = x + y + [CO_2]_{Out}$
FEED CH_4 CO
$[CO_2]_{Out} = 1.1772 - x - y$ (1)

Oxygen Balance

IN =OUT
$1.07 + 4.6036 = y + 2*(1.1772 - x - y) + [H_2O]_{Out}$
$[CO_2]$ Steam [CO] $[CO_2]$
$[H_2O]_{Out} = 4.6036 - y - 3.544 + 2x + 2y$
$= 1.059587 + 2x + y$ (2)

Hydrogen Balance

IN = OUT

$2.131 + 4.6036 = 1.059587 + (2x + y) + [H_2]_{Out}$

FEED Sam

$[H_2]_{Out} = 5.675 - 4x - y$ (3)

Equilibrium Constants

The equilibrium constants for reforming (R1) and shift (R4) reactions can be written as

$$K_1 = \frac{(pH2)^{\wedge}3 \ x \ (pCO)}{(pCH_4) \ x \ (pH_2O)} \quad (4)$$

$$K_2 = \frac{(pCO_2) x (pH_2)}{(pCO) x (pH_2O)} \quad (5)$$

For the solver solution, the work sheet from Table 3 is extended to include products as well as the material balance equations, as shown in Table 4.

The decision variables are x and y, whose values are used in computing the formulas for equilibrium rate constants for reforming and shift reactions as per equations (4) and (5). The computed values of equilibrium constants are as per cell values G2 and G3. The solver adjusts the values in the decision variable cells C3 and C4 to satisfy the limits on constraints cells F2 and F3 to produce results for objective cells. In simple terms, Solver determines the optimum (maximum or minimum) value of one cell by changing other cell values.

The optimal solver solution is shown in column H, Table 4. The complete reformed gas composition at the reformer exit is H_2 = 70.6353%m, CH_4 = 16.0957%m, CO_2 = 6.9933%m, CO=6.2060%m, and N_2 (inert) = 0.0697%m (dry basis). The criteria for coke deposition in the primary reformer are shown in Table 5.

Coke Deposition

Equilibrium constants for coke formation through Boudouard coking, CO reduction and methane cracking were computed as per the procedure described by Mearns. The computed values are compared with the literature values of equilibrium constants for coke forming reactions (R7, R8) as shown in Figure 1.

Table 4. Work sheet showing solver solution to material balance equations

File Home Insert Page Layout Formulas Data Review View Help

J18

Constraints cells

Variable cells

Decision Variables

	A	B	C	D	E	F	G	H
1		t= 785.0000		°C		Required	Currently	
2		P_{Total} = 28.0000		Atm	$K_{Reforming}$	60.5900	60.5900	
3		x = 0.6468			K_{Shift}	1.2290	1.2290	
4		y = 0.2494						
5		P° = 1.0000		Atm				
6								
7	Species	In, Kmol	Out,Kmol		y_i	P_i	P_i / P°	% m [Dry]
8	CH_4	0.8470	x	0.6468	0.0977	2.7353	2.7353	16.0957
9	C_2H_5	0.0859	0.0000		0.0000	0.0000	0.0000	
10	C_3H_8	0.0339	0.0000		0.0000	0.0000	0.0000	
11	i- C_4H_{10}	0.0033	0.0000		0.0000	0.0000	0.0000	
12	n-C_4H_{10}	0.0037	0.0000		0.0000	0.0000	0.0000	
13	i-C_5H_{12}	0.0006	0.0000		0.0000	0.0000	0.0000	
14	n-C_5H_{12}	0.0005	0.0000		0.0000	0.0000	0.0000	
15	n- C_6H_{14}	0.0003	0.0000		0.0000	0.0000	0.0000	
16	CO_2	0.0214	1.1772-x-y	0.2810	0.0424	1.1884	1.1884	6.9933
17	N_2	0.0028	0.0028	0.0028	0.0004	0.0118	0.0118	0.0697
18	H_2O	4.6036	1.059587 +2x +y	2.6026	0.3931	11.0062	11.0062	
19	CO	0.0000	y	0.2494	0.0377	1.0546	1.0546	6.2060
20	H_2	0.0000	5.675-4x-y	2.8384	0.4287	12.0036	12.0036	70.6353
21	$\sum$		7.914587 - 2x	6.6210	1.0000	28.0000	28.0000	100.0000
22	$\sum$ Removing H_2O			4.0184				

Table 5. Check for Criteria for coke deposition in primary reformer

Coke formation reactions in the reformer					
(R6) CO $\rightleftharpoons CO_2 + C$		(R7) CO $+H_2 \rightleftharpoons C + H_2O$		(R8) $CH_4 \rightleftharpoons C + 2H_2$	
Boudouard coking		Carbon monoxide reduction		Methane cracking	
$K_{1,2} = \frac{(pCO2)}{(pCO)^2}$		$K_{1,3} = \frac{(pH2O)}{(pCO)(pH2)}$		$K_{1,4} = \frac{(pH2)^2}{(pCH4)}$	
Calculated	Literature	Calculated	Literature	Calculated	Literature
1.0675	0.1350	0.8694	0.1500	52.6800	~18.0
For all the reactions, the calculated values of rate constants are higher than graphical, hence no credible scenario for coke formation is envisaged under the operating conditions of reformer					

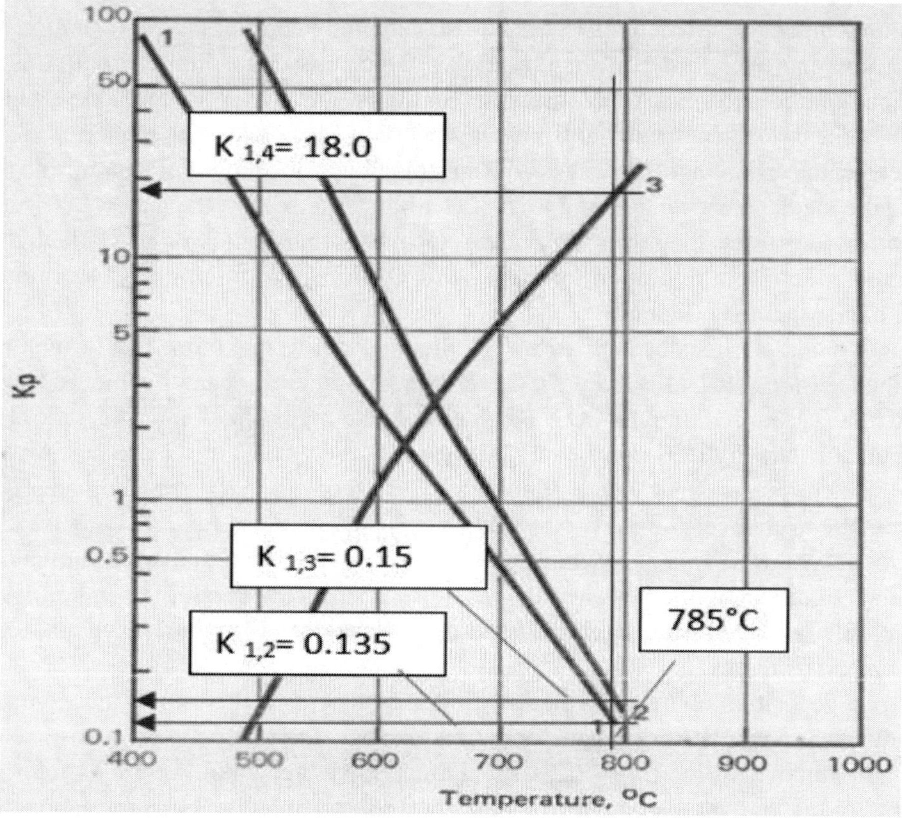

Figure 1. Rate constants for carbon-forming reactions [7]. (Fig 1.30, p 82, Catalyst Handbook, Spencer M.S., 2nd Ed Edited by M.V.Twigg, Copyright (1989). (Reproduced by permission from Taylor & Francis Group).

Results and Discussion

Starting with initial values of x = 0.4 and y=0.3, the converged values of x and y are 0.6468 and 0.2494, respectively. Using these values, equations 1, 2 and 3 are solved to obtain mole fractions of H_2, CO_2, and H_2O. A change in the initial guess had no effect on the convergence.

For the case studied, the outlet gas composition of the mixture of gases from the exit of the primary reformer was calculated using Belton's method. The complete reformed gas composition on a dry basis is H_2 = 70.6353%m, CO_2 = 6.9933%m, CO = 6.206%m, CH_4 = 16.0957%m, and inert (N_2) = 0.0697%m.

3.1. Effect of Changes in the Operating Parameters of the Primary Reformer

Outlet pressure: Operating at lower pressure levels reduces feedstock consumption and lowers the risk of carbon laydown. The optimum pressure is mainly decided based on the pressure level required in the CO purification stage.

Outlet temperature: Reactions (R1) and (R3) are endothermic; hence, operation at higher temperatures improves feed conversion. Higher temperatures are limited by the mechanical constraints of the reformer tube material. Similarly, at higher temperatures, the carbon deposition due to methane cracking is higher.

Inlet steam/carbon ratio: An excess of the stoichiometric amount of steam is necessary to improve the methane conversion and to avoid carbon lay-down on the catalyst. The condition of carbon lay-down sets the lower limit of the steam to carbon ratio. For a CO plant, the excess steam used is kept to a minimum to maximize CO formation, minimize CO_2 formation and achieve higher thermal efficiency.

CO_2/C ratio: CO_2 is added at the inlet of the reformer to maximize CO formation. where CH_4 is the feedstock, and a 1:3 ratio of CO_2/H_2 at the reformer exit can be achieved by recycling the entire CO_2 produced. Import CO_2 can be used for further reduction of this ratio. However, the carbon utilization of CO_2 significantly decreases as the CO_2/C ratio increases. An excessive CO_2/C ratio increases fired heater duty and CO_2 recovery and recycling costs, thereby increasing the overall cost of CO production.

Carbon laydown: This is a vital consideration in the operation of reformers. Carbon deposition significantly deactivates the catalyst and reduces reformer tube life due to overheating. Carbon deposition is a trade-off between equilibrium reactions for carbon formation and its removal.

Carbon deposition on the reformer tube internals occurs if the rate of reaction(s), having thermodynamic potential for carbon formation, exceeds the rate of the (reverse) reaction(s), which will remove carbon, which exists during operating conditions in a reformer. The reforming of higher hydrocarbons particularly poses the problem of coke formation due to the increased ease of thermal cracking or pyrolysis. Coking is favored at a low steam-to-carbon ratio and high temperatures.

Conclusion

The composition of the reformed gas from the primary reformer exit has been determined for a specific case. Although major contracting companies already use their patented software to solve this kind of problem, the proposed method and solution method should be useful to the plant operators and design engineers who are confronted with estimating the outlet gas composition of the reformer for evaluating performance for selecting the optimum operating conditions of the plant. The proprietary software is access controlled.

Graduating engineers who are tackling reforming problems in their project work can find the method useful since the original article was written by Hampson [4, 5] for students who use the analytical method as a starting point in estimating gas composition and later are given access to the computer program which looks at complete mass and energy balance of the ammonia manufacture.

The results obtained need to be validated with data from actual plants. This will help in the refinement of the model to represent the actual scenario. It is found that the reformer is operating under no-coking conditions, a conclusion that is only qualitative in nature and has no practical significance for the process design of steam-NG reforming.

References

[1] Chopey N., (2004) Equilibrium composition for simultaneous known chemical reactions, "*Handbook of Chemical Engineering Calculations,*" Part-1, Sections 4-7 & 4-8.

[2] Deshpande V. V., Saha S., (2013) Steam Reforming Natural Gas with Higher Hydrocarbons, *PTQ Journal*, *GAS*, April-2013,1-5.

[3] Deshpande V., (2017) Solution to Steam Reforming Natural Gas for Primary Reformer Process Design Equations, *International Journal of Research in Science & Engineering*, Vol 3, Issue 1, (e-ISSN: 2394-8299; p-ISSN: 2394-8280).

[4] Hampson, G., (1979) Design Procedures: A Simple solution to Steam Reforming equations (Part-1), *The Chemical Engineer*, 523.

[5] Hampson, G., (1979) Design Procedures: A Simple solution to Steam Reforming Equations (Part-2), *The Chemical Engineer*, 621.

[6] Belton D. (2018), Steam Reformer Material balance, Parts 1,2,3 & 4, *Chemical Engineering Tutor*, YouTube, Oct 30, 2018.

[7] Mearns A. M. (1973), *"Chemical Engineering Process Analysis,"* Oliver & Boyd, 94-95.

[8] Smith J. M., Van Ness H. C., Abbott M. M., (1996) *Introduction to Chemical Engineering Thermodynamics*, 5th Edition, McGraw Hill International, 538-602.

Appendix I: Checking the Analytical Solution Proposed by Hampson [4, 5]

Example: A primary reformer has natural gas feed consisting of CH_4 90%$_m$, C_2H_6 6%$_m$ and N_2 4%$_m$. Using a steam-to-carbon ratio of 3, an outlet temperature of 800^0 C and a pressure of 28 atmospheres, the reformed gas composition was calculated. Reforming reaction $K_1 = 118.2$ @ 785^0C, Shift Reaction $K_2 = 1.136$ (Assuming 15^0 C Approach to equilibrium) @ 800°C.

Steam/Total Carbon = 3.0

Total carbon in the system = [carbon in CH_4] + [carbon in C_2H_6] = 0.9 x 1 + 0.06 x 2 =1.02

Total Steam added = 1.02 x 3 = 3.06 kmol

Material Balance Equations across primary reformer

Carbon Balance

In = Out

$[CH_4] + [C_2H_6] = [CO_2] + x + y$

$0.9x1 + 0.06x3 = [CO_2]_{out} + x + y$

$[CO_2]_{out} = 1.02 - x - y$

Oxygen Balance

In = Out

$[O]_{Steam} = [O]_{CO2} + [O]H_2O,_{out} + [O]\,CO$

$3.06 = 2 \times [1.02 - x - y] + [O]_{H2O,\,out} + y$

$3.06 = 2.04 - 2x - 2y + y + [O]\,H_2O$, Out

$[O]_{H2O,\,out} = 1.02 + 2x + y$

Hydrogen Balance

In = Out

$[H]_{CH4} + [H]_{C2H6} + [H]_{H2O,\,Steam} = [H]_{CH4} + [H]_{H2O} + [H]_{H2}$

$0.9 \times 2 + 0.06x3 + 3.06 = 2x + 1.02 + 2x + y + [H]_{H2}$

$1.8 + 0.18 + 3.06 = 1.02 + 4x + y + [H]_{H2}$

$[H]_{H2} = 4.02 - 4x - y$

Total K mol gases out = 1.02-x-y+1.02+2x+y+4.02-4x-y
Total K mol gases out = 6.06 - 2x

Work Sheet Showing Solver Solution to Material Balance Equations

	t=	800°C	°C	Hampson K_1 & K_2		Required	Currently
	T_{otal} =	28.0000	bar	$K_{Reforming}$		118.2000	118.2000
	x=	0.3754		K_{Shift}		1.1360	1.1360
	y=	0.3112	$P^{\varnothing}$ =				
	t=	800°C	°C	1.0000		bar	Standard
Component	Kmol, In	Kmol, Out*	y_i	P_i, Bar	$P_i/P^{\varnothing}$	%m (Dry)	Hampson's Result
CH_4	0.9	x	0.0704	1.9723	1.9723	11.4893	11.4600
C_2H_6	0.06	0	0.0000	0.0000	0.0000	0.0000	0.0000
N_2	0.04	0.04	0.0626	1.7519	1.7518	10.2052	10.2
CO	0	y	0.0075	0.2102	0.2101	1.2243	1.2200
CO_2	0	1.02- x - y	0.3907	10.9387	10.9386	0.0000	0.0000
H_2O	3.06	1.02 + 2 x + y	0.0584	1.6350	1.6349	9.5242	9.5400
H_2	0	4.02 -4 x - y	0.4142	11.5971	11.5971	67.5570	67.5800
		6.1-2 x	1.0000	28.0000	28.0000	100.0000	100.0000

There is a satisfactory agreement between calculated gas composition using Belton's method and published result as seen from the last two columns of the above table.

Appendix II: Solution by Gibb's Free Energy Minimization Method [1, 8]

For the present case, in the initial unreacted state, from material balance, there are 877.8246 moles of CH_4 and 3846.9273 moles of H_2O. It is necessary to determine the equilibrium composition of the final system @758°C and 28.85 bar that contains CO, CO_2 H_2, CH_4, and H_2O.

Values of ΔG_f^0 at 758°C and 28.85 bar for hydrocarbon species using Aspen+ (7.3 version) with Peng-Robinson EOS are $\Delta G_{CH_4}^0$ = + 2.87 Kcal/mole (+12008.08 Joule.mole^{-1}),$\Delta G_{fCO_2}^0$ = -93.58 Kcal/mole (- 391538.72 J. mole^{-1}), ΔG_{CO}^0 = - 45.46 Kcal/mole (- 190204.64 J. mole^{-1}), and $\Delta G_{H_2O}^0$ = - 44.64 Kcal/mole (- 186773.76 J. mole^{-1}).

Determining the number of gram atoms of each atom present in the system

One gram of CO contains 1 g atom each of C and O, and 1 g mol of H_2contains 2 g atoms of H

The required values of A_k (i.e., the total number of atomic masses of the kth element in the system, as determined by the initial constitution of the system, are determined from the initial numbers of moles, and the values of $a_{i,k}$ come directly from chemical formulas of the species) are tabulated below.

A_k = Number of atomic masses of k in the system						
Species *i*	Element, k					
	Carbon		Oxygen		Hydrogen	
	A_C =1040.2632		A_O= 4104.9092		A_H =11895.5127	
	$a_{i,k}$ = No of atoms of k per mole of *i*					
CH_4	$a_{CH4,C}$	1	$a_{CH4,O}$	0	$a_{CH4,H}$	4
H_2O	$a_{H2O,C}$	0	$a_{H2O,O}$	1	$a_{H2O,H}$	2
CO	$a_{CO,C}$	1	$a_{CO,O}$	1	$a_{CO,H}$	0
CO_2	$a_{CO2,C}$	1	$a_{CO2,O}$	2	$a_{CO2,H}$	0
H_2	$a_{H2O,C}$	0	$a_{H2O,O}$	0	$a_{H2O,H}$	2

Writing equations for minimization of total Gibbs free energy,

$$\Delta G_{f_i}^0 + RT \ln\left(y_i \hat{\emptyset}_i' \frac{P}{P^0}\right) + \Sigma_k \lambda_k a_{i,k} = 0, (i = 1,2,3\ldots\ldots N) \qquad (A)$$

R= Gas constant, T = Temperature, y = mole fraction, $\hat{\emptyset}_i'$ = fugacity coefficient, P = Pressure, P^0= Vapor pressure, λ_k = Lagrange undetermined multiplier for element k within substance i. Writing material balance and mole-fraction equations

Assuming that at 1031 K and 28.85 bar, the assumption of ideal gas is justified, and the $\hat{\emptyset}_i$ values are all unity.

Since P= 28.85 Bar, $\frac{P}{P^0}$ = 28.85,

For five species, CH_4,H_2O,CO,CO_2,H_2. Five material balance equations can be written using equation (A) as follows:

$$CH_4: \quad \frac{12008.08}{8571.734} + 3.3621 + \ln\left\{\frac{n_{CH_4}}{\Sigma_i n_i}\right\} + \frac{\lambda_C}{8571.734} + \frac{4\lambda_H}{8571.734} = 0$$

$$8571.734 \ln \boldsymbol{x_1} + \boldsymbol{x_6} + 4\boldsymbol{x_8} = 32585.4468 \qquad \text{Eq. (1)}$$

$$H_2O: \quad -\frac{186773.76}{8571.734} + 3.3621 + \ln\left\{\frac{n_{H2O}}{\Sigma_i n_i}\right\} + \frac{\lambda_O}{8571.734} + \frac{2\lambda_H}{8571.734} = 0$$

$$8571.734 \ln \boldsymbol{x_2} + \boldsymbol{x_7} + 2\boldsymbol{x_8} = 231359.6343\ldots \qquad \text{Eq. (2)}$$

$$CO: \quad -\frac{190204.64}{8571.734} + 3.3621 + \ln\left\{\frac{n_{CO}}{\Sigma_i n_i}\right\} + \frac{\lambda_C}{8571.734} + \frac{\lambda_O}{8571.734} = 0$$

$$8571.734 \ln \mathbf{x_5} + \mathbf{x_6} + \mathbf{x_7} = 234790.5144 \qquad \text{Eq. (3)}$$

$$CO_2: \quad -\frac{391538.72}{8571.734} + 3.3621 + \ln\left\{\frac{n_{CO_2}}{\Sigma_i n_i}\right\} + \frac{\lambda_C}{8571.734} + \frac{2\lambda_O}{8571.734} = 0$$

$$8571.734 \ln \mathbf{x_3} + \mathbf{x_6} + \mathbf{2x_7} = 436124.5944 \qquad \text{Eq. (4)}$$

$$H_2: \; 3.3621 * \ln\left\{\frac{n_{H_2}}{\Sigma_i n_I}\right\} + \frac{2\lambda_H}{8571.734} = 0$$

$$RT \ln\left(\frac{n_{CO_2}}{\Sigma_i n_I} * 28.85\right) + 2\lambda_H = 0$$

$$8571.734 \ln \boldsymbol{n_{H_2}} + \mathbf{2}\,\boldsymbol{\lambda_H} = 42497.2627 \quad \text{Eq. (5)}$$

Material balance equation for each element of species based on values in the above table

$\sum_i n_i a_{i,k} = A_k$ (k = 1,2,3 .... w) (B)

Three material balance equations from equation (B) considering where w=3 is the total number of elements comprising the system

C: $n_{CH_4} + n_{CO} + n_{CO_2} = 1040.2632$ Eq. (6)

H: $4\, n_{CH_4} + 2\, n_{H_2O} + 2\, n_{H_2} = 11895.5127$ Eq. (7)

O: $n_{CO} + 2\, n_{CO_2} + n_{H_2O} = 4104.9092$ Eq. (8)

$\sum_i n_i = n_{CH_4} + n_{CO} + n_{CO_2} + n_{H_2O} + n_{H_2} = 5232.3699$ Eq. (9)

Additionally, the sum of mole fractions must be equal to = 1.0

$y_{CO} + y_{CO_2} + y_{H_2} + y_{CH_4} + y_{H_2O} = 1.0$ Eq. (10)

Simultaneous equations from (1) to (9) are solved for the mole fractions of all species present at equilibrium, namely, y_{CH_4} , y_{H_2O} , y_{CO_2}, y_{H_2}, y_{CO}, λ_C, λ_H, and λ_O, n and implicit constraints,

$0 \le y_{CH_4} \le 1,\ 0 \le y_{H_2O} \le 1,\ 0 \le y_{CO_2} \le 1,\ 0 \le y_{CO} \le 1,\ 0 \le y_{H_2} \le 1$

A suitable numerical method such as the Lagrange multiplier technique can be used to solve nonlinear equations. The Lagrange multipliers have no physical significance and are eliminated from the calculation scheme.

The solution was attempted using ROMeo software from Invensys. Its capability is similar to that of Aspen. It has the inbuilt solver Opera, originally developed by Shell. It has a solver that solves the nonlinear algebraic equations using an open equation approach. Basically, it uses sequential quadratic programming (SQP) an iterative method for nonlinear optimization.

Table 6. Solution to the equations

	C1	X1	C2	X2	C3	X3	C4	X4	C5	X5	C6	X6	C6	X6	C7	X7	C8	X8	$R.H.S._{calcd}$	$R.H.S._{Given}$	Error
Eq 1	8571.734	6.24	0	3097.72	0	477.54	0	1828.82	0	52.12	1	18735.65	0	18735.65	0	182256.5	4	-9900.04	32585.447	32585.447	0.00
Eq 2	0	510.61	8571.731	8.04	0	477.54	0	1828.82	0	52.12	0	18735.65	1	18735.65	1	182256.5	2	-9900.04	231359.609	231359.609	0.00
Eq 3	0	510.61	0	3097.72	0	477.54	0	1828.82	8571.734	3.95	1	18735.65	1	18735.65	1	182256.5	0	-9900.04	234880.514	234880.515	0.00
Eq 4	0	510.61	0	3097.72	8571.734	6.17	0	1828.82	0	52.12	1	18735.65	0	18735.65	2	182256.5	0	-9900.04	436124.594	436124.5994	0.00
Eq 5	0	510.61	0	3097.72	0	477.54	8571.734	7.51	0	52.12	0	18735.65	0	18735.65	0	182256.5	2	-9900.04	44585.874	44585874	0.00
Eq 6	0	510.61	0	3097.72	1	477.54	0	1828.82	1	52.12	0	18735.65	0	18735.65	0	182256.5	0	-9900.04	1040.263	1040.263	0.00
Eq 7	0	510.61	2	3097.72	0	477.54	2	1828.82	0	52.12	0	18735.65	0	18735.65	0	182256.5	0	-9900.04	11895.513	11895.513	0.00
Eq 8	0	510.61	1	3097.72	2	477.54	0	1828.82	1	52.12	0	18735.65	0	18735.65	0	182256.5	0	-9900.04	4104.909	4104.909	0.00

Appendix III

The analytical method originally proposed by Hampson [4, 5] can be programmed for a computer solution using a suitable numerical technique such as Newton Raphson. For the present case, the derivative of pressure "P" with respect to methane gas content "d" in the reformer exit gases has been shown. It is proposed to solve the problem using MATLAB.

A = Total carbon in the system, kmole = 117.8610
B = Total hydrogen in the system, kmole = 672.8586
C = Total oxygen in the system, kmole = 231.0343
Inerts, Kmole = 0.28
Moles of Steam 3.9102 x 117.861= 459.8786
Moles of Hydrogen in steam = 459.8786
Moles of Oxygen in Steam = 229.9393
Intermediate calculation parameters θ & Ø

$$\Theta = 2A + B - 2C = 446.01 \quad (1)$$

$$Ø = A + B\,K2 + 3\,d\,(1-2K2) - \theta\,(1-K2) = 1046.6894 - 4.374*d \quad (2)$$

Equations for "a" and "P"

$$a = \frac{-Ø \pm \sqrt{Ø^2 + 4K2(1-K2)(\Theta-4d)(B-2d)}}{2(1-K2)} \quad (3)$$

$$a = \frac{(1046.6894 - 4.374 * d) - \sqrt{(1046.6894 - 4.374 * d)^2 + 4.916(-0.229)(446.01 - 4d)(672.8586 - 2d)}}{0.458} \quad (4)$$

$$P = \frac{A+B-2d+f}{a}\sqrt{K_1 \frac{d}{a}\frac{(B-a-2d)}{(\Theta-a-4d)}} \quad (5)$$

$$P = \frac{790.7486-2d}{a}\sqrt{60.59\frac{d}{a}\frac{(672.8586-a-2d)}{(446.01-a-4d)}} \quad (6)$$

It is observed that a = a (d) and P = P (a, d), using the chain rule of differentiation

$$\frac{dP}{dd} = \frac{\partial P}{\partial a}\frac{\partial a}{\partial d} + \frac{dP}{dd} \quad (7)$$

The final expression for the derivative is

$$\frac{dP}{dd} = \frac{(790.4087-2d)}{a} \times \frac{1}{2}\sqrt{\frac{(a)*(446.35-a-4d)}{(60.59d)(673.4087-a-2d)}}\,\Big[\left\{\frac{60.59d}{a}\right\}\frac{227.058}{(446.35-a-4d)^2} -$$

$$\left\{\frac{(673.4087-a-2d)}{(446.35-a-4d)}\right\}*\frac{1}{2}*\frac{60.59d}{a^2}\Big]$$

$$+\sqrt{\frac{60.59d}{a}\frac{(673.4087-a-2d)}{(446.35-a-4d)}} \times \frac{(2d-790.4087)}{a^2} \times - [9.5502 +$$

$$\frac{(22.1086-6694.3644)}{\sqrt{(10.1258d^2-6132.055d+758995.0625}}]$$

$$+\frac{(790.4087-2d)}{a} \text{ x } [\frac{1}{2}\sqrt{\frac{(a)*(446.35-a-4d)}{(60.59d)(673.4087-a-2d)}}[\left\{\frac{60.59d}{a}\right\} x \frac{1800.9348-2a}{(446.35-a-4d)^2}$$

$$+\left\{\frac{(673.4087-a-2d)}{(446.35-a-4d)}\right\} \text{ x } \left\{\frac{60.59}{a}\right\}]] - \{ \sqrt{\frac{60.59d}{a}\frac{(673.4087-a-2d)}{(446.35-a-4d)}} \text{ x } \frac{2}{a} \} \quad (8)$$

(Symbols and notations from the original article are retained for the sake of ease.)

Chapter 10

The Effects of the Direction of Mass Transfer and Channel Diameter on Solvent Extraction in Core-Annular Flow (CAF) in a Microchannel

R. K. Chaurasiya[1,2,*]
K. K. Singh[1,2]
S. Chakraborty[1,3]
and S. Mukhopadhyay[1,2]

[1]Homi Bhabha National Institute, Anushaktinagar, Mumbai, India
[2]Chemical Engineering Division, Bhabha Atomic Research Centre, Trombay, Mumbai, India
[3]Radiopharmaceutial Division, Bhabha Atomic Research Centre, Trombay, Mumbai, India

Abstract

Core-annular flow (CAF) is one of the liquid-liquid flow patterns observed in microchannels when used for solvent extraction. The liquid that preferentially wets the wall of the microchannel forms the annulus, while the other liquid forms the core. Thus, depending on wettability, different liquid–liquid systems may have different directions of mass transfer (either from core to annulus or annulus to core). The direction of mass transfer will also be different for extraction and stripping modes. The diameter of the microchannel also affects the rate of mass transfer in the CAF. In this study, CFD modeling of liquid-liquid mass transfer in CAF is performed to study the effect of the direction of mass transfer and channel diameter. Despite the change in the concentration profile in the core and annulus due to a change in the direction of mass transfer, mass transfer variables such as the concentration boundary layer thickness (δ), local mass transfer coefficient (K_L) and Sherwood number (Sh) of the core and annulus do not change significnatly. However, on interchanging the properties of the liquid in the core and liquid in the annulus in addition to the change in the direction of mass transfer, the change in mass transfer variables is significant. A decrease in microchannel diameter increases the mass transfer coefficient for the core (K_{Lc}) and the annulus (K_{La}), but the Sherwood number for the core (Sh_c) and the annulus (Sh_a) remain the same. This means that the convective flux and diffusive flux increase in the same proportion when the diameter of the microchannel is increased.

---

* Corresponding Author's Email: chaurasiyarajnesh@yahoo.co.in.

In: Sustainability in Chemical Processes through Digitalization and Green Chemistry Approaches
Editors: Susarla V. A. R. Sastry, Shravan Kumar and Ashish Kapoor
ISBN: 979-8-89113-257-3

Keywords: computational fluid dynamics (CFD), core-annular flow, mass transfer, microchannel, Sherwood number

1. Introduction

Solvent extraction is a commonly used separation technique for the recovery of desired metals or moieties. It typically involves a solvent (an organic phase) that is used to extract the desired metal(s)/moiety(ies) from the feed (an aqueous phase). Solvent extraction in microchannels leads to process intensification due to higher specific interfacial area for mass transfer, shorter diffusion length, shorter contact time, etc. [1]. When flowing solvent and feed (immiscible liquids) come in contact at a microfluidic junction, different flow patterns, such as droplet, slug, parallel, and core-annular flow (CAF), arise. CAF is observed at higher Weber numbers, which means that CAF is observed when inertial forces dominate over surface tension forces, which is typically the case for higher throughputs. Due to differences in the wettability of phases toward the microchannel wall, one phase that preferentially wets the walls of the microchannel forms the annulus, whereas the other phase forms the core. The radius of the core depends on the flow rate ratio, viscosity ratio and radius of the microchannel. During the extraction of species from the aqueous phase, the direction of mass transfer will be from the core to the annulus when the organic phase preferentially wets the wall of the microchannel and the annulus to the core when the aqueous phase preferentially wets the wall of the microchannel. The extraction is followed by stripping of metals from the organic phase. Now, the direction of mass transfer will be from the annulus (core) to the core (annulus) when the organic phase (aqueous phase) preferentially wets the wall. Thus, there are scenarios in which mass transfer is from core to annulus as well scenarios in which mass transfer is from annulus to core. The velocity profiles are different in the core and the annulus of the CAF, with maximum velocity observed at the center of the core and zero velocity at the wall of the microchannel. The local velocity profiles may affect the rate of mass transfer when the direction of mass transfer is changed. In our previous work [2], CFD modeling of the mass transfer of CAF in microchannels was reported. The previous work considered the direction of mass transfer from the core to the annulus only. The case of mass transfer from the annulus to the core was not studied. Additionally, in the previous work, the effect of microchannel diameter on mass transfer was not explored. This study is an extension of the previously reported work and aims to address the above-mentioned gap areas.

2. Model

2.1. Computational Model

The computational model is 2-D axisymmetric and contains two domains: core and annulus, one for each phase. The length of the computational domain is 0.05 m, whereas the width of the computational domain varies from 100 μm to 500 μm according to the radius of the microchannel. Fluids are assumed to be incompressible and Newtonian. The physical and thermodynamic properties of the fluids are given in Table 1. Mass transfer of a single component as dilute species transport is considered in the model. The effect of gravity is neglected due to flow in micro dimensions.

Table 1. Physical and thermodynamic properties of the fluids

Parameter	**Value**
Viscosity of the aqueous phase (μ_{aq})	0.001 Pa·s
Viscosity of the organic phase (μ_{org})	0.003 Pa·s
Density of the aqueous phase (ρ_{aq})	1000 kg m^{-3}
Density of the organic phase (ρ_{org})	800 kg m^{-3}
Diffusivity of solute in the aqueous phase (D_{aq})	2×10^{-9} m^2 s^{-1}
Diffusivity of solute in the organic phase (D_{org})	1×10^{-9} m^2 s^{-1}
Distribution coefficient (K_D)	1

2.2. Computational Methodology

First, the velocity profile is obtained by solving Navier–Stokes equations for each phase. Then, the concentration profile is obtained by solving the convection-diffusion equation for the same velocity profile. Local mass transfer coefficients (K_L) for each phase are evaluated using diffusive flux and concentration at different positions along the liquid–liquid interface. The concentration boundary layer thickness (δ) is the radial distance from the interface to the point at which the concentration gradient becomes negligible. The governing equations, boundary conditions and further details of the computational model are discussed in our previous work [2] and hence omitted here.

3. Results and Discussion

3.1. Effect of the Direction of Mass Transfer

To study the effect of the direction of mass transfer in the CAF, three different independent cases are considered, as shown in Figure 1. In case 1, the aqueous phase forms the core, whereas the organic phase forms the annulus, and extraction takes place from core to annulus. In case 2, the aqueous phase forms the core, whereas the organic phase forms the annulus, and stripping of the solute takes place from the annulus to the core. In case 3, the aqueous phase forms the annulus, the organic phase forms the core, and extraction takes place from the annulus to the core. The diameter of the microchannel is 400 μm, and the flow rate ratio is unity in all cases. Figure 2 compares the dimensionless concentration profiles at the center of the core, at the interface and the wall of microchannel for different flow rates along the dimensionless length of the microchannel. Q_{aq} and Q_{org} are the flow rates of the aqueous and organic phases, respectively.

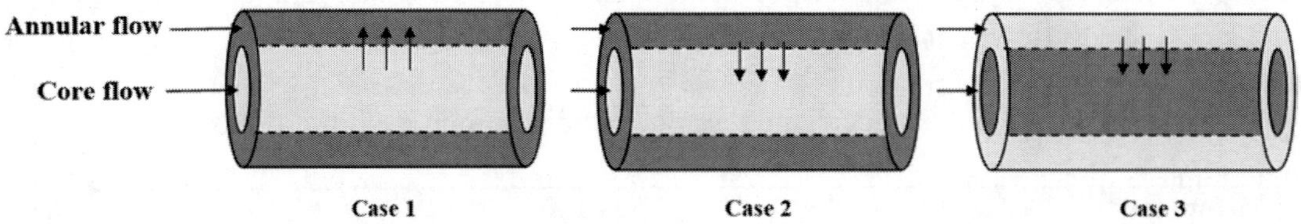

Figure 1. Schematic diagram of CAF for different cases (arrows at the interface show the direction of mass transfer; dark shade represents the organic phase whereas light shade represents the aqueous phase).

3.1.1. Comparison of Case 1 and Case 2

When the direction of mass transfer is interchanged from core to annulus (case 1) to annulus to core (case 2) while keeping the same velocity profiles in the core and the annulus (i.e., the mass transfer direction is changed without changing the flow rates and without interchanging the liquids in the core and the annulus), the obtained concentration profiles are different. The average velocity in the annulus is always lower than the average velocity in the core. The fluid in the annulus moves slowly and has a greater residence time compared to the fluid in the core. Thus, the interfacial concentration for the case of mass transfer from the annulus to the core (case 2) is lower along the length of the microchannel compared to the case of mass transfer from the core to the annulus (case 1). Thus, it is concluded that the direction of mass transfer affects the concentration profiles.

3.1.2. Comparison of Case 2 and Case 3

When fluids are interchanged from the aqueous phase in the core and organic phase in the annulus (case 2) to the organic phase in the core and aqueous phase in the annulus (case 3) (direction of mass transfer remains the same, i.e., annulus to core), all concentration profiles change for $Q_{aq} = 0.1$ mL/min, but only the interfacial concentration changes for 5 mL/min. The concentration boundary layer will be very thin for a higher flow rate, and the effect of mass transfer is not observed at the points away from the interface, such as the wall of the microchannel and the center of the core. The interfacial concentration is lower for case 2 than for case 3. The average velocity of the liquid phase containing the solute flowing in the annulus is lower in case 2 since case 2 has a smaller core diameter due to the effect of the viscosity ratio. Thus, in case 2, the liquid phase containing the solute flowing in the annulus has more residence time for mass transfer compared to case 3, which results in a greater decrease in the interfacial concentration for case 2. This is also observed for a higher flow rate, $Q_{aq} = 5$ mL/min.

3.1.3. Comparison of Case 1 and Case 3

In addition to interchange of direction of mass transfer, when liquids are also interchanged from aqueous phase in the core and organic phase in the annulus (case 1) to organic phase in the core and aqueous phase in the annulus (case 3), the obtained concentration profiles are different. Despite the liquid initially containing the solute having a higher residence time for case 3 compared to case 1, the interfacial concentration is higher for case 3. This is because of the combined effect of changes in the direction of mass transfer and hydrodynamics. The hydrodynamics, mass transfer variables (diffusivities) and direction of mass transfer are all changed when comparing case 1 and case 3. Thus, the change in interfacial concentration when switching from case 1 to case 3 is the result of several factors. Thus, the higher interfacial concentration for case 3 compared to case 2 cannot be explained solely on the basis of the residence time of the liquid initially containing the solute.

Although the concentration profiles are changing, it would be interesting to check the profiles of the local Sherwood number for the different cases. Therefore, the values of the Sherwood number of the core (Sh_c) and Sherwood number of the annulus (Sh_a) are evaluated and results are shown in Figure 3. It is found that both Sh_c and Sh_a are exactly the same for case 1 and case 2. Sh_c for case 3 is higher than that for case 1/2, whereas Sh_a for case 3 is lower than that for case 1/2. It should be noted that K_{Lc} for case 3 is less than that for case 1/2 due to a decrease in D_c (diffusivity in the core) and the average velocity of the core, but Sh_c increases due to a decrease in D_c. Similarly, Sh_a is decreased when switching from case 1/2 to case 3. The results thus show that the local Sherwood number remains invariant with changes in the direction of mass transfer, keeping hydrodynamics the same (case 1 and case 2), but it changes when hydrodynamics is changed owing to the interchange of liquids flowing in the core and the annulus (case 1/2 and case 3).

The resistance for the rate of mass transfer is usually represented in terms of the concentration boundary layer thickness (δ) in accordance with film theory [3]. The profiles of concentration boundary layer thickness of the core (δ_c) and concentration boundary layer thickness of the annulus (δ_a) in dimensionless form are shown in Figure 4a & 4b, respectively. When case 1 and case 2 are compared, there is no variation in the profiles of dimensionless δ. δ_c decreases and δ_a increases when switching from case 1 or case 2 to case 3. The physical process of mass transfer by diffusion in the CAF does not change the δ profiles when the direction of mass transfer is reversed, keeping the hydrodynamics the same (case 1 and case 2). But the interchange of fluids, which leads to changes in hydrodynamics and mass transfer variables (diffusivities of solute) (case 1 and case 2 or case 2 and case 3), does change the δ profiles. Thus, it can be said that the resistances in the core and annulus remain the same when only the direction of the mass transfer changes but the resistances change when there is interchange of fluids due to changes in hydrodynamics and mass transfer variables (diffusivities of solute).

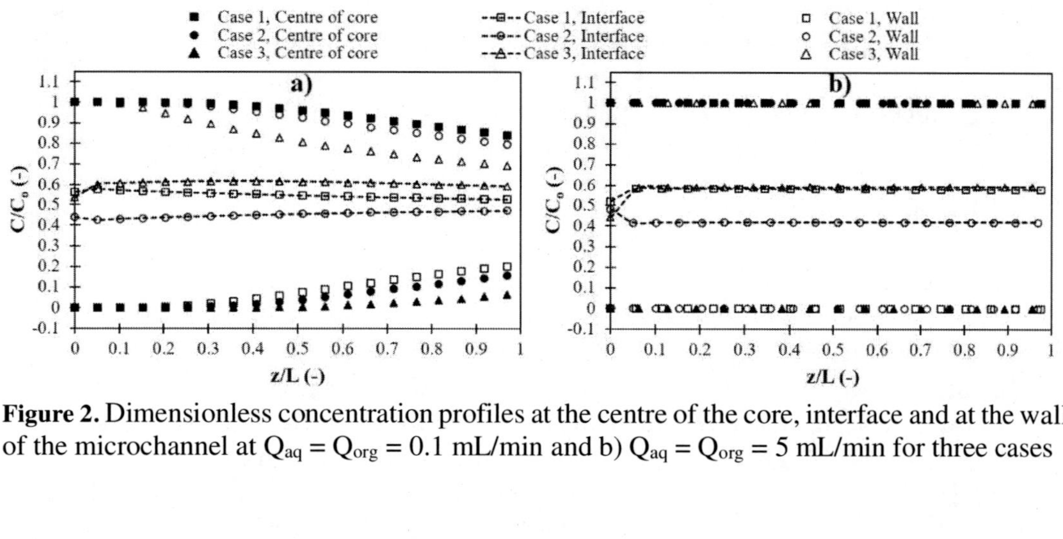

Figure 2. Dimensionless concentration profiles at the centre of the core, interface and at the wall of the microchannel at $Q_{aq} = Q_{org} = 0.1$ mL/min and b) $Q_{aq} = Q_{org} = 5$ mL/min for three cases

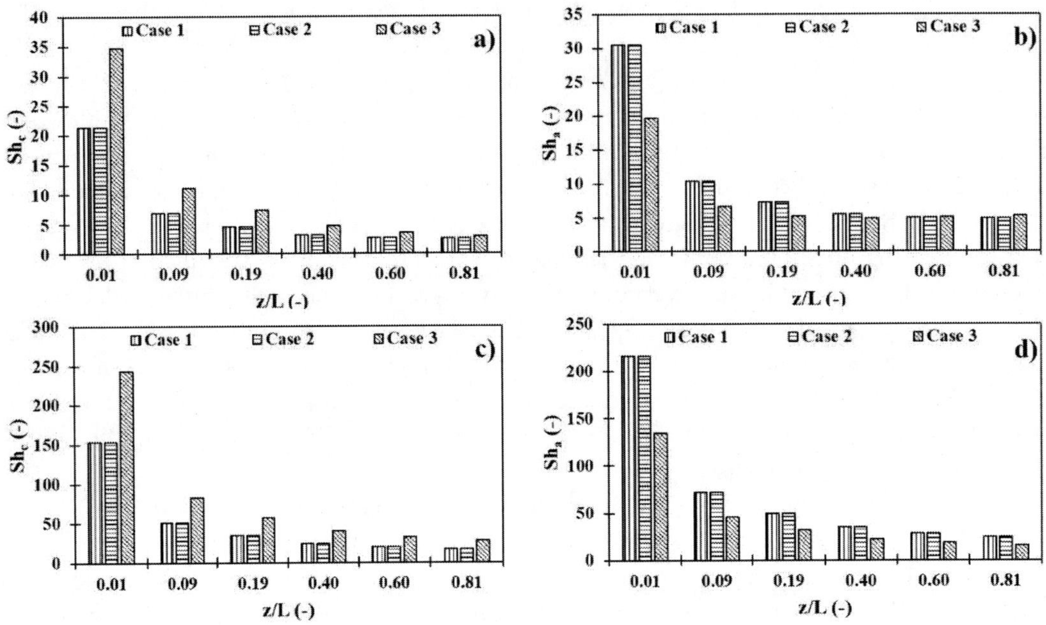

Figure 3. Variation of Sh_c and Sh_a for different flow rates (a,b: $Q_{aq} = Q_{org} = 0.1$ mL/min; c,d: $Q_{aq} = Q_{org} = 5$ mL/min)

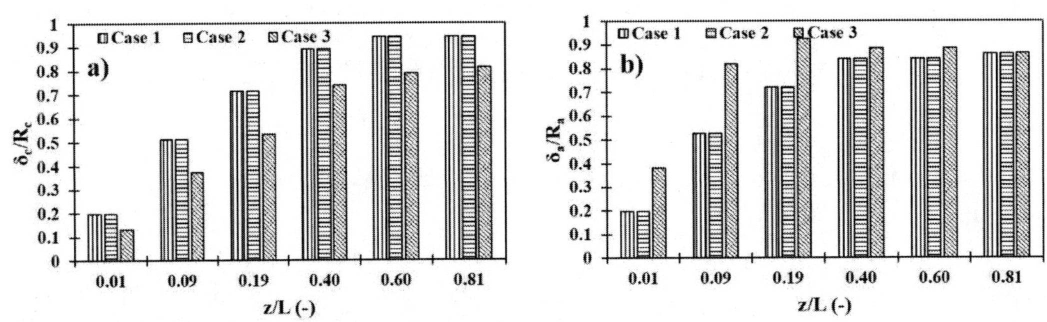

Figure 4. Variation of dimensionless δ_c and δ_a at $Q_{aq} = Q_{org} = 0.1$ mL/min

3.2. Effect of Microchannel Diameter

The diameter of the microchannel is varied from 200 μm to 1000 μm while keeping the same flow rate in the core (Q_{aq} = 0.1 & 5 mL/min) and annulus (Q_{org} = 0.1 & 5 mL/min). When the diameter of the microchannel is changed for constant flow rates, the average flow velocity is changed accordingly. Therefore, the effect of flow velocity is also encountered along with the effect of microchannel diameter. Figure 5 shows the effect of microchannel diameter on K_{Lc} and K_{La}. When diameter is decreased for a constant flow rate, the flow velocity is increased which leads to thinner δ and greater K_L, as seen in Figure 5. Concentration profiles change when the microchannel diameter is changed while keeping the flow rates the same. However, it is interesting to observe that Sh_c and Sh_a remain invariant with respect to the change in diameter of the microchannel, as shown in Figure 6. This means that the ratio of diffusive flux and convective flux remains the same despite the change in microchannel diameter.

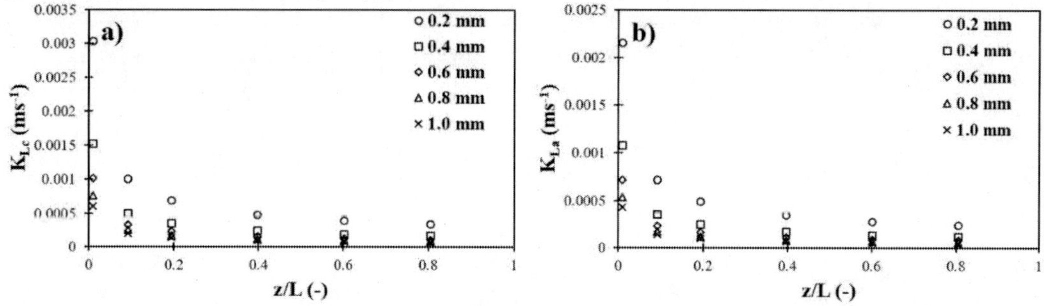

Figure 5. Effect of microchannel diameter on for a) K_{Lc} and b) K_{La} at $Q_{aq} = Q_{org} = 5$ mL/min

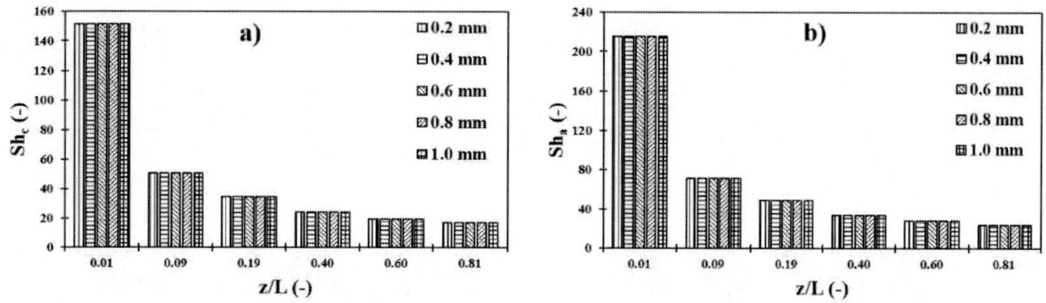

Figure 6. Effect of microchannel diameter on a) Sh_c and b) Sh_a at $Q_{aq} = Q_{org} = 5$ mL/min

Conclusion

CFD modeling is used to study the effect of the direction of mass transfer and channel diameter on liquid–liquid mass transfer in core-annular flow (CAF) in microchannels. Changes in the direction of mass transfer without altering hydrodynamics affect the concentration profiles, but local Sherwood number values remain invariant. The local Sherwood number also remains invariant when the diameter of the microchannel is increased without changing the liquids in the core and the annulus. When hydrodynamics is changed due to the interchange of liquids flowing in the core and the annulus, the local Sherwood numbers of the core and annulus change.

References

[1] Ciceri, D., Perera, J. M., & Stevens, G. W. (2014). The use of microfluidic devices in solvent extraction. *J. Chem. Technol. Biotechnol.*, *89*(6), 771-786.

[2] Chaurasiya, R. K., & Singh, K. K. (2022). CFD modeling of mass transfer in liquid-liquid core-annular flow in a microchannel. *Chem. Eng. Sci.*, *249*, 117295.

[3] Lewis, W. K., & Whitman, W. G., (1924). Principles of gas absorption. *Ind. Eng. Chem.*, *16*, 1215-1220.

Chapter 11

A Comparison of the Drop Formation at a Flat-Tip Nozzle, Sharp-Tip Nozzle and a Hole in a Plate: A Numerical Study

Swati Agarwal[1,2*]
Nirvik Sen[1,2]
K. K. Singh[1,2]
and G. Sugilal[1,2]
[1]Homi Bhabha National Institute, Anushaktinagar, Mumbai, India
[2]Bhabha Atomic Research Centre, Trombay, Mumbai, India

Abstract

This chapter presents the results of numerical simulations aimed at understanding the drop formation phenomena at nozzles or hole in a plate submerged in a quiescient continuous phase. Three different inlet geometries, i.e., a flat tip nozzle, a sharp tip nozzle and a sieve plate hole, are used in the studies. The phase-field algorithm implemented in COMSOL Multiphysics 5.3 is used to track the interface of the drop being formed. The impacts of flow velocity and dispersed phase inlet diameter on the drop formation phenomenon are predicted using the developed model.

Keywords: CFD; flat tip nozzle; sharp tip nozzle; sieve plate; drop, phase-field

1. Introduction

In the nuclear industry, solvent extraction is the most preferred separation process. Among other factors, the efficiency of the process also depends on the size of the smallest unit generated during the process, i.e., the drop. Drop size is a crucial parameter, as small drops are required for high specific interfacial area, but extremely small drops cause settling issues. Therefore, an optimum drop size is needed, and understanding the drop formation phenomenon at a fundamental level becomes important. The solvent extraction equipment used in the nuclear

* Corresponding Author's email: swatia@barc.gov.in.

In: Sustainability in Chemical Processes through Digitalization and Green Chemistry Approaches
Editors: Susarla V. A. R. Sastry, Shravan Kumar and Ashish Kapoor
ISBN: 979-8-89113-257-3

industry involves complex flow conditions [1, 2]. The first step in comprehending such complex cases is to understand drop-level hydrodynamics for simple flow conditions. Further complexities, including flow effects (cocurrent, counter-current and cross-current), pulsation effects, mass transfer effects, etc., can be gradually incorporated to reach the final goal. The studies presented in the paper focus on the first step, i.e., dispersion of a liquid phase (organic) in a quiescent continuous (aqueous) phase [3–5].

The physical properties of the two-phase system, dispersed phase flow rate, wetting characteristics of liquids and inlet geometry of the dispersed phase are some crucial factors that affect the formation of drops in the quiescent continuous phase. Extensive studies, both experimental and numerical, have been carried out to understand the drop formation process at a fundamental level [6–7]. However, studies evaluating the impact of dispersed phase inlet geometry on the formation of drops are very few [7, 8].. In this work, we report a 2D axisymmetric CFD model that captures drop formation for three different dispersed phase inlet geometries, i.e., flat tip nozzle, sharp tip nozzle and sieve plate hole. The key variables of drop formation process (i.e., equivalent drop diameter 'd_d', drop detachment height 'h_d', and drop detachment time 't_d') for all three cases are compared, and then the process of drop formation is examined in relation to the flow velocity and nozzle diameter.

2. Numerical

The numerical approach used for simulation involves the coupled solution of Navier–Stokes equations and the Cahn-Hilliard equation. Eqs. 1 and 2 represent continuity and momentum conservation equations, respectively, where $\vec{U}$ represents velocity vector. The Cahn-Hilliard equation (Eq. 3) is a partial differential equation for the conservation of dimensionless phase-field variable (Φ) in two-phase flow [8]. Φ varies in the range of -1 to 1 in the diffuse interface region.

$$\nabla.\vec{U} = 0 \tag{1}$$

$$\rho(\frac{\partial\vec{U}}{\partial t} + \vec{U}.\nabla\vec{U}) = -\nabla p + \rho\vec{g} + \overrightarrow{F_{st}} + \nabla.\{\mu(\nabla\vec{U} + \nabla\vec{U}^T)\} \tag{2}$$

$$\frac{\partial\Phi}{\partial t} + \vec{U}.\nabla\Phi = \nabla.[\frac{\gamma\lambda}{\varepsilon^2}\nabla\{-\nabla.(\varepsilon^2\nabla\Phi) + (\Phi^2 - 1)\Phi\}] \tag{3}$$

Eqs. 4 and 5 define the average values of density (ρ) and viscosity (μ) that are used for the calculation in the two-phase region, and volume fractions of fluids (V_f) are given by Eqs. 6 and 7.

$$\rho = \rho_1 + (\rho_2 - \rho_1)V_{f2} \tag{4}$$

$$\mu = \mu_1 + (\mu_2 - \mu_1)\,V_{f2} \tag{5}$$

$$V_{f1} = \frac{1-\Phi}{2} \tag{6}$$

$$V_{f2} = \frac{1+\Phi}{2} \tag{7}$$

The relation between the interface thickness parameter (ε), mixing energy density (λ) and interfacial tension (σ) is given by Eq. 8.

$$\sigma = \frac{2\sqrt{2}}{3}\frac{\lambda}{\varepsilon} \tag{8}$$

The typical values of the mobility parameter (γ) and interface thickness parameter are set to $\gamma = \varepsilon^2$ and $\varepsilon = hc/2$, respectively. Here, hc represents the characteristic mesh size in the interface region, and γ determines the time scale of Cahn-Hilliard diffusion.

The surface tension force $(\overrightarrow{F_{st}})$ in the momentum conservation equation (Eq. 2) is evaluated by Eqs. 9 and 10. G represents chemical potential.

$$\overrightarrow{F_{st}} = G\nabla\Phi \tag{9}$$

$$G = \lambda(-\nabla^2\Phi + \frac{\Phi(\Phi^2-1)}{\varepsilon^2}) \tag{10}$$

These equations are implemented for numerical simulation using the phase-field method of COMSOL Multiphysics 5.3. Figure 1 displays the mesh and boundary conditions utilized in the simulation as well as the 2D axisymmetric computational domain. The geometry used for simulation has a 20 mm radius and a 40 mm height. The inner diameter of the dispersed phase inlets is 2 mm, while the wall thickness is 0.6 mm for the flat tip and sharp tip nozzles.

It is assumed that the geometry is initially filled with the continuous phase (3 N nitric acid) and that the dispersed phase (30% TBP in dodecane) is fed through different inlet geometries. The physical properties of the two-phase system used for the studies are listed in Table 1. The numerical technique to simulate drop formation in an initially quiescent continuous phase has been validated previously [9]. The same methodology is used in this study.

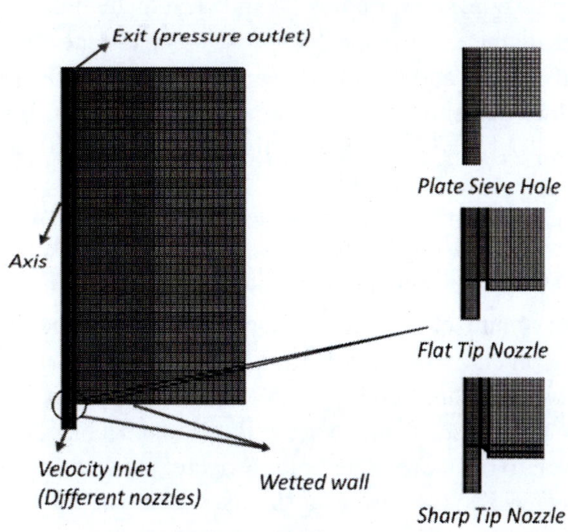

Figure 1. Computational domain and boundary conditions.

Table 1. Physical properties of the phases used in the simulation (at 25 °C)

Phase	Density (kg/m^3)	Viscosity (mPa s)	Density difference (kg/m^3)	Surface tension (mN/m)
Continuous (3 N Nitric acid)	1094	0.91		10.99
Dispersed (30% TBP in dodecane)	832	1.70	262	

In the study, drop size is evaluated by equivalent drop diameter, and Eqs. 11 and 12 are used for its calculation. Here, Q_d is the volumetric flow rate of the dispersed phase, d_d is the equivalent drop diameter, U is the velocity of the dispersed phase, t_d is the drop detachment time and d_n is nozzle/hole diameter.

$$Q_d = \pi \frac{d_n^2}{4} U \tag{11}$$

$$d_d = \left(\frac{6 Q_d t_d}{\pi}\right)^{1/3} \tag{12}$$

3. Results and Discussion

3.1. Visualization of the Drop Formation Phenomenon

Figure 2 presents the temporal evolution of the liquid–liquid interface for drop formation in all three cases, i.e., flat tip nozzle, sharp tip nozzle and sieve plate hole. The values of velocity of dispersed phase, inner diameter of dispersed phase inlet and nozzle wall thickness used in simulation are 0.03 m/s, 2 mm and 0.6 mm, respectively. It is evident from the figure that for the sieve plate hole, drop detachment is seen after 2.2 s; however, for the flat tip nozzle, the drop detaches nearly at 2.2 s, and for the sharp tip nozzle, the drop detaches well before 2.2 s. Thus, as the wettable area of dispersed phase inlets decreases (largest for the sieve plate hole and smallest for the sharp tip nozzle), the drop formation rate becomes faster.

Figure 3 quantifies the effect of dispersed phase inlet geometry on drop formation. Figure 3(a) shows the variation of d_d with dispersed phase inlet geometry. Figure 3(b) shows the variation in h_d (the maximum elongation attained by the drop before its detachment) with dispersed phase inlet geometry, and Figure 3(c) shows the variation in t_d with dispersed phase inlet geometry. It is observed that as the wettable area is decreased, drops of smaller sizes are formed, i.e., approximately 12% smaller for the flat tip nozzle and 15% smaller for the sharp tip nozzle in comparison to the sieve plate hole. A decrease in t_d is observed with a decrease in wettable area (33% decrease in the case of a flat tip nozzle and 39% in the case of a sharp tip nozzle), while a marginal difference is observed in h_d (less than 3%). This is because an increase in wettable area increases the extent of spreading of the dispersed phase (which wets the surface to a finite degree), leading to a larger drop volume and longer detachment times.

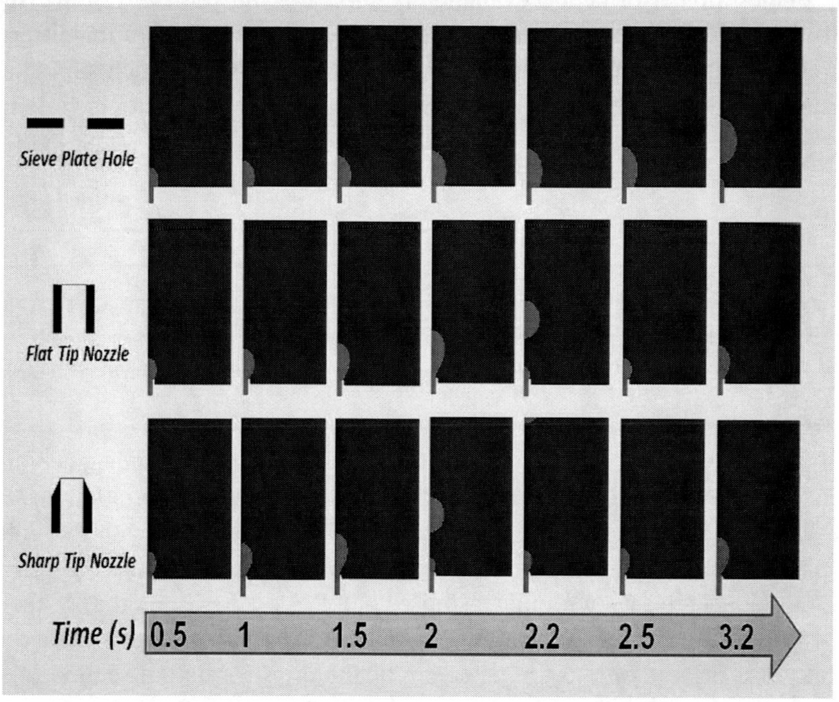

Figure 2. Simulation results of the drop evolution for different nozzle geometries.

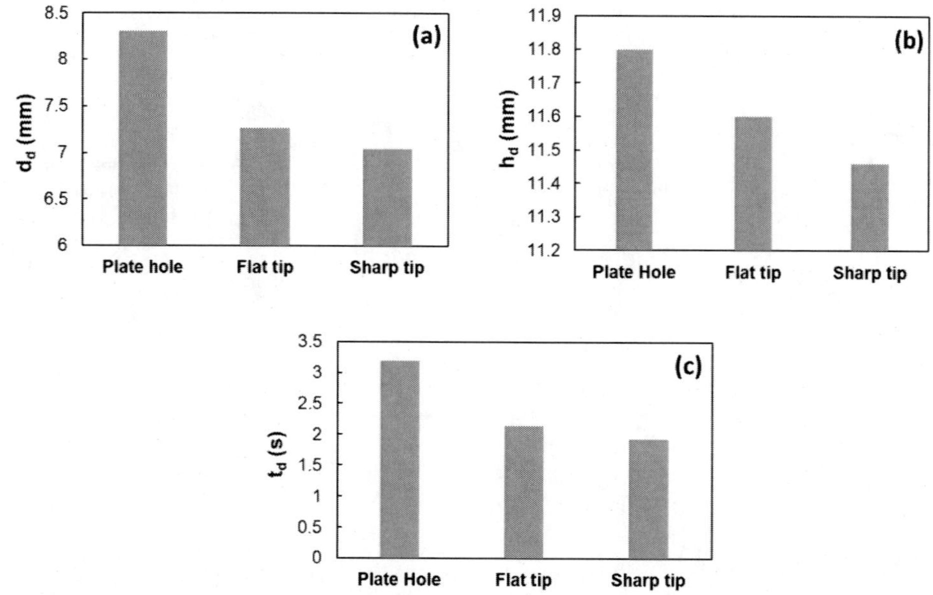

Figure 3. Variation of (a) d_d (b) h_d (c) t_d for different types of dispersed phase inlets of 2 mm diameter at a dispersed phase velocity of 0.03 m/s.

3.2. Parametric Analysis

Parametric studies are performed to evaluate the impact of the velocity of the dispersed phase (U) and dispersed phase inlet diameter (d_n) on the formation of drops for different types of dispersed phase inlets. Table 2 shows the range of parameters for which simulations are run. The underlined parameter value is kept constant when another parameter is varied.

Table 2. Range of parameters used for study

U (m/s)	d_n (mm)
0.03	1
0.06	2
0.1	3

3.2.1. Impact of Dispersed Phase Velocity

Figure 4 highlights the impact of the velocity of the dispersed phase on the generation of drops for different dispersed phase inlet geometries. Figure 4(a) shows the variation in d_d, Figure 4(b) shows the variation in h_d and Figure 4(c) shows the variation in t_d with the velocity of the dispersed phase for all three dispersed phase inlet geometries. For all three geometries, d_d and t_d decrease while h_d increases with increasing velocity. As velocity increases, the momentum force, a disruptive force, increases, resulting in rapid drop detachment and hence a smaller drop size. The increase in h_d is due to the tendency of the drop formation regime to switch from dripping to jetting. As observed above, d_d, h_d and t_d are the largest for a plain hole, followed by a flat tip nozzle and are the smallest for a sharp tip nozzle.

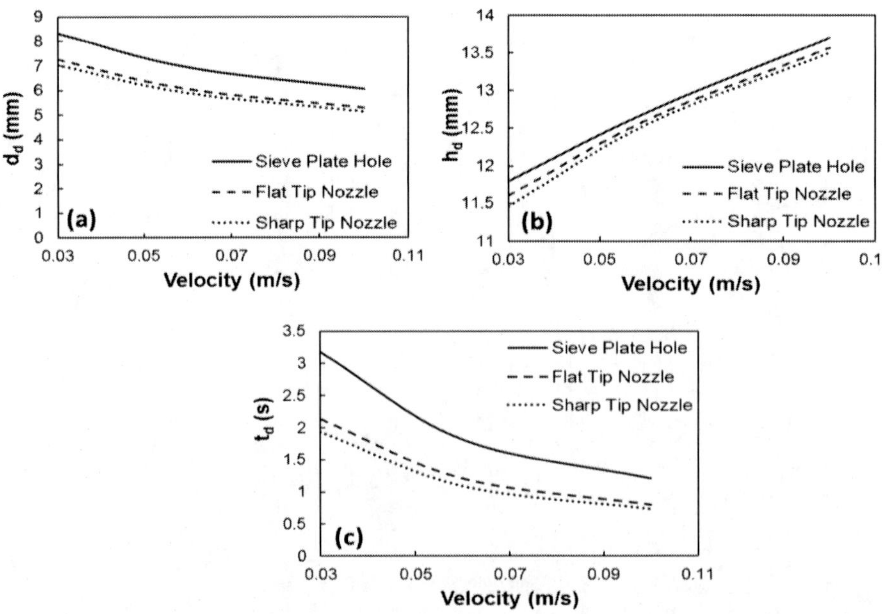

Figure 4. Variation of (a) d_d (b) h_d (c) t_d with dispersed phase velocity for different types of dispersed phase inlets of 2 mm diameter.

3.2.2. Effect of Nozzle Diameter

Figure 5 highlights the impact of nozzle diameter on drop formation phenomena for different dispersed phase inlet geometries. Figure 5(a) shows the variation in d_d, Figure 5(b) shows the variation in h_d, and Figure 5(c) shows the variation in t_d with nozzle diameter for all three dispersed phase inlet geometries. It is observed that d_d and h_d increase while t_d decreases as the dispersed phase inlet diameter increases. This occurs due to an increase in the interfacial tension force, a restoring force that is proportional to the inlet diameter [7, 8]. Thus, the formed drop holds a larger volume and is larger in size.

Additionally, the thickness of the base of the liquid thread connecting the drop being formed to its bulk phase grows along with the dispersed phase inlet diameter. The thicker thread takes a longer distance to thin out and break, thereby increasing the height of drop detachment. A decrease in t_d is observed because, with increasing dispersed phase inlet diameter, the volumetric flow rate increases for constant velocity. Thus, larger volume of dispersed phase can be accommodated in a short time. Additionally, from Figure 5, the effect of nozzle or hole diameter is observed to diminish as diameter of the nozzle or hole increases. This is due to the ratio of the inner radius to the outer radius approaching 1 as the nozzle diameter increases. Thus, the wall thickness effect and therefore the wettable area become the same for all three cases.

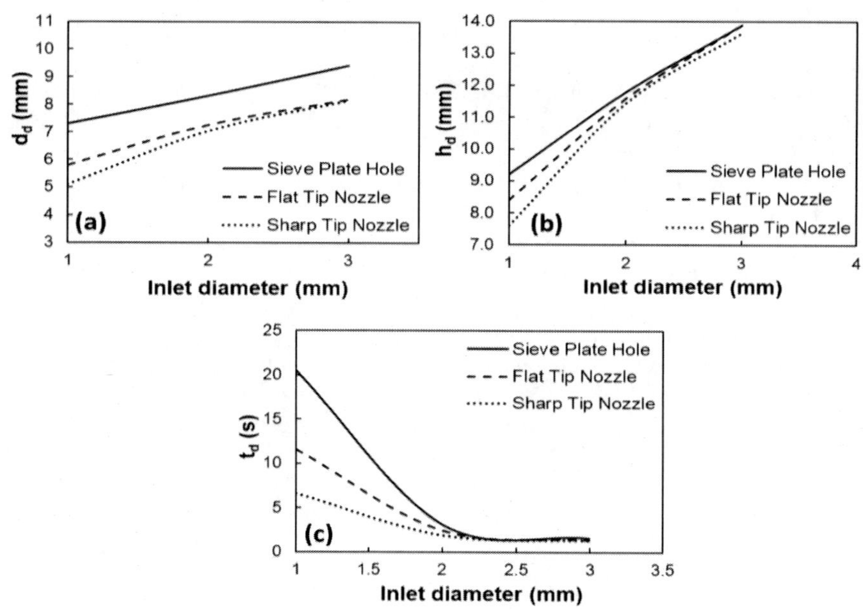

Figure 5. Variation of (a) d_d (b) h_d (c) t_d with dispersed phase inlet diameter for different types of dispersed phase inlets at a dispersed phase velocity of 0.03 m/s.

4. Conclusion

The drop formation phenomenon is studied numerically for three different types of dispersed phase inlets to analyze the effect of inlet geometry on drop formation. A flat tip nozzle, a sharp tip nozzle and a sieve plate hole are used in the studies. The geometries of dispersed phase inlets affect the wettable area, which, in turn, affects the extent of dispersed phase spreading. Thus, an increase in the wettable area of the dispersed phase inlet results in increased d_d, h_d and t_d.

The model developed for all three dispersed phase inlet geometries is extended for parametric analysis to quantify the effect of dispersed phase velocity and inlet diameter on d_d, h_d and t_d. It is observed that d_d and t_d decrease while h_d increases with an increase in the velocity of the dispersed phase for a constant dispersed phase inlet diameter. d_d and h_d increase while t_d decreases with increasing dispersed phase inlet diameter for a constant dispersed phase velocity.

References

[1] Harkins W D, Brown F E (1919) The determination of surface tension (free surface energy), and the weight of falling drops: The surface tension of water and benzene by the capillary height method. *J. Am. Chem. Soc.* 41:499–524.

[2] Hayworth C B, Treybal R E (1950) Drop formation in two-liquid-phase systems. *Ind. Eng. Chem.* 42:1174–1181.

[3] Narasinga Rao EVL, Kumar R, Kuloor N R (1966) Drop formation studies in liquid—liquid systems. *Chem. Eng. Sci*. 21:867–880.

[4] Null H R, Johnson H F (1958) Drop formation in liquid–liquid systems from single nozzles. *AIChE Journal 4*:273–281.

[5] Roy A, Darekar M, Singh K K, Shenoy K T, Grover R B (2018) Drop formation at nozzles submerged in quiescent continuous phase: An experimental study with TBP-dodecane and nitric acid system. *Nuclear Science and Techniques* 29:88.

[6] Scheele G F, Meister B J (1968) Drop formation at low velocities in liquid–liquid systems: Part I. Prediction of drop volume. *AIChE Journal* 14:9–15.

[7] Soleymani A, Laari A, Turunen I (2008) Simulation of drop formation in a single hole in solvent extraction using the volume-of-fluid method. *Chemical Engineering Research and Design* 86:731–738.

[8] Sen N, Singh K K, Mukhopadhyay S, Shenoy K T, Rao H, Ghosh S K (2012, December 10-12). Numerical simulation of drop formation at a hole in quiescent continuous phase [Paper presentation]. *International Congress on Computational Mechanics and Simulation, IIT Hyderabad, India.*

[9] Sen N, Singh K K, Mukhopadhyay S, & Shenoy K T, (2019). Drop formation at a hole in a plate submerged in quiescent continuous phase : comparison of plain hole and nozzle hole. *Chemical Engineering Communications*, 206(10), 1317-1336.

Chapter 12

Water Absorption Behavior of Bamboo Fiber-Reinforced Polymer Composites

N. Shukla*
and G. L. Devnani
[1]Chemical Engineering Department, Harcourt Butler Technical University, Kanpur, U. P., India

Abstract

Bamboo fiber comes into the category of natural fiber and has more advantages over synthetic fibers. Lower cost, renewability and eco-friendliness make natural fibers more suitable than synthetic fibers. When it is mixed with matrix, it forms a reinforced composite. Composites have many applications in the automotive industry, construction applications, furniture applications, etc. Fibers are water-loving, so the absorption of water is high, resulting in lower compatibility between the fiber and matrix. The surface treatment method of fiber results in an increase in the strength of composites. To understand the water absorption behavior of composites, the kinetic behavior is established, and the diffusivity is calculated.

Keywords: natural fiber, bamboo fiber, composites, water absorption

1. Introduction

Natural fibers are thread-like formations that are inherited from plants and animals. These are either cellulose or proteins. These are quite a bit longer than wider. The main resources to inherit natural fibers are plant parts such as bast, leaf, stem and fruit, such as hemp, kenaf, abaca, cotton, etc., and animal fur such as wool, silk, etc. The easier the availability is, the cheaper the cost. The main advantages of selecting natural fiber over synthetic fiber are its cheaper price, availability and highly specific properties, such as moisture sensitivity, thermal sensitivity and recyclability [1]. Natural fibers are eco-friendly by nature, which is an important

---

* Corresponding Author's Email: snidhishukla26@gmail.com.

In: Sustainability in Chemical Processes through Digitalization and Green Chemistry Approaches
Editors: Susarla V. A. R. Sastry, Shravan Kumar and Ashish Kapoor
ISBN: 979-8-89113-257-3

step for green technology. The various industrial applications where natural fiber can be used with additives are ceramics, automobiles, aerospace, construction sites, packaging materials, furniture, roofing sheets, etc. There are diverse applications where it has further scope [2]. Natural fiber-reinforced composites are matrices in which natural fibers are dispersed into additives and matrix materials such as thermosetting materials and thermoplastics [3].

The nature of natural fiber is hydrophilic (hydro means water and philic means loving). Their moisture absorption competency is very high. The hydrophilic nature of natural fibers can be understood through diffusion theory. The increasing moisture content results in low mechanical properties, such as hardness, resilience, toughness and stiffness [4]. The main drawback of natural fiber is its moisture absorption capability, which results in fiber swelling. The swelling behavior of polymers due to their moisture absorption properties makes them less reliable. Bamboo fiber is bast fiber. The demand for bamboo is increasing as the population is increasing [5].

Bamboo fibers have water-loving properties, so water absorption is high in these types of fibers. If we prepare composites, the bonding between the fiber and matrix is not as good. To improve the bond strength between the fiber and matrix, it is necessary to make the fiber hydrophobic in nature [6]. Alkali treatment of bamboo fiber gives it better strength [7].

Moisture absorption depends on the content of cellulose and fiber volume fractions. Different fiber-reinforced composites show different water absorption behaviors, such as hemp fiber reinforcement with polyester composites, which show lower mechanical properties due to the absorption of water in the fiber [8]. Coconut fiber absorbs more water than synthetic fiber. The water absorption simply shows the hydrophilicity. Hydrophilic materials absorb water through different processes, such as diffusion and absorption [9]. The moisture content absorbed results in the swelling of the fiber. Due to swelling, microbreaks occur, and continuous absorption occurs due to bulk diffusion. The water absorption property of fiber causes a reduction in mechanical properties. Hence, to reduce this, we use different treatments to improve the mechanical properties and reduce the swelling behavior of fibers [10]. We can understand the water absorption property of fiber through the diffusion theory from Figure 1, which explains the moisture absorption and bulk diffusion causing swelling.

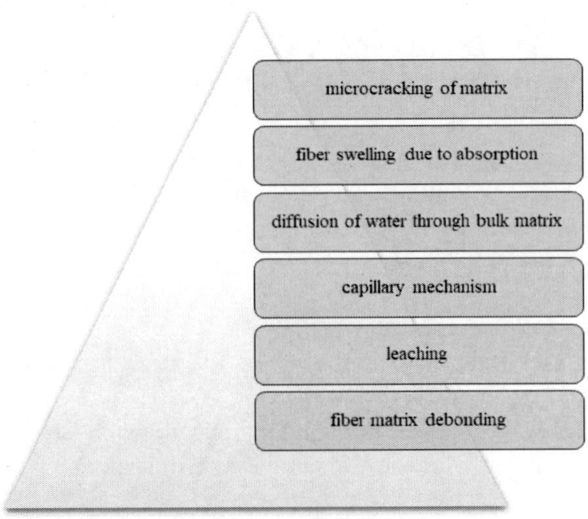

Figure 1. Steps of water absorption in composites.

2. Material and Methods

2.1. Materials

- The bamboo fiber was purchased from Vruksha Composites and Services in Chennai.
- Epoxy resin, hardener and all other chemicals were purchased from the local market.

2.2. Method

Fibers were dried under ambient conditions and cut to 60–70 mm before use. The fiber is treated with a 5 wt% alkali treatment method to make it hydrophobic in nature. The fiber was then washed to remove moisture at 70°C. Araldite (AW106) and hardener (HV 953IN) are used as the matrix to prepare a composite [11]. The hand layup method is used for the preparation of untreated and treated composites. The liquor to fiber ratio was maintained at 20:1. A homogeneous mixture of fiber and matrix is kept in clean polythene to make a proper shape, and a heavy roller is used to reduce voids and bubbles. Finally, 25 kg of weight is put on a mold for 24 hours. For uniform preparation of sheets, the final sheet is used for characterization purposes [12].

2.2.1. Preparation of Untreated & Treated Fiber Composites

AW 106 is selected as an epoxy resin, and HV 953 IN is selected as a hardener. Both are mixed in a ratio of 5:4. Different fiber loadings are used for the preparation of untreated and treated fibers. The matrix and fiber are mixed in a high-speed stirrer for 10–15 minutes to make a homogeneous mixture. Different percentages (5, 10, 15, and 20) of fiber and matrix loading are used for preparations of untreated composites, and 15% optimal loading is used for treated composites [13]. The hand layup method is used for the preparation of untreated and treated composites. For the removal of voids and bubbles, a heavy roller is used. It made the surface uniform and put approx. 25 kg of weight on it. The final prepared sheet was used for characterization purposes.

2.2.2. Water Absorption Kinetics

ASTM D570 is used for water absorption characteristics [14].

Equation used for analysis:

$$\% \mathrm{M} = \frac{M_f - M_i}{M_i} * 100 \quad \text{(i)}$$

Fick diffusion and non-Fickian diffusion are two types of methods used for water absorption behavior.

For the diffusion behavior equation used:

$$F_s = \frac{M_t}{M_m} = kt^n \quad \text{(ii)}$$

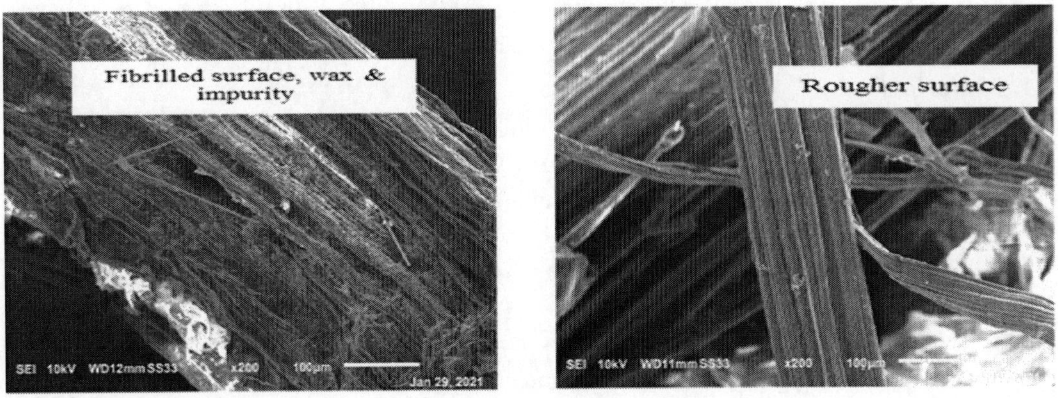

Figure 2. (a) Untreated fiber. (b) 5 wt% NaOH-treated bamboo fiber.

The diffusion coefficient can be calculated by:

$$F_s = \frac{M_t}{M_m} = \frac{4}{h}(\frac{Dt}{\pi})^{1/2} \quad \text{(iii)}$$

where

M_i: Dry initial weight, M_f : Weight after immersion in water; % M: water absorption rate, M_t: Percentage of the water that is absorbed in the sample at time t; h: thickness of the sample.

M_m: Maximum percentage of the water absorbed; K and n: kinetic parameters.

3. Results and Discussion

3.1. SEM

SEM is scanning electron microscopy. It is used to study the morphological behavior of untreated and treated composites. JSM-6010LA was used for scanning electron microscopy. The analysis was performed in detection mode with 10X–300,000X, followed by gold coating [15]. Figures 2 (a) and (b) show the morphology of the untreated fiber, and the next adjacent image shows the treated fiber. In the case of untreated fiber impurities, wax and a fibrillated surface are present, as shown in the image. To improve bonding between the fiber and matrix, a chemical treatment method is used, and the surface becomes rougher.

3.2. Water Absorption Behavior

To understand the moisture absorption behavior of composites, water absorption kinetics play a vital role. Different percentages (5, 10, 15, and 20) of loading are used for the preparation of untreated and treated composites. It is concluded from both figures that water absorption increases with increments in fiber loading. In the case of treated composites, there is less water absorption compared to untreated ones, just as in treated composites, bonding between fiber and matrix is better compared to untreated ones. From Figures 3 and 4, it is concluded that water absorption increases as loading increases, and with the chemical treatment effect, water absorption is lower in the case of treated fiber composites with 15% loading.

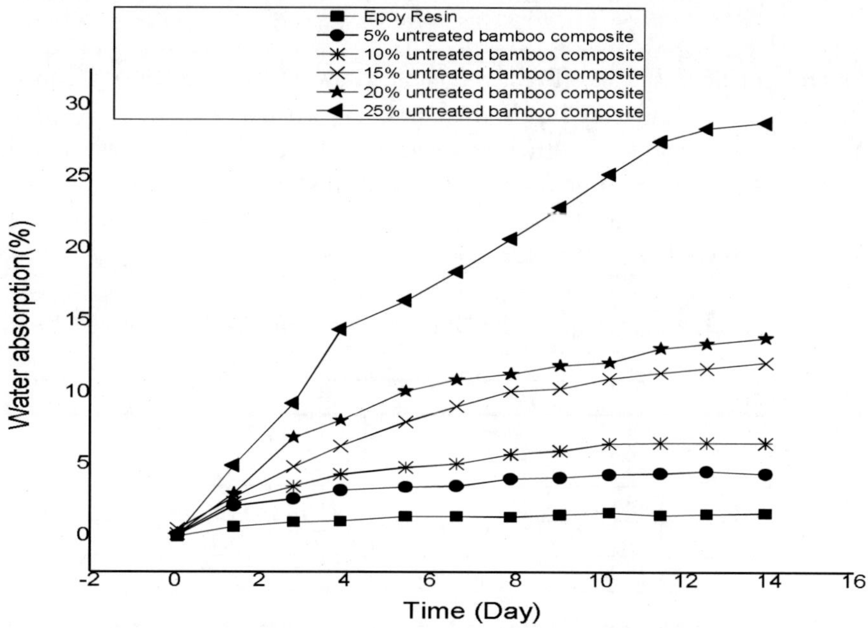

Figure 3. Water absorption behavior of the untreated bamboo fiber-reinforced composite.

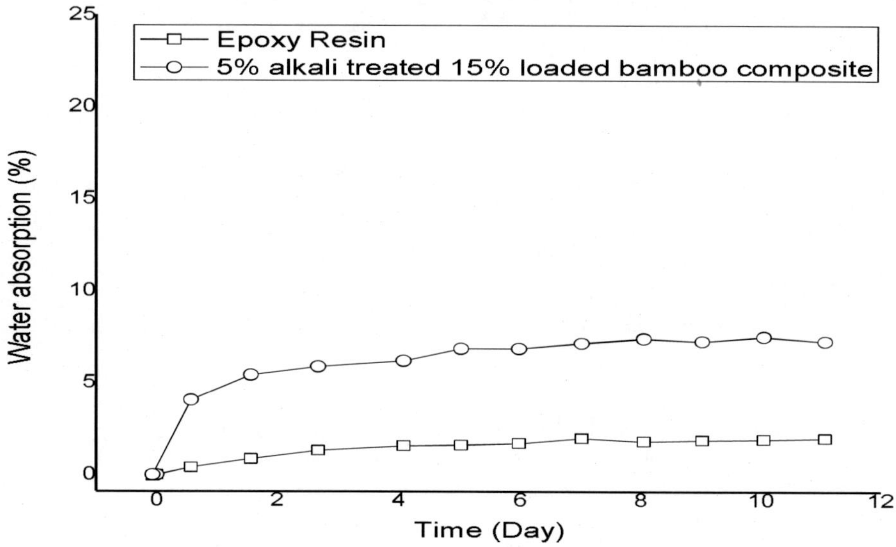

Figure 4. Water absorption behavior of the treated bamboo fiber-reinforced composite.

3.3. Mathematical Modeling of Water Absorption Behavior

From equation (ii), by taking the log of both sides of the equation, we obtain where

$$logF_s = \log\left(\frac{M_t}{M_m}\right) = \log k + n\, logt \qquad \text{(iv)}$$

n and k are statistical analyses. Figures 5 and 6 show a fitting curve for diffusion, and the values of n and k can be calculated from the curve. A statistical technique is applied to both graphs. The values of n and k are different for different loadings, as shown in Table 1. The value of n is close to 0.5, i.e., it obeys the Fickian diffusion mechanism. From Table 1, it is clear that the value of diffusivity is lower for treated bamboo fiber-reinforced composites, so water absorption is lower in treated composites.

Table 1. Diffusion Parameters for the Untreated and Treated Reinforced Composite

Untreated composite			
Loading %	n	k	D (m^2/s) *10^{12}
5	.349	-0.79	4.50
10	.434	-.98	5.7
15	.478	-1.01	7.03
20	.359	-.82	8.02
Treated composite			
15	.37	-.84	6.02

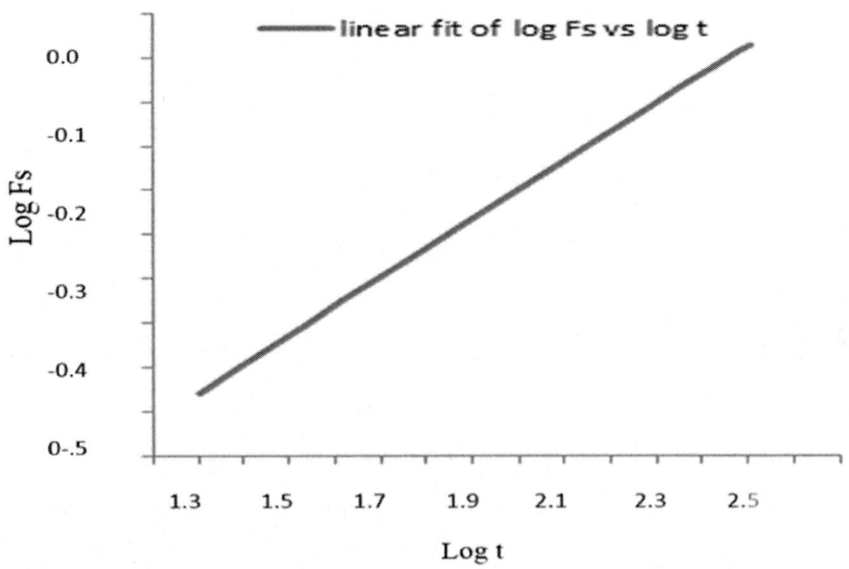

Figure 5. Diffusion curve fitting plot of the untreated bamboo fiber-reinforced epoxy composites.

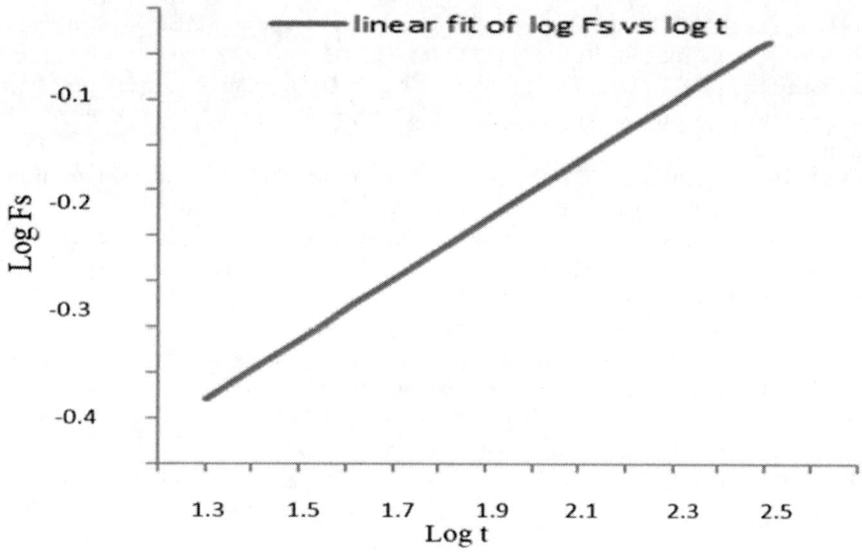

Figure 6. Diffusion curve fitting plot of the treated bamboo fiber-reinforced composite.

Conclusion

Surface treatment of the fiber improves adhesion between the fiber and matrix. This results in good bonding between the fiber and matrix. With increasing fiber loading, water absorption also increases in composites. Alkali treatment is used as a surface treatment method. The fiber is treated with a 5-wt% alkali treatment method. The value of n is close to 0.5, which shows that the Fickian model is being followed, and it is clear from the table that diffusivity is lower in the case of treated composites, so water absorption is also lower compared to the untreated ones.

References

[1] Akil H. M., Cheng L. W., Mohd Ishak Z. A., Abu Bakar A., Abd Rahman M. A. Water absorption study on pultruded jute fiber reinforced unsaturated polyester composites. *Compos Sci Technol* 2009; 69:1942–8.

[2] Alix S., Lebrun L., Morvan C., Marais S. Study of water behavior of chemically treated flax fibers-based composites: A way to approach the hydric interface. *Compos Sci Technol* 2011;71:893–9. https://doi.org/10.1016/j.compscitech.2011.02.004.

[3] Devnani G. L., Mittal V., Sinha S. *Science Direct Mathematical modeling of water absorption behavior of bagasse fiber reinforced epoxy composite material* 2018;5:16912–8.

[4] Devnani G. L., Sinha S. Epoxy-based composites reinforced with African teff straw (Eragrostis tef) for lightweight applications. *Polym Polym Compos* 2019;27:189–200. https://doi.org/10.1177/0967391118822269.

[5] El Hachem Z., Célino A., Challita G., Moya M. J., Fréour S. Hygroscopic multiscale behavior of polypropylene matrix reinforced with flax fibers. *Ind Crops Prod* 2019;140:111634. https://doi.org/10.1016/j.indcrop.2019.111634.

[6] Gupta M. K. Water absorption and its effect on mechanical properties of sisal composite. *J Chinese Adv Mater Soc* 2018;6:561–72.

[7] Holbery J., Houston D. Natural-fiber-reinforced polymer composites in automotive applications. *Jom* 2006;58:80–6. https://doi.org/10.1007/s11837-006-0234-2.

[8] Hosseinihashemi S. K., Arwinfar F., Najafi A., Nemli G., Ayrilmis N. Long-term water absorption behavior of thermoplastic composites produced with thermally treated wood. *Meas J Int Meas Confed* 2016;86:202–8.

[9] Hu R. H., Sun M. young, Lim J. K. Moisture absorption, tensile strength and microstructure evolution of short jute fiber/polylactide composite in hygrothermal environment. *Mater Des* 2010;31:3167–73.

[10] Oke S. R., Omotoyinbo J. A., Alaneme K. K. Water absorption characteristics of polyester matrix composites reinforced with oil palm ash and oil palm fiber. *Usak Univ J Mater Sci* 2013;2:109–109.

[11] Kumar P. S. S., Allamraju K. V. A review of natural fiber composites [Jute, Sisal, Kenaf*]. Mater Today, Proc* 2019;18:2556–62. https://doi.org/10.1016/j.matpr.2019.07.113.

[12] Masoodi R., Pillai K. M. A study on moisture absorption and swelling in biobased jute-epoxy composites. *J Reinf Plast Compos* 2012;31:285–94.

[13] Muñoz E., García-Manrique J. A. Water absorption behavior and its effect on the mechanical properties of flax fiber reinforced bioepoxy composites. *Int J Polym Sci* 2015;2015:16–8.

[14] Sanjay M. R., Madhu P., Jawaid M., Senthamaraikannan P., Senthil S., Pradeep S. *Characterization and properties of natural fiber polymer composites: A comprehensive review.* vol. 172. Elsevier B.V.; 2018. https://doi.org/10.1016/j.jclepro.2017.10.101.

[15] Thyavihalli Girijappa Y. G., Mavinkere Rangappa S., Parameswaranpillai J., Siengchin S. Natural Fibers as Sustainable and Renewable Resource for Development of Eco-Friendly Composites: A Comprehensive Review. *Front Mater* 2019;6:1–14. https://doi.org/10.3389/fmats.2019.00226.

Chapter 13

Experimental Studies on Mixed Micelle Concentrations and Gemini Surfactants

Sara Fatima*
Department of Chemical Engineering, Aligarh Muslim University, Aligarh, U. P., India

Abstract

This research paper presents the data from experiments conducted in the laboratory on CTAB and mixed surfactants. Surfactants find a wide range of applications in domestic and industrial areas. These include soaps, detergents, dispersions, paints, emulsions, foams and anti-foams. Recently, there has been an interest in biosurfactants due to the adverse effects of conventional surfactants on the environment. More recently, there has been considerable research on certain dimeric surfactants, which have two hydrophilic tails and two hydrophobic tails attached together. These are called Gemini surfactants.

Keywords: surface tension, surface active agents, micelle, bio surfactants, mixed surfactants, gemini surfactants

1. Introduction

Surface tension is the tendency of fluid surfaces to shrink to the minimum surface area possible. Intermolecular forces, such as the van der Waals force, hold liquid particles close by. One characteristic feature of surfactants is their tendency to absorb at interfaces. Another fundamental property of surfactants is that monomers in solution tend to form aggregates called micelles. Micelle formation or micellization can be viewed as an alternative mechanism to adsorption at the interfaces for removing hydrophobic groups from contact with water. Surfactant molecules behave differently when present in micelles than as free monomers in solution. The SI unit of surface tension is Newton/meter (N/m), and in CGS, the unit is dyne/cm.

* Corresponding Author's Email: fsara4788@gmail.com.

In: Sustainability in Chemical Processes through Digitalization and Green Chemistry Approaches
Editors: Susarla V. A. R. Sastry, Shravan Kumar and Ashish Kapoor
ISBN: 979-8-89113-257-3

The surface tension phenomenon is responsible for the shape of liquid droplets. Small insects can work on water, as their weight is considerably less than that needed to penetrate the water surface. There are several such surface tension examples found in nature. Some cases are provided below:

1. Clinical test for jaundice
2. Insects walking on water
3. Rainproof tent materials
4. Round bubbles where the surface tension of water provides the wall tension for water bubble formation.
5. Washing with cold water
6. Cleaning clothes with soaps and detergents lowers the surface tension of water.
7. Needle floating on the water surface
8. Raindrops are spherical due to surface tension and the presence of cohesive forces between liquid molecules.

Some methods of measuring surface tension include the spinning drop method, the Du-nong ring method, the pendant drop method, the session drop method, and the bubble pressure method.

When a cloth is washed using a surfactant, the hydrophobic group or branch attaches itself to the dirt and oil of the cloth, and the hydrophilic end of the micelle exposes itself to water. The polar hydrophilic head group is attached at one end to the alkyl chain (8–18 C atoms). This polar, hydrophilic part may be ionic or nonionic. On the other hand, the hydrophobic part may be branched or linear. When the fluid is water, the surfactant has a hydrophobic (water-repelling) part and a hydrophilic (water-loving) part as the tail. The simple micelle chemistry of soap is shown in Figure 1.

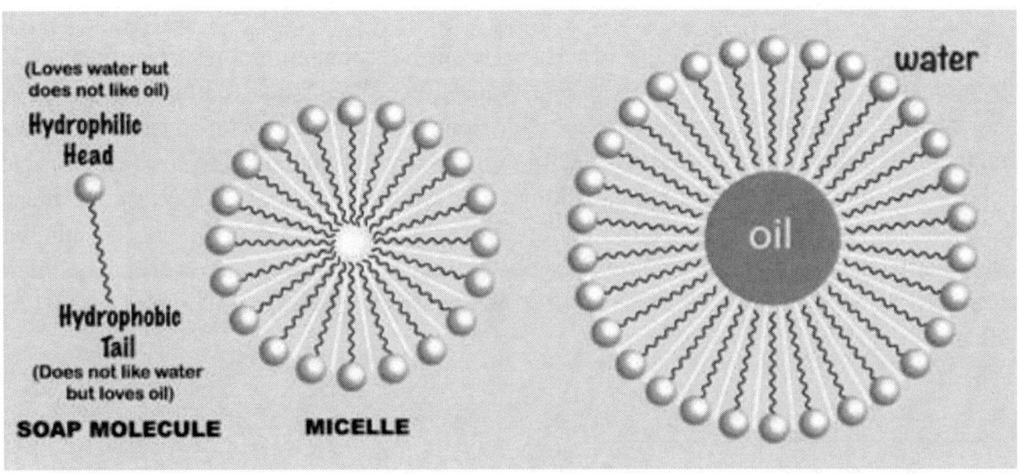

Figure 1. Micelle chemistry of soap as a surfactant.

A surfactant is called amphiphilic because it has two parts. One soluble and the other insoluble.

Table 1 presents how the two parts of amphiphilic surfactant work.

Table 1. Two parts of an amphiphilic surfactant

Micelle Chemistry			
#	Surfactant End	Water/Oil Repelling	Physiochemical Characteristics
1	Hydrophilic end	Water loving	Ionic/Nonionic polar group
2	Hydrophobic end	Oil loving	Branched/Linear chain nonpolar group

This research paper investigates an important property of conventional and mixed surfactants called CMC. Micelles begin to form at very low surfactant concentrations in water. Adding more drops of surfactant continues to increase the surface tension of the water-surfactant solution. However, when more surfactant is added, it does not cause any further increase in the surface tension of the solution. In fact, upon further increasing the surfactant concentration, the surface tension remains constant and is still at that value. This is because micelle formation has reached saturation. This constant value where micelle saturation occurs is called the critical micelle concentration (CMC). A series of 1 mM ionic surfactants indicates that the monomer concentration will not exceed 1 mM, even if more surfactant is added to the solution. Figure 2 illustrates the CMC graph of surface tension vs. micelle concentration.

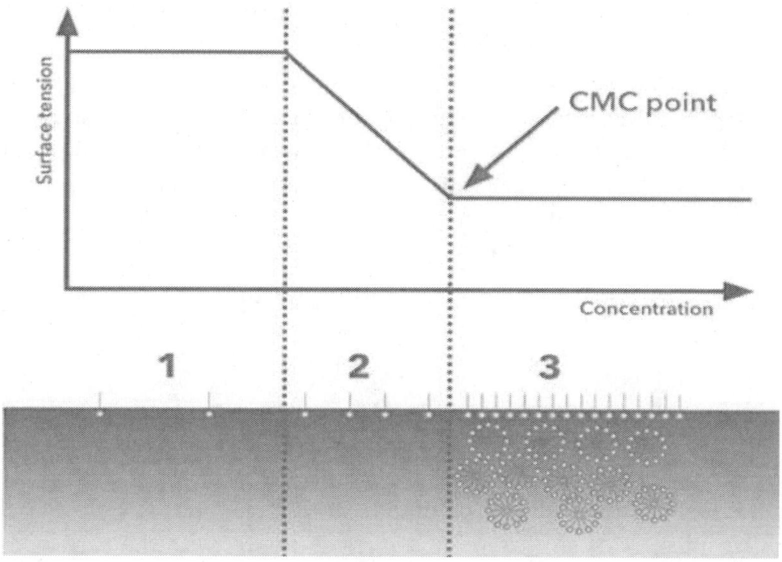

Figure 2. Surface tension vs. concentration graph.

A series of experiments were conducted by the author on conventional CTAB surfactant and CTAB-Gemini mixed surfactants in the laboratory. The CMC determination of CTAB was done using manual tentiometer and CMC determination of CTAB-Gemini surfactant was done using an automatic tentiometer in laboratory by the author.

2. Literature Review

Conventional surfactants pose the most challenging hazard today because they are continuously discharged into wastewater streams. Surfactants pose a severe environmental and public threat to ecosystems and humans. Ionic surfactants, predominantly linear alkylbenzene sulfonates (LAS), cause adverse biochemical and physiological impacts on terrestrial and aquatic ecosystems [1]. These cause irritation of the skin and respiratory problems [2], reduce the resistance of aquatic species and have a harmful impact on reproduction and growth processes. Surfactants facilitate the solubility of contaminants and thus increase water eutrophication [3]. Furthermore, the hydrophobic properties of the surfactants increase their toxicity. These adverse effects raise public health and environmental concerns about surfactants. Some surfactants have a severe impact on humans through the ingestion or drinking of contaminated food items. For instance, surfactants react with liver proteins and serum, thus causing long-term metabolic effects and disruption of the human endocrine system [4]. Some surfactants cause human skin burning, eye irritation, and respiratory problems. Nonylphenolethoxylates have estrogenic effects on amphibians, mammals, and fish. LAS surfactants damage the cell membrane and alter its structure. Agrochemicals such as biocides, herbicides, pesticides, and industrial products such as cosmetic products or emulsifiers, wetting agents, detergents, coatings, fabric softeners, paper and carpet are continuously prone to environmental degradation. Effluent streams and land are treated by primary, secondary, and tertiary treatments. Primary treatments of effluent streams with the removal of floatable objects reduce biological oxygen demand (BOD) and total suspended solids (TSS) in the wastewater. Secondary treatments are the main processes to remove constituents that fall into the category of detergents and surfactant products. These secondary techniques include flocculation-adsorption, filtration, ultrafiltration, membrane filtration, coagulation, and electro-oxidation [5–6]. Thus, municipal wastewater treatment plants usually receive a high concentration of surfactants and byproducts upon secondary treatment. Secondary treatments transform surfactants into less toxic sulfophenyl carboxylic acids and phenol ethylates.

Hence, due to the numerous disadvantages of conventional surfactants, a need is being felt for more biodegradable alternative surfactants. Biosurfactants, mixed surfactants and gemini surfactants are prospective alternatives as clean surfactants.

3. Method

Critical micelle concentration is the concentration of surfactants (e.g., CTAB) in water solution at which the first micelles of surfactant appear. If one starts with the surface tension of pure water and slowly adds drops of surfactant, the surface tension decreases until it reaches a minimum value. Further adding extra surfactants ends up as micelles in water and does not cause any further reduction in surface tension. At this point, the micelle concentration is called the CMC. Then, curves are plotted for surface tension versus concentration. These curves are called "isotherms" and can be fitted to various equations. Below are the CMC value tables as reported in the literature for conventional and biosurfactants. Table 2 presents the CMC values of synthetic surfactants, and Table 3 presents the CMC values of biosurfactants reported in the literature at ambient temperature.

CTAB CTAB and SDS find wide applications in the soap, detergent and shampoo industries. CTAB (Cetrimonium Bromide) is a low-CMC surfactant employed in DNA extraction and isolation and in the manufacturing of novel gene therapies, polysaccharide bacterial vaccines and viral vaccines for influenza. Surface tension methods were conducted over a period of eight months in 2016 by the author on two tensiometers installed in the lab.

Table 2. CMC values of synthetic surfactants as reported in the literature at T = 25°C

CMC values of synthetic surfactants as reported in literature			
#	Type	Surfactant	CMC (mM)
1	Anionic	Sodium dodecylsulfate (SDS)	5.2
2	Cationic	Dodecyltrimethylammonium bromide (DTAB)	14.6-16.0
3		Tetradecyltrimethyl ammonium bromide (TTAB)	3.6 3.72
4		Cetyltrimethyl Ammonium bromide (CTAB)	0.92-1.00
5	NonIonic	Triton X-100 (TX-100)	0.24-0.27
6		Octaethylene glycol monododecyl ether	0.11
7		Octoethylene glycol monododecyl ether	0.064

Table 3. CMC values of bio surfactants as reported in the literature at T = 25°C

CMC values of bio surfactants as reported in literature					
#	Surfactant	Producing Organism	C-source	Charge	CMC (mg/L)
1	Cellobioselipids	Ustilago sp.	Vegetable Oil	anionic	20
2	Conynomycolates	Arthobacter sp.	Different sugars	nonionic	1 to 19
3	Mannosylerythriol lipids	Candida sp.	Glucose soyabean oil	"	10 to 200
4	Rhamnolipids	Psuedomonas sp.	n-alkanes, glycerol	anionic	60 to 82
6	Lipopeptides	Bacillus Lichen i-formis	Glucose	"	0.02-10
7	Surfactin	Bacillus subtilis	Glucose	"	5

Table 4 presents the CMC values of certain surfactants reported in the literature at ambient temperature.

Table 4. CMC values of surfactants in distilled water

CMC values of surfactants in pure water					
#	Surfactant	CMC (mM) at 298.15K	303.15K	318.15K	Literature
1	CTAB	0.98	1.1	1.2	0.92-1
2	SDS	8	9	10	6.0-8.0
3	16-5-16	0.037	0.04	0.045	0.035
4	16-6-16	0.038	0.04	0.05	0.04

For this, a first stock solution of surfactant (10 mM CTAB) was prepared, and the following formula was used to determine the CTAB weight content present in it:

Weight of CTAB (g) = (MW of CTAB (g) x volume of water (mL) x molarity (mMol) x 10^{-3} mL)/volume of water solution = (364.6 g x 25 mL x 10 M x 10^{-3} mM)/1000 L = 0.09115 g.

Hence, approximately 0.009115 g of CTAB was weighed using a chronometer. Then, this CTAB powder was added to 50 liters of distilled water to prepare a surfactant solution. Meanwhile, a manual tensiometer was calibrated to prepare it for taking surface tension readings of this water-CTAB solution. Ring parameters from the panel of the manual tensiometer were noted (C = 6.11986 cm and R/r = 55.685714) to begin the readings. Every time 0.1 or 0.2 mL of water-CTAB solution was added to 25 mL of distilled water, a corresponding surface tension reading was taken. The tabulated data from surface tension experiments conducted using a manual tensiometer are presented below.

Table 5 presents the surface tension vs. log C data from Experiment 1 for 10 mM CTAB molarity in water. Experimental graph ϒ vs. log C plots are shown in Figures 3, 4, 5, 6, 7 and 8. Experimental data on ϒ vs. log C obtained by experiments are given in Tables 5, 6, 7, 8, 9, 10, 11 and 12.

Table 5. ϒ vs. log C data from Experiment 1 for 10 mM CTAB in water

#	V(mL)	C(mM)	log C	ϒ (MN/m)
1	0	0	-	72
2	0.2	0.0398406	-1.39967372	43.5
3	0.4	0.0793651	-1.10037055	42
4	0.6	0.1185771	-0.92599927	41.5
5	0.8	0.1574803	-0.80277373	41
6	1	0.1960784	-0.70757018	40.5
7	1.2	0.234375	-0.63008871	40
8	1.4	0.2723735	-0.56483508	38
9	1.6	0.3100775	-0.50852972	38
10	1.8	0.3474903	-0.45905725	38
11	2	0.3846154	-0.41497335	39
12	2.2	0.4214559	-0.37524782	38
13	2.4	0.4580153	-0.33912005	38
14	2.8	0.530303	-0.27547589	38
15	3.2	0.6015038	-0.22076165	37
16	3.6	0.6716418	-0.17286229	36.5
17	4	0.7407407	-0.13033377	37

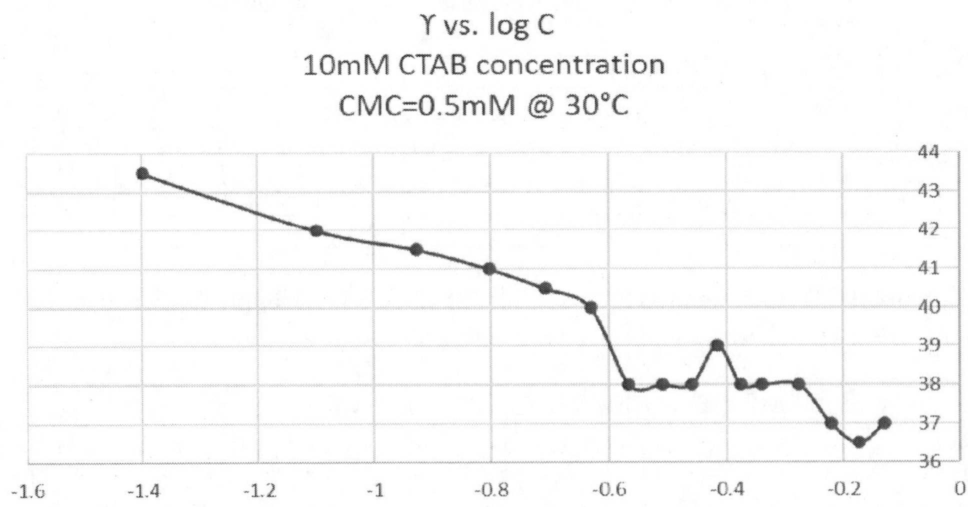

Figure 3. ϒ vs. log C plot from Experiment 1 for 10 mM CTAB in water.

Figure 3 shows the ϒ vs. log C plot from Experiment 1 for 10 mM CTAB in water.

3.1. Determination of the Comic of Mixed Surfactants

Gemini surfactants show great promise as clean, biodegradable and low-CMC surfactants. They increase wetting, promote emulsification of oil in water, enhance dispersion of solids and have high foaming stability. Furthermore, they have better solubilizing properties, stronger tolerance to multivalent metal ions, antimicrobial activity, and mildness to the skin and are safe for the environment.

Mixed micellization was achieved in the laboratory by mixing Sigma–Aldrich Gemini surfactant (16-5-16) and CTAB in different ratios. The gemini surfactant used in this experiment was 16-5-16, also called Gemini-5. Table 6 presents the Gemini-5 and CTAB mixing ratios.

Table 6. G5-CTAB ratios used in five experiments conducted on an automatic tensiometer

#	G5-CTAB ratio
1	10:90
2	30:70
3	50:50
4	70:30
5	90:10

Table 7. ϒ vs. log C data from experiment 2 for 20 mM-20 mM G5-CTAB (10:90) in water

#	V(mL)	C(mM)	log C	ϒ (mN/m)
1	0	0		65.1
2	0.1	0.00797	2.09864	
3	0.2	0.01587	1.79934	32.37
4	0.3	0.02372	1.62497	31.99
5	0.4	0.0315	1.50174	31.36
6	0.5	0.03922	1.40654	30.67
7	0.6	0.04688	1.32906	28.04
8	0.7	0.05447	1.26381	31.36
9	0.8	0.06202	1.2075	31.36
10	1	0.07692	1.11394	30.67
11	1.2	0.0916	1.03809	28.04
12	1.4	0.10606	0.97445	30.75
13	1.6	0.1203	0.91973	29.97
14	1.8	0.13433	0.87183	31.05
15	2	0.14815	0.8293	30.28
16	2.2	0.16176	0.79112	31.36
17	2.4	0.17518	0.75651	30.59

Sample calculations that were performed for the G5-CTAB solution (2 mM each) are given below:

CTAB: 346.6 g x 50 mL x 2 mM x $10^{\wedge-6}$ = 0.03466 g

G5 (16-5-16): 786.22 g x 50 mL x 2 mM x $10^{\wedge-6}$= 0.0786 g

Hence, the tabulated data for mixed micellization experiments performed on an automatic tensiometer are presented below.

Table 7 presents surface tension vs. log C data from experiment 2 for a 20 mM-20 mM G5-CTAB (10:90) composition in water using an automatic centimeter for mixed micellization.

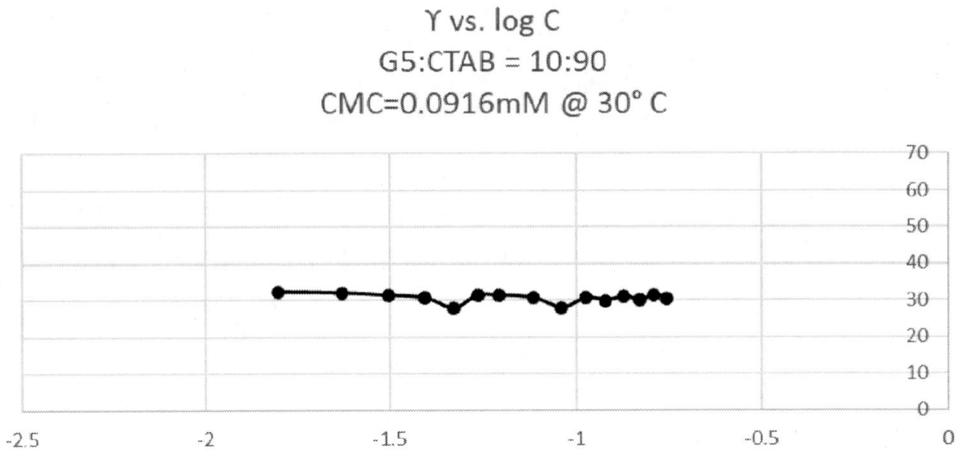

Figure 4. ϒ vs. log C plot from experiment 2 for 20 mM-20 mM G5-CTAB (10:90) in water.

Figure 4 shows the ϒ vs. log C plot from experiment 2 for 20 mM-20 mM G5-CTAB (10:90) in water.

Table 8 presents the data of ϒ vs. log C from experiment 3 for 20 mM-20 mM G5-CTAB (30:70) in water.

Table 8. ϒ vs. log C data from experiment 3 for 20 mM-20 mM G5-CTAB (30:70) in water

#	V(mL)	C(mM)	log C	ϒ (mN/m)
1	0	0		65.7
2	0.1	0.007968	2.09864	49.97
3	0.2	0.015873	1.79934	43.79
4	0.4	0.031496	1.50174	40.86
5	0.6	0.046875	1.32906	38.8
6	0.8	0.062016	1.2075	37.15
7	1	0.076923	1.11394	36.68
8	1.2	0.091603	1.03809	36.05
9	1.4	0.106061	0.97445	36.13
10	1.6	0.120301	0.91973	35.81
11	1.8	0.134328	0.87183	34.95
12	2	0.148148	0.8293	34.33
13	2.2	0.161765	0.79112	33.86
14	2.4	0.175182	0.75651	34.33
15	2.6	0.188406	0.72491	26.36
16	2.8	0.201439	0.69586	28.97
17	3	0.214286	0.66901	26.59

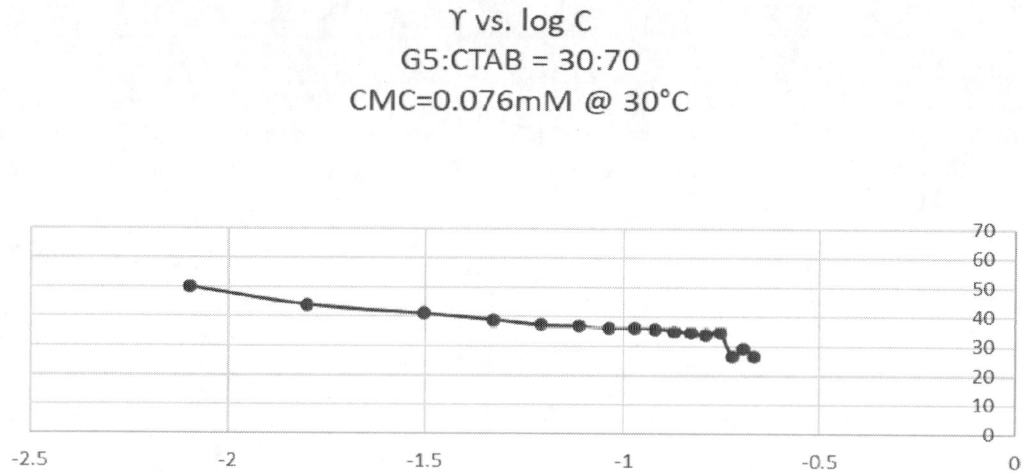

Figure 5. ϒ vs. log C plot from experiment 3 for 20 mM-20 mM G5-CTAB (30:70) in water.

Figure 5 shows the ϒ vs. log C plot from experiment 3 for 20 mM-20 mM G5-CTAB (30:70) in water.

Table 9 presents ϒ vs. log C data from experiment 4 for 20 mM-20 mM G5-CTAB (50:50) in water.

Figure 6 shows the ϒ vs. log C plot from experiment 4 for 20 mM-20 mM G5-CTAB (50:50) in water.

Table 10 presents ϒ vs. log C data from experiment 5 for 20 mM-20 mM G5-CTAB (70:30) in water.

Table 9. ϒ vs. log C data from experiment 4 for 20 mM-20 mM G5-CTAB (50:50) in water

#	V(mL)	C(mM)	log C	ϒ (mN/m)
1	0	0		69.36
2	0.1	0.007968	2.09864	51.35
3	0.2	0.015873	1.79934	43.88
4	0.4	0.031496	1.50174	40.78
5	0.6	0.046875	1.32906	39.04
6	0.8	0.062016	1.2075	38.88
7	1	0.076923	1.11394	38.72
8	1.2	0.091603	1.03809	38.17
9	1.4	0.106061	0.97445	38.01
10	1.6	0.120301	0.91973	37.46
11	1.8	0.134328	0.87183	37.15
12	2	0.148148	0.8293	37.62
13	2.2	0.161765	0.79112	38.33
14	2.4	0.175182	0.75651	37.25
15	2.6	0.188406	0.72491	38.33
16	2.8	0.201439	0.69586	38.17
17	3	0.214286	0.66901	38.25

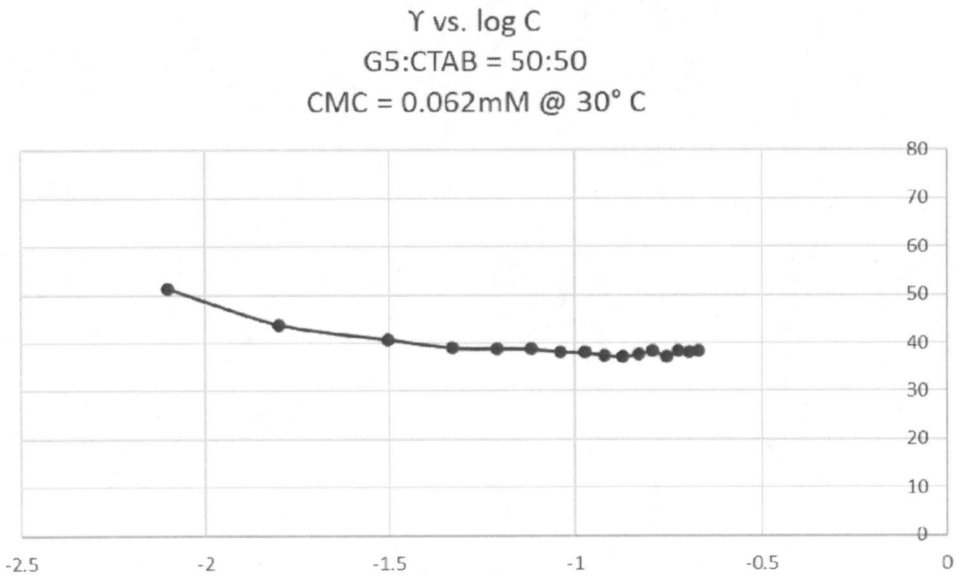

Figure 6. ϒ vs. log C plot from experiment 4 for 20 mM-20 mM G5-CTAB (50:50) in water.

Table 10. ϒ vs. log C data from experiment 5 for 20 mM-20 mM G5-CTAB (70:30) in water

#	V(mL)	C(mM)	log C	ϒ (mN/m)
1	0	0		68.19
2	0.1	0.007968	2.09864	
3	0.2	0.015873	1.79934	49.73
4	0.4	0.031496	1.50174	42.84
5	0.6	0.046875	1.32906	42.84
6	0.8	0.062016	1.2075	41.81
7	1	0.076923	1.11394	41.73
8	1.2	0.091603	1.03809	41.33
9	1.4	0.106061	0.97445	40.86
10	1.6	0.120301	0.91973	40.22
11	1.8	0.134328	0.87183	39.83
12	2	0.148148	0.8293	39.59
13	2.2	0.161765	0.79112	39.35
14	2.4	0.175182	0.75651	39.59
15	2.6	0.188406	0.72491	39.35
16	2.8	0.201439	0.69586	39.12
17	3	0.214286	0.66901	39.19

Figure 7 shows the ϒ vs. log C plot from experiment 5 for 20 mM-20 mM G5-CTAB (70:30) in water.

Table 11 presents ϒ vs. log C data from experiment 6 for 20 mM-20 mM G5-CTAB (90:10) in water.

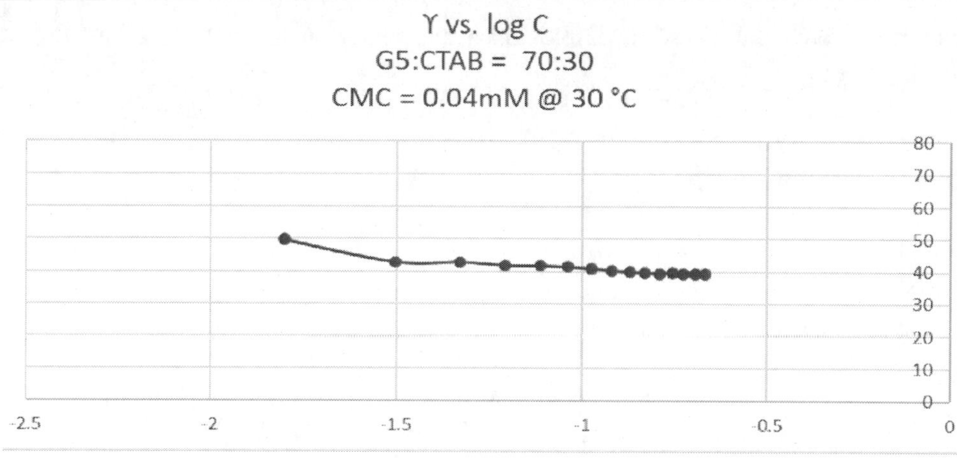

Figure 7. ϒ vs. log C plot from experiment 5 for 20 mM-20 mM G5-CTAB (70:30) in water.

Table 11. ϒ vs. log C data from experiment 6 for 20 mM-20 mM G5-CTAB (90:10) in water (automatic thermometer)

#	V(mL)	C(mM)	Log C	ϒ (mN/m)
1	0	0		68.69
2	0.1	0.007968	2.09864	
3	0.2	0.015873	1.79934	52.4
4	0.4	0.031496	1.50174	41.65
5	0.6	0.046875	1.32906	40.22
6	0.8	0.062016	1.2075	39.83
7	1	0.076923	1.11394	39.83
8	1.2	0.091603	1.03809	39.59
9	1.4	0.106061	0.97445	39.43
10	1.6	0.120301	0.91973	39.19
11	1.8	0.134328	0.87183	39.19
12	2	0.148148	0.8293	39.19

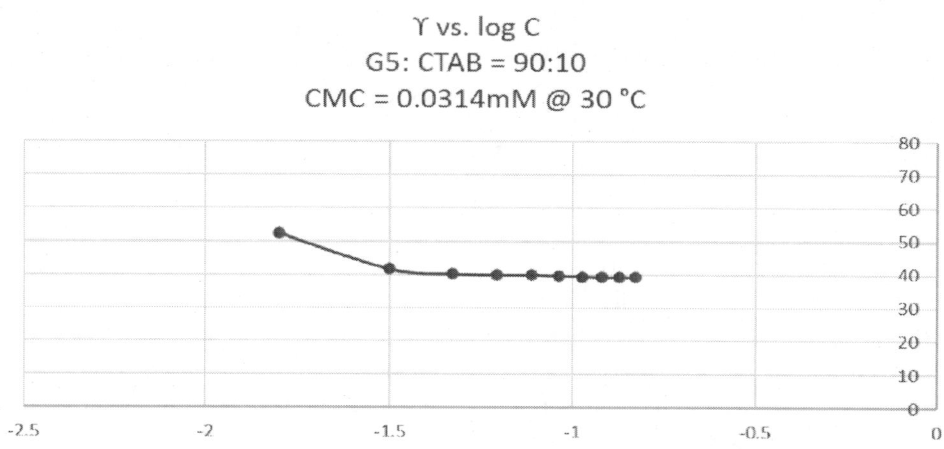

Figure 8. ϒ vs. log C plot from experiment 6 for 20 mM-20 mM G5-CTAB (90:10) in water (automatic thermometer).

Figure 8 shows the ϒ vs. log C plot from experiment 6 for 20 mM-20 mM G5-CTAB (90:10) in water.

4. Results and Discussion

Mixed micelle systems are encountered in nearly all practical applications of surfactants. These mixtures arise from several sources. First is the natural polydispersity of commercial surfactants, which results from impurities in starting materials and variability in reaction products during their manufacture. These are less expensive to produce than isomerically pure surfactants and often provide better performance. Second is the deliberate mixture formulation of mixtures of different surfactant types with different performance characteristics in a single formulation.

Mixtures of different surfactant types often exhibit synergism in their effects on the properties of the system. The observed synergism can be attributed to nonideal mixing effects in the aggregates and results in substantially lower CMCs and interfacial tensions than would be expected based on the properties of the unmixed surfactants alone. This situation can be exploited in applications such as detergency, enhanced oil recovery and mineral flotation. Monomer-micelle equilibrium exerts a key controlling influence on mixed surfactant systems. This equilibrium property of mixed surfactant systems was employed in experiments no. 2, 3, 4, 5 and 6. Among the experimental techniques commonly used for measuring mixed CMCs are surface tension measurements, conductivity, dye solubilization and light scattering. This research utilized the surface tension property of mixed surfactants, specifically the G5-CTAB solution.

The first CTAB surface tension readings were taken using a manual tensiometer at room temperature. Surface tension (ϒ) vs. log C (CTAB concentration in water solution) was tabulated in MS Excel, and a graph was generated therewith. The CMC found by this experiment was 0.5 mM, which is very close to the 0.92-1 value of the CMC of CTAB reported in the literature. Table 12 presents the comparison between the CMC experimental value obtained and the literature value reported.

Table 12. CMC value obtained from experiment 1 using a manual tensiometer (CTAB in water solution)

Experiment #	Experimental CMC (mM)	Experiment #
1	0.5	0.92-1.0

Then, Gemini-5 (G5 or 16-5-16) surfactant was mixed in five different ratios with CTAB to obtain a mixed surfactant and observe the mixed micellization behavior. These five experiments on an automatic tensiometer showed that as the G5 ratio increased, the CMC value of the mixed surfactant decreased and became closer to the value of 0.03 reported in the literature for 16-5-16. The CMC values obtained from the experiments are shown in Table 13.

Table 13. CMC values obtained from Experiments 2, 3, 4, 5 and 6 using an automatic tensiometer (G5-CTAB in water solution)

Experiment #	G5-CTAB ratio	CMC (mM)
reading for G5-CTAB mixed micellization		
2	10:90	0.0916
3	30:70	0.0760
4	50:50	0.0620
5	70:30	0.0468
6	90:10	0.0314

Conclusion

As the gemini surfactant percentage increased in the mixed micelle experiment of the G5-CTAB solution, the CMC value decreased in the lab experiments.

Acknowledgments

This research work is a result of experimental work undertaken by the author while working as a Junior Research Fellow in the Department of Petroleum Studies, Aligarh Muslim University, as part of the DST-PURSE project in 2016. This DST-PURSE project was procured by the Department of Petroleum Studies, ZHCET, AMU, in 2015-16 from the Department of Science, New Delhi, India.

References

[1] Bhattarai, A, Ghenshyam S. 2014 Study of critical micelle concentration of cetyl trimethyl ammonium bromide(CTAB) in pure water in presence and absence of magnesium sulfate and sodium sulphate, *Bibechana*: 18, 123-127.

[2] Banipal, T. S, Kaur, H, Banipal, PK and Sood, A. K. 2014. Effect of head groups, temperature and polymer concentration on surfactant polymer interaction, *Journal of surfactants and detergents*. 17, 1181-1191.

[3] Ali, M. J, Suhail, M, Ghosh, G and Kamil, M Din. Interaction between cationic Gemini or conventional surfactants with polyvinyl pyrolidone: specific conductivity and dynamic light scattering studies, *Colloids and surfactants: Physico chemical Engineering aspects,* 350, 51-56.

[4] Holmberg krister Johnson Bo, Kronberg Bengt and Lindman Bjorn. 2002. Chapter 1: Introduction to Surfactants. *Surfactant and Polymers in Aqueous Solution*, John Wiley and sons Ltd, 1-30.

[5] Abott Steven. 2015. Surfactant Science: Principles and Practice. *Creative Commons* BY-ND, 1-15.

[6] Badmus O. Suaibu, Amusa k. Hossein, Oyehen A. Tajudeen, Salih A Tawfiq. 2021. Environmental risks and toxicity of surfactants: Overview of analysis, assessment and remediation techniques. *Environmental science and pollution research international.* 28(44).

Chapter 14

Maleic Anhydride-Based Surfactants and Their Properties

Smriti Dwivedi*
P. K. S. Yadav
and Gaurav Singh
Department of Oil Technology, School of Chemical Technology, Harcourt Butler Technical University, Kanpur, Uttar Pradesh, India

Abstract

A novel class of surfactants derived from maleic anhydride has incredible properties. Maleic anhydride is an acid anhydride of maleic acid that is a colorless organic compound that is mostly used as an additive in the paint and coating industries. These surfactants provide excellent solubilization, dispersion, low foaming, and synergic properties. In addition, maleic anhydride-based surfactants are less toxic, biodegradable and can be produced from green synthesis. These surfactants are produced by reacting maleic anhydride with triglycerides in the presence of acid catalysts, which impart certain enhanced characteristics to the surfactant. These properties can be further demonstrated with the help of NMR, IR and elemental analysis. Additionally, maleic anhydride-based surfactants are widely used in the manufacturing of laundry detergents, hard surface cleaners, fabric softeners, cement setters, unsaturated polyester resins, coatings, pharmaceuticals and pesticides. The present paper addresses the routes of synthesis, properties and applications of maleic anhydride-based surfactants.

Keywords: maleic anhydride-based surfactants (MAS), surface active agents, oleochemicals, detergents, surface tension, low foaming, textiles

1. Introduction

Maleic anhydride-based surfactants are a type of surfactant that is synthesized from maleic anhydride, a reactive intermediate in the production of various products. These surfactants have

* Corresponding Author's Email: smritidwivedi05@gmail.com.

In: Sustainability in Chemical Processes through Digitalization and Green Chemistry Approaches
Editors: Susarla V. A. R. Sastry, Shravan Kumar and Ashish Kapoor
ISBN: 979-8-89113-257-3

unique properties that make them useful in a range of applications, including personal care, cleaning and industrial products. Maleic anhydride-based surfactants are known for their high solubility in both water and oil, making them ideal for use in emulsions and formulations that require a stable mix of oil and water [1, 2]. They are also effective at reducing the surface tension of liquids and are used as foaming agents, dispersants and wetting agents. Basically, surfactant is a term that is defined as a surface-active agent that, when present in a certain concentration, has the ability to adsorb onto the surface interface and lower the surface tension of the solvent phase in which it is dissolved. Surfactants are compounds that have both hydrophobic (water-fearing) and hydrophilic (water-loving) properties [3]. These surfactant molecules have a wide range of applications in the detergent, pharmaceutical, food, mineral processing, cosmetics, hydrometallurgy, paint and coating industries [4, 5]. The hydrophilic group is the polar group that comprises the head of the surfactant and is responsible for the surface activity, as shown in Figure 1. On the other hand, the hydrophobic part is a nonpolar group that could be a long hydrocarbon chain, comprises the tail and is responsible for solubility. Due to this property, these molecules tend to concentrate at the interfacial surfaces and modify the interfacial tensions.

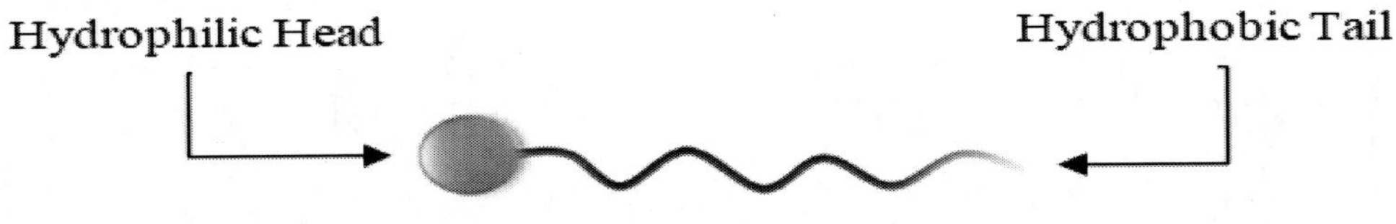

Figure 1. General Surfactant Structure.

The hydrophobic part of the surfactant molecule tends to dissolve in oils and fats, while the hydrophilic part is attracted to water. This allows surfactants to reduce the surface tension of liquids and help mix oil-based and water-based substances [6]. This is why they are often used in products such as shampoo, which needs to clean both oil and dirt from hair while also remaining in liquid form.

There are four different types of surfactants, including anionic, cationic, nonionic and amphoteric surfactants, as mentioned below. Each type has unique properties and is used for specific purposes, depending on the desired outcome [7, 8].

- Anionics: Surfactants that carry a negatively charged head group are known as anionics. The most common class of surfactants used in detergents and many other industries. Some popular anionics are sodium dodecyl sulfate and sodium dodecylbenzene sulfonate.
- Cationics: A surfactant with a positively charged head group falls under this category. These are very efficient with the surfaces, which are negatively charged and adsorb on them by strong electrostatic interactions. Some commonly used cations are laurylamine hydrochloride and cetyl trimethylammonium bromide.
- Nonionics: These surfactants have no charge groups in their heads. Most of the newer biodegradable and nontoxic products are produced from nonionic surfactants. These are excellent for detergents; they have moderate foaming, wetting and emulsifying properties. Some of the common examples are polyoxyethylene alcohol and alkylphenol ethoxylate.

- Amphoterics: These have both positively and negatively charged groups. Some examples are lauramidopropyl betaine and cocoamido-2-hydroxypropyl sulfobetaine.

Surfactants can have both positive and negative impacts on the environment, depending on their composition and how they are used [9]. Some surfactants can be toxic to aquatic life, while others are biodegradable and have a low environmental impact.

Many efforts toward the formulation of new structures of surfactants have been made. These surfactants include gemini surfactants, polymeric surfactants, methylester ethoxylates, and polymerisable surfactants [10].

One of the key advantages of maleic anhydride-based surfactants is their biodegradability and low toxicity. They are typically made from renewable resources, such as vegetable oils, and are designed to breakdown into nontoxic byproducts after use, reducing their impact on the environment [11, 12]. Maleic anhydride surfactants also fall under the category of novel surfactants. Novel surfactants show synergistic interactions, enhanced surface and aggregation properties, less toxicity and high biodegradability [12, 13]. The last two characteristics are due to the more restrictive regulations, especially for personal care products and house detergents. In aqueous systems, the addition of a surfactant enables two immiscible liquids to become soluble in one another. This striking phenomenon is attributed to the unique structure of the surfactant molecules.

In particular, maleic anhydride was first produced in 1872 by dehydrating maleic acid. Currently, it is widely used in many industries. In addition to acetic anhydride and phthalic anhydride, it is the second most commercially important compound. Until 1940, only 4.5 million lbs of maleic anhydride were commercially produced [14]. Undoubtedly, maleic anhydride suddenly started gaining importance in 1978, when the demand hyped with a production rate of 341 million lbs. The effects of the 1974 oil embargo may contribute to the high demand for maleic anhydride. Typically, the foregoing research and advancement in technologies are apparently intensifying the behavior and properties of maleic anhydride-based products and successfully inventing a variety of compounds that are commercially considered to have enormous importance and potential. The surfactant produced from maleic anhydride and its derivatives is easy to handle, store, use and transport. As maleic anhydride is an organic compound, it is less toxic in nature [15]. Therefore, surfactants produced using maleic anhydrides are biodegradable and can be handled safely. It is obvious that the raw materials for these novel surfactants are from renewable sources such as oleochemicals, petrochemicals, and certain maleic anhydride derivatives.

Moreover, as a result of the unique functionality of these surfactants, they have a wide range of applications in many areas. Maleic anhydride can also undergo a number of reactions, such as addition, polymerization, and malenization. Some important reactions also include Diels-Alder and Friedel-Crafts reactions [16].

The attachment of unsaturated sites of maleic anhydride with triglycerides of oils and fats produced mainly from renewable sources such as vegetable oils and oleochemicals can be used in the production of a unique class of eco-friendly and biodegradable surfactants, monomers and macromonomers. Recently, this area of research and development has gained momentum with a precise number of different applications. The versatile nature, chemistry and need for renewable and biodegradable systems are some of the reasons for the routes of synthesis for these novel surfactants. Maleimides and their derivatives are important building blocks in chemical synthesis, and some of them demonstrate biological activity [17]. A special feature of the reactivity of maleimides is their susceptibility to additions across the double bond, either by Michael additions or via Diels-Alder reactions. The carbon–carbon double bond in maleimide is also capable of free-radical or anionic polymerization or copolymerization to produce functional polymers that are utilized in high-temperature applications, such as adhesives in the semiconductor industry. Biological applications of maleimide derivatives result from their high and specific reactivity and binding toward the thiol groups in biological substrates (e.g., chemical probes), site-directed modification of proteins and peptides, bioconjugates and prodrugs.

2. Novel Surfactants

2.1. Definition

Novel surfactants are surfactants that are newly developed or have unique properties compared to conventional surfactants. These surfactants can have improved performance, biodegradability, or reduced toxicity [16, 17]. Moreover, these surfactants can be manufactured using green synthesis, which makes them more extraordinary than convectional surfactants.

2.2. Types and Properties of Novel Surfactants

Novel surfactants possess different properties depending on their nature and route of synthesis. Therefore, some of the unique examples with properties are mentioned below [18, 19].

- Amphiphilic copolymers: These are copolymers that have both hydrophobic and hydrophilic segments, which give them unique surfactant properties. They can be used as emulsifiers, dispersants or wetting agents [20].
- Biobased surfactants: These are surfactants derived from renewable resources, such as plants or microorganisms. They are often biodegradable and eco-friendly, making them an attractive alternative to conventional surfactants.
- Gemini surfactants: These are two surfactant molecules linked together by a spacer group. They have unique properties, such as enhanced stability and lower critical micelle concentrations, compared to traditional surfactants [21].
- Ionic liquids: These are salts that are liquid at room temperature. They can be used as surfactants and have unique properties, such as high thermal stability and low volatility.
- Sulfosuccinates: These are anionic surfactants that are made from sulfo-succinic acid and an alcohol [22, 23]. They are commonly used as emulsifiers and have good solubility in both water and oil.
- Sugar-based surfactants: These are surfactants made from sugar derivatives, such as glucosides [24]. They are biodegradable and have low toxicity, making them a promising alternative to traditional surfactants.

- Polypeptide-based surfactants: These are surfactants made from polypeptides, which are chains of amino acids. They are biodegradable, have low toxicity and can be used as emulsifiers or dispersants.

These are a few types of novel surfactants that are mentioned above. There are many other surfactants that have been used in different applications. The field of surfactant research is constantly evolving, and new surfactants with unique properties are being developed.

2.3. Polymerizable Surfactants

Polymerizable surfactants are a type of surfactant that is made up of long chains of repeating monomer units. They are larger and more complex than traditional surfactants and can have unique properties and applications [24]. Some of the key benefits of these surfactants are as follows:

- Improved stability: These surfactants can have improved stability compared to traditional surfactants. This means they are less likely to degrade or lose their effectiveness over time.
- Better emulsifying properties: These surfactants can be more effective emulsifiers, which makes them useful for applications including the formulation of emulsions, suspensions and gels.
- Increased biodegradability: Some polymerizable surfactants can be biodegradable, which means they breakdown into natural substances after use, reducing their impact on the environment.
- Enhanced Solubility: Polymerizable surfactants can have enhanced solubility in both water and oil, making them useful for a range of applications, including the formulation of surfactant solutions, emulsions and suspensions.

2.3.1. Structure of Polymerizable Surfactants

The structure of a polymerizable surfactant can vary depending on its chemical composition, but in general, it is composed of long chains of repeating monomer units. The monomer units can be linked together through covalent bonds, and the overall structure of the surfactant can be linear, branched, or crosslinked [25]. The monomer units in a polymerizable surfactant can contain both hydrophobic and hydrophilic segments. The hydrophobic segments are nonpolar and have a tendency to associate with oil, while the hydrophilic segments are polar and have a tendency to associate with water [26]. The balance of hydrophobic and hydrophilic segments in the surfactant determines its overall surface activity and its ability to stabilize emulsions, suspensions and gels.

The term graft homopolymer is associated with polymeric surfactants, as it is a type of polymer that consists of a main chain (the backbone) and a series of side chains (the branches) that are attached to the backbone through chemical bonds [27]. However, unlike graft copolymers, which are composed of two or more different types of repeating units, the side chains in a graft homopolymer are made up of the same type of repeating unit.

These homopolymers can be synthesized through various methods, such as polymerization techniques or physical methods such as melting and solidification. The size and length of the side chains can be controlled, which can affect the properties of the material, such as its thermal stability, solubility and mechanical strength [28]. Some common applications of graft homopolymers include use in the food industry as gelling agents and thickening agents, as well as in materials science, such as in the development of emulsions, surface coatings and adhesives. A generalized structure of graft homopolymers is shown in Figure 2.

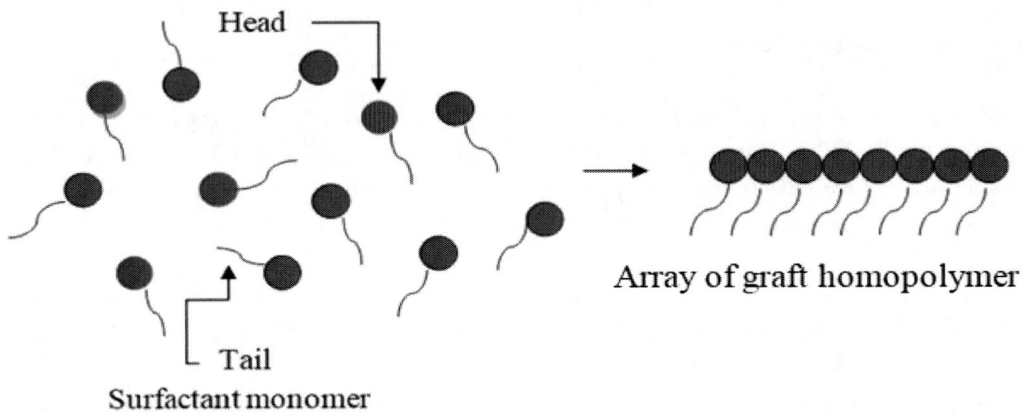

Figure 2. Structure of graft homopolymers.

The size and molecular weight of a polymeric surfactant can also play a role in its properties and applications. Lower molecular weight surfactants can be more soluble and have better emulsifying properties, while higher molecular weight surfactants can be more stable and provide better film-forming properties [29].

Maleic anhydride-based surfactants fall under the category of polymerizable surfactants. Some common examples of polymerizable surfactants also include polyvinyl alcohol, polyethylene oxide, and polyvinylpyrrolidone [30]. These polymers can be used in a range of industries, including cosmetics, pharmaceuticals and food, as well as in oil and gas, agriculture and other industrial applications [31]. Therefore, polymeric surfactants are a promising alternative to traditional surfactants and can offer a range of benefits and applications, making them a valuable addition to the field of surfactants.

3. Maleic Anhydride Surfactants: Definition and Function

The unsaturation of maleic anhydride is actuated by conjugation with the carboxyl group, due to which maleic anhydride reacts with olefins. Furthermore, it undergoes a nucleophilic addition reaction [32]. The electrophilic nature of the double bond generated by conjugation is an important aspect that leads maleic anhydride to respond to a number of reactions, such as the Diels-Alder reaction. The internal functioning of anhydride makes maleic anhydride very reactive at the sites of unsaturation, which makes maleic anhydride superior to its other anhydride derivatives, such as phthalic anhydride and succinic anhydride [33].

Due to the unique functionality of maleic anhydride (as a hydrophilic functional group) and long-chain fatty acids (as a hydrophobic part), when micelles form in aqueous solution above the CMC value, the surfactant monomers aggregate with the tails inside the micelle shielded from water and the heads at the micelle surface in contact with water, which is depicted in Figure 3 [34, 35].

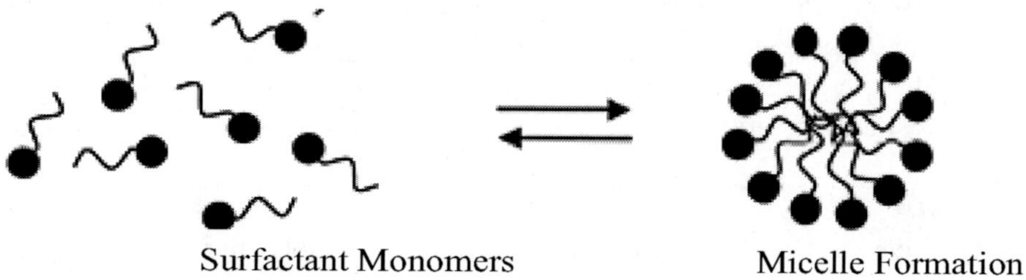

Figure 3. Micelle formation in aqueous phase.

The importance of oleochemicals as olefins is still difficult to determine completely because of the extraordinarily complex reactions in both monomeric and polymeric functionality of products that are produced from maleic anhydride. It is wise to claim that unsaturation plays a major role in various products for many industries, such as surfactants, varnishes, paints, lubricants, coatings, agricultural products, etc. [35]. Since the sites of unsaturation similar to maleic anhydride are also available in some of its derivatives, maleic anhydride derivatives such as maleic acid, esters or amides are being used for manufacturing many products [36]. These derivatives are used because of their easy availability and obtainability, but their lower reactivity empowers maleic anhydride. In addition, many factors, such as tranquillity, availability, convenience and routes for the synthesis of the end products, take precedence [37]. For this reason, anhydride-functionalized reactions could be useful since esters, amic acids or imides are produced by anhydride functionalization.

4. Raw Materials

4.1. Maleic Anhydride

The chemical names of maleic anhydride are 2,5-furandione and cis-butene-dioic anhydride. It is also known as maleic acid anhydride or malic acid anhydride. It is a white-colored solid with a hygroscopic nature [38]. It has an acrid odor. It forms an orthorhombic-crystalline-needle-like structure. Commercially, it is available in two to three forms: liquid or briquettes. The molecular weight is 98.06 g/mol, and the melting point of maleic anhydride is 53.58°C. Its boiling point is 202°C [39, 40]. In addition to its solubility in water, one last property to make special note of is its density, which is 1.48 g/cm^3. Maleic anhydride, being an anhydride, is more prone to hydrolysis reactions during storage. There are many uses for maleic anhydride, ranging from a chemical reagent in laboratories to a component in resin products. One particular use that is quite interesting is maleic anhydride's role in the modification of food and beverages. Yes, it sounds very odd to have an acid anhydride in our food and beverages, but it is commonly used as an additive [41]. Specifically, when maleic anhydride is converted to maleic and/or fumaric acid, it is a great preservative or flavor enhancer.

Therefore, for the pure maleic anhydride sample, one required operation is purification. Distillation is used in large-scale industries for purification, whereas in small-scale laboratories, sublimation or crystallization could be used [42]. It is soluble in various solvents, including benzene, toluene, alcohols, etc. Chloroform or aromatic solvents such as benzene, xylene and toluene are appropriate as crystallizing solvents. Large batches of maleic anhydride can be purified by first dehydrating a sample with solvents having a high boiling point, such as tetrachloroethane, toluene, and xylene, to facilitate azeotropic water removal. After dehydrating, these solvents are removed, and maleic anhydride is distilled under less pressure [43]. Again, the decomposition of maleic anhydride in its vapor phase is possible above 350°C.

4.1.1. Structure

Maleic anhydride (MA) is an organic compound with the molecular formula $C_4H_2O_3$. It is a white, crystalline solid that is soluble in water, alcohol and other polar solvents.

The molecular structure of maleic anhydride is based on a planar, five-membered ring with two double bonds and two single bonds. The molecule contains two carbonyl (C=O) groups and a carboxylic acid (-COOH) functional group [44]. The double bonds in maleic anhydride are conjugated, which gives the molecule its reactive properties. This reactive structure of maleic anhydride makes it useful as an intermediate in the synthesis of a wide range of organic compounds, including surfactants, polymers and other materials.

Maleic anhydride is formed by heating maleic acid, which has a different structure. However, maleic anhydride is cyclic, and a dicarboxylic group is attached to it. This conversion of malic acid to maleic anhydride is shown in Figure 4.

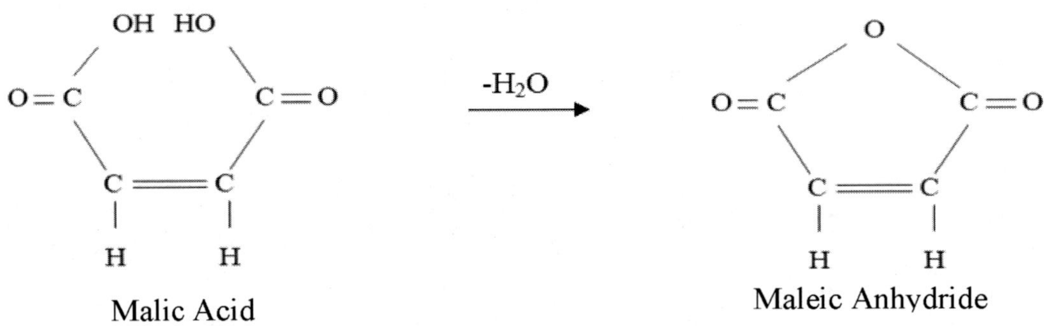

Figure 4. Formation of Maleic Anhydride.

4.1.2. Properties of MA

- Solubility: Maleic anhydride is highly soluble in water and alcohol ether, but it is insoluble in nonpolar solvents such as benzene and toluene.
- Melting and boiling points: Maleic anhydride has a melting point of 52.8°C and a boiling point of 202.5°C.
- Reactivity: Maleic anhydride is highly reactive and can undergo a variety of chemical reactions, including addition reactions with alkenes, nucleophilic addition reactions with amines and alcohols and hydrolysis reactions with water.
- Polymerization: Maleic anhydride can undergo polymerization reactions to form polymers such as polymaleic anhydride, which are used in a variety of applications.

- Toxicity: Maleic anhydride is a corrosive and irritant substance that can cause skin, eye and respiratory irritation. Dust and vapor irritate the skin, eyes, nose, throat and upper respiratory tract. It has a faint, acrid odor noticeable at 0.49 ppm [45]. It should be handled with care, and appropriate safety precautions should be taken.

4.2. Vegetable Oils and Fats

Oils and fats can be used as raw materials in the production of maleic anhydride-based surfactants. Maleic anhydride can be reacted with various types of oils and fats, such as fatty acids, glycerides and fatty alcohols, to produce surfactants with different properties and applications. The type of oil or fat used can affect the properties of the resulting surfactant, such as its hydrophilic-lipophilic balance (HLB), surface tension and stability. For example, the reaction of maleic anhydride with fatty acids can produce fatty acid esters, which can be used as surfactants in cleaning and personal care products [46]. The reaction of maleic anhydride with glycerides such as glycerol or triglycerides can produce glycerol esters, which can be used as surfactants in food and pharmaceutical applications. The reaction of maleic anhydride with fatty alcohols can produce alcohol esters, which can be used as surfactants in the textile, paper and paint industries.

It is important to note that the production of maleic anhydride-based surfactants from oils and fats can produce byproducts and waste that may be reused in other applications. It is also necessary to consider the sustainability and environmental impact of the production process when selecting oils and fats as raw materials for the production of maleic anhydride-based surfactants.

4.3. Acid-Base Catalyst

Maleic anhydride-based surfactants are typically produced through the reaction of maleic anhydride with a compound containing a nucleophilic functional group, such as an amine or an alcohol. The reaction is catalyzed by a suitable catalyst to promote the reaction. Several catalysts can be used in the production of maleic anhydride-based surfactants. One common catalyst is a strong acid, such as sulfuric acid, which can protonate the nucleophilic functional group on the reacting compound and activate it toward nucleophilic attack by the maleic anhydride. Another catalyst that can be used is a Lewis acid, such as boron trifluoride or aluminum chloride, which can coordinate with the maleic anhydride and activate it toward nucleophilic attack by the reacting compound. This type of catalyst is often used when the reacting compound is not very nucleophilic, such as in the case of alcohols.

In addition to these catalysts, other types of catalysts can be used depending on the specific reaction conditions and the desired product properties. For example, some surfactant production processes may use a base catalyst, such as sodium hydroxide, sodium bisulphite, disodium maleate, DMF (dimethyl formaldehyde), DMM (dimethylmaleate), or hydrochloric acid, to promote the reaction of maleic anhydride with an amine-containing compound. Overall, the choice of catalyst depends on the specific reaction conditions and the desired product properties, and different catalysts can be used to optimize the reaction and the surfactant properties.

5. Routes of Synthesis

5.1. Esterification Reaction

The primary addition of alcohols to maleic anhydride can be performed by the esterification reaction. In the presence of sodium alcoholate, which is a strong base, under certain conditions, maleic anhydride or its derivatives have the ability to add an alcohol group at the sites of unsaturation. The fabrication of alkoxy succinic acids is shown in R1.

Alcohol Maleic Anhydride Alkoxysuccinic Acid

R1: Reaction of MA with alcohol

This study shows that ethanol yielded almost 60% of the final product. Similarly, isobutyl alcohol and methanol can also react, but the yield would be lower compared to ethanol. The production of the final product, which is monomethyl maleate, tends to be only 19%. Dialkyl maleate reacts with excessive amounts of alkyl alcohol to produce other monomers [47]. There are many advantages to the reaction of sodium salts in the production of tri-sodium carboxymethyloxy succinate. This is produced by reacting disodium maleate with sodium glycolate, which is shown in R2 [48].

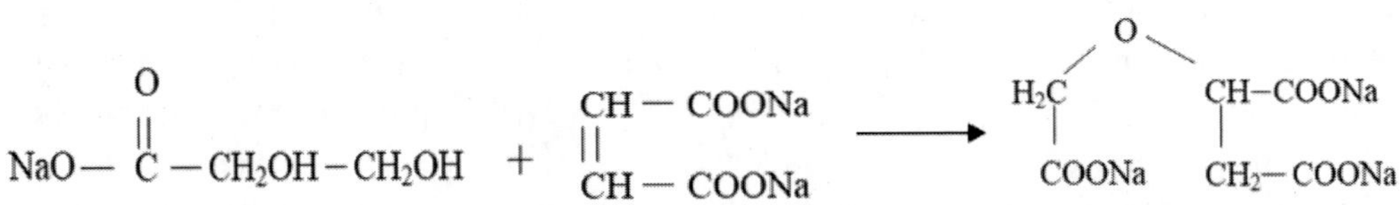

Alcohol Dialkylmaleate Carboxymethyloxy Succinic acid

R2: Formation of tri-sodium salts

This is a very important reaction, as the products so formed have a number of applications in synthetic detergents and monomers. Carboxymethyloxy succinic acid contains sequestering characteristics for calcium and magnesium ions. In the late 1960s, phosphates were being used as builders in many detergents, but later, claims were made that phosphates were not environmentally friendly, as they contributed to the eutrophication process of water bodies, which was very hazardous for marine life. Therefore, the need for new organic products that could replace phosphates resulted in a number of these esterified products. Moreover, the other anhydrides obtained from the dehydration of alcoholic reactions have been widely used in the food industry, mainly for flavoring food items, in detergents such as builders, coatings, surfactants, gasoline additives, etc. The products made from dodecyl alcohol and decyl alcohol are best used as detergents and emulsifying agents in the latex and paint industries.

5.2. Malenization of Oils & Fats

The reaction of fatty acids with maleic anhydride in the presence of an acid–base catalyst under certain controlled conditions. Many efforts are being made to organize newer surfactants for the application of detergents, dyes, etc., from this reaction. This reaction chemically modifies vegetable oils through Diels-Alder or conjugated dienes and "ene" reaction compounds of allylic hydrogens with maleic anhydride [46].

The main modifications that have been made to olefinic compounds with the aim of increasing their hydrophilicity are "ene" reactions, and many adducts can be prepared through Diels-Alder reactions of unsaturated fatty acids with maleic anhydride. These products and their derivatives are employed as dry oils, water-soluble paints and surfactants for metals; some applications of these reactions with vegetable oils are incredible in the era of novel surfactants.

Conclusion

Surfactants play a crucial role in many consumer and industrial products and have a wide range of applications due to their ability to reduce surface tension, emulsification, wetting, etc. However, it is important to consider their potential environmental impacts while choosing and using them. Maleic anhydride-based surfactants are a versatile type of surfactant that offer a range of benefits, including high solubility, low toxicity, biodegradability and much more. These properties make them an attractive option for a variety of applications, including personal care, detergent, coating, textile, plastic and industrial products. In addition, the structure also plays an important role in determining the properties and applications of a particular surfactant. The balance of hydrophobic and hydrophilic segments, the size and molecular weight of the surfactant, and the overall chemical composition play a role in determining the properties and behavior of maleic anhydride-based surfactants; therefore, more innovation is yet to be done in this area.

References

[1] Augustyn A. (2023); Surfactant, *Encyclopedia Britannica.*

[2] Dave N., Joshi T. (2017); A Concise Review on Surfactants and its Significance, *International Journal of Applied Chemistry*,13,663-672.

[3] Kissa E. (1994); Fluorinated Surfactants: Synthesis, Properties and Applications, *M. Dekker New York*,13-14.

[4] Kosarik N. (1995); Biosurfactants: Production, Properties and Applications, *M. Dekker, New York*,221-223.

[5] Wasan D. T. (1998); Surfactants in Chemical Engineering, *M. Dekker, New York*, 1994,37-40.

[6] JoÈnsson B., Lindman B., Holmberg K., Kronberg B. J.; *Surfactants and Polymers in aqueous solutions.* John Wiley &sons Ltd., 24-25.

[7] Zana R. (2002); Dimeric Gemini Surfactants: Effect of the Spacer Group on the Association Behavior in Aqueous Solution, *The Journal of Colloid and Interface Science*, 203.

[8] Gajbhiye P. G, Gogte B. B. (2011); Ecofriendly liquid detergents based on malenised vegetables oils, *International Journal of Chemical Science*,40(5), 92.

[9] Kharkate S. K., Gogte B. B. (2011), Testing of ecofriendly detergent powder and liquid detergent, *International Journal of Research in Pharmacy & Chemistry*,1(1).

[10] Irja, Piirma (1992). Polymeric Surfactants, *Marcel Dekker Inc, New York*,42,230.

[11] Pravin A. Dhakite, Gogte B. B., Phate B. W. (2011); Ecofriendly liquid detergents based on malenised vegetables oils, *The Asian Journal of Experimental Chemistry*, 6 (1), 35-37.

[12] Pravin A. Dhakite, Gogte B. B., Phate B. W. (2013); Malenised vegetables oils - a substitue for linear alkyl benzene sulphonate and alpha olefienic sulponate for ecofriendly powder detergents, *ElixirApplied Chemistry*,13274-13277.

[13] Irja, Piirma (1992), *Polymeric Surfactants, Surfactant Series*, Marcel Dekker Inc, New York,

[14] Garrelt H. E. (1972); Surface active chemicals programmer press, *Marcel Dekker Inc, New York.*

[15] Nabel A. Negm, Sayed A. Ahmed, Ali A. Abd-Elaal, Tarek A. (2016); Synthesis and Surface Activity of Nonionic Surfactants Derived from Gallic Acid, *Arabian Journal for Science and Engineering*, 41,67-73.

[16] Seddon E. J., Friend C. L., Roski J. P. (2010): *Detergents and Dispersants chemistry and Technology of Lubricants,* Springer, 3, 213–236.

[17] Infante M., Pinazo A., Seguer J. (1997); Nonconventional surfactants from amino acids and glycolipids: structure, preparation and properties. *Colloid. Surf. A. Physicochem. Eng. Asp.*, 123(124), 49–70.

[18] Rondel C., Portet B., Alric I., Mouloungui Z., Blanco J. F., Silvestre F. (2011); Green production of anionic surfactant obtained from pea protein. *J. Surfact. Deterg.,* 14, 535–544.

[19] Adewuyi A., Oderinde R. A., Rao B. V., Prasad R. B. (2012): Synthesis of alkanolamide: a nonionic surfactant from the oil of Gliricidia sepium. *J. Surfact. Deterg.,* 15, 89–96.

[20] Leal J. S., Gonzalez J. J., Kaiser K. L., Palabrica V. S., Comelles F., Garcıa M. T. (1994); On the toxicity and biodegradation of cationic surfactants. *Acta Hydrochim. Hydrobiol.,* 22, 13–18.

[21] Do L. D., Withayyapayanon A., Harwell J. H., Sabatini D. A. (2009); Environmentally friendly vegetable oil microemulsions using extended surfactants and linkers. *J. Surfact. Deterg.,* 12, 91—99.

[22] Negm N. A., El Farargy A. F. M., Mohammed D. E., Mohamad H. N. (2012): Environmentally friendly nonionic surfactants derived from tannic acid: synthesis, characterization and surface activity, *J. Surf. Deterg.,*2012, 15, 433–443.

[23] Andondonskaya B. L., Ezrielev A. T., Roskin E. S., Vysokomol, Soedin B. (1971), *Maleic anhydride bookmatter,* 9(13), 648.

[24] Kato M., Nakano Y. (1972), Influence of crystallization temperature, molecular weight, and additives on the crystallization kinetics of poly(ethylene terephthalate), *J. Polym. Sci.*, 10, 157.

[25] Trivedi B., Shulyndin S. V., Sanatullin K. G., Gubanov E. F., Ivanov B. E., *Vysokomol. Soedin.* (1973), *Chern. Abstr*, 14(9), 664.

[26] Trivedi B., Shulyndin S. V., Sanatullin K. G., Gubanov E. F., Ivanov B. E., (1973); *Chern. Abstr.*, 4 0,15668.

[27] Negm N. A., El-Tabl A. S., Aiad I. A., Zakareya K., Moustafa A. H. (2013); Synthesis, characterization, biodegradation and evaluation of the surface active properties of nonionic surfactants derived from Jatropha oil., *J. Surfact. Deterg,* 16, 857–863.

[28] Naylor C. G., Williams J. B., Varineau P. T., Yunick R. P., Serak K., Cady C., Severn D. J. (1991); *19th Annual Meeting of the Society of Environmental Toxicology and Chemistry.*

[29] Negm N. A., Tawfik S. M. (2013); Isoxazolium cationic schiff base surfactants: surface and antimicrobial properties. *Chimica Oggi/Chemistry Today,*5–8.

[30] Syed N. A., Mad I., Helmy S. M., Azab W. I. M., (2003); Aqueous solution properties, biodegradability, and antimicrobial activity of some alkyl polyglycosides surfactants. *Tenside Surf Deterg.*, 46,311–316.

[31] Azum N., Abdul-Rub M., Asiri A. M., Khan A. P., Khan A. (2013), Physico-chemical investigations of mixed micelles of cationic Gemini and conventional surfactants: a conductometric study. *J Surf Deterg*., 16,77–84.

[32] Griffin W. C. (1949); Classification of surface-active agents by HLB. *J Soc Cosm Chem*., 1,311.

[33] Vaughan C. D., Dennis A. (1990); Predicting O/W emulsion stability by the required HLB equation, *J Dispers Sci Technol.,*11,83–91.

[34] Negm N. A., Aiad I. A. (2007); Synthesis and characterization of multifunctional surfactants in oil-field protection applications, *J. Surfact. Deterg*., 10, 87–92.

[35] Dianno D., Talmon Y., Zana R. (1995); Alkanediyl.-, Alph.-, Omega-bis(dimethyl ammonium bromide) surfactants (dimeric surfactants), *Langmuir,* 11,1448–1453.

[36] Negm N. A., Mahmoud S. A. (2005): Synthesis, surface and thermodynamic properties of carboxymethyl (di-2-ethanol) ethyl alkanoate ammonium bromide amphoteric surfactants, Egypt. *J. Petrol.,* 14, 79–86.

[37] John A. M., Christopher J. M., Ingrid K. M. (2004); Interactions of surfactants with a derivatized low molecular weight styrene maleic anhydride copolymer-differences between acetylenic diol-based wetting agents and other ethoxylates, *Colloids and Surfaces A:Physicochem. Eng. Aspects,* 234, 63–69.

[38] McCutcheon's (2002), *Emulsifiers and Detergents,* vol. 1. MC Publishing Co., Glen Rock, NJ.

[39] Zhang C., Valsaraj K. T., Constant W. D., Roy D. (1998); Surfactant screening for soil washing, *J. Environ. Sci. Health*, 1249–1273.

[40] Nelson R. D. (1997); Dispersing powders in liquids: *Handbook of Powder Technology,* Elsevier, 7.

[41] Sato S., Kishimoto H. (1985); Interactions between dyes and surfactants in inkjet ink used for textiles, *Bull. Chem. Soc. Jpn*., 58, 282–287.

[42] Wang S., Zeng Z., Yang S., Weng S., Philip C. L. (2000); Controlled Sulfonation of Poly(N-vinylcarbazole). *Macromolecules*, *33* (9),3232-3236.

[43] Osama M. M. (*2016), Handbook of maleic anhydride based materials: Syntheses, properties and applications,* 151-184.

[44] Anna M. P., Alberto D., Dinuccio D., Federica C. (2011); 2-Methoxy Aniline Grafted Poly(maleic anhydride-alt-butyl vinyl ether) Hemiester: A New Biocompatible Polymeric Free Radical Scavenger, *Macromolecules,*44, 848–856.

[45] Chiellini F., Piras A., Gazzarri M., Bartoli C., Ferri M., Paolini L., Chiellini E. (2011); Bioactive polymeric materials for targeted administration of active agents: synthesis and evaluation, *Macromol Biosci*.,8,516–525.

[46] Tonge S. (2009); *Compositions comprising a lipid and copolymer of styrene and maleic acid,* US 8,623,414 B2.

[47] Bertz S., Miksza F., Zucker E. (2001); *High purity adduct of castor oil and maleic anhydride,* US6,225,485 B1.

[48] Coats C. (1996); *Electrocoating compositions comprising of aromatic amine amidated drying oil copolymer-maleic anhydride adducts,* US 3,428,589.

Chapter 15

Nitrogen-Doped Porous Carbon Electrodes with Ionic Gel Electrolytes for Super Capacitive Energy Storage

Mercy Jacquline B.[1]
Ameelia Roseline A.[2]
and Kalaiselvam S.[1,3,4,*]
[1]Department of Applied Science and Technology, Anna University, Chennai, India
[2]Department of Electronics and Communication Engineering, Panimalar Engineering College, Chennai, India
[3]Centre for Industrial Safety, Anna University, Chennai, India
[4]Department of Mechanical Engineering, Anna University, Chennai, India

Abstract

With the diminishing range of fossil fuel resources, there is an urgent need for sustainable alternative resources, so researchers are focused on developing sustainable energy resources from a renewable source; developing energy from waste is more advantageous. This work focuses on using biomass waste resources to make porous carbon electrodes with efficient ion-conducting electrolytes for supercapacitor applications. The synthesis methodology involves simple steps: the *Artocarpus hirsutus* dried leaves were activated using phosphoric acid with a 1:1 impregnation ratio, followed by carbonization at 650°C and then pyrolysis at 850°C in a tubular furnace in a nitrogen atmosphere; the porous carbon was finally mixed in melamine and heated for nitrogen doping in this process. The polymer gel electrolyte is made of a mixture of neutral salt and polymer, i.e., KOH and cationic polyacrylamide. The electrochemical performance was analyzed using a three-electrode system; this gave a high specific capacitance value of 353.3 F g^{-1} at a current density of 0.5 A g^{-1}, an energy density of 24.54 Wh kg^{-1} and a power density of 36 kW kg^{-1}. Thus, a dual benefit of waste being used as feedstock for energy and waste management is possible.

Keywords: renewable source, biomass, electrode, electrolyte, supercapacitor

---

[*] Corresponding Author's Email: kalai@annauniv.edu.

In: Sustainability in Chemical Processes through Digitalization and Green Chemistry Approaches
Editors: Susarla V. A. R. Sastry, Shravan Kumar and Ashish Kapoor
ISBN: 979-8-89113-257-3

Nomenclature

EDLC	Electric Double Layer Capacitors
AC	Activated carbon
N-AHPAC	Nitrogen doped *Artocarpus hirsutus* leaves based activated carbon
SC	Supercapacitor
AH	*Artocarpus hirsutus*
SEM	Scanning Electron Microscope
XRD	X-ray diffraction
D	Defect band
G	Graphite band
I_D/I_G	Intensity ratio of D and G band
FT-IR	Fourier Transform Infrared Spectroscopy
BET	Brunauer–Emmett–Teller
CV	Cyclic Voltammetry
GCD	Galvanostatic Charge/Discharge
EIS	Electrochemical Impedance Spectroscopy
C_{sp}	Specific capacitance of the N-AHPAC sample (F g^{-1})
v_s	Scan rate (V s^{-1})
I	Current (A)
m	Mass of the active material (g)
ΔV	Applied potential window (V)
E_g	Energy density (Whkg^{-1})
P_g	Power density (W kg^{-1})
R_s	Series resistance (Ω)
R_{ct}	Charge transfer resistance (Ω)
W	Warburg resistance (Ω)

1. Introduction

With rapid urbanization, industrialization and population growth, there is an enormous demand for clean, sustainable and pollution-free energy resources [1]. Super capacitors are considered the best alternate energy storage device for the current situation due to their high-power density, fast charging and discharging, and long cycle life; they are widely used in electric vehicles, power systems and other portable electronic devices [2]. Based on the storage mechanism, supercapacitors are classified into electric double layer capacitors (EDLCs), pseudocapacitors, and hybrid capacitors. EDLC is an electric energy storage system based on a charge–discharge process in an electric double layer on porous electrodes [3]. An "electric double layer" is created when one of the electrodes receives a potential and the number of ions from the opposing signal that gather on its surface increases proportionally to the applied voltage [4], [5]. Since pure EDLC is nonfaradaic, energy storage is entirely electrostatic, and neither charge nor mass transfer occurs at the electrode-electrolyte contact during charging or discharging [6], [7].

Electrostatic interaction poses no threat to the electrode's stability or integrity; hence, EDLC may carry through 100,000 charge–discharge cycles with a 10% degradation rate in organic electrolytes [8], [9], [10]. They are advantageous due to fast charge/discharge, long cycle life because no chemical reactions occur; only adsorption and desorption of electrolyte ions occur; high efficiency for charge/discharge cycles; no heavy metals are used, thereby being environmentally friendly; it is possible to discharge at high current density; and they work effectively over a wide range of temperatures. Literature studies have shown that supercapacitor electrodes from carbonaceous materials have exceptional properties in terms of surface area, porosity, electrical conductivity and surface functionalities. To achieve both economic and sustainability goals, efforts have been made to produce supercapacitor electrodes from biomass-derived activated carbons. Activated carbons (AC) are generally renowned for their high surface area, large microporosity, excellent conductance, low cost and adsorption capacity, which allow them to be employed as electrode materials. The synthesis process of AC generally involves two phases. In the first stage, the precursors are carbonized at a relatively low temperature to synthesize the charcoal, and in the next step, the black-colored char is activated at a high temperature. Activation may be performed by a physical or chemical method. Physical activation encompasses the precursors to be activated in the presence of steam or CO_2 at high temperatures. Chemical activation encompasses the precursor's impregnation with different materials, for example, NaOH, KOH, $ZnCl_2$ and H_3PO_4 in an inert atmosphere. The presence of heteroatoms such as nitrogen (N), phosphorous (P), sulfur (S), and oxygen (O) has been proven to improve electrochemical performance to some extent. Therefore, heteroatom doping promotes the hydrophilicity and pseudocapacitance of carbon materials [11], [12]. Researchers are focused on fabricating and developing carbon materials with interconnected pores of high porosity decorated with heteroatoms to improve the hydrophilicity and pseudocapacitance of the carbon materials. The electrolyte also affects the energy storage performance; the aqueous electrolyte has high ionic conductivity but limitations of easy leakage and decomposition voltage, so an obviously better choice will be a polymer gel electrolyte [13].

In this work, AC materials with interconnected and evenly dispersed micro, meso- and macropores were produced by phosphoric acid (H_3PO_4) activation. The practical performance of the supercapacitor (SC) device was studied using a three-electrode system in a polymer gel electrolyte, and the electrochemical performance was examined.

2. Materials and Methods

2.1. Preparation of Porous Activated Carbon

The dried waste leaves of *Artocarpus hirsutus*(AH) were collected from the southern part of India. The leaves were completely washed in distilled water to remove contaminants and sand particles. They were then sun-dried for a few days, followed by oven drying for complete removal of moisture. The moisture-free leaves were size-reduced and chemically activated using H_3PO_4 as an activating agent with an impregnation ratio of 1:1 [14]. The mixture was subjected to a carbonization temperature of 650°C for 1 h in a box furnace, followed by 850°C for 1 h in a tubular furnace in a nitrogen atmosphere. The obtained porous carbon was washed using double-distilled water to attain a neutral pH of 7. For nitrogen doping, melamine [15] of equal weight was mixed with the obtained carbon and heated for a few hours to obtain nitrogen-doped *Artocarpus hirsutus*leaf-based porous activated carbon, which was named N-AHPAC.

2.2. Material Characterization

To analyze the morphological aspect of the N-AHPAC sample, scanning electron microscopy (SEM) analysis was carried out. The crystal structure and planes of the N-AHPAC sample were analyzed using X-ray diffraction (XRD). Raman spectra of sample N-AHPAC were taken, with spectra recorded ranging from 4000 to 400 cm^{-1}. The functional groups of the N-AHPAC sample were analyzed using Fourier transform infrared spectroscopy (FT-IR). The surface area of the N-AHPAC sample was analyzed using the Brunauer–Emmett–Teller (BET) method. Electrochemical characterization was performed using a three-electrode system.

2.3. Preparation of the Electrode and Electrolyte

The N-AHPAC was mixed with the ethanol solvent and Nafion binder [16] to form an electrode slurry, which was finally coated onto the glassy carbon electrode. The electrolyte was a mixture of the strong base KOH, neutral salt KCl and polyacrylamide, thus forming a gel-type electrolyte.

3. Results and Discussion

3.1. SEM Morphological Analysis

The surface morphology of the N-AHPAC sample contains meso- and micropores, proving the influence of the activating agent (H_3PO_4) and activating temperature (650°C/850°C), which facilitated the formation of pores on the surfaces of the original pores, thereby disrupting the internal structure of the material. The SEM image of N-AHPAC is shown in Figure 1. The lignin and cellulose inside the N-AHPAC sample were hydrolyzed by H_3PO_4 during impregnation, and the surface lignin and cellulose groups were polymerized by H_3PO_4 during activation [17]. The regular pores favor the adsorption of the electrolyte onto the surface of the electrode for better ion transfer and conductivity.

3.2. XRD Structural Analysis

The XRD patterns of the N-AHPAC sample exhibit broad peaks that prove that the sample is more amorphous in nature [18]. The X-ray diffraction pattern peaks were observed at 2θ = 25.5° and 43.6°, corresponding to the (002) and (101) planes, respectively, as shown in Figure 2 (b).

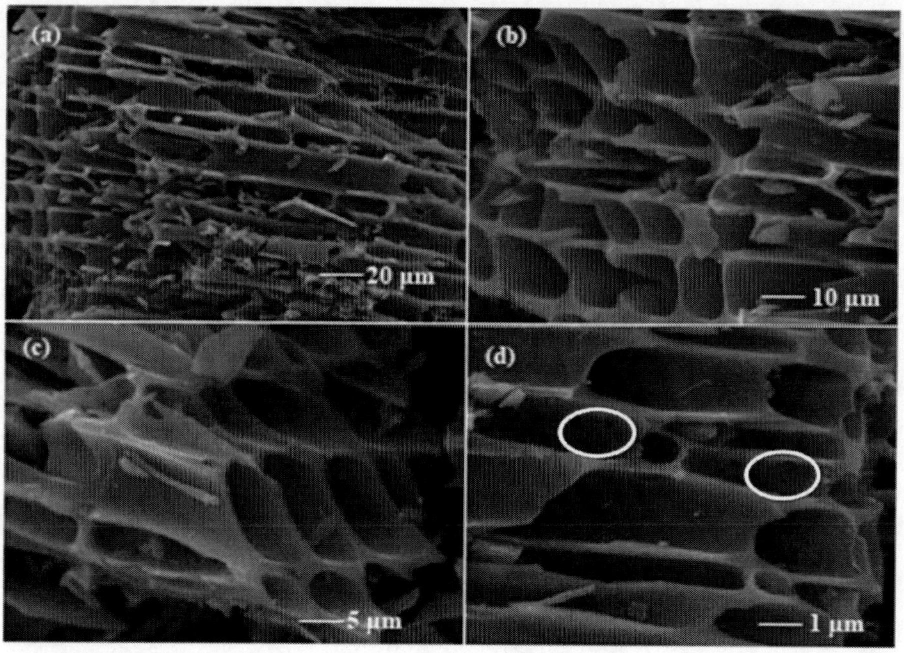

Figure 1. SEM image of N-AHPAC.

3.3. Raman Analysis

It is observed in Figure 2(c) that the Raman spectra show two sharp peaks at 1397 cm^{-1} that correspond to the D (defect) band and 1667 cm^{-1} that correspond to the G (graphite) band [19]. The vibrational modes of these two bands can be attributed to domains of disordered microcrystalline carbon atoms that are sp^2 bonded. The G band, in particular, can be attributed to the stretching of bonds between pairs of sp^2 carbon atoms (either in aromatic chains or rings). The I_D/I_G intensity ratio is typically used to calculate the degree of graphitization or disorder; this is attributed to a degree of graphitization of approximately 0.84, which shows the presence of a low graphitization degree or highly disordered arrangement in the N-AHPAC sample [20]. The peak arising in the range of 1620–1690 cm^{-1} confirms the presence of amide with a functional group of =C=O with NH deformation, and the peak near 1397 cm^{-1} confirms the CH_3 symmetric deformation.

3.4. FTIR Functional Group Analysis

FTIR spectra were recorded in a Bruker ALPHA spectrometer using a KBr pellet technique from 400 to 4000 cm^{-1}. The functional group of the sample, N-AHPAC, gives a peak at 3742.9 cm^{-1}, indicating the presence of O-H groups. The peak at 3384.7 cm^{-1} is due to stretching vibrations of hydrogen-bonded groups such ascarboxyls, phenols or alcohols. The peaks at 2923 cm^{-1} and 2853 cm^{-1} indicated C-H stretching; the 1570.8 cm^{-1} peak corresponds to C-C vibrations in aromatic rings; and the 1322.8 cm^{-1} to 1093.1 cm^{-1} peaks indicate oxidized carbon due to C-O stretching in acids, alcohols, phenols, ethers and/or ester groups, as shown in Figure 2 (a). The peak at 809.2 cm^{-1} is due to out-of-plane C-H deformation, and 525.7 cm^{-1} indicates the disulfide stretch of C-S, S=O and S-O due to surface groups HSO_4 and SO_4. The N-H stretching falls between 3500 cm^{-1} and 3100 cm^{-1}, and the C-N stretching falls between 1400 cm^{-1} and 1000 cm^{-1} [21].

3.5. BET Surface Area Analysis

The N_2 adsorption/desorption isotherm curves of N-AHPAC exhibit the characteristic mesoporous and microporous nature of the sample with a specific surface area of 320 $m^2\,g^{-1}$, as observed in Figure 2 (b, c, and d). Both the activation temperature and impregnation ratio have ideal values because the activation process has two components: the formation of pores and their expansion. The initial stage will see an increase in activation temperature and impregnation ratio, which will encourage the formation of pores. However, an excessive activation temperature or impregnation ratio will cause excessive collapse or expansion of pores, slowing down the formation of pores and reducing the specific surface area [22].

3.6. Electrochemical Performance Analysis

The electrochemical characterization was carried out using a three-electrode system, with N-AHPAC as the electrode material, Nafionas a binder and aqueous gel polymer as the electrolyte.

3.7. Cyclic Voltammetry Analysis

The cyclic voltammetry (CV) measurements of the N-AHPAC electrode are carried out with a scan rate ranging from 50 mV s^{-1} to 10 mV s^{-1} at a potential range of 0 to 1 V. The typical curve proves EDLC-type material; the shape of the curve does not change with change in scan rate, proving the stability of the fabricated electrode; no redox peaks are observed, proving the purity of the material. The specific capacitance (C_{sp}) is calculated from Eq. (1).

$$C_{sp} = \frac{1}{m \times v_S \times \Delta V} \int_{V1}^{V2} I dV \tag{1}$$

where C_{sp} of the N-AHPAC sample is measured in F g^{-1}, v_s is the scan rate (V s^{-1}), I is the current (A cm^2) for the applied scan rate, m is the mass of the active material (g) and ΔV is the applied potential window (V) [23]. This gave a high specific capacitance value of 353.3 F g^{-1} at a current density of 0.5 A g^{-1}.

3.8. Galvanostatic Charge/Discharge Analysis

The GCD curve forms an approximately triangular shape with a good charge–discharge rate; the current density of the curves was maintained from 0.5 A g^{-1} to 2.5 A g^{-1}in the potential range between 0 and 1.0 V. The energy density (E_g) was calculated from Eq. (2).

$$E_g = \frac{1}{2}\frac{C_{sp}}{4} \Delta V^2 \frac{1}{3.6} \tag{2}$$

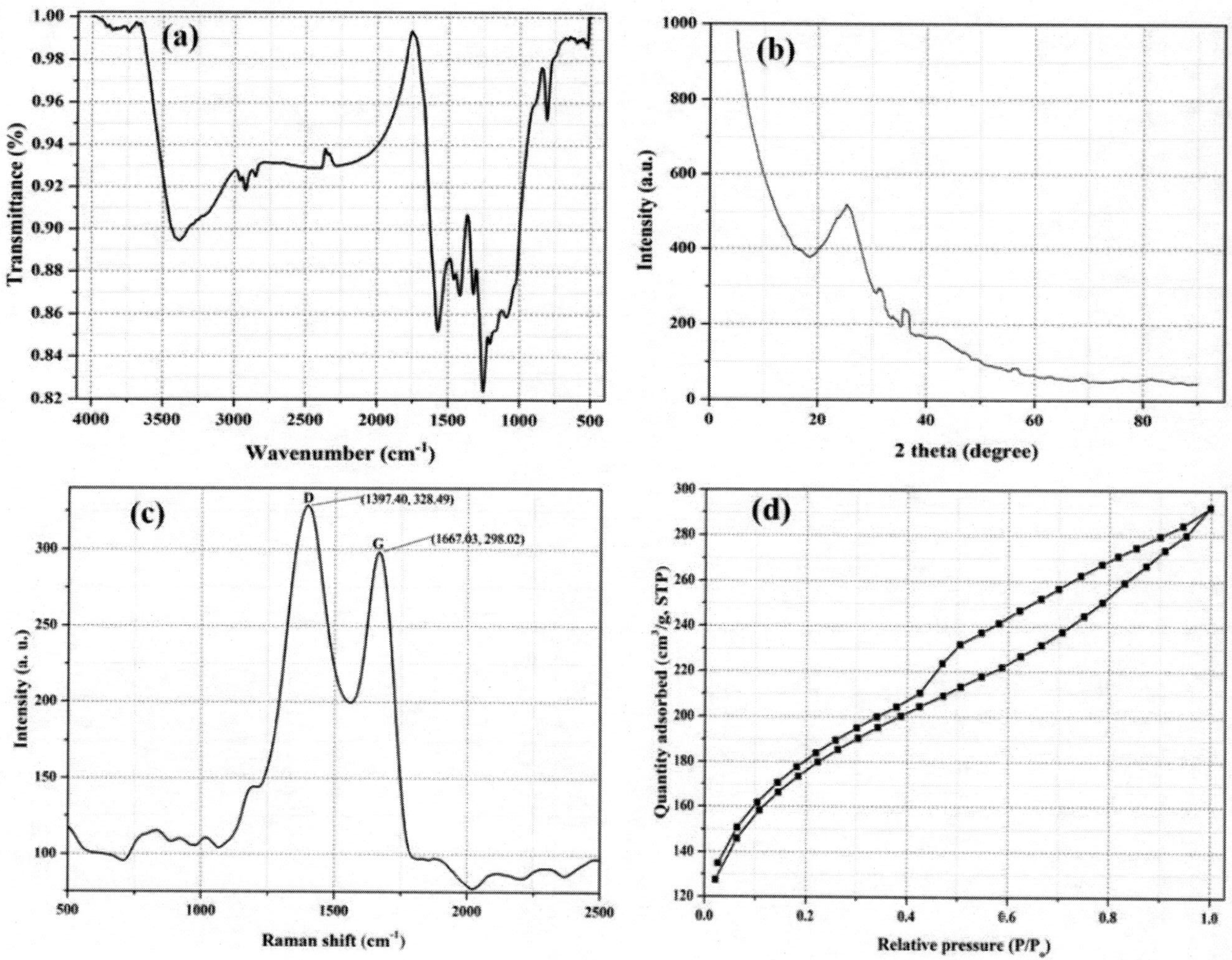

Figure 2. Characterization with (a) FTIR (b) XRD (c) Raman spectra and (d) BET for N-AHPAC.

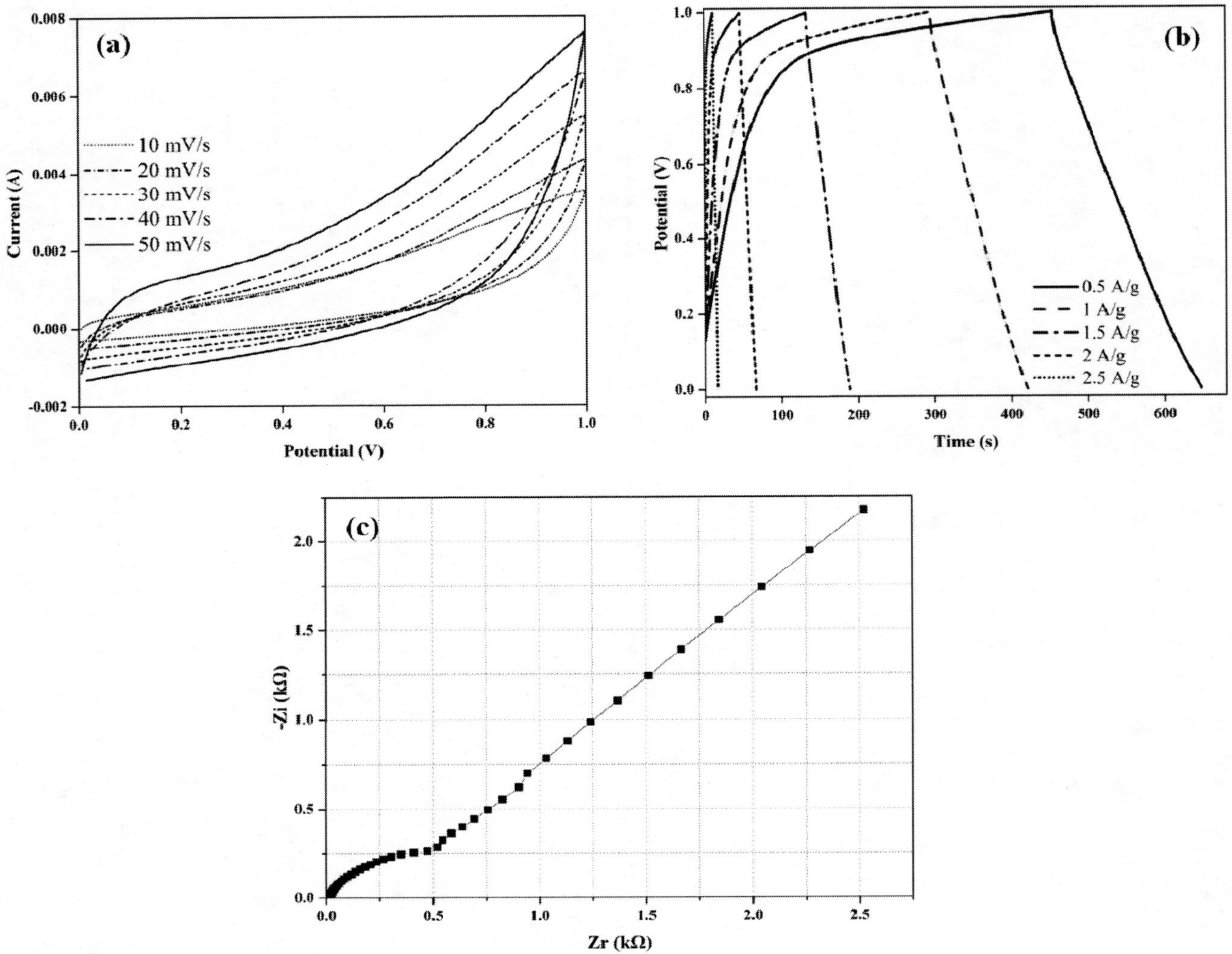

Figure 3. Electrochemical characterization of N-AHPAC sample (a) CV (b) GCD and (c) EIS.

Power density P_g is calculated from Eq. (3)

$$P_g = \frac{E_g}{t} \tag{3}$$

where t is the discharge time (h) (energy density of 24.54 Wh kg^{-1} and power density of 36 kW kg^{-1} for the N-AHPAC sample).

3.9. Electrochemical Impedance Spectroscopy

Electrochemical impedance spectra (EIS) measurements were carried out to investigate the ionic diffusion and transport kinetics of electrolytes in the N-AHPAC sample [24]. The Nyquist plots of the N-AHPAC systems measured at an open circuit voltage (frequency range: 0.01 Hz–100 kHz) display two principal parts. The intercept value on the horizontal axis and the size of the semicircle at high frequency refer to the equivalent series resistance (R_s) and charge transfer resistance (R_{ct}), respectively. The straight line with a 45° slope in the middle frequency corresponds to the Warburg resistance (W), which is related to the diffusion of electrolyte ions into the pores of the electrode material. From the graph, the semicircles represent the charge transfer resistance of the systems, and it was suggested that the ion diffusion resistance was high due to its poor crystalline nature [25].

Electrochemical tests of the three-electrode system show that N-AHPAC possesses excellent capacitive performance based on its outstanding hierarchical porous structure, suitable graphitization degree and trace amount of heteroatom doping. Figure 3 (a, b and c) shows theelectrochemical characterization of the N-AHPAC sample with CV, GCD and EIS, respectively.

4. Conclusion

This work provides an efficient route for synthesizing AC-based electrodes and polymer gel electrolytes for high-performance supercapacitors. N-AHPAC with an interconnected hierarchical structure was successfully synthesized by H_3PO_4 activation, carbonization, and nitrogen doping using melamine. The material exhibited favorable features such as pore internal connectivity and pore size distribution (micro, meso and macropores). In a three-electrode system with the prepared ionic polymer gel as the electrolyte and N-AHPAC as the electrode, a high specific capacitance of 353.3 F g^{-1} was observed at a current density of 0.5 A g^{-1}, an energy density of 24.54 Wh kg^{-1} and a power density of 36 kW kg^{-1}. This shows that the N-AHPAC electrode material has superior ion adsorptivity and that the polymer gel electrolyte has enhanced ion conductivity.

Funding Details

The authors gratefully acknowledge the Science and Engineering Research Board (SERB), a statutory body of the Department of Science and Technology (DST), Government of India, for supporting the research work under EEQ (SERB Sanction Order No. EEQ/2018/000694). Furthermore, the authors are grateful to Department of Science and Technology (DST) for sustainable support for the research work under FIST Sanction Order No. SR/FST/ET-1/2017/44.

References

[1] Vinayagam, M., Babu, R. S., Sivasamy, A., and de Barros, A. L. F. (2020). Biomass-derived porous activated carbon from Syzygiumcumini fruit shells and Chrysopogonzizanioides roots for high-energy density symmetric supercapacitors. *Biomass and Bioenergy*, 143, 105838.

[2] Zhang, L., Wang, Y., Yang, S., Zhao, G., Han, L., Li, Y., and Zhu, G. (2022). Biomass-derived S, P, Cl tri-doped porous carbon for high-performance supercapacitor. *Diamond and Related Materials*, 126, 109061.

[3] Muthulakshmi, G., Mohamed Ismail, M., Arivanandhan, M., and Bhaskaran, A. (2022). Effect of calcination on structural, morphological and electrochemical properties of SnO2 nanoparticles. *Journal of Materials Science: Materials in Electronics*, 33(8), 5534-5551.

[4] Cai, N., Cheng, H., Jin, H., Liu, H., Zhang, P., and Wang, M. (2020). Porous carbon derived from cashew nut husk biomass waste for high-performance supercapacitors. *Journal of Electroanalytical Chemistry*, 861, 113933.

[5] Gao, Y., Zhang, Y., Huang, H., Deng, C., and Cheng, Y. (2022). Low-cost carbon derived from coal-coke for high-performance supercapacitors. *Journal of Electroanalytical Chemistry*, 921, 116678.

[6] Wang, B., Wu, X., Yu, Y., Wang, N., and Zhou, Z. (2022). Simultaneously, tuning the hierarchical porous structure and graphitization degree of biomass derived carbon for supercapacitors. *Electrochimica Acta*, 432, 141219.

[7] Chen, Z., Peng, X., Zhang, X., Jing, S., Zhong, L., and Sun, R. (2017). Facile synthesis of cellulose-based carbon with tunable N content for potential supercapacitor application. *Carbohydrate polymers*, 170, 107-116.

[8] Jiang, G., Senthil, R. A., Sun, Y., Kumar, T. R., and Pan, J. (2022). Recent progress on porous carbon and its derivatives from plants as advanced electrode materials for supercapacitors. *Journal of Power Sources*, 520, 230886.

[9] Bi, R., Pang, S. K., Yung, K. C., and Yin, L. K. (2022). Comprehensive study of used cigarette filters-derived porous activated carbon for Supercapacitors: From biomass waste to sustainable energy source. *Journal of Electroanalytical Chemistry*, 925, 116915.

[10] Luo, L., Lan, Y., Zhang, Q., Deng, J., Luo, L., Zeng, Q.,and Zhao, W. (2022). A review on biomass-derived activated carbon as electrode materials for energy storage supercapacitors. *Journal of Energy Storage*, 55, 105839.

[11] Du, W., Zhang, Z., Du, L., Fan, X., Shen, Z., Ren, X., and Wei, S. (2019). Designing synthesis of porous biomass carbon from wheat straw and the functionalizing application in flexible, all-solid-state supercapacitors. *Journal of Alloys and Compounds*, 797, 1031-1040.

[12] Yeletsky, P. M., Lebedeva, M. V., and Yakovlev, V. A. (2022). Today's progress in the synthesis of porous carbons from biomass and their application for organic electrolyte and ionic liquid based supercapacitors. *Journal of Energy Storage*, 50, 104225.

[13] Xu, X., Allah, A. E., Wang, C., Tan, H., Farghali, A. A., Khedr, M. H.,and Yamauchi, Y. (2019). Capacitive deionization using nitrogen-doped mesostructured carbons for highly efficient brackish water desalination. *Chemical Engineering Journal*, 362, 887-896.

[14] Sylla, N. F., Ndiaye, N. M., Ngom, B. D., Mutuma, B. K., Momodu, D., Chaker, M., and Manyala, N. (2020). Ex situ nitrogen-doped porous carbons as electrode materials for high performance supercapacitor. *Journal of colloid and interface science*, 569, 332-345.

[15] Dai, H., Zhang, G., Rawach, D., Fu, C., Wang, C., Liu, X.,and Sun, S. (2021). Polymer gel electrolytes for flexible supercapacitors: Recent progress, challenges, and perspectives. *Energy Storage Materials*, 34, 320-355.

[16] Mathangi, J. B., Sharma, M. S., Jacquline, B. M., and Kalavathy, M. H. (2018). Development of carbon-based material from biomass for the removal of Ni2+ and CO2 from fluid phase. *Vacuum*, 158, 236-248.

[17] Sun, J., Zhang, J., Shang, M., Zhang, M., Zhao, X., Liu, S., and Yi, X. (2023). N, O codoped carbon aerogel derived from sodium alginate/melamine composite for all-solid-state supercapacitor. *Applied Surface Science*, 608, 155109.

[18] Hossain, R., Nekouei, R. K., Mansuri, I., and Sahajwalla, V. (2021). *In situ* O/N-heteroatom enriched activated carbon by sustainable thermal transformation of waste coffee grounds for supercapacitor material. *Journal of Energy Storage*, 33, 102113.

[19] Zeng, F., Meng, Z., Xu, Z., Xu, J., Shi, W., Wang, H.,and Tian, H. (2023). Biomass-derived porous activated carbon for ultrahigh performance supercapacitor applications and high flux removal of pollutants from water. *Ceramics International*.

[20] Liu, Y., Liu, P., Li, L., Wang, S., Pan, Z., Song, C., and Wang, T. (2021). Fabrication of biomass-derived activated carbon with interconnected hierarchical architecture via H3PO4-assisted KOH activation for high-performance symmetrical supercapacitors. *Journal of Electroanalytical Chemistry*, 903, 115828.

[21] Xue, B., Xu, J., and Xiao, R. (2023). Ice template-assisting activation strategy to prepare biomass-derived porous carbon cages for high-performance Zn-ion hybrid supercapacitors. *Chemical Engineering Journal*, 454, 140192.

[22] Ahmed, S., Ahmed, A., and Rafat, M. (2019). Investigation on activated carbon derived from biomass Butneamonosperma and its application as a high performance supercapacitor electrode. *Journal of Energy Storage*, 26, 100988.

[23] Li, D., Huang, Y., Yu, C., Lu, Y., Tang, C., and Lin, J. (2022). Pore structure regulation of hierarchical porous agaric-derived carbon via boric acid activation for supercapacitors. *Diamond and Related Materials*, 130, 109432.

[24] Zhou, H., Shu, R., Guo, F., Bai, J., Zhan, Y., Chen, Y., and Qian, L. (2021). NOP codoped porous carbon aerogel derived from low-cost biomass as electrode material for high-performance supercapacitors. *Diamond and Related Materials*, 120, 108614.

[25] Yan, B., Feng, L., Zheng, J., Zhang, Q., Dong, Y., Ding, Y., and He, S. (2023). Nitrogen-doped carbon layer on cellulose derived free-standing carbon paper for high-rate supercapacitors. *Applied Surface Science*, 608, 155144.

Chapter 16

The Design and Development of Molten Salt Batteries

Aswin Muraleedharan[*]
and Jyoti Sahu
Department of Chemical Engineering, National Institute of Technology, Tiruchirappalli, Tamil Nadu, India

Abstract

Energy storage systems have been of great interest since the advent of the industrial revolution. Energy storage systems are defined by the amount of power and energy they can store. It basically stores energy and uses it later. The presence of an energy storage device in the system reduces the imbalance between power demand and production. A large variety of energy storage systems are available in the industrial sector, which includes energy discharge on both short-term and long-term scales.

Research regarding a novel molten battery has been in progress for the past few years. Research regarding state-of-the-art *lithium-ion batteries* has been ongoing for even longer. The replacement of the metal lithium with a more abundant metal such as Sodium has been under discussion for the past two decades. In comparison to its counterpart, *the lithium-Sulphur battery*, the *Sodium-Sulphur* battery has seen very little exposure to the world. The construction of this battery involves the use of Sodium and Sulphur as the electrodes and a solid electrolyte. The *electrical circuit model* and *electrochemical model* are considered for modeling. The parameters analyzed using the model are the state *of charge, voltage* and *current* generated by the battery, which are then compared to its more successful counterpart, the *lithium-Sulphur* battery. The purpose of this research is to model a Sodium Sulphur battery that can be used to store large amounts of energy by combining multiple cells and to reduce the operating temperature of the original model.

Keywords: energy storage systems, Sodium-Sulphur battery, lithium-ion battery, electrical circuit model

1. Introduction

Developing the technology to store energy so that it is available when needed to meet demand represents a breakthrough in energy distribution around the world. Energy management

[*] Corresponding Author's Email: aswinmuraly@gmail.com.

In: Sustainability in Chemical Processes through Digitalization and Green Chemistry Approaches
Editors: Susarla V. A. R. Sastry, Shravan Kumar and Ashish Kapoor
ISBN: 979-8-89113-257-3

systems provide tremendous support in supplying energy not only to our homes but also to large consumers such as industry. Utilizing such systems encourages the prudent use of renewable sources of power. For enterprises that use a large amount of energy, energy storage devices can help provide a more dependable power supply [1]. The most prevalent energy storage system in the industrial sector is an electrochemical method of energy storage using a common rechargeable battery, which uses a mechanism for electrical energy conversion from chemical energy. The modern industrial sector uses battery power plants that combine multiple batteries to store energy and use it when needed. Battery technology can be used during power fluctuations, but only for short periods of time. The electrochemical process shows how single units can be combined to build a large single-unit battery energy storage system (BESS). The system has an initial cost but is reusable.

Electrochemical processes use a combination of electrochemical cells to store energy for later use. It is known as a secondary battery because of its rechargeable nature. The rationale behind electrochemical methods of energy storage is to store energy using reversible chemical reactions. The basic model of an electrochemical cell consists of a combination of electrodes and electrolyte. Several combinations of electrodes and electrolytes have been tested and used on an industrial scale. Batteries made of lead-acid, nickel-cadmium, lithium, and Sodium are some of the most commonly used types [2]. Research into the formation of electrochemical cells has been conducted for decades. Researchers have worked to improve lithium-ion batteries, which have come a long way since their invention, with the only downside being the availability of lithium. Therefore, new metal electrodes have to be proposed.

Due to the difficulty of extracting lithium and its limited availability, new materials have tried to replace it [3]. Research into new molten salt batteries has been ongoing for a long time. Lithium depletion was a major factor in this process. This required the development of a new type of liquid salt battery. Attempts to replace the lithium electrode with more common metals, such as Sodium, have been successful. Compared to their lithium-Sulphur counterparts, Sodium-Sulphur batteries have little exposure to the world. Research on this new molten-salt battery is currently underway, and great results have been achieved. Due to the abundance of Sodium and the challenge of obtaining lithium, Sodium-Sulphur batteries have recently received much attention. Sodium is readily available, which raises the question of whether to use lithium metal. For this reason, researchers started with the idea of using Sodium instead of lithium as the electrode material. Modern Sodium-Sulphur batteries are not the focus of many researchers who are working on lithium-Sulphur batteries. Figure 1 shows a schematic of a Sodium Sulphur battery.

The innovative energy storage device known as Sodium-Sulphur batteries (Na-S) uses Sodium and Sulphur as the active electrode components. Since the middle of the 1980s, the Na-S battery development program has been collaboratively carried out by NGK Insulators (NGK) and Tokyo Electric Power Company (TEPCO). Batteries are charged and discharged by reversible electrochemical reactions [4]. The battery voltage is approximately 2 volts. The main characteristics of the battery are high energy density, high coulombic efficiency, no self-discharge, long service life, low production cost, low operating cost and fully recyclable raw materials.

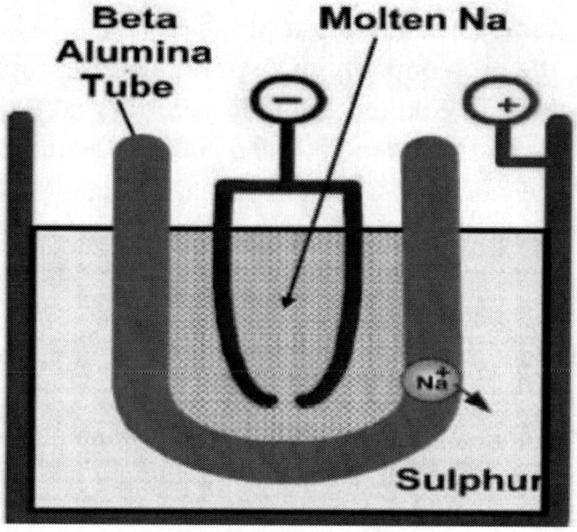

Figure 1. Sodium Sulphur battery [4].

The performance of the Sodium Sulphur battery is significantly influenced by the electrolyte used. The main characteristic should be the selective passing of Sodium ions through the battery module. The other essential characteristics of the electrolyte are high electronic conductivity, high electrical resistivity, high density without porosity and long-term durability. The anode material must be composed of Sodium or its composites such that the Sodium ions can easily dissociate from it and move freely through the electrolyte. The necessary conditions that the anode must satisfy are that the anode must be in contact with the electrolyte throughout the charging and discharging phases and that there must be very little or no interfacial polarization at the surface of the electrolyte used. Interfacial polarization causes an increase in resistance at the anode, reducing the voltage produced. This usually happens due to the quality of the Sodium used in the anode and the quality of the electrolyte used.

In the cathode, the discharge is produced by the production of Sodium polysulfides. The polysulfides formed have x values ranging from 2 to 5. Initially, only polysulfide is formed. On further discharge, higher polysulfides appear. The higher polysulfides aid in voltage dissipation and exist in a liquid state at the operating temperature. The polysulfide is usually formed at the end of the discharge. This polysulfide has a higher melting point and tends to exist in a solid-state at the operating temperature. The solid tends to reduce the voltage dissipation as it forms a layer over the cathode material, reducing the cathode capacity and decreasing the life of the battery. Thus, extreme care must be taken while operating the battery, as the polysulfides can also be corrosive to the electrodes. The operating temperature is an issue that must be addressed in battery design. Na-S batteries are divided into three categories based on operating temperature: high-temperature Na-S batteries, intermediate-temperature Na-S batteries and room-temperature Na-S batteries. High-temperature Sodium-Sulphur batteries are ordinary Na-S batteries developed by NGK in the 1980s that operate at temperatures above 300°C. The high temperature of the module keeps the electrodes in a molten state. This means that the ionic resistance of the module is low enough to allow electrons to pass through it easily. This battery's electrolyte needs to be highly Na^+ conductive, have a low electron transfer number, have a high specific gravity and be extremely strong mechanically. They are commercialized for energy storage systems due to their high energy density, high charge/discharge efficiency and prolonged life. This type of battery typically has a large surface area, so it does not cool down quickly [5]. Intermediate-temperature Sodium-Sulphur (IT-Na-S) batteries were discovered as an immediate solution to the problem of high operating temperatures in high-temperature Na-S batteries. Thus, research on IT Sodium-Sulphur batteries and RT Sodium-Sulphur batteries began. IT Na-S batteries operate between 100° and 200°C. The increased interest in this battery system is evidenced by the numerous papers that have been published on IT-Na-S energy storage systems using nonaqueous electrolytes or polymer electrolytes [6]. A lower operating temperature guarantees a higher level of safety, but it also has some drawbacks. Because Sodium cations carry the majority of the current and because melts with homogeneous composition have very limited sulphide ion mobility, a high Sodium concentration is crucial. The next drawback is the decrease in conductivity of the electrolyte due to the decrease in temperature. Next is the decrease in polysulfide solubility. Melts of higher-order polysulfides have a density that increases with decreasing temperature, whereas the irreversible formation of lower-order polysulfides (and Sulphur) leads to capacity loss and cell failure. A final perceived drawback is the inability to take full advantage of the Sulphur electrode. This is because the produced polysulfide solidifies on the Sulphur electrode and lowers the electrode's capacity. Developing a room-temperature battery is the next step in this process [7]. Due to their similar performance to HT-Na-S batteries and increased safety factor, RT-Na-S batteries are well received. This particular battery has a different working principle, as the electrodes are in a solid state. A major drawback of this battery is the formation of unwanted deposits on the electrodes [8]. The solid nature of Sodium electrodes can lead to dendrite formation, which reduces battery capacity and life. Another drawback is that the Sodium electrode and electrolyte form a solid-electrolyte interface with good ionic-conducting properties but poor electronic-conducting properties. The Na-S battery now uses regular Sodium and Sulphur as electrodes and a beta-alumina solid electrolyte. This is a hot battery. The battery showed good energy efficiency and an improved lifespan, but high operating temperatures are always a disadvantage for this model, and electrolyte degradation has also been discussed. A study on thermal management, safety and cell reliability [9]. This study showed that the electrolyte exhibited dendrite formation after several days of runtime, which caused an increase in cell resistance. Various electrolyte and electrode materials have been used to improve their properties. In 1998, a 64 MW, 240-module Na-S system for Tokyo Electric Power Company was designed and has been working on load balancing applications ever since. The structure and operating performance of Sodium-Sulphur battery cells, modules and systems [10].

Research into new Sodium-Sulphur batteries has been ongoing for decades. As mentioned in the previous paragraph, the main drawbacks of using this battery are the excessive polysulfides that cause corrosion and the high operating temperature of the battery. Several battery modification studies have been conducted to remedy these shortcomings. The design should consider electrode stability and performance characteristics. This particular proposal discusses in detail the problems associated with the above drawbacks. Research has shown that the voltage loss depends on the electrolyte concentration. Note that ether-based electrolytes perform similarly to their lithium counterparts. Other studies used the improved electrodes to fabricate IT Na-S batteries and RT Na-S batteries. The IT-Na-S battery has the major drawback that the formed polysulfides do not dissolve in the electrode, and the Sulphur electrode cannot be fully utilized. The use of ether- and carbonate-based electrodes tends to aid in the design of such batteries. Research based on the polysulfide shuttle phenomenon is also being conducted around the world. The interactions between polysulfides and MOFs in the discharge/charge process, sulfide adsorption and fast conversion kinetics can be tuned by changing the dynamic electronic states of Ni centers in 2D MOFs, according to synchrotron X-ray diffraction, electrochemical measurements and theoretical calculations. The resulting battery is believed to be the most stable of the reported batteries. Understanding electrode chemistry and structure is a key component of battery design.

Extensive research on electrode nanostructures discusses new methods. The research results of Shanghai Ceramic University were very promising [11]. They published the first state-of-the-art work on this Na-S battery. It taught us how easy it is to make and use a battery. Their work gave me the idea to work on battery electrodes. Delhi University has been working on this for some time [12]. The primary negative of using this particular battery is its high temperature, which poses a risk to the user. This strong interest has led to a move toward batteries that operate at room temperature. The RT Na-S battery developed by them provides an excellent solution to this problem. Although this work is still in its early stages, it provides an excellent backbone for the development of Sodium-Sulphur batteries. A key study by Arumugam Manthiram presented work on Sodium-Sulphur batteries at room temperature [13]. This piece shows his current RT Na-S battery. It highlights the strengths and weaknesses of previous works. This work proposed an improvement for room-temperature Na-S batteries. Another analysis by Indian researchers demonstrates that a battery with a microporous carbon-Sulphur composite cathode and a liquid carbonate electrolyte that incorporates the ionic liquid 1-methyl-3-propylimidazolium chlorate bonded to SiO_2 nanoparticles at room temperature can be calibrated and operated at that temperature. [14]. The use of these materials increases energy density and improves battery cycle life and stability. This work helped develop a low-cost rechargeable Na-S battery using nanoparticles. A study demonstrating the use of Sulphur/soot nanocomposites decorated with aluminum oxyhydroxide nanosheets as a cathode is also a highly decorated work by Indian researchers [15]. Using this material, RT-Na-S batteries have been shown to be highly stable and have excellent specific capacities. This study demonstrated the use of Lewis acid–base catalysts to improve the results. The Gujarat Institute of Technology has also shown interest in this topic. Recently, the use of additives such as ionic liquids in electrolytes to improve their performance has been discussed. The use of gel polymers is very well described in research, as the use of solid electrolytes has proven inefficient due to their low conductivity [16]. The catalysis of the free radical mechanism, which uses stabilized free radicals to improve battery performance, is also discussed. The polymerization of benzoxazines using a Sulphur electrode and reduced graphene oxide has been investigated to optimize the use of the Sulphur electrode. The nanostructure of the electrodes has been thoroughly studied, and improved techniques are discussed in a review by Yun-Xiao Wang. We discuss the use of different electrolytes in the electrodes and the use of composites of real electrode materials instead of standard electrodes to improve electrode properties. The Shanghai Ceramic Research Institute conducted a review of the Na-S battery's application as energy storage. The basic characteristics of Sodium-Sulphur batteries are presented in this paper, which also provides a summary of recent advancements in Sodium-Sulphur batteries and their use in stationary energy storage [17]. Tera ethylene glycol dimethyl ether is used as the base solvent [18]. According to a study, ether-based electrolytes help reduce Sulphur more completely, allowing for greater electrochemical reversibility [19]. On the other hand, carbonate solvents show irreversible liquid-phase reactions. This study provides insight into the achievement of the high energy potential of the RT Na-S battery and illustrates the solvent-reliant Na-S reaction pathway of the battery at that temperature. The cyclability of the Na-S system was greatly improved by using ether-based electrolytes for the first time. Improvements in stability and voltage loss have opened the opportunity to experiment with batteries using more types of ether-based electrolytes that can help reduce battery operating temperatures. In a study by Reddy, B.S. gives an example of his use of KOH as the electrolyte in room-temperature batteries. Conductivity, surface area, and porosity are all improved in this study by KOH activation [20]. The use of Sulphuric acid as an electrolyte is discussed in a previous study [21].

Applications for batteries include heavy equipment operation, space applications, and load balancing in grid systems. Na-S batteries are employed in stand-alone renewable energy applications as well as to support electrical grids. Na-S batteries can provide value under specific market circumstances by engaging in energy transfer and voltage control [22]. Na-S batteries have the potential to support the production of renewable energy, particularly solar and wind power. Batteries for wind farms store energy at times of strong winds but low electricity demand. During periods of maximum load, this stored energy may be drained from the battery. In addition to this power shift, Sodium-Sulphur batteries can be employed to stabilize the power output of wind farms during wind fluctuations. Where alternative storage choices are not practical, these kinds of batteries provide a storage option for energy. Compressed air storage (CAES) requires specific geological features, such as salt caves, whereas pumped storage operations require considerable space and water resources [23]. Na-S batteries have been proposed for use in space applications due to their high energy density. Batteries composed of Sodium and Sulphur might be used in space. In reality, an orbital test battery made of Sodium and Sulphur was launched. In Na-S flight testing, a battery operating at 350°C with 150 W h/kg of specific energy—three times the energy density of Ni-MH batteries—was demonstrated. On the STS-87 mission in November 1997, it was launched and operated experimentally for ten days.

This kind of battery is also being considered for the Venus Land Sailing Rover Mission Concept. This is so that the rover and its payload can function for approximately 50 days without a cooling system on Venus' heated surface [24]. It has been challenging to deploy Sodium-Sulphur batteries in electric vehicles due to their high operating temperatures. By lowering the operating temperature of the battery, it can be used as a car battery. A lower operating temperature also ensures operator safety. This work builds on results analyzed in previous work performed on Na-S batteries. The shortcomings of conventional Na-S batteries have been shown in several papers. The purpose of this project is to modify the basic battery model to cover the shortcomings of the traditional model. They include the following goals: the development and design of new molten salt batteries to replace conventional lithium-ion batteries. Since lithium is difficult to extract and the concentration of lithium is low compared to Sodium, using the more abundant Sodium instead of lithium is satisfactory. Consider combining and improving conventional Sodium electrodes with other materials to achieve better performance. Performance improvements must also be commercially viable. The new battery aims to act as a perfect replacement for traditional energy storage, designing the battery for high operating temperature conditions to drop to mid-temperature and room temperature capabilities. Electrodes and electrolytes should be used in the form of alloys in combination with other materials, or doped materials should also be tested. The battery is designed to reduce the polysulfide shuttle phenomenon. The formation of polysulfides and their deposition on electrodes causes corrosion and reduces electrode capacity. This also reduces the number of battery cycles.

1.1. Theory

A Sodium-Sulphur battery consists of Sodium as the anode and Sulphur as the cathode. The standard model was developed by NGK Insulators Ltd. It is crucial to choose the right electrolyte for Sodium-Sulphur batteries. The main criterion for choosing an electrolyte is the Sodium ion conductivity of the material. The best choice of electrolyte is a beta-alumina solid electrolyte. According to the tests carried out, this electrolyte showed excellent Sodium ion conductivity. An important parameter to discuss for Sodium-Sulphur battery performance is the operating temperature of the module. The electrode material was kept at a temperature of 300°C. The electrodes are guaranteed to be in a molten state at this temperature. The ionic conductivity of Sodium ions increases with temperature. This facilitates the movement of the ions to the cathode to form polysulfides. It is also in a liquid state. The generated voltage of a Na-S battery is approximately 2 volts.

The Na-S battery is a rechargeable battery based on electrochemical oxidation and reduction reactions.

CELL REACTION: $2Na + xS \leftrightarrow Na_2S_x$

The discharge phase shows the reaction between the Sodium ions and the Sulphur cathode to form polysulfides. Sodium is first ionized within the anode material, forming Na^+ ions. The electrolyte is subsequently traversed by these ions to reach the Sulphur-based cathode material. The electrons given off by the anode are allowed to move through an electrical load and back into the battery at the cathode. At the cathode, an electron reacts with Sulphur to form Sulphur. This ion, in combination with the positive Sodium ions, forms polysulfides where x ranges from 2 to 5. During the charging period, the polysulfides formed during discharge dissociate back into Na^+ ions and ions. The ions then return to their respective electrodes. The electrons that were initially injected into the positive electrode during the discharge phase are returned to the Sodium electrode by applying an external current. Ions must traverse from the positive electrode to the negative electrode for the battery to store energy. The amount of Sulphur in the cathode determines the discharge voltage of the battery [17].

ANODE REACTION: $2Na \leftrightarrow Na^+ + 2e^-$

CATHODE REACTION: $S + 2e^- \leftrightarrow S_x^{2-}$

It is crucial to choose the right electrolyte for Sodium-Sulphur batteries. The main criterion for choosing an electrolyte is the Sodium ion conductivity of the material. The other essential characteristics of the electrolyte are high electronic conductivity, high electrical resistivity, high density without porosity and long-term durability. The anode material must be composed of Sodium or its composites such that the Sodium ions can easily dissociate from it and move freely through the electrolyte. The necessary conditions that the anode must satisfy are that the anode must be in contact with the electrolyte throughout the charging and discharging phases and that there must be very little or no interfacial polarization at the surface of the electrolyte used. Interfacial polarization causes an increase in resistance at the anode, reducing the voltage produced. This usually happens due to the quality of the Sodium used in the anode and the quality of the electrolyte used.

In the cathode, the discharge is produced by the production of Sodium polysulfides. The polysulfides formed have x values ranging from 2 to 5. Initially, only polysulfide is formed. On further discharge, the and appear. The higher polysulfides aid in voltage dissipation and exist in a liquid state at the operating temperature. The polysulfide is usually formed at the end of the discharge. This polysulfide has a higher melting point and tends to exist in a solid-state at the operating temperature. The solid tends to reduce the voltage dissipation as it forms a layer over the cathode material, reducing the cathode capacity and decreasing the life of the battery. Thus, extreme care must be taken while operating the battery, as the polysulfides can also be corrosive to the electrodes.

2. Model

The design and modeling of batteries have seen a wide range of approaches. The computational model and the electrochemical model show the details of how the actual Na-S battery works. The models can be designed to be one-dimensional, two-dimensional or three-dimensional. The computational model can be simulated in MATLAB or Simulink using the governing equations of the battery. The electrochemical aspect of the battery was also considered in this research by modeling the 1-D model and 2-D in COMSOL software.

2.1. Electrical Circuit Model

Electrical circuit models are probably the most common and effective way to mathematically model batteries. A mathematical description or an electrical equivalent circuit diagram is used to represent the system. Electrical circuit models are created by analyzing various analogies between currents, voltages and various other physical parameters. These relationships are used to find other properties of the system. This modeling approach makes it easy to design very complex energy management systems. The various electrical circuit models used for design simulation are simple models, Thevenin models, and modified battery models.

i. Simple Battery Model: This is probably the simplest and most common model used when modeling batteries. This model uses only a simple resistor placed in series with the open-circuit voltage. This model assumes that the combined internal resistance of the anode, cathode and electrolyte is constant. This is a shortcoming of the model. Assuming that the battery's resistance is constant is incorrect because it changes depending on how the battery is charged and the ambient conditions. Therefore, we recommend using this model for modeling and learning purposes only. Therefore, it is considered unsuitable for battery modeling.

ii. Thevenin Model: This model consists of a resistor, a voltage source and a parallel combination of an overvoltage resistor and a capacitor. Similar to the simple model above, this model also has an ideal open-circuit voltage. This model is perhaps one of the most common modeling techniques used in battery design and development. The principle behind the Thevenin model is the Thevenin theorem. According to Thevenin's Theorem, any complicated linear circuit can be reduced to a simpler design by adding a series resistor and a single voltage source. The Venin model offers a simple method for streamlining circuits with dynamic loads. Without having to recalculate the complete circuit, this modeling technique offers an effective way to calculate the voltage and current across the load. Common secondary batteries, such as nickel-cadmium, lithium-ion and lead-acid batteries, were evaluated using this theorem. However, this approach has the drawback of not accounting for the voltage drop across the battery. Therefore, the model is not suitable for modeling Sodium-Sulphur batteries.

iii. Modified Battery Model: This model has proven effective in this case study. This model has a significant advantage for modeling purposes compared to other models, as it consists of different internal resistances and no-load loads. A significant advantage of the modified model is that it takes into account changes in model resistance as the battery discharges. This model is predicated on the notion that when a battery drains, the resistance and voltage will fluctuate according to the battery's level of charge. We can thus create the fundamental design of a Sodium-Sulphur battery using this model. The Sodium-Sulphur battery is modeled using MATLAB or Simulink. Mathematical modeling is performed using equations derived from references and papers and integrated into MATLAB and Simulink software.

$$V_{bat} = E_0 + V_{pol} - R \times i - R_{pol} \times i + V_{exp} - V_{act} \qquad (1)$$

where represents the voltage produced by the battery, represents the open circuit voltage, represents the polarization voltage, represents the exponential voltage, represents the activation potential, represents the polarization resistance, represents the internal resistance of the battery and represents the current generated by the battery.

$$State\ of\ Charge\ (SOC) = SOC_0 - \frac{1}{Q} \times \int i(t)dt \qquad (2)$$

where is the initial state of charge of the system and is the charge capacity of the system in ampere hours.

$$V_{pol} = K \times \frac{Q}{Q-it} \times it \qquad (3)$$

where is the polarization constant in V/Ah

$$V_{exp} = A \times e^{-B \times it} \qquad (4)$$

where is the exponential zone amplitude (V) and is the exponential zone time constant inverse (Ah^{-1})

$$R_{pol} = K \times \frac{Q}{Q-it} \qquad (5)$$

$$V_{act} = \frac{2\alpha RT}{nF} \sinh^{-1} \frac{i}{2i_0} \qquad (6)$$

where is the current, is the current density, is the universal gas constant, is the stabilizing temperature of the battery, is Faraday's constant, and α is the transfer coefficient.

By combining these mathematical equations in MATLAB or Simulink, we can obtain the real-time parameter values of the Sodium Sulphur battery, which include the state of charge, voltage and current. Thus, the mathematical model of the battery can be obtained using MATLAB or Simulink. These parameters are plotted against time and compared with their counterpart, the lithium-ion battery.

The lumped battery interface is used to define a battery model based on a minimal set of lumped characteristics, necessitating no prior knowledge of the internal composition, design or material selection of the battery electrodes. The status of the charge and voltage response of a battery during a load cycle may often be tracked using models developed with the Lumped Battery interface. To represent battery cooling and thermal management, the interface additionally defines a battery heat source that may be connected to a heat transfer interface. To study the internal structure and design of the battery electrodes, other physics of the battery must also be studied. The physics that must be studied for this include the transport of species inside the module, the electrostatics of the module, the tertiary current distribution and the electroneutrality of the cell.

2.2. Electrochemical Model

Based on electrochemical processes that take place in the cell compartments, batteries are charged and discharged. Therefore, modeling a cell based on its electrochemical behavior can also be done by calculating the number of ions that move from one electrode to the other. Batteries can be described by finding the relationship between the potential of an electrochemical cell and the free enthalpy. The electrochemical function of the cell is well described by the Nernst and Butler-Volmer equations. The fundamental tenet of this model is that the greatest amount of work that a cell or battery may produce is dependent upon the variation in the Gibbs free energy of the internal processes. Sodium-Sulphur batteries have been found to have activation potential. The kinetics of the electrochemical reactions that occur in this cell are very slow. A light push is required to move the battery. The extra energy expended to keep the cell running is the activation potential. The cell activation potential is detected by applying electrochemical modeling to the modified model. The concept of electrochemical modeling tells us how different ions pass through the cell and how their concentration affects the voltage of the cell.

The concentration field of diluted solutes in solvents is calculated using the dilute species transport interface. The movement and reactions of chemical species dispersed in gases, liquids or solids can be calculated. Diffusion, in accordance with Fick's law, convection, when linked with fluid flow, and migration, when paired with an electric field, are the factors that propel transport. When the charge distribution is explicitly provided, the electrostatic interface is utilized to calculate the electric, displacement and potential distributions in dielectrics. Although the formulation is stationary, it also supports natural frequencies in all spatial dimensions, the frequency domain, tiny signal analysis and time domain modeling for usage in conjunction with other physics. The scalar potential is used as the dependent variable in the physics interface to solve Gauss' law of the electric field. When describing the distribution of currents and potentials in an electrochemical cell, the cubic current distribution, or Nernst–Planck interface, takes into consideration the individual movements of charged species (ions) in the electrolyte. Either random equations or the preset Butler-Volmer and Tafel equations can be used to describe the electrode kinetics of charge transfer reactions. The electroneutrality condition of the electrolyte must be true for the charge retention option to work. Keep in mind that the electroneutral condition requires that the model contain all important current-carrying ions. Charge balance and Ohm's law are used to describe the current flow in the electrodes.

The model equation used in COMSOL is as follows:

The mass and charge transport are modeled using the Nernst–Planck equations.

$$Ji = -\frac{1}{kT} Di(x)A(x)\rho_i(x)\frac{d}{dx}[\mu i(x)+z_i eV(x)] \quad (7)$$

where Ji is the particle flux, Di is the chemical potential profile, A, zis the ionic valence, T is the temperature, k is the Boltzmann constant and e is the elementary charge.

You can calculate the transport and reactions of chemical species dissolved in gases, liquids or solids. Diffusion, in accordance with Fick's law, when linked with fluid flow and migration, when paired with an electric field, is the factor that propels transport. When the charge distribution is explicitly provided, the electrostatic interface is utilized to calculate the electric, displacement and potential distributions in dielectrics.

The Lumped Battery Interface is used to define battery models based on a small set of lumped parameters that do not require knowledge of the internal structure and design of battery electrodes or material selection. An electrochemical model designed in COMSOL uses the physics of how electrode materials behave during discharge. Figure 2 shows the COMSOL model of a Sodium Sulphur battery.

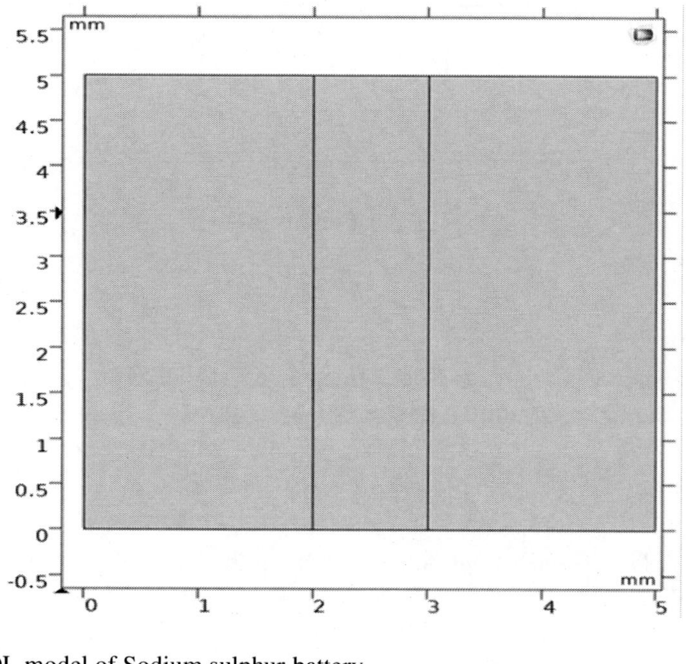

Figure 2. COMSOL model of Sodium sulphur battery.

Input parameters	Input values
Diffusion coefficient Na^+	1.5×10^{-12} m^2/s
Diffusion coefficient S_8	0.88×10^{-11} m^2/s
Diffusion coefficient S_5^{2-}	$3.5 \times 10^{-12} m^2/s$
Diffusion coefficient S_4^{2-}	3.5×10^{-12} m^2/s
Diffusion coefficient S_3^{2-}	1.75×10^{-12} m^2/s
Diffusion coefficient S_2^{2-}	0.88×10^{-12} m^2/s
Reference concentration S_8	19 mol/m³
Reference concentration S_5^{2-}	0.18 mol/m³
Reference concentration S_4^{2-}	0.32 mol/m³
Reference concentration S_3^{2-}	0.02 mol/m³
Reference concentration S_2^{2-}	5.23×10^{-7} mol/m³
Reference concentration Electrolyte	1000 mol/m³
Reference concentration Anode	1000 mol/m³

3. Results and Discussions

3.1. Simulink Model Results

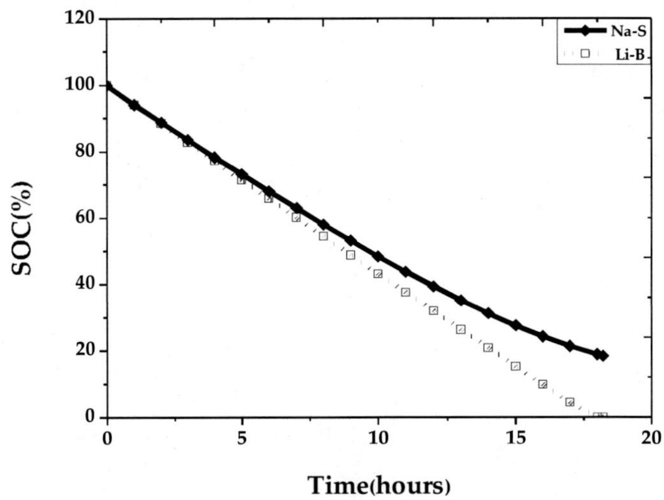

Figure 3. State of Charge v/s Time.

Figure 3 is the state of charge vs. time graph of the Na-S battery and the lithium-ion battery. The mathematical model of the battery performed using MATLAB or Simulink gives the above result. From the graph, it is clear that the Na-S battery provides better performance compared to its counterpart. We can see that the lithium-ion battery is clearly depleted, while the Na-S battery still has charge left in it. The comparison of the voltage dissipated and the current is also given in the results below. From the voltage vs. time graph shown in Figure 4, we can see that a voltage of approximately 2 V is produced by the cell. Figure 5 shows the variation in the current with time.

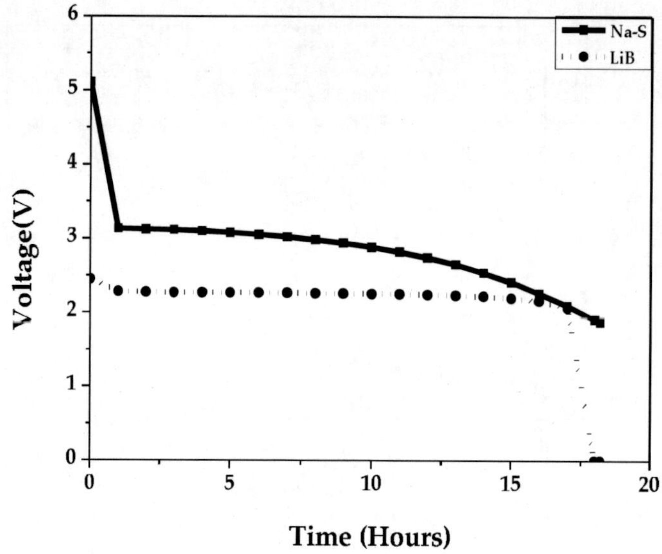

Figure 4. Voltage v/s Time.

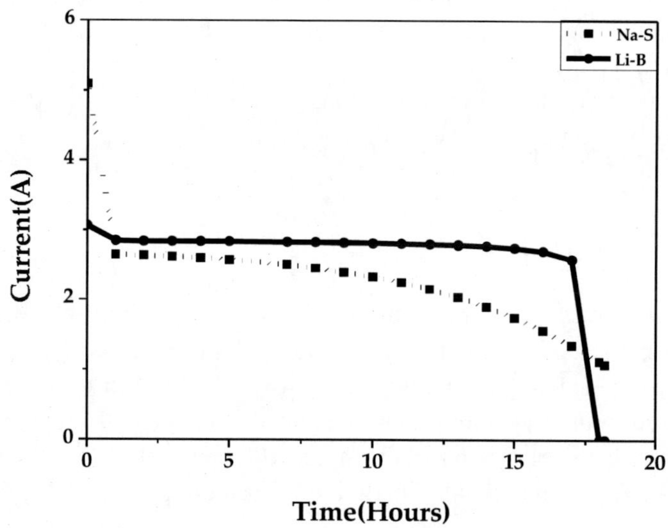

Figure 5. Current v/s Time.

One of the major criteria in cell performance is the resistance of the cell. The major contributor to cell resistance is the resistance due to the polarization effect. It is noted that during discharging of the cell, the resistance due to polarization will increase, which in turn increases the internal resistance of the cell. The graph given below in Figure 6 shows the change in resistance due to polarization with time.

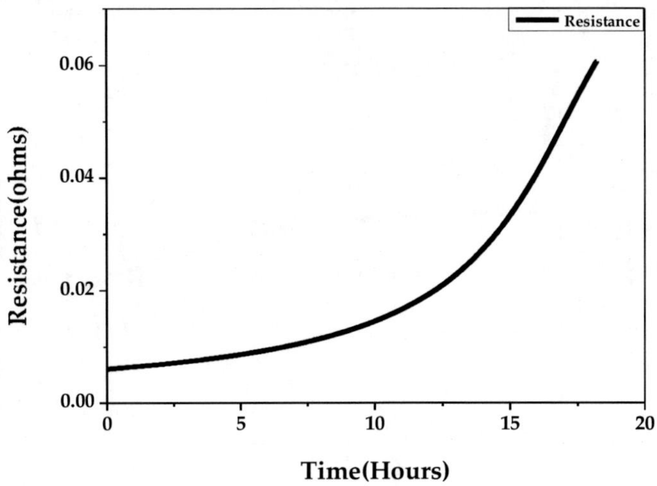

Figure 4. Voltage v/s Time.

There is an activation loss occurring at the surface of the electrode, where the electron transfers occur. The reactions occurring on the surface of the electrode are very slow. Therefore, these reactions require an extra push. Thus, extra power is required to force the reaction to proceed at the speed required by the current. This minimum energy must be present if a collision between two atoms causes a chemical reaction. This loss is known as the activation voltage. The variation in activation potential with time is given below.

3.2. COMSOL Results

The discussed graph gives us the information that Sulphuric acid electrolyte gives the highest voltage dissipation for different capacities, followed by potassium hydroxide. The lowest voltage is observed when using the polymer electrolyte $LiPF_6$. For the ether-based electrolyte, the voltage is found to increase along with the capacity of the cell. The concentration change of each ion and the electrodes dissipated during the discharge of the battery for each electrolyte are discussed in the following results. The voltage dissipation depends on the concentration of the cathode. As observed from the results above, it can be seen that when Sulphuric acid is used, more of the Sulphur electrode is depleted faster, and it also produces the highest volume compared to the other electrolytes. The voltage produced by each electrolyte is proportional to the amount of cathode dissociation. From the graphical results, we can observe that the cathode dissociation for each electrolyte is different and that the voltage produced is directly proportional to the amount of cathode dissociated. The production of S_4^{2-} and S_5^{2-} ions is also vital to voltage dissipation. The formation of S_4^{2-} and S_5^{2-} is good for high voltages, and the formation of S_2^{2-} ions reduces the voltage dissipation. The formation of higher polysulfides is vital to voltage production. The formation of lower polysulfides is found to be a degrading factor. Dendrite formation seems to be an issue in this process. The lower polysulfides are found to exist in a solid-state even while being operated at higher temperatures. These lower polysulfides degrade the capacity of the cathode, which in turn affects the thermal dissipation capacity of the battery. The formation of various polysulfide ions and depletion of electrolyte concentration while discharging are also mentioned in the results below. From the graphs, it is observed that the $LiPF_6$ polymer electrolyte produces the least voltage, and we can observe that this electrolyte supports the formation of a lower polysulfide.

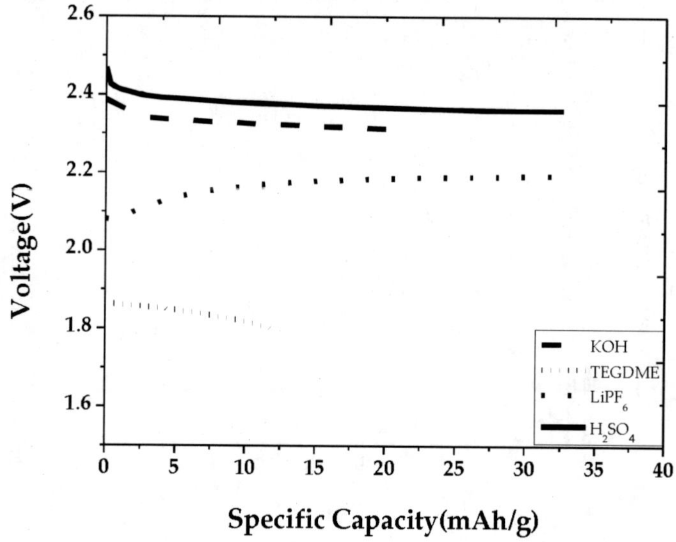

Figure 7. Voltage v/s Specific Capacity.

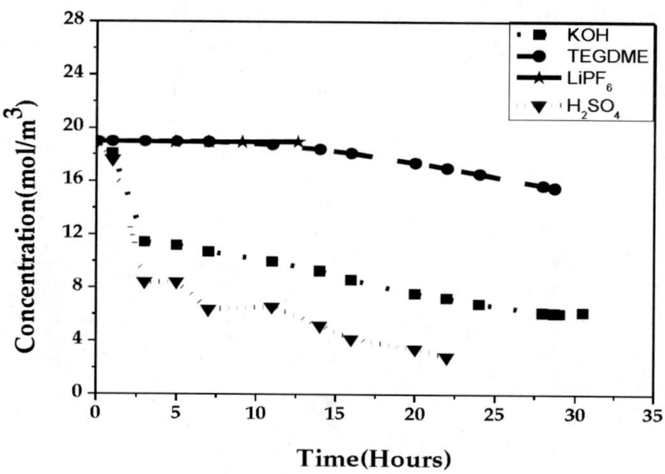

Figure 8. Concentration of Sulphur cathode v/s Time.

Figures 7 and 8 show the change in concentration of the Sulphur electrode with time as the discharge proceeds. The graph shows that the Sulphur electrode becomes depleted at a faster rate when the electrolyte used is Sulphuric acid due to the strong dissociative power of Sulphuric acid. This produces the highest voltage in comparison to other electrolytes. The Sulphuric acid electrolyte is followed by the KOH electrolyte, which dissociates the cathode material. The least dissociation is observed with the TEGDME-based electrolyte and $LiPF_6$ electrolyte. The choice of the model of the battery should be based on its sulphur dissociation capacity, which is associated with the voltage dissociation capacity. The next important parameter is the operating temperature of the model. The operating temperatures of the models are discussed in further sections.

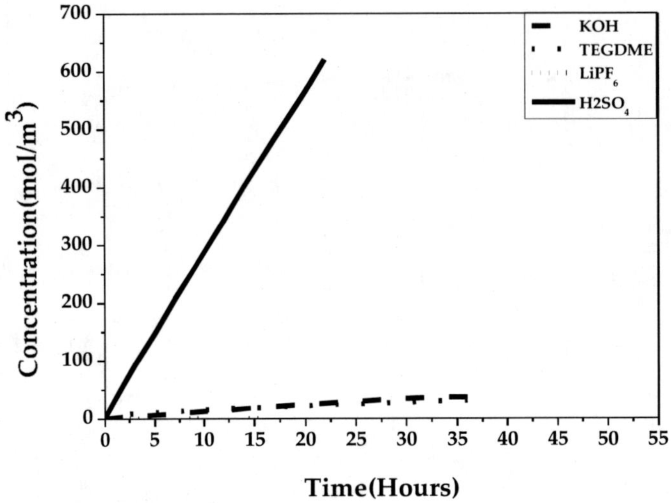

Figure 9. Formation of S_5^{2-} v/s Time.

The formation of S_5^{2-} polysulfide ions is essential for higher voltage dissipation, as shown in Figure 9. The formation of more polysulfide ions favors higher voltage dissipation because more polysulfide ions tend to have better ionic conductivity at lower temperatures than less polysulfide ions. The above graph shows that the concentration of the higher polysulfide ions is the highest when using Sulphuric acid, followed by potassium hydroxide, TEGDME and $LiPF_6$. The concentration profile follows the same order as that of the voltage dissipation capacity of the battery.

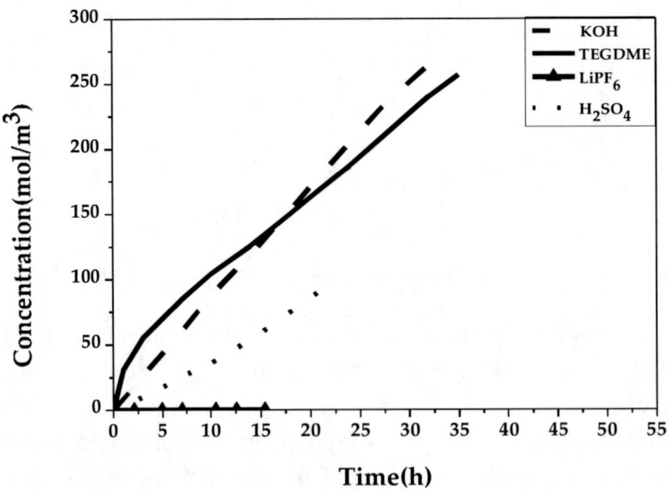

Figure 10. Formation of S_4^{2-} v/s Time.

Figure 10 shows the formation of S_4^{2-} ions for the various models formed. The formation of S_4^{2-} is lowest when using the $LiPF_6$-based electrolyte. Since most of the cathode dissociated while using the Sulphuric acid–based electrolyte is used up in the formation of S_5^{2-} ions, the amount of S_4^{2-} ions formed is lower than that formed when using the KOH-based electrolyte and TEGDME-based electrolyte. The highest concentration is observed when using the KOH-based electrolyte, followed by the TEGDME-based electrolyte. Therefore, we can imply that the higher polysulfide formation is only present in the Sulphuric acid-, KOH- and TEGDME-based electrolytes.

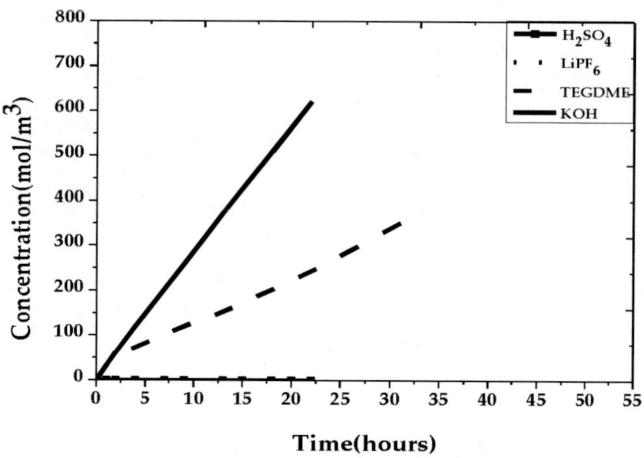

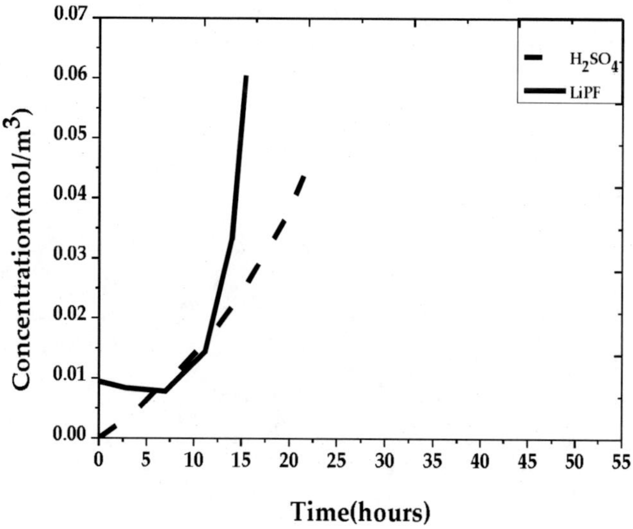

Figure 11. Formation of S_3^{2-} v/s Time.

The S_3^{2-} ions can be considered higher polysulfide ions, as shown in Figure 11. The formation of more polysulfide ions is followed by higher voltage dissipation. The highest S_3^{2-} ions are formed with the KOH-based electrolyte, followed by the tetraglyme-based electrolyte. The least is observed with the Sulphuric acid–based electrolyte and lithium hexaflurophosphate electrolyte. The lower concentration when using the lithium hexafluorophosphate electrolyte is explained by the low kinetics of reaction with the higher sulphide ions leading to only the formation of lower sulphide ions. The reaction with the sulphuric acid–based electrolyte causes only very low formation of S_3^{2-} ions. The reason for this is that the Sulphur electrode is already used up in making the S_5^{2-} ions and the S_5^{2-} ions. Therefore, the kinetics for the production of lower polysulfides are lower compared to those of the higher polysulfides.

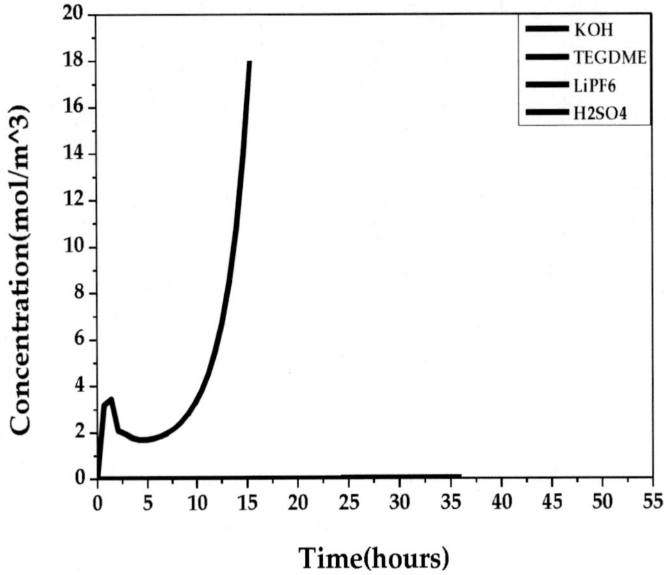

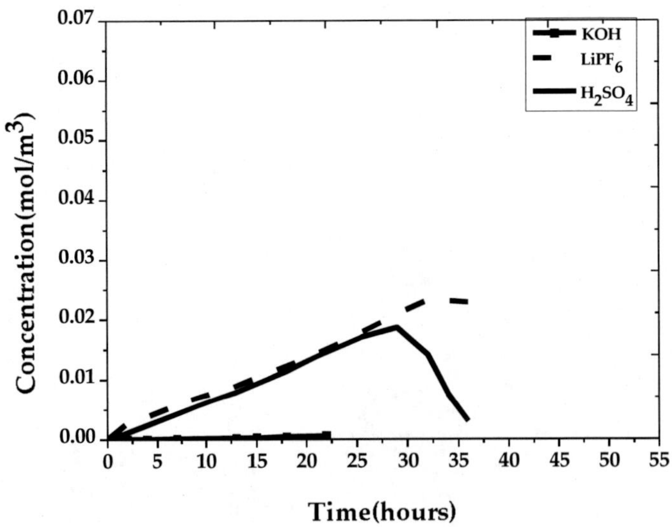

Figure 12. Formation of S_2^{2-} v/s Time.

Figure 12 shows the formation of S_2^{2-} ions. One of the major drawbacks of using the Sodium Sulphur battery is that at lower temperatures, dendrite formation reduces the cathode capacity. The reason for dendrite formation is the formation of lower polysulfides, namely, S_2^{2-} ions. These ions cause a layer of polysulfide to be formed on the cathode. This causes a decrease in the movement of Sulphur ions and reduces the capacity and life of the cathode. The formation of S_2^{2-} ions is highest with lithium hexafluorophosphate due to its low voltage dissipation. The other electrolytes show very low S_2^{2-} ion formation, which is very low in comparison to the lithium hexafluorophosphate electrolyte.

3.3. Temperature Profile

Since Sulphuric acid–based electrolytes can be strictly avoided because they cause high deterioration of the electrodes, as seen from the above graphs, we can clearly see the deterioration of the Sulphur cathode. The lithium hexafluorophosphate-based electrolyte causes very low deterioration of the Sulphur cathode. Since the deterioration is very low, it also leads to low voltage dissipation while using this electrolyte. Therefore, from these inferences, we can clearly avoid the use of these electrolytes. Research regarding the use of KOH-based Sodium Sulphur batteries and tetraglyme-based batteries has been performed in further research. The temperature profile while using the above-given electrolytes is given below.

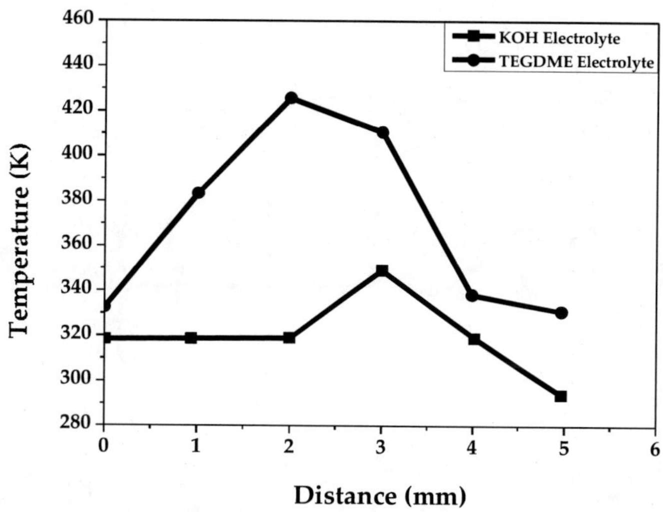

Figure 13. Temperature profile along the length of the battery modeule.

Electrolyte	Peak Temperature	
	Simulation	Research paper
KOH	76°C	30°C-80°C
TEGDME	168°C	150°C-200°C

Figure 14. Comparison of temperatures with research paper data.

Figure 13 is a representation of the temperature profile of the battery along the length of the battery while using the KOH-based Sodium Sulphur battery and the tetraglyme-based Sodium Sulphur battery. According to the experimental research conducted, it is noted that the KOH-based battery works in the room temperature range and the intermediate temperature battery works in the intermediate temperature range.

The temperature distribution curve shows the peak temperature achieved by the battery while using the two electrolytes. The peak temperature achieved by the battery while using the KOH battery is 76°C, while the actual range should be between 30°C and 80°C. The peak temperature observed while using the tetraglyme-based electrolyte in the simulation is 168°C. The actual temperature of operation was 150°C. Therefore, we can conclude that the temperatures are in range and in accordance with the experimental values. Figure 14 shows the comparison of temperatures with research paper data.

3.4. Discharge Profiles at Different Discharge Rates of TEGDME-Based Sodium Sulphur Batteries

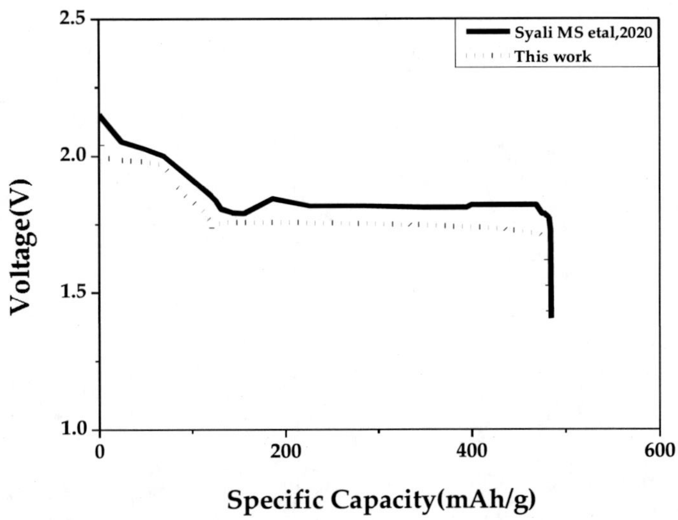

Figure 15. Voltage v/s specific capacity for TEGDME electrolyte at 0.1 C.

Figure 15 shows the discharge voltage of the battery while using the TEGDME electrolyte at a discharge rate of 0.1 C. The data obtained using simulation were compared to those obtained using experimental research in the reference papers. The capacity of the battery increases with the amount of electrode material present. Therefore, although they may have the same open-circuit voltage, a small battery will have less capacity than a larger cell. Figure 18 depicts the discharge cycle number and the capacity of the cell. The simulated value shows an initial capacity of 795 mAh/g, while the experimental results show 855 mAh/g. The simulation is found to run over 110 cycles, while the experiment is stopped at 104 cycles, as shown in Figure 16.

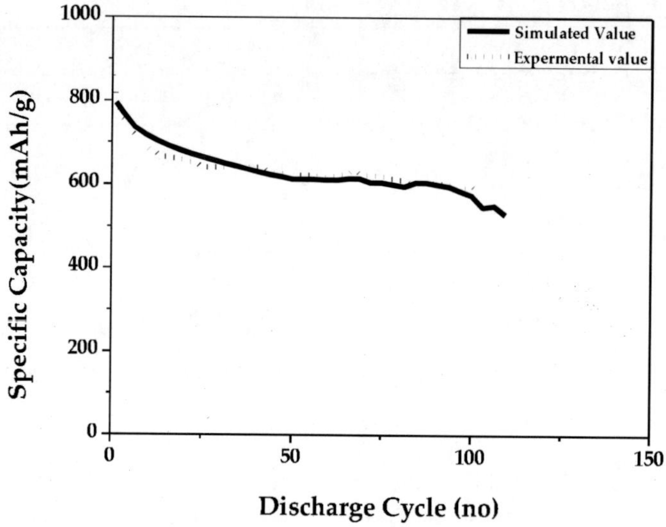

Figure 16. Specific capacity v/s Discharge Cycle for TEGDME based electrolyte at 0.1 C.

3.5. KOH-Based Sodium Sulphur Battery

The voltage dissipation of the KOH-based battery at different discharge rates is shown in Figure 17. The discharge circumstances, including the amount of current, the battery's permitted terminal voltage, temperature, and other factors, affect a battery's capacity. The pace of battery drain determines how much capacity is accessible. A battery's usable capacity will be lower than anticipated if it is depleted quite quickly. The battery's energy is delivered more effectively when discharging at a low rate compared to a higher discharge rate.

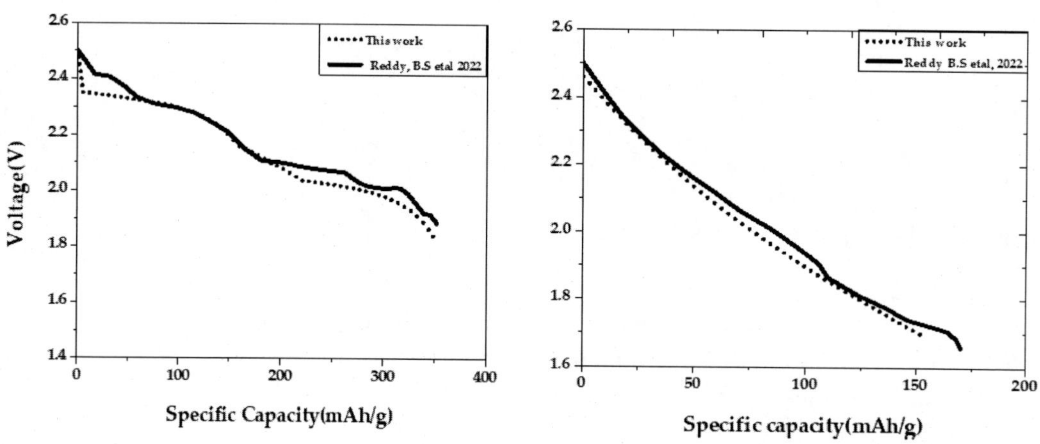

a

b

Figure 17. (Continued)

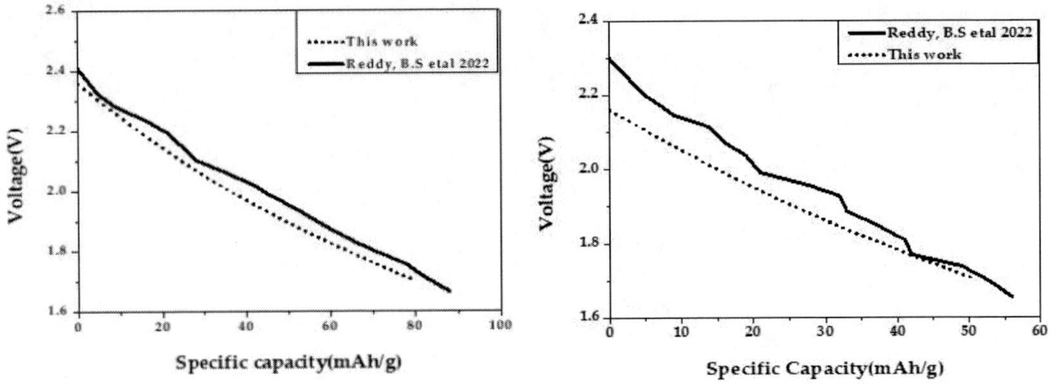

c

d

Figure 17. Voltage v/s Specific capacity for KOH electrolyte at different discharge rates a) 0.1 C b) 0.5 C c)1 C d) 2 C.

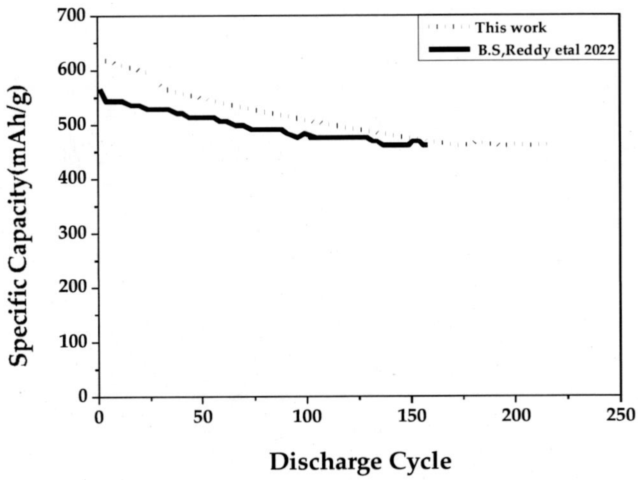

Figure 18. Specific Capacity v/s cycle number.

Figure 18 depicts the discharge cycle number and the capacity of the cell for KOH electrolyte. The simulated value shows an initial capacity of 620 mAh/g, while the experimental results show 592 mAh/g. The simulation is found to run over 200 cycles, while the experiment is stopped at 158 cycles.

Conclusion

Research regarding the development of Sodium Sulphur batteries was conducted, and the results were simulated. The simulations were conducted by varying the different parameters of the battery and changing the electrolytes on which the battery is based. The electrolytes that were used for the experiments were potassium hydroxide, Sulphuric acid, TEGDME and lithium hexafluorophosphate. The following conclusions were drawn from the observed simulations:

1. The use of Sulphuric acid–based electrolytes caused the deterioration of the cathode with an immediate effect. Even though the electrolyte gave very high voltages, Sulphuric acid–based electrolytes can be discarded due to the corrosion caused to the electrodes.
2. A lithium hexafluorophosphate-based electrolyte, which is mostly used alongside lithium-ion batteries, is tested with a Sodium Sulphur battery. The battery produced a very low voltage when used due to the poor kinetics of the Sulphur-based electrode while using the electrolyte.
3. The TEGDME-based electrolyte and KOH-based electrolyte yielded good voltages with minimal deterioration of the electrodes and electrolytes. The temperature of operation of the batteries with these electrolytes is simulated. The KOH-based electrolyte was found to operate at room temperature in accordance with the reference paper. The operating temperature of TEGDME-based electrolytes operated in an intermediate temperature range, which is in accordance with the research paper.
4. The batteries with the abovementioned electrolytes are run at different discharge rates, and the voltages and capacities are calculated and validated.
5. Further research on the battery can be conducted by improving the current-collecting material properties and the capacity of the electrodes by doping the electrode materials with good conductors. Thus, we can improve the capacity of the Sulphur cathode, thus improving the charge capacity of the battery module.

References

[1] Ibrahim, H., Ilinca, A. and Perron, J., 2008. Energy storage systems—Characteristics and comparisons. *Renewable and sustainable energy reviews*, *12*(5), pp.1221-1250.

[2] Buckley, D. N., O'Dwyer, C., Quill, N. and Lynch, R. P., 2018. Electrochemical energy storage. *Energy storage options and their environmental impact*, *46*, p.115.

[3] Vikström, H., Davidsson, S. and Höök, M., 2013. Lithium availability and future production outlooks. *Applied Energy*, *110*, pp.252-266.

[4] Oshima, T., Kajita, M. and Okuno, A., 2004. Development of Sodium-sulphur batteries. *International Journal of Applied Ceramic Technology*, *1*(3), pp.269-276.

[5] Nikiforidis, G., Van de Sanden, M. C. M. and Tsampas, M. N., 2019. High and intermediate temperature Sodium–sulphur batteries for energy storage: development, challenges and perspectives. *RSC advances*, *9*(10), pp.5649-5673.

[6] Lu, X., Kirby, B. W., Xu, W., Li, G., Kim, J. Y., Lemmon, J. P., Sprenkle, V. L. and Yang, Z., 2013. Advanced intermediate-temperature Na–S battery. *Energy & Environmental Science*, *6*(1), pp.299-306.

[7] Hueso, K. B., Palomares, V., Singh, G., Armand, M. and Rojo, T., 2015. Sodium–Sulphur Batteries, Handbook of Clean Energy Systems, ed. J. Yan.

[8] Kohl, M., Borrmann, F., Althues, H. and Kaskel, S., 2016. Hard Carbon Anodes and Novel Electrolytes for Long-Cycle-Life Room Temperature Sodium-Sulphur Full Cell Batteries. *Advanced Energy Materials*, *6*(6), p.1502185.

[9] Sudworth, J. L. and Tilley, A. R., 1983. Sodium-sulphur battery with beta-alumina electrolyte. *Sulphur Electrode, Fused Salts and Soild Electrolytes*, pp.235-325.

[10] Kodama, E. and Kurashima, Y., 1999. Development of a compact Sodium sulphur battery. *Power Engineering Journal*, *13*(3), pp.136-141.

[11] Wang, Y. X., Zhang, B., Lai, W., Xu, Y., Chou, S. L., Liu, H. K. and Dou, S. X., 2017. Room-temperature Sodium-sulphur batteries: a comprehensive review on research progress and cell chemistry. *Advanced Energy Materials*, *7*(24), p.1602829.

[12] Kumar, D., Rajouria, S. K., Kuhar, S. B. and Kanchan, D. K., 2017. Progress and prospects of Sodium-sulphur batteries: A review. *Solid State Ionics*, *312*, pp.8-16.

[13] Manthiram, A. and Yu, X., 2015. Ambient temperature Sodium–sulphur batteries. *Small*, *11*(18), pp.2108-2114.

[14] Wei, S., Xu, S., Agrawral, A., Choudhury, S., Lu, Y., Tu, Z., Ma, L. and Archer, L. A., 2016. A stable room-temperature Sodium–sulphur battery. *Nature communications*, *7*(1), pp.1-10.

[15] Tadayon, N., Ramazani, A., Torabi, M. and Seyyedin, S.T., 2020. Using of various metal species for improvement of electrochemical performances of lithium sulphur batteries. *Journal of Electroanalytical Chemistry*, *878*, p.114652.

[16] Syali, M. S., Kumar, D., Mishra, K. and Kanchan, D. K., 2020. Recent advances in electrolytes for room-temperature Sodium-sulphur batteries: A review. *Energy Storage Materials*, *31*, pp.352-372.

[17] Wen, Z., Cao, J., Gu, Z., Xu, X., Zhang, F. and Lin, Z., 2008. Research on Sodium sulphur battery for energy storage. *Solid State Ionics*, *179*(27-32), pp.1697-1701.

[18] Liu, D., Li, Z., Li, X., Cheng, Z., Yuan, L. and Huang, Y., 2019. Recent Advances in Cathode Materials for Room-Temperature Sodium− Sulphur Batteries. *ChemPhysChem*, *20*(23), pp.3164-3176.

[19] Zhang, H., Diemant, T., Qin, B., Li, H., Behm, R. J. and Passerini, S., 2020. Solvent-dictated Sodium sulphur redox reactions: investigation of carbonate and ether electrolytes. *Energies*, *13*(4), p.836.

[20] Reddy, B. S., Reddy, N. S., Nam, S. Y., Ahn, H. J., Ahn, J. H. and Cho, K. K., 2022. Potassium hydroxide activated carbon derived from albumen as an efficient sulphur host for room temperature Sodium-sulphur batteries. *Journal of Energy Storage*, *45*, p.103666.

[21] Zhang, M. F., Zhu, C. F., Yang, C. P. and Lv, Y. N., 2018. Preparation of amorphous Ni-W coating for the current collector of Na/S battery by electrodeposition. *Surface and Coatings Technology*, *346*, pp.40-47.

[22] Walawalkar, R., Apt, J. and Mancini, R., 2007. Economics of electric energy storage for energy arbitrage and regulation in New York. *Energy Policy*, *35*(4), pp.2558-2568.

[23] Stahlkopf, K., 2015. Taking wind mainstream. *IEEE Spectrum, retrieved Mar*, *21*, p.4.

[24] Koenig, A. A. and Rasmussen, J. R., 1990. Development of a high specific power Sodium sulphur cell. In *Proceedings of the 34th International Power Sources Symposium* (pp. 30-33). IEEE.

Chapter 17

The Facile Synthesis of an Amino-Functionalized Nanocellulosic Fibrous Mat (NH_2/CA-EC) from a Plant Source for the Enhanced Removal of Organo Arsenic Contaminants

B. Uma Maheswari
V. M. Sivakumar*
and M. Thirumarimurugan
Department of Chemical Engineering, Coimbatore Institute of Technology, Coimbatore, India

Abstract

Roxarsone (ROX) and P-arsanilic acid (PA) are persistent, toxic, bio accumulative Organo arsenic contaminants emerging from feeding additives that pose high potential risks to the environment. In the present study, a novel biodegradable NH_2-functionalized nano cellulosic fibrous mat was synthesized using the waste aquatic weed *Eichhornia crassipes* (NH_2/CA-EC) and employed to adsorb ROX and PA. The chemical and morphological structures of as-synthesized NH_2/CA-EC were investigated by scanning electron microscopy (SEM) and Fourier transform infrared spectroscopy (FTIR). The results of NH_2/CA-EC exhibit ample nanofiber structures and abundant functional groups, such as -NH_2, C=O and -OH. Various parameters, such as adsorbent loading (0–5 g/L), solution pH (1–12), initial concentration (ROX, PA) (10–100 mg/L), and contact time (0–120 min), on the adsorption behavior of the fabricated NH_2/CA-EC cellulosic fibrous mat was investigated.

High removal efficiencies of 97.62% (ROX) and 95.23% (PA) and adsorption capacities of 96.23 mg/g (ROX) and 103.25 mg/g (PA) were achieved under the optimum conditions.

Keywords: adsorption, amino-functionalized, NH_2/CA-EC, cellulosic fiber mat, organo arsenic contaminants, remediation

---

* Corresponding Author's Email: vmsivakumar@cit.edu.in.

In: Sustainability in Chemical Processes through Digitalization and Green Chemistry Approaches
Editors: Susarla V. A. R. Sastry, Shravan Kumar and Ashish Kapoor
ISBN: 979-8-89113-257-3

1. Introduction

Arsenic has become a major concern worldwide because it is highly toxic and poses a series of threats to human health and ecosystems [1]. Generally, organ arsenic compounds, such as roxarsone (ROX) and p-arsanilic acid (PA), have been widely used in the poultry and swine industries as feed additives. It is mainly used to rheostat intestinal parasites and improve feed efficiency [2]. The archetypal levels of ROX and PA in poultry and swine feeds are found to be 20–50 gt^{-1} and 45–100 $g\,t^{-1}$, respectively [3]. Although they have substantial benefits, nearly 90% of these compounds are not metabolized properly and are released directly into the environment as solid waste through animal excrement. Consequently, owing to their high-water solubility nature, these compounds are quickly leached from soils, thereby increasing their prevalence in both aquatic and terrestrial ecosystems [4]. This in turn subsequently develops an environmental risk, as they possess carcinogenic properties and might slowly transmute into highly toxic inorganic arsenic compounds such as arsenate and arsenite via the biological degradation process [5]. To protect the environment, it is urgent to develop treatment methods to remove organoarsenic contaminants before they are converted into their inorganic form.

Recent research on organoarsenic removal has mainly focused on adsorption and degradation. Although the degradation method of removal shows exemplary results (nearly 99%) on removal, the major drawback is the formation of secondary pollutants. To overcome this problem, the adsorption process came into practice. To date, ample studies have been reported for organ arsenic compound removal through various adsorbents, such as metal oxides [6], multiwalled carbon nanotubes [7], iron oxides [8] and metal organic frameworks [9]. However, while these adsorbents have shown promising removal efficiency, the disposal of the generated sludge is the main disadvantage. Hence, the choice of biodegradable and inexpensive materials has been considered for our present research.

Cellulose, a natural and renewable polymer, has attracted attention owing to its wide range of applications, as it is thermally stable and biodegradable in nature [10]. Cellulose contains abundant hydroxyl functional groups; however, they are not very effective owing to their lack of stability. Hence, functionalization of cellulose has been adopted to improve its stability. As a substitute, the incorporation of specific functional groups into raw cellulose creates good mechanical strength, high surface area and pore volume [11]. In recent times, amino-functionalized cellulose adsorbents have exhibited high potential for the removal of contaminants from wastewater [12]. As NH_2 groups are more prone to acidic environments, ion exchange and electrostatic interaction mechanisms play a major role in the removal process. Similarly, functionalizing the surface of cellulose with amine groups hosts more cationic charges on the surface and can also be used to conjugate biomolecules on the surface [13]. Previous studies have investigated the potential use of amino-functionalized biopolymers such as cellulose for various organic dyes and heavy metal ions [14, 15].

However, exploiting physical as well as chemical routes for the synthesis process may substantially lead to increased energy consumption, high costs, low production yield and damage to the environment by retaining harmful reducing agents. In addition, the utilization of microorganisms is also much riskier owing to pathogenic issues and high maintenance. Hence, in the present work, we preferred a greener method for the synthesis of nanofibrous mats. In our previous work, *Eichhornia crassipes* was utilized as a greener source to synthesize nano adsorbents, and we found that it has more potential for removing heavy metal ions from wastewater [16]. In the present study, we develop and optimize a greener approach that involves the functionalization of cellulose acetate with amines in aqueous solution to synthesize a nano cellulosic fibrous mat (NH_2/CA-EC) by a simple reaction protocol. The removal efficiency of ROX and PA using NH_2/CA-EC was evaluated in batch adsorption experiments. Further isotherm, kinetic, mechanism, and regeneration studies of the process by NH_2/CA-EC are discussed subsequently.

2. Materials and Methods

2.1. Raw materials and Reagents

Eichhornic crassipes (EC) was collected near ponds in and around Coimbatore district. Cellulose acetate (CA), acetone, ammonium hydroxide solution, hydrochloric acid, N, N-dimethylacetamide (DMA), ethyl alcohol, and Millipore-ionized (D.I.) water were used for all sample preparations.

2.2. Synthesis of Amine-Functionalized Nanocellulosic Mat (NH_2/CA-EC)

Eichhornia crassipes utilized for the preparation of adsorbents were washed thoroughly with distilled water and retained in an oven at 120 °C for 10 h. They were then ground to a fine powder and sieved. The finely powdered weed sample was treated with 0.5 mL of HCl, 25 mL of acetone, 20 mL of tetraethylorthosilicate (TEOS), and 5 mL of distilled water. The reaction mixture was stirred in a magnetic stirrer for 60 minutes, followed by 30 minutes of sonication. Finally, it was dried at 60 °C for 3 h, ground, and stored in desiccators for further experimental purposes.

First, 0.5 g of EC was added to 0.8 g of CA in 5 mL of acetone and 2.5 mL of DMA, and the reaction mixture was stirred at room temperature for 2 h to obtain a homogenous solution. The prepared solution mixture was further transferred to a syringe with a stainless-steel needle of 0.82 mm inner diameter and fed at a flow rate of 0.8 mLh^{-1}. During the electrospinning process, aluminum foil paper was utilized to receive the fibrous material. The obtained fiber mats were kept in a fume hood for 12 hours to remove the remaining solvents. Finally, the dried mats (NH_2/CA-EC) were shed from the foil for further experiments.

2.3. Characterization of Amine-Functionalized Nano cellulosic Mat (NH_2/CA-EC)

The surface morphologies of NH_2/CA-EC were observed by scanning electron microscopy (SEM, Nova Nano SEM 230, FEI, USA). A Fourier transform infrared (FTIR) spectrometer (400 to 4000 cm^{-1}) was utilized to determine the functional groups of NH_2/CA-EC using AVAT-AR 360, Nicolet, USA.

2.4. Adsorption Studies

Adsorption studies were conducted to examine equilibrium parameters such as pH, dosage, metal ion concentration and contact time. Batch adsorption experiments were carried out using 0.1 g of NH_2/CA-EC added to 50 mL of ROX and PA solutions at an initial concentration of 10 mgL^{-1}. The pH of the prepared solution was adjusted by adding a 1 M HCl solution. The mixture was kept in a shaking incubator (NB-205 L, N-BIOTEK) at 200 rpm for the predetermined time interval. Then, the mixture was filtered, and the filtrate was analyzed using an atomic absorption spectrometer (AAS) to determine the ROX and PA concentrations. By varying a particular parameter and keeping the remaining constant of ROX and PA uptake at equilibrium, q_e (mgg^{-1}) and its percentage removal (%), it can be optimized. Once equilibrium was achieved, the adsorbent was filtered and measured. The amount was determined by using equation (1):

$$q_e = (C_i - C_e)\frac{V}{W} \qquad (1)$$

Percentage removal was calculated as per Equation (2):

$$\text{Percentage Removal (\%)} = \frac{C_i - C_e}{C_i} \times 100 \qquad (2)$$

where C_i (mgL^{-1}) and C_e (mgL^{-1}) are the initial and final concentrations of ROX and PA, respectively. V (L) is the volume of the test solution, and W (g) is the mass of NH_2/CA-EC.

Adsorption isotherms were tailored with the Langmuir, Freundlich and Temkin models. Similarly, the kinetic study results were fitted with nonlinear pseudo first-order and pseudo second-order models. The reusability efficiency of NH_2/CA-EC was evaluated at 0.1% HCl for 24 h. The regeneration process was repeated three times to completely remove the adsorbed species.

3. Results and Discussion

3.1. Characterization of NH_2/CA-EC

The synthesized NH_2/CA-EC samples were characterized by FESEM and FT-IR spectroscopy. SEM micrographs of electro spun CA-EC nanofibers and amine-functionalized CA-EC nanofibers are presented in Figures 1(a, b), respectively. These fibers were uniform, long, continuous and in nonwoven mat form. These fibers were even, elongated and of nonwoven mat structure. Furthermore, after the deacetylation process, the morphology of the fibers remained unbroken. From Figure 1 (b), it could be observed that a little winding distortion and enlargement take place in the nanofibers. This was attributed to the formation of hydrophilic OH groups present in the cellulose chain that would instantly form hydrogen bonds, thereby adsorbing more water molecules. This would in turn increase the space between the cellulose chains, resulting in the subsequent expansion and bending of nanofibers [17].

After functionalization with amino groups, the nanofibers exhibit obvious crosslinking, and certain interspaces between the nanofibers are further blocked. The functionalized amino groups on the surface of CA-EC reacted with COO groups on CHO cellulose membranes (Schiff base reaction), which rigorously intensified the cross-linking effect amid the nanofibers.

Furthermore, the EDS mapping of NH_2/CA-EC is depicted in Figure 1(c). The presence of C and O elements was consistently dispersed on the external surface of the fiber, and the presence of nitrogen was clearly observed, indicating that the amino group was successfully functionalized onto the surface of the CA-EC.

FTIR analysis was used to detect the prominent peaks of CA-EC and NH_2/CA-EC, which are represented in Figure 1 (d). The evident C=O widening peak at 1730 cm^{-1} was due to the presence of esters in CA-EC. Similarly, the corresponding stretch at 1241 cm^{-1} indicates a C-O stretch, followed by a peak at 1376 cm^{-1} C–CH_3 bend [18].

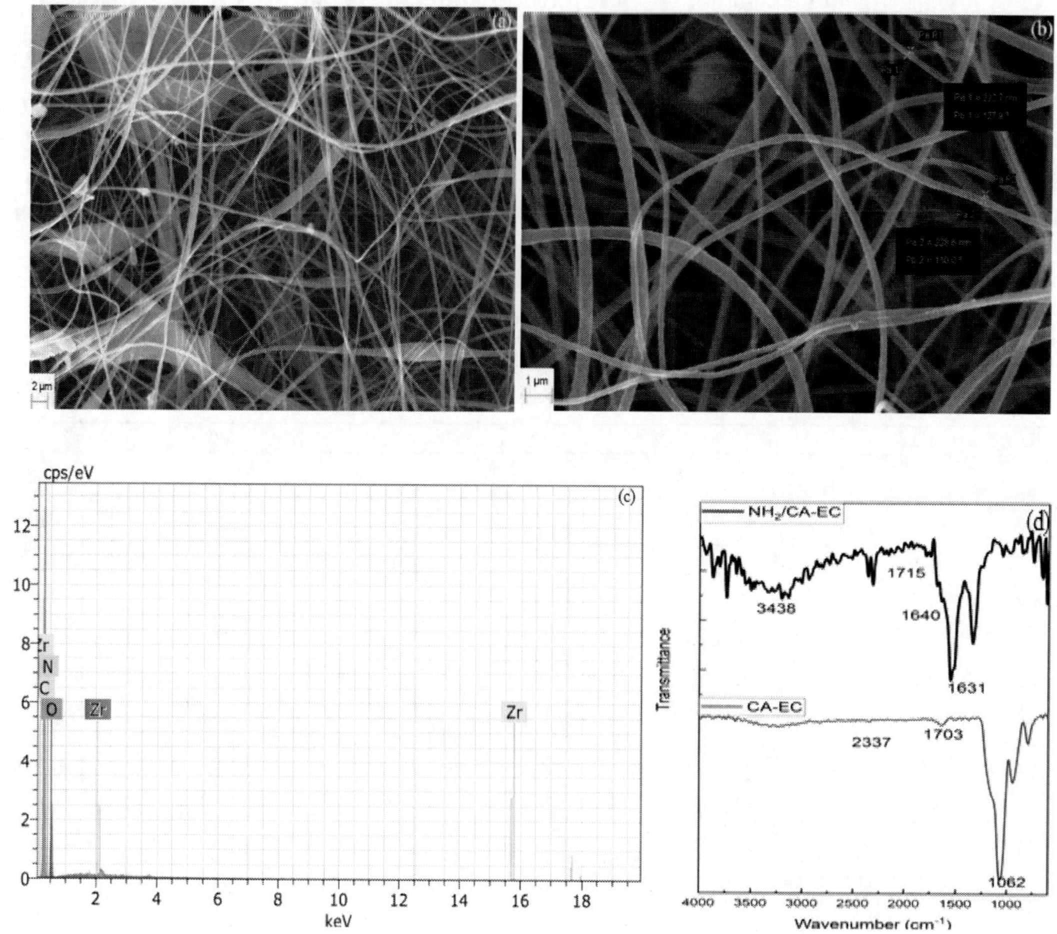

Figure 1. (a) SEM images of CA-EC (b) SEM images of NH_2/CA-EC (c) EDX of NH_2/CA-EC d) FTIR of NH_2/CA-EC.

The spectrum of a deacetylated sample displays stretching at 1062 cm^{-1} owing to C-OH molecules. The functionalization of amine groups (-NH_2) on the surface of CA-EC was confirmed by the peaks present at 1631 cm^{-1} and 3438 cm^{-1} [19, 20]. Similarly, the peaks at 1715 cm^{-1} (C = O stretching) and 1640 cm^{-1} (C = N stretching) indicate that NH_2 was successfully functionalized on the CA surface [17].

3.2. Effect of pH

To determine the adsorption behavior of the synthesized NH_2/CA-EC, the effect of pH varying from 1–12 on ROX and PA was analyzed, as shown in Figure 2(a). Initially, the adsorption rate was high and increased with increasing pH. The maximum removal efficiency occurred at pH 8. However, when pH values increased from 8 to 12, a sharp decline in adsorption rate was observed. The adsorption capacities of ROX and PA were found to be 51.23 mgg^{-1} and 29.23 mgg^{-1} at pH 8 and 7.7, respectively, which decreased to 21.36 mgg^{-1} at pH 10. From these results, it was evident that neither very low nor very high pH favors the adsorption process.

Furthermore, in the zeta potential analysis, the isoelectric point of NH_2/CA-EC was found to be 10.5. Hence, the surface charge of NH_2/CA-EC was found to be positive, owing to the protonation of the amine group at pH <10.5. However, as the pH decreases, the charge of the adsorbent will increase further. Instead, when pH > 10.3, upon further increasing pH < 10.5, a high negative charge dominates on the NH_2/CA-EC surface, thereby causing deprotonation of amino groups. The surface charges at various levels of pH are tabulated (Table 1) as follows:

Table 1. Surface charge of NH_2/CA-EC values at various pH values

pH range	NH_2/CA-EC	ROX	PA
$1 \leq pH \leq 3$	+	-	-
$3.5 \leq pH \leq 10.5$	+	-	-
$10.5 \leq pH \leq 12$	-	-	-

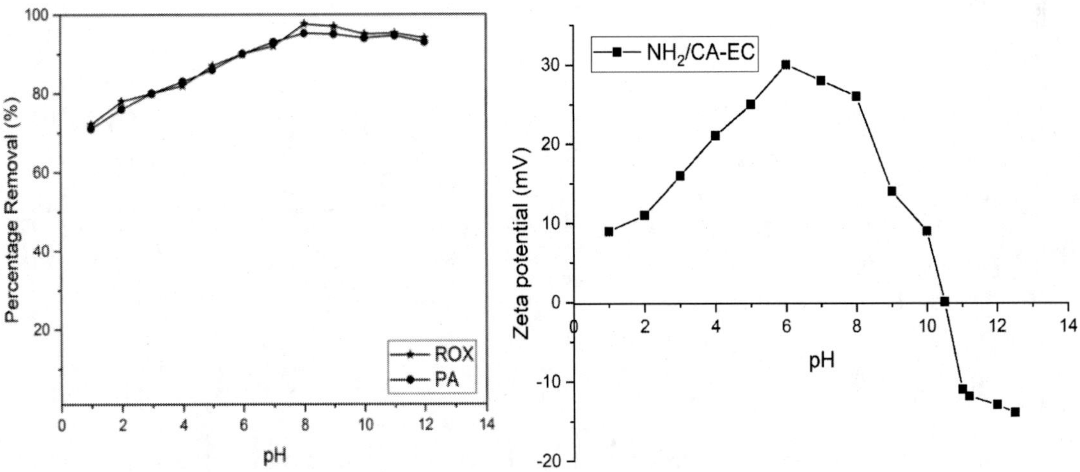

a) b)

Figure 2. a) Effect of pH b) Zeta potential of NH_2/CA-EC.

3.3. Adsorption Isotherm Analysis

To determine the interaction between ROX and PA, isotherm studies were carried out to elucidate the interaction between the solid and aqueous surfaces. The adsorption equilibrium data were analyzed using Langmuir, Freundlich and Temkin models [21–23]. Nonlinear regression analysis was performed using Origin 2017 software to determine the values and regression coefficient (R^2). Based on the results of the analysis, a suitable adsorption isotherm model was chosen for the study. The schematic representations of the models are depicted in Figures 3 (a, b). All the constants and parameter values are presented in Table 2. As depicted in Table 2, the Langmuir isotherm model designated better adsorption of ROX and PA onto NH_2/CA-EC than the Freundlich and Temkin isotherm models, with higher regression coefficients of 0.99, 0.95, and 0.90, respectively. Hence, it could be concluded that a uniform energy was exhibited on the surface of NH_2/CA-EC and that the adsorption of ROX and PA fit monolayer adsorption.

Table 2. Isotherm Model Parameters for Adsorption of ROX and PA

Isotherm model	M-WH	M-DW	Equation
Langmuir	q_m= 96.23 K_L =0.56 R^2=0.98	q_m= 103.25 K_L =2.6 R^2=0.99	$ROX: q_e = \frac{53.88\, C_e}{1+0.56C_e}$ $PA: q_e = \frac{268.45\, C_e}{1+2.6C_e}$
Freundlich	K_F=2.1 n=2.3 R^2=0.98	K_F=2.16 n=1.64 R^2=0.98	$ROX: q_e = 2.19\, C_e^{0.43}$ $PA: q_e = 3.37\, C_e^{0.60}$
Temkin	A=19.25 B=6.21 R^2=0.89	A=31.73 B=2.66 R^2=0.88	$ROX: q_e = 6.21 \ln(19.25\, C_e)$ $PA: q_e = 2.66 \ln(31.73\, C_e)$

Table 3. Comparison of the Maximum Monolayer Adsorption Capacity of Various Adsorbents

S.No	Heavy Metal	Adsorbents	q_m (mg/g)	References
1	ROX	NH_2/CA-EC	96.23	Present Study
2	PA	NH_2/CA-EC	103.25	Present Study
3	ROX	Fe/La-Mt	31.07	[24]
4	ROX	Goethite-cellulose composite	17.4	[25]
5	ROX	Chitin nano fibers	69.15	[17]
6	PA	Nanohydrated zirconium oxide	24	[26]
7	PA	Fe-Ti-Mn composite oxide	45.6	[27]
8	PA	MgO	86.8	[28]

Based on the Langmuir model, the maximum monolayer adsorption capacity (q_m) was found to be 96.23 mg/g (ROX) and 103.25 mg/g (PA). The higher q_m values of ROX and PA on NH_2/CA-EC clearly indicate that the introduction of NH_2 groups onto CA-EC effectively improved the adsorption capacity and affinity of CA-EC toward ROX and PA. Table 3 displays the value of q_m for the present study along with the list of adsorbents used for the removal of ROX and PA published in the literature. From the tabulated values, NH_2/CA-EC clearly displays better adsorption capacity than most other adsorbents used for the removal of ROX and PA [24–26].

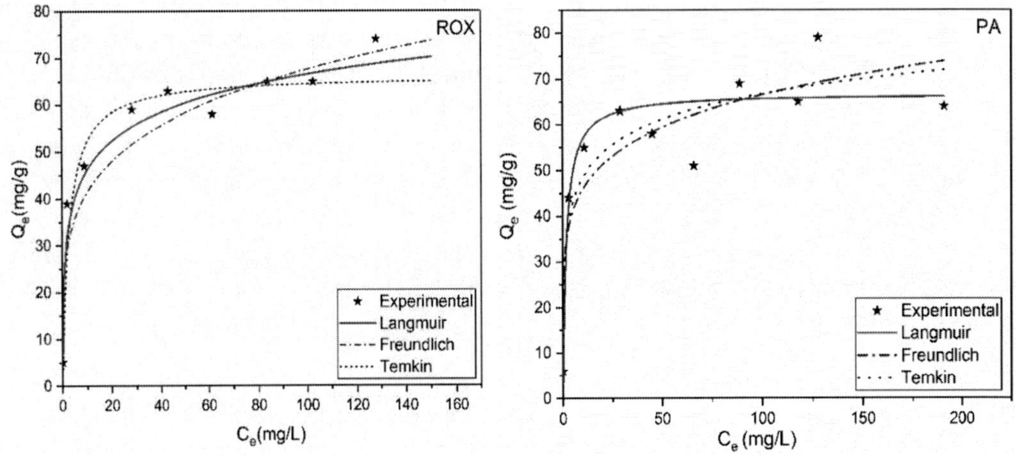

a) b)

Figure 3. Isotherm plots of a) ROX and b) PA.

3.4. Kinetic Study

Experimental data were fitted to pseudo first order [27–29], pseudo second order [30] and intraparticle diffusion models [31] to analyze the rate-governing mechanism of the process. A nonlinear analysis was performed, and the corresponding images fitting the models are illustrated in Figure 4 (a, b). The corresponding parameter results are mentioned in Table 4. In Figure 3 (a, b), NH_2/CA-EC exhibited relatively rapid adsorption within 80 min of equilibrium for ROX and PA, respectively.

Table 4. Kinetic Model Parameters for Adsorption of ROX and PA

Model name	Parameters &Units		ROX	PA
Pseudofirst order	k_1	(mg/min)	0.21	0.14
	q_e	(mg/g)	66.33	101.62
	R^2		0.70	0.85
Pseudosecond order	k_2	(mg/min)	3.16	1.46
	q_e (exp)	(mg/g)	114.02	167.09
	q_e(cal)	(mg/g)	112.39	166
	R^2		0.99	0.99
Intra particle diffusion	k_d	(mg/g $min^{0.5}$)	11	13
	C	(mg/g)	75.3	99.32
	R^2		0.90	0.88

As illustrated in Figure 3 (a, b), NH_2/CA-EC exhibited relatively rapid adsorption within 80 min of equilibrium for ROX and PA. Furthermore, as tabulated in Table 4, the pseudo second-order model displayed higher R^2 values for ROX and PA than the pseudo first-order model. From the experimental findings, it is evident that the ROX and PA adsorption process on NH_2/CA-EC was a chemisorption process in which electron transference between ROX and PAA with respect to NH_2/CA-EC was the main controlling mechanism during the process.

In addition, the data were fitted to an intraparticle diffusion model to designate the diffusion of ROX and PA inside NH_2/CA-EC. In the intraparticle model, the adsorption of contaminants on the adsorbent is divided into three linear regions. The initial stage was the boundary layer diffusion process, followed by void diffusion. In the initial stage, ROX and PA are randomly diffused from the aqueous solution to the surface of NH_2/CA-EC, resulting in an increase in adsorption capacity. Then, ROX and PA persistently diffuse into the adsorption site in the pores of NH_2/CA-EC, thereby increasing the adsorption capacity gradually in this stage.

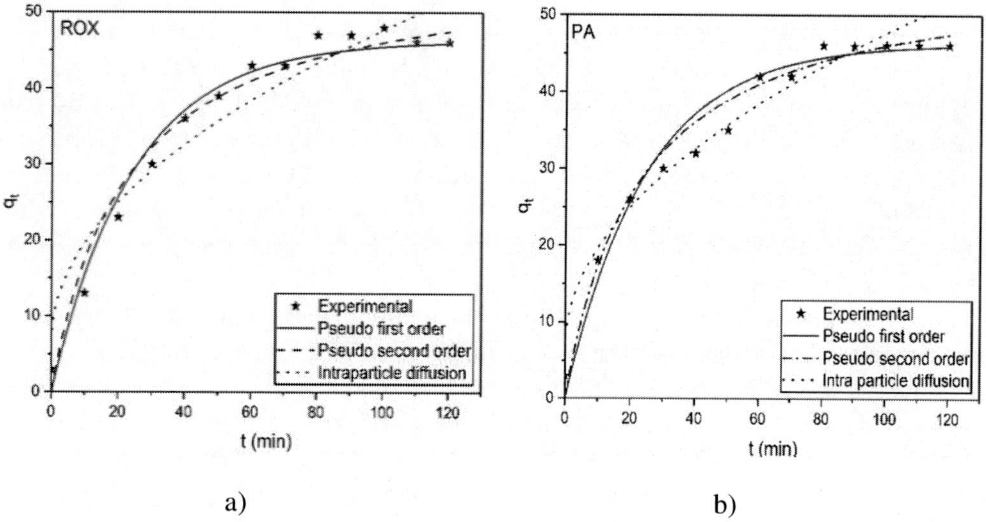

Figure 4. Kinetic plot a) ROX b) PA.

Finally, the last stage was the adsorption reaction stage, where the adsorption remained constant, indicating that adsorption equilibrium was attained.

3.5 Synergistic Adsorption Mechanism of NH_2/CA-EC

The main mechanisms involved in ROX and PA adsorption over NH_2/CA-EC are electrostatic interactions and hydrogen bonding. Being a base, amine groups strongly attract protons, thereby having a much higher isoelectric point (10.3) than the original (3.3). This indicates that amino-functionalized compounds have a higher positive charge even at high pH. Thus, the NH_2/CA-EC nano cellulosic mat exhibits excellent ROX and PA removal performance in the pH range of 7-8 due to electrostatic interactions. Another important and yet commonly occurring mechanism is hydrogen bonding. Generally, oxygen atoms accept hydrogen atoms easily and form hydrogen bonds easily [9]. However, due to the formation of nitrogen atoms, they tend to possess a strong affinity toward protons owing to their lone pair of electrons [32].

Based on the FT-IR results, it is evident that there were more –OH, C = O, COO and -NH_2 species on the surface of NH_2/CA-EC and the ROX (4-Hydroxy-3-nitrophenylarsonic acid), PA (4-Aminophenyl) arsenic acid, molecules possess a number of nitrogen and oxygen-containing functional groups. This is in turn coupled with hydrogen atoms to form hydrogen bonds such as O···H-O, O-H···O, C-OO···H, N···H-N, and O···H-N. The formation of these bonds highly enhances the adsorption process, thereby increasing the removal efficiency.

However, -OH, C=O, and $-NH_2$ will tend to decrease further. When amine groups are introduced into CA-EC, more hydrogen acceptors are enhanced by these positively charged protons, which stimulate hydrogen bonding, resulting in a significant enhancement in ROX and PA adsorption over CA-EC.

Conclusion

In this work, an NH_2/CA-EC nanocomposite was synthesized for the effective removal of ROX and PA, which is an incipient organo arsenic pollutant. Amino-functionalized green synthesized nanocomposites from waste aquatic plants were utilized as efficient adsorbents for contaminant removal. The adsorption capacity of ROX and PA obtained from NH_2/CA-EC was analyzed using four operational parameters, such as contact time, NH_2/CA-EC dosage, and initial ROX and PA concentration, followed by pH. Based on the investigations, the optimal conditions were as follows: pH: 8, contact time: 80 min, adsorbent dosage of 2 g/L, and initial concentration of 20 mg/L.

In addition, the ROX and PA equilibrium adsorption data fit well to the Langmuir model, and the maximum monolayer adsorption capacities were 96.23 mg/g and 103.25, respectively. Additionally, adsorption kinetic data followed the pseudo second-order model, where the experimental data coincided well with the theoretical data. NH_2/CA-EC exhibited excellent reusability during the adsorption of ROX and PA for up to four cycles. The main mechanisms involved are the synergistic effects of electrostatic interactions and hydrogen bonding, which are the main mechanisms prevailing between adsorbents and adsorbates. Therefore, this study provides a good experimental approach for the applications of green synthesized cellulose-based functionalized nanomaterials for the remediation of organic contaminants.

Acknowledgment

The authors would like to acknowledge the support and grant received through the Research Promotion Scheme (grant no. 8-97/FDC/RPS (POLICY-1)/2019-20) from the All-India Council for Technical Education for their financial support.

References

[1] Zubair Y. O., S. Fuchida, and C. Tokoro, "Insight into the Mechanism of Arsenic(III/V) Uptake on Mesoporous Zerovalent Iron–Magnetite Nanocomposites: Adsorption and Microscopic Studies," *ACS Appl. Mater. Interfaces*, vol. 12, no. 44, pp. 49755–49767, Nov. 2020, doi: 10.1021/acsami.0c14088.

[2] Xu Y., "Efficient removal of low-concentration organoarsenic by Zr-based metal–organic frameworks: cooperation of defects and hydrogen bonds," *Environ. Sci. Nano*, vol. 6, no. 12, pp. 3590–3600, 2019, doi: 10.1039/C9EN00923J.

[3] Liu X., "Arsenic pollution of agricultural soils by concentrated animal feeding operations (CAFOs)," *Chemosphere*, vol. 119, pp. 273–281, Jan. 2015, doi: 10.1016/j.chemosphere.2014.06.067.

[4] Liu X., W. Zhang, Y. Hu, and H. Cheng, "Extraction and detection of organoarsenic feed additives and common arsenic species in environmental matrices by HPLC–ICP–MS," *Microchem. J.*, vol. 108, pp. 38–45, May 2013, doi: 10.1016/j.microc.2012.12.005.

[5] Wang L. and H. Cheng, "Birnessite (δ-MnO 2) Mediated Degradation of Organoarsenic Feed Additive p -Arsanilic Acid," *Environ. Sci. Technol.*, vol. 49, no. 6, pp. 3473–3481, Mar. 2015, doi: 10.1021/es505358c.

[6] Joshi T. P., "Adsorption of aromatic organoarsenic compounds by ferric and manganese binary oxide and description of the associated mechanism," *Chem. Eng. J.*, vol. 309, pp. 577–587, Feb. 2017, doi: 10.1016/j.cej.2016.10.084.

[7] Mahaninia M. H.and L. D. Wilson, "A Kinetic Uptake Study of Roxarsone Using Cross-Linked Chitosan Beads," *Ind. Eng. Chem. Res.*, vol. 56, no. 7, pp. 1704–1712, Feb. 2017, doi: 10.1021/acs.iecr.6b04412.

[8] Depalma S., S. Cowen, T. Hoang, and H. A. Al-Abadleh, "Adsorption Thermodynamics of p-Arsanilic Acid on Iron (Oxyhydr)Oxides: In-Situ ATR-FTIR Studies," *Environ. Sci. Technol.*, vol. 42, no. 6, pp. 1922–1927, Mar. 2008, doi: 10.1021/es071752x.

[9] Lv Y., "Removal of p-arsanilic acid by an amino-functionalized indium-based metal–organic framework: Adsorption behavior and synergetic mechanism," *Chem. Eng. J.*, vol. 339, pp. 359–368, May 2018, doi: 10.1016/j.cej.2018.01.139.

[10] Suhas, V. K. Gupta, P. J. M. Carrott, R. Singh, M. Chaudhary, and S. Kushwaha, "Cellulose: A review as natural, modified and activated carbon adsorbent," *Bioresour. Technol.*, vol. 216, pp. 1066–1076, Sep. 2016, doi: 10.1016/j.biortech.2016.05.106.

[11] Stephen M., N. Catherine, M. Brenda, K. Andrew, P. Leslie, and G. Corrine, "Oxolane-2,5-dione modified electrospun cellulose nanofibers for heavy metals adsorption," *J. Hazard. Mater.*, vol. 192, no. 2, pp. 922–927, Aug. 2011, doi: 10.1016/j.jhazmat.2011.06.001.

[12] Shen J., "Amino-functionalized cellulose: a novel and high-efficiency scavenger for sodium cholate sorption," *Cellulose*, vol. 27, no. 7, pp. 4019–4028, May 2020, doi: 10.1007/s10570-020-03066-0.

[13] Akhlaghi S. P., "Synthesis of amine functionalized cellulose nanocrystals: optimization and characterization," *Carbohydr. Res.*, vol. 409, pp. 48–55, May 2015, doi: 10.1016/j.carres.2015.03.009.

[14] Wang Y., B. Wang, Q. Wang, J. Di, S. Miao, and J. Yu, "Amino-Functionalized Porous Nanofibrous Membranes for Simultaneous Removal of Oil and Heavy-Metal Ions from Wastewater," *ACS Appl. Mater. Interfaces*, vol. 11, no. 1, pp. 1672–1679, Jan. 2019, doi: 10.1021/acsami.8b18066.

[15] Wang R. J., L. W. Zhang, R. Liu, L. Liu, and J. M. Yao, "Ultrafast and probe-free cellulose biosensor for visual detection of Cu2+ ions in biological samples," *Carbohydr. Polym.*, vol. 223, p. 115117, Nov. 2019, doi: 10.1016/j.carbpol.2019.115117.

[16] U. M. Balasubramanian, S. Vaiyazhipalayam Murugaiyan, and T. Marimuthu, "Enhanced adsorption of Cr(VI), Ni(II) ions from aqueous solution using modified Eichhornia crassipes and Lemna minor," *Environ. Sci. Pollut. Res.*, vol. 27, no. 17, pp. 20648–20662, Jun. 2020, doi: 10.1007/s11356-019-06357-7.

[17] K. Liu, "Fabrication of amino-modified electrospun nanofibrous cellulose membrane and adsorption for typical organoarsenic contaminants: Behavior and mechanism," *Chem. Eng. J.*, vol. 382, p. 122775, Feb. 2020, doi: 10.1016/j.cej.2019.122775.

[18] Araga R. and C. S. Sharma, "Amine Functionalized Electrospun Cellulose Nanofibers for Fluoride Adsorption from Drinking Water," *J. Polym. Environ.*, vol. 27, no. 4, pp. 816–826, Apr. 2019, doi: 10.1007/s10924-019-01394-2.

[19] Booth H., R. M. Silverstein, G. C. Bassler and T. C. Morrill. "Spectrometric identification of organic compounds. Wiley, Chichester, 1991, pp. x + 419, £50.25 (cloth), ISBN 0 471 63404 2," *Magn. Reson. Chem.*, vol. 30, no. 4, pp. 364–364, Apr. 1992, doi: 10.1002/mrc.1260300417.

[20] Jin L., Q. Sun, Q. Xu, and Y. Xu, "Adsorptive removal of anionic dyes from aqueous solutions using microgel based on nanocellulose and polyvinylamine," *Bioresour. Technol.*, vol. 197, pp. 348–355, Dec. 2015, doi: 10.1016/j.biortech.2015.08.093.

[21] Langmuir I., "The Adsorption of Gases on Plane Surfaces of Glass, Mica and Platinum.," *J. Am. Chem. Soc.*, vol. 40, no. 9, pp. 1361–1403, Sep. 1918, doi: 10.1021/ja02242a004.

[22] Hi F., "Over the Adsorption in solution," *J Phy Chem*, vol. 57, pp. 385–470.

[23] Temkin M.J., Pyzhev. "Recent Modifications to Langmuir Isotherms," *Acta Physicochim. URSS*, vol. 12, pp. 12217–225.

[24] Wang Y.-J., F. Ji, W. Wang, S.-J. Yuan, and Z.-H. Hu, "Removal of roxarsone from aqueous solution by Fe/La-modified montmorillonite," *Desalin. Water Treat.*, vol. 57, no. 43, pp. 20520–20533, Sep. 2016, doi: 10.1080/19443994.2015.1108875.

[25] Kong D. and L. D. Wilson, "Synthesis and characterization of cellulose-goethite composites and their adsorption properties with roxarsone," *Carbohydr. Polym.*, vol. 169, pp. 282–294, Aug. 2017, doi: 10.1016/j.carbpol.2017.04.019.

[26] Zhao Z., P. Wu, Z. Fang, and X. Zhang, "Selective sequestration of p-arsanilic acid from water by using nanohydrated zirconium oxide encapsulated inside hypercross-linked anion exchanger," *Chem. Eng. J.*, vol. 391, p. 123624, Jul. 2020, doi: 10.1016/j.cej.2019.123624.

[27] Zhang W. "Simultaneous adsorption and oxidation of para arsanilic acid by a highly efficient nanostructured Fe-Ti-Mn composite oxide," *Chem. Eng. J.*, vol. 407, p. 127142, Mar. 2021, doi: 10.1016/j.cej.2020.127142.

[28] Peng Y., "Iron Humate as a Novel Adsorbent for p -Arsanilic Acid Removal From Aqueous Solution," *J. Dispers. Sci. Technol.*, vol. 37, no. 11, pp. 1590–1598, Nov. 2016, doi: 10.1080/01932691.2015.1120219.

[29] Lagergren S., "About the Theory of So-called Adsorption of Soluble Substances," *K. Sven. Vetenskapsakademiens Handl.*, vol. 24, pp. 1–39.

[30] Ho YS, G. McKay. "Pseudosecond order model for sorption processes," *Proc Biochem*, vol. 34, pp. 451–465.

[31] Morris J.C., W.J. Weber, "Kinetics of adsorption on carbon from solution," *J. Sanit. Eng. Div. Am. Soc. Civ. Eng.*, vol. 89, pp. 31–60, 1963.

[32] Liu B., F. Yang, Y. Zou, and Y. Peng, "Adsorption of Phenol and p -Nitrophenol from Aqueous Solutions on Metal–Organic Frameworks: Effect of Hydrogen Bonding," *J. Chem. Eng. Data*, vol. 59, no. 5, pp. 1476–1482, May 2014, doi: 10.1021/je4010239.

Chapter 18

The Effect of Microwave Blanching as a Green Chemistry Technique on the Quality of Lotus Rhizome Slices

Singh Akanksha[1,*]
and Kumar Vivek[2]
[1]School of Biochemical Engineering, IIT BHU, Varanasi, India
[2]Department of Food Technology, Harcourt Butler Technical University, Kanpur, India

Abstract

Microwave radiation is a clean alternative to conventional energy and hence is used in green chemistry. Blanching is an effective pretreatment technique used to inactivate browning enzymes in vegetables. The effect of microwave blanching at 360 W, 540 W and 900 W was studied in comparison to hot water blanching at 90°C and 50°C and steam blanching on lotus rhizome slices. This vegetable crop was selected because it is an amalgamation of good taste, rich nutrient content and crisp texture. The time of all the blanching methods was optimized to minimize nutritional and quality losses. Blanching efficacy was checked by the inactivation of peroxidase and catalase enzymes. Several quality parameters, such as ascorbic acid content, hardness, residual activity of the enzymes peroxidase and polyphenol oxidase, browning index and microstructural analysis, were determined. An increase in ascorbic acid content was observed after microwave blanching. Similarly, texture and microstructure were also found to be intact in microwave-blanched lotus rhizome slices compared to steam- and hot-water-blanched samples. Higher nutritional loss was observed in the case of hot water blanching due to leaching, making it an inefficient method of blanching. Browning decreased under all three treatments equally. Microwave blanching can therefore be regarded as the best and most ecologically sustainable pretreatment method.

Keywords: microwave radiation, green chemistry, optimized, steam and hot water blanching, ecologically sustainable

---

* Corresponding Author's Email: singhakanksha13@gmail.com.

In: Sustainability in Chemical Processes through Digitalization and Green Chemistry Approaches
Editors: Susarla V. A. R. Sastry, Shravan Kumar and Ashish Kapoor
ISBN: 979-8-89113-257-3

1. Introduction

Microwaves are electromagnetic radiation lying at the lower end of the electromagnetic spectrum. It lies in a frequency range of 3×10^2 to 3×10^5 MHz. This energy source is nonionizing, unlike X-rays and gamma rays. Several studies to date have proven the successful utilization of microwave radiation to accelerate organic biochemical reactions and synthesis [1]. Microwave chemistry is regarded as a green chemistry technique, as there is very minimal or no generation of hazardous or harmful substances that can harm the environment or organisms. The 12 principles of green chemistry are thoroughly followed by microwave chemistry. It increases the efficiency of biochemical reactions and, at the same time, remains eco-friendly. It is one of the best examples of green chemistry [2].

The lotus rhizome is a modified subterranean stem of the lotus plant [2]. In India, tropical lotus is found with a whip-like rhizome and a longer green period and flowering time. Its length varies from 10–20 cm, and its diameter varies from 5–10 cm. The rhizomes appear as sausage networks that are connected to one another through smaller rhizomes in underground water. They have a firm texture with whitish to light purplish skin that changes to a brown-beige hue with dark brown speckles when fully mature. Below the skin lies the starchy, crispy and whitish edible flesh. Several air pockets are present in the flesh throughout the length, which represent a wheel-like structure. Along with its crispy and crunchy texture, the taste of lotus rhizome is sweeter, resembling that of taro or water chestnut. Mature rhizomes are used in extensive cooking applications, whereas younger rhizomes are used for culinary purposes, being juicier and more tender (Speciality Produce, USA). The largest exporter of lotus rhizome in 2021 was the Netherlands, with an export value of USD 133 million, while the largest importer was Germany, with an import value of USD 81.4 million (Tridge Intelligence, 2021). India does not rank among the top ten exporters or importers. The lotus rhizome is exported from India in either fresh form, frozen form, individually quick-frozen form, canned, brined or vacuum packed. The major exporting ports from India are Nhava Sheva (29.0%) and Mundra (43.0%) (Connect2India, 2021).

Lotus rhizome is an amalgamation of rich nutrient content and crisp texture. Its nutrient richness can be justified by its proximate composition, vitamin composition and mineral composition [3]. The rich composition of lotus rhizome makes it an efficient competitor in health foods. They are mostly consumed fresh owing to enzymatic degradation upon storage. Several bioactive compounds have been identified, such as isoquercetin, catechin and rutin [4]. Despite its rich nutritional composition, lotus rhizome is one of the most underutilized vegetables, especially in India. The main reason for its underutilization is the high rate of enzymatic browning due to the action of the enzyme polyphenol oxidase (PPO). Traditionally, blanching has been used as an effective pretreatment method to curb this enzymatic activity, which is followed by processing techniques such as drying, freezing, canning or frying [5]. It basically involves heating the fruit or vegetable for a few seconds or minutes at a predetermined temperature and then rapidly cooling it or passing it on to the next process. A variety of advantages of blanching have been reported, such as the inactivation of quality-deteriorating enzymes such as polyphenol oxidase (PPO) and peroxidase (POD), decreasing the microbial load, increasing the extraction efficiency of bioactive compounds, removing toxic substances and pesticide residues, reducing the browning reaction, and enhancing the drying rate and overall quality of fruits and vegetables.

Several conventional methods, such as hot water blanching and steam blanching, have been utilized profoundly. However, these thermal methods result in quality and water-soluble nutrient loss along with substantial environmental pollution due to the generation of a large amount of wastewater. Increasing consumer awareness regarding health and diet has compelled the food industry to search for mild processing techniques that could yield a better result along with being environmentally friendly [6]. The majority of these techniques are based on dry blanching. Microwave blanching has previously been used for the blanching of several fruits and vegetables, such as star fruit, broccoli, mushroom, and apple, but never for lotus rhizome. This study is the first such attempt to do so. This study tries to compare the effect of microwave blanching to conventional blanching methods on lotus rhizome slices. The availability of information on postharvest treatments, such as various optimized methods of blanching for keeping the fresh lotus rhizome intact and how this impacts the sensory quality and shelf life of the unique aquatic crop, is a limitation to the food industry's utilization of this horticultural crop to its full potential. Therefore, the objective of this research was to determine the most efficient blanching method for lotus rhizome, which could be regarded as a green chemistry technique. In this study, blanching of lotus rhizome slices was performed with the help of a microwave technique as a green chemistry method. The results obtained were compared with those obtained in conventional blanching methods of steam and hot water blanching.

2. Material and Methods

2.1. Material

The work deals with different methods and materials that are used to obtain and accomplish the desired objectives. Lotus rhizome was purchased in special order from vegetable shops in Nawabganj Market, Kanpur. All chemicals and reagents were of 99% purity and were used in the completion of the work. The chemical reagents were mostly Thomas Baker's. The experimental work was performed in 2021–2022 on the "effect of blanching methods on quality of lotus rhizome slices" in the Department of Food Technology, Harcourt Butler Technical University, Kanpur (U.P.).

2.2. Methods

Lotus rhizomes were refrigerated at 4°C after being purchased from the market. The lotus rhizomes used in the experimental analysis had a length of 20–30 cm each. It was ensured that the rhizome purchased was of the same cultivar and subjected to equal storage conditions every time for reproducibility. The samples were then selected for size uniformity, quality, and lack of mechanical damage and then washed with tap water to remove mud. After that, they were cut into slices of 5 mm thickness each using a slicer and subjected to different blanching treatments. Figure 1 depicts a typical lotus rhizome slice used in this study.

Figure 1. Lotus rhizome slice with 5 mm uniform thickness.

2.3. Experimental Setup

2.3.1. Steam Blanching

The slices, after weighing 20 g, were leaned on a wire grid placed in a pressure cooker after the generation of steam with 200 ml of water. The lid was kept closed. The slice-to-water ratio was kept fixed at 1:10 in each case. Blanching was performed for 15 s, and the time was varied to 30, 45, 60, 75, 90, 105 s, 120 s, 135 s, 150 s and so on, until the peroxidase and catalase enzymes were inactivated and confirmed by blanching efficacy tests [3].

2.3.2. Hot Water Blanching

Two temperature combinations were chosen for hot water blanching: 90°C and 50°C, representing high- and low-temperature blanching, respectively [7]. Twenty grams of lotus rhizome samples were placed in a temperature-controlled hot water bath at the two temperatures. The blanching time varied from 15 s to 60 min for both temperatures until the inactivation of POD and catalase enzymes was achieved [6].

2.3.3. Microwave Blanching

Three power combinations were chosen for microwave blanching: 360 W, 540 W and 900 W. Twenty grams of rhizome slices were placed in a domestic microwave oven at a frequency of 2450 MHz. Samples were placed in a 500-ml beaker and covered with a Pyrex plate (to prevent water loss). Dry blanching was performed at different power levels. The time varied from 5 s to the inactivation of enzymes [8].

2.3.4. Optimization of Blanching Methods

The optimization of blanching time is the process of determining the minimum blanching time at which enzymes are fully inactivated. Qualitative inactivation of peroxidase and catalase enzymes was performed using the Iwase method [9]. Peroxidase (POD) and catalase, according to evidence, are key endogenous source enzymes involved in the browning process [10]. As a result, POD and catalase inactivation are crucial for increased color retention in plants.

One milliliter of hydrogen peroxide was added to 1 ml of juice from the blanched lotus rhizome slice. If strong effervescence was observed, the catalase enzyme was present; otherwise, it was absent. This effervescence was due to the release of oxygen gas. In this test tube, 0.5 ml of guaiacol solution was added and kept for 2 s. If the color changed to red, it confirmed the presence of the peroxidase enzyme; otherwise, the enzyme had been inactivated. This process was repeated for optimization of all the blanching methods. The time taken from zero seconds to the time when the red color is no longer observed is called the blanching time for the vegetable under study.

2.3.5 Quality Tests

2.3.5.1 .Residual Activity of the Peroxidase Enzyme

The POD residual activity was found using the method described by Severini et al. (2015) [11], where guaiacol was used as a hydrogen donor and hydrogen peroxide. Slices of blanched rhizome (2 g) were crushed in a mortar and pestle after being added to a 5 mL phosphate buffer solution (pH 6.5; 0.1 M). Using a homogenizer, samples were homogenized at 18000 rpm for 30 seconds. After centrifugation of the slurry at 7000 rpm at 4°C for 20 minutes, filtration was performed using filter paper (Whatman no. 40). Phosphate buffer solution (pH 6.5; 0.1 M), guaiacol (0.1% v/v), and hydrogen peroxide (0.1% v/v) were combined to create the POD substrate solution. A total of 4.7 ml of buffer, 1 ml of enzymatic solution, 0.1 ml of guaiacol, and 0.2 ml of hydrogen peroxide were reacted at 30°C for 5 min. The UV–VIS analyst program was used to quantify absorbance at 470 nm in a spectrophotometer (LAB UV 3000 $^{+}$ double beam). The enzyme activity was determined using an equation. Subsequently, the percent residual activity was determined using the equations by Pfluck et al. (2018) (Equations 1 and 2). The control samples were considered indicative of 100% residual activity. All determinations were carried out in triplicate. The unit of POD activity was mol min^{-1} mL.

$$\text{POD Activity} = \frac{A}{\Delta E} \times \frac{V1}{V} \times \frac{1}{t} \times 10^{6} \quad (1)$$

$$\text{POD Residual Activity} = \frac{\textit{Enzymatic activity of blanched sample}}{\textit{Enzymatic activity of raw sample}} \times 100 \quad (2)$$

2.3.5.2. Residual Activity of Polyphenol Oxidase Enzyme

Polyphenol oxidase activity was assayed according to Jiang et al.'s 1999 method using catechol. To the 5 g of crushed blanched sample, 1.5 ml of phosphate buffer (pH 6.5-7) was added and mixed in a shaker for 2 min. The constituents were filtered using filter paper and centrifuged at 75,000 rpm for 5 minutes. The supernatant was collected, and 0.5 ml of 0.1 M catechol and 2 ml of buffer were added. PPO residual activity was determined using a spectrophotometer (LAB UV 3000 $^{+}$ double beam) at 425 nm. The rest procedure and calculation (Equations 3 and 4) were the same as in POD by replacing the value for the molar extinction coefficient in the formula of guaiacol with catechol [12]. The unit of PPO activity was mol min^{-1} mL [13].

$$\text{PPO Activity} = \frac{A}{\Delta E} \times \frac{V1}{V} \times \frac{1}{t} \times 10^{6} \quad (3)$$

$$\text{PPO Residual Activity} = \frac{\textit{Enzymatic activity of blanched sample}}{\textit{Enzymatic activity of raw sample}} \times 100\ \% \quad (4)$$

2.3.5.3. Ascorbic Acid Loss

Vitamin C or ascorbic acid is an antiscorbutic. Citrus fruits, gooseberries, bitter gourds, etc., all contain significant amounts of it. In general, it can be found in all fresh fruits and vegetables. It is a highly heat-labile and water-soluble vitamin. The procedure to determine it is simple, quick and allows for the quick analysis of numerous samples. In this case, it was estimated by the method suggested by the FSSAI Manual (2106). By macerating the blanched sample with stabilizing agents such as 3% metaphosphoric acid, the ascorbic acid content in the lotus rhizome was measured. Metaphosphoric acid (3%) was prepared and added to 100 mg L-ascorbic acid, and the volume was made up to 100 ml. Subsequently, 2,6-dichloroendophenol dye was standardized using prepared L-ascorbic acid and 3% metaphosphoric acid after titration. The dye factor value was used to calculate (using equation 5) the ascorbic acid concentration.

$$\text{Ascorbic Acid} = \frac{\text{Titre value} \times \text{dye factor} \times \text{volume make up} \times 100}{\text{Aliquot} \times \text{volume of sample}} \quad (5)$$

2.3.5.4. Texture Profile Analysis

The texture profile was measured using a texture profile analyzer (TA-XT Plus). The texture analyzer was equipped with a probe with a 5-kg load cell. The shear force (N) of the samples was measured using a single-blade probe with thickness P/6. The blanched samples were allowed to acclimate to room temperature (approximately 20°C) before measurement, which took approximately 30 minutes. The blade was kept perpendicular to the samples so that they were carefully positioned on the support base. The distance was 5 mm, the testing parameter was 1 mm/s, the compression ratio was 30%, the plotting parameter was 1 mm/s and the posttest speed was 1 mm/s. Computer-generated force–time graphs were captured, and Texture Expert for Windows (version 1.15, Stable Micro Systems Ltd.) was used to evaluate them. The force–time profile's peak height or shear force was measured. Readings were taken in triplicate, and hardness was defined as the most vigorous during the first compression.

2.3.5.5. Browning Index

It is the term that is used for denoting the purity of brown color and is reported as an important parameter in processes where enzymatic or nonenzymatic browning takes place. It was calculated according to the given equations 6 and 7 by Zhang et al. (2017). The L*, a* and b* values were determined using a Hunter Lab colorimeter.

$$\text{B.I} = \frac{[100(x-0.31)]}{0.17} \quad (6)$$

$$x = \frac{a* + 1.75\, L*}{5.645\, L* + a* - 0.3012\, b*} \quad (7)$$

2.3.5.6. Microstructural Analysis

The blanched rhizome and fresh slices were fixed using 3% glutaraldehyde solution for 18-24 hours and rinsed in buffer solution at pH 6.8. Serial dehydration was performed with ethanol for 15 min at concentrations of 15%, 30%, 45%, 60%, 75% and 90%. The samples were freeze dried until the moisture content of the samples fell below 5%. Finally, gold coating was performed, and the samples were passed through a scanning electron microscope. Micrographs were taken at 500x magnification using a scanning electron microscope (SEM). This method was taken from [14].

3. Results and Discussion

3.1. Optimization of Steam Blanching, Hot Water Blanching and Microwave Blanching Methods

The blanching method optimization was based on the inactivation of both catalase and peroxidase enzymes. The steam blanching time was varied from 15 s to 150 s. The catalase and peroxidase enzymes were inactivated at 105 s and 150 s, respectively (Table 1). The inactivation time of lotus rhizome slices under steam blanching was in line with the optimized time of 2 minutes in steam blanched carrot slices, as determined by [6], and 1.5 minutes in steam blanched broccoli [11].

Two different blanching temperature combinations were chosen in the case of hot water blanching: 90°C and 50°C. The blanching time was varied from 1 min to 5 min in the case of 90°C and 10 min to 45 min in the case of 50°C (Table 1). [6] reported a 5-minute optimized blanching time for hot water blanching at 95°C in carrot slices.

Similarly, three different power combinations of 360 W, 540 W and 900 W were chosen for the microwave blanching of lotus rhizome slices. The blanching time varied in each case. The microwave blanching time varied from 10 s to 80 s at 360 W, 15 s to 75 s at 540 W and 4 s to 24 s at 900 W. The catalase and peroxidase enzymes were inactivated at 60 s and 70 s at 360 W, 45 s and 60 s at 540 W and 16 s and 20 s at 900 W, respectively (Table 1).

Table 1. Optimization of steam, hot water and microwave blanching time

Blanching Method	Optimized Blanching Time
Steam	150 s
Hot Water at 90	4 min
Hot Water at 50	40 min
360 W	70 s
540 W	60 s
900 W	20 s

3.2. Residual Activity of POD and PPO during Different Blanching Methods

POD is known to be one of the most heat-stable enzymes and is frequently used as a blanching index. One hundred percent residual activity was considered for the control. At least three separate experiments were conducted for each assay, and the findings clearly demonstrate that the increase in temperature and power (in the case of the microwave) leads to a decrease in blanching time and a higher inactivation of POD and PPO enzymes.

Considering the different optimized blanching times under different methods, the highest rate of enzymatic inactivation was achieved in the case of hot water blanching at 90°C, followed by steam blanching. This is clearly evident in Figures 2 and 3. The residual activity of POD under both of these methods was calculated to be 3.9% and 5.2%, respectively. Similarly, the residual activity of PPO under both of these methods was calculated to be 0.6% and 3.98%, respectively. The highest residual activity of POD and PPO was determined to be in the case of hot water blanching at 50°C, with values of 20.6% and 17.8%, respectively. This clearly depicts the inefficacy of this method at this temperature–time combination to blanch the lotus rhizome slices. The results obtained are in carrot slices, stating total inactivation of POD and PPO in steam blanching and hot water blanching at 95 °C for 4 minutes [6]. Similarly, the highest residual activity of 40.9% was obtained in hot water blanching at 60°C for 40 min.

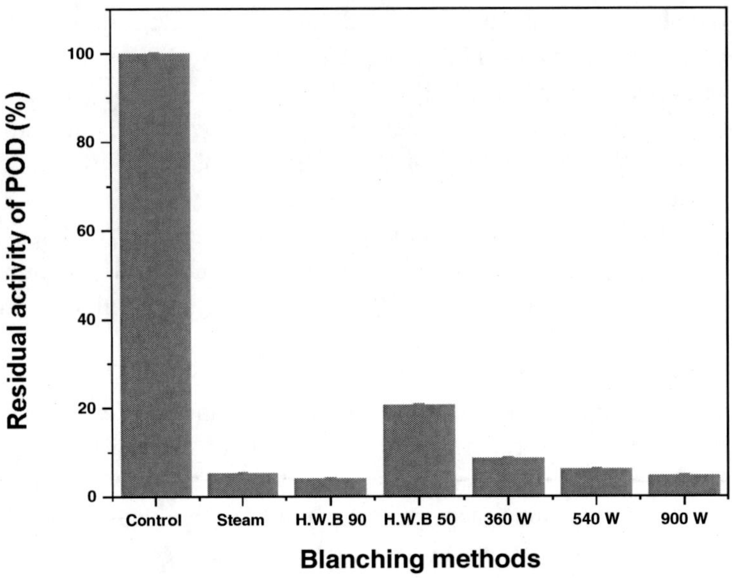

Figure 2. Relation between % residual activity of POD and blanching methods.

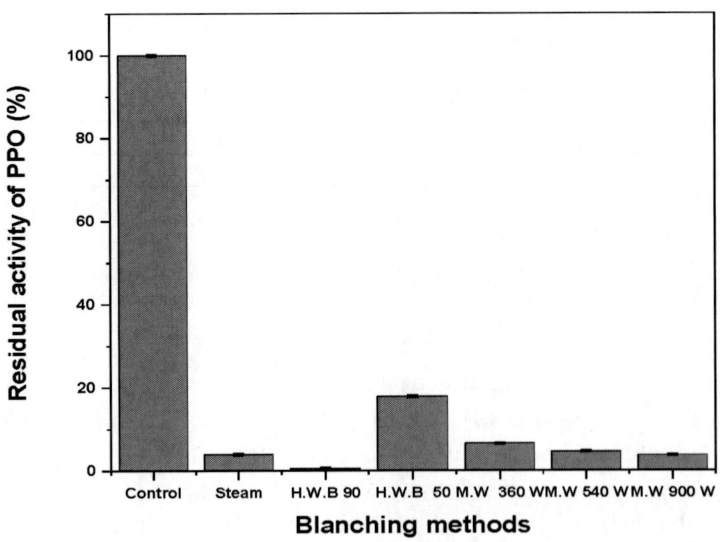

Figure 3. Relation between % residual activity of PPO and blanching method.

Among dry microwave blanched lotus rhizome slices, the lowest residual activity of POD and PPO, with values of 4.7% and 3.7%, respectively, was found at 900 W. With an increase in microwave power, the blanching time decreases and the blanching efficacy increases, as here in the case of microwave blanching at 360 W and 540 W. The values of residual activities obtained are in synchrony in broccoli [11]. Total inactivation of POD was reported at 900 W for 80 seconds of blanching time. According to the microwave, blanching is an efficient way to blanch vegetables such as broccoli with no water or energy waste. As per the existing research reports and the current research findings, we can conclude that the residual activity of POD and PPO does not solely depend on blanching time but is also drastically affected by blanching temperature, as was observed in the case of hot water blanching at 90°C and 50°C, respectively [11]. Additionally, an inactivation of 90% can be regarded as a benchmark for efficient blanching, as seen in most of the research literature.

3.3. Effect of Blanching Methods on Ascorbic Acid Loss of Lotus Rhizome

The content of ascorbic acid varies greatly depending upon rhizome maturity at harvest, growing conditions, soil quality and climatic conditions [14–15]. It is generally in the range of 15 to 79.39 mg/100 g [16]. The amounts of ascorbic acid present in the control and blanched lotus rhizome slices used in this investigation are displayed in Table 2. The ascorbic acid content in the control was found to be 77.751 mg/100 g. This quantity was higher than that of the USDA (2016; 44 mg/100 g). The amount of ascorbic acid in lotus rhizomes was considerably reduced after blanching, except in the case of microwave blanching. This is evident from Figure 4. In the case of steam blanching, 26.4% ascorbic acid loss was observed, while in the case of hot water blanching at 90°C, another 26.4% loss was observed. Dramatic loss was observed in the case of hot water blanching at 50°C, with a 76.3% decrease in ascorbic acid [17–18].

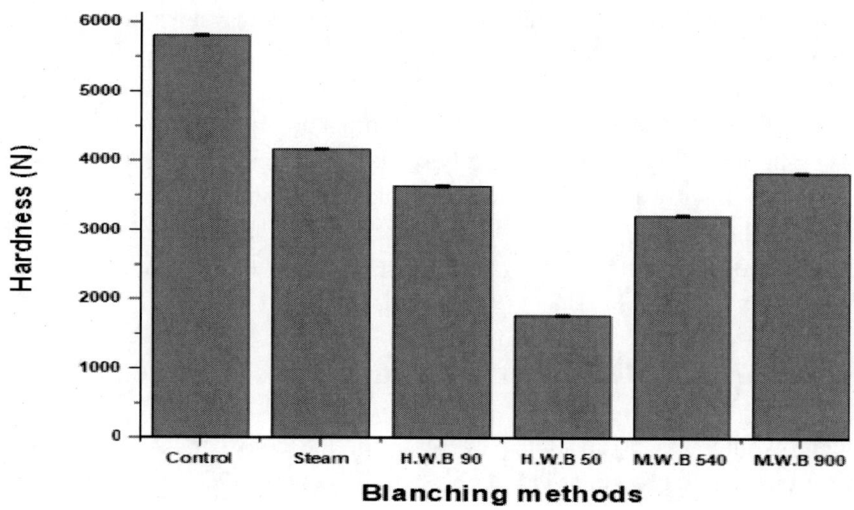

Figure 4. % Loss in Ascorbic Acid content (mg/100 g).

Due to long exposure to water and heat, ascorbic acid (which is water-soluble) leaches out in water, thereby leading to a loss in antioxidant activity. Unlike hot water blanching, the ascorbic acid value increases in the case of microwave blanching. This was observed at all three powers: 360 W, 540 W and 900 W. This increase can be attributed to enhanced chemical extractability after microwave heating [19–21]. This value slightly decreases as the power is increased. This can be due to the heat degradation of ascorbic acid. Therefore, in terms of ascorbic acid retention, the best blanching treatment is microwave, followed by steam, and then hot water blanching.

Table 2. Ascorbic acid content in blanched lotus rhizome slices

Blanching method	Time of Blanching	Ascorbic Acid (mg)
Control	0	77.751 ± 0.3
Steam	150s	57.219 ± 0.3
H.W 90	4 min	37.533 ± 0.4
H.W 50	40 min	18.411 ± 0.1
M.W 360 W	70s	78.139 ± 0.2
M.W 540 W	60s	76.936 ±0.1
M.W 900 W	20s	75.487 ± 0.2

3.4. Effect of Blanching Methods on the Texture of Lotus Rhizome

Figure 6 clearly depicts the results obtained upon texture profile analysis. The graph signifies that the hardness of lotus rhizome slices decreased with an increase in blanching time, with the exception of microwave blanching. In this case, instead of decreasing, the hardness property slightly increased. Hence, hardness was lowest in the case of hot water blanching at 50°C, with a value of 1769.6 N. This was far less than the hardness of the control sample, which had a value of 5810.44 N. Steam blanched lotus rhizomes (hardness value 4169 N) were far better in texture compared to hot water blanched rhizomes at 90°C with values of 3637.5 N. However, the best texture was retained in the case of microwave-blanched samples. The highest hardness was retained at 900 W with a value of 5600.2 N. This can be explained by the increase in hardness and cell wall firmness due to the loss of moisture content in lotus rhizome slices, as also observed in okra [20]. Therefore, the best blanching method to retain hardness in lotus rhizome slices is microwave followed by steam blanching. Hot water blanching is an inappropriate method for blanching in terms of texture retention. The variation in texture is presented in Figure 5.

3.5. Effect of Blanching Methods on Browning of Lotus Rhizome

Enzymatic browning was taken as a parameter for the color of lotus rhizome slices. The condition for enzymatic browning is a high phenolic content [22–24]. In fresh-cut fruits and vegetables, the dynamic balance between phenol and quinone is broken during processing and storage, and phenols are oxidized to o-quinone by related enzymes (e.g., PPO, POD, and CAT). Due to its high degree of oxidation, quinine eventually gives rise to hydroxyquinone by undergoing secondary hydroxylation. After further polymerization, the hue of hydroxyquinone changes from red to brown to black.

The enzymatic browning decreased with an increase in blanching treatment. The browning index is evident in Figure 6. The browning index for the control was determined to be 4.1172. The lowest value of browning was in the case of steam, microwave blanching at 900 W and hot water blanching at 90°C.

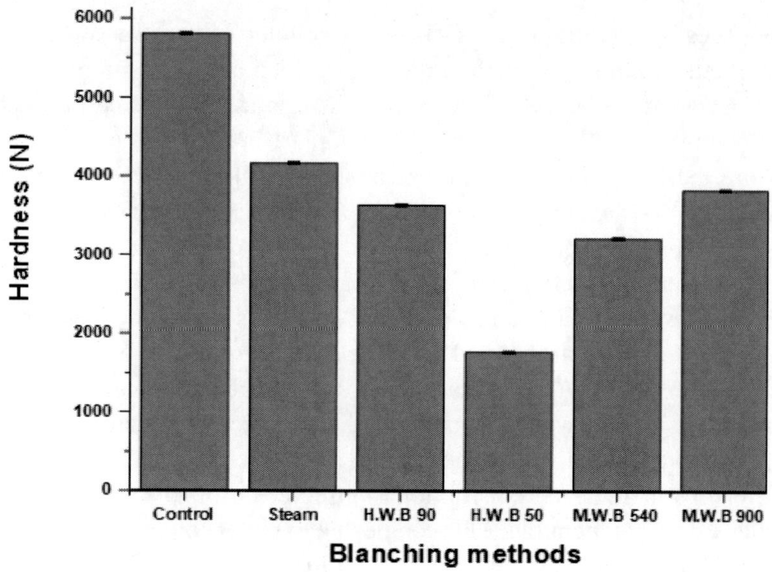

Figure 5. Effect of blanching methods on the hardness of lotus rhizome slices.

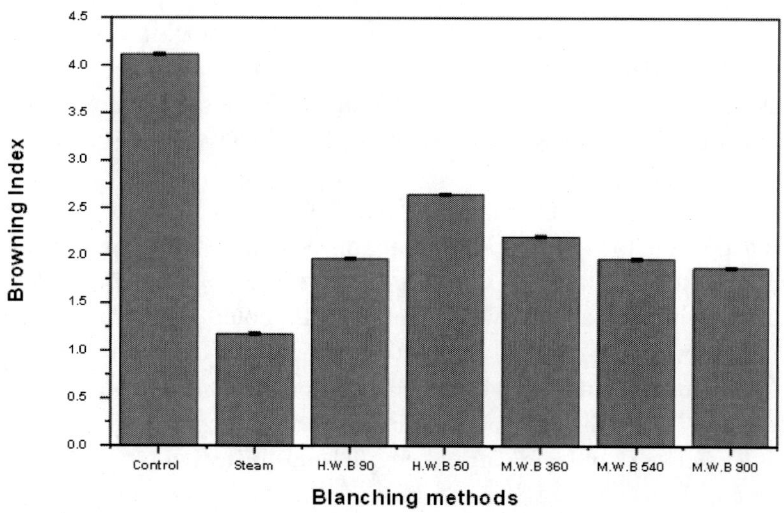

Figure 6. Effect of blanching on the browning index of lotus rhizome slices.

3.6. Microstructural Analysis Using Scanning Electron Microscopy (SEM) of the Optimized and Best Methods

According to SEM observations conducted at 500 magnifications, as shown in Figure 7(a), fresh lotus rhizome tissue (control) is composed of well-joined cells with intact cell walls, demonstrating the characteristic eutectic artifact that results from sample preparation for SEM observation. The solute aggregation phenomenon, also known as the eutectic artifact, is caused by the buildup of soluble solutes during sample etching and manifests as a network. No intercellular spaces or pores could be seen in the matrix, which supports the freshness of the lotus rhizome slices used in the trials. Cell walls, cellular cements and plasmalemma are all generally unharmed. Additionally, cell protoplasts, which are made up of the cell membrane, plasmalemma and cell content, are preserved in close proximity to the cell wall and are difficult to distinguish from one another. Most of the protoplast is enclosed inside a sizable vacuole. The observations made are in line with those in sliced [22] unblanched carrots. Additionally, the SEM micrographs of the control lotus rhizome exhibit a polyhedrical form and smooth cell walls of the phloem parenchyma cells.

In the case of steam blanching, the micrographs with a magnification of 500 show cells with good cell adhesion, no intercellular gaps, and no separation between cell walls and membranes, similar to the control sample. The cell walls are intact, but only in a few cells, as can be clearly seen in Figure 7(b). Distorted cell walls can be observed in a few cells, thereby leaving organelles visible. However, throughout the tissue, cellular cements, which are directly related to the durability of cell walls, appeared to be unaffected. As per observations made on the phloem parenchyma tissue, less disruption is observed during the steam blanching processes compared to the xylem parenchyma. This can be due to better solute retention. The membranes of various organelles that were not readily apparent in the tissue of the control samples can now be seen in the blanched samples. It can be concluded that steam blanching does not have much adverse effect on the microstructure of lotus rhizome slices.

Figure 7(c) reports the micrograph of a lotus rhizome slice that was blanched in hot water at 90°C. Comparing this sample to that of fresh (control) rhizome, the eutectic artifact reveals larger cells and a lower cell density, reflecting the lower solid and higher water content brought on by the heat treatment. The membranes of various organelles that were not readily apparent in the tissue of the control sample were observed in the hot water-blanched samples. An increased prevalence of detachments between adjacent cell walls and impending plasmolysis were also noticed. Specifically, the blanching treatment led to a considerable retraction of the plasmalemma and/or the tonoplasts to the center of the cell, resulting in a separation between the plasmalemma and the cell wall in some regions in the samples treated at 90°C for 4 min. In this sample, distortion of the cellular wall edges was also noted. This process may be related to the loss of soluble solids as a result of the solubilization of the cell wall polymers [23].

Compared to lotus rhizome slices blanched in water or steam, slices exposed to microwave blanching have a noticeably different appearance, as shown in Figure 7(d). The tissue magnified at 500 exhibits a patchwork of layers of sunken cells and clusters of well-preserved cell groups. Surprisingly, the cells were preserved and intact, similar to those in the control sample. The lotus rhizome shows only minor cell wall deterioration, and the structure remains mostly unaltered. Microwave blanching produced a heterogeneous structure made up of both intact cell areas and collapsed cell areas. Cell damage may be partially explained by the high internal vapor pressure that results from the microwave conversion of water into steam. In the current experiment, very few fissures and cracks can be seen after microwave heating. It is observed that the xylem parenchyma tissue's inner core is harder compared to a water-blanched sample. Additionally, it was also discovered that microwave heating caused the light-microscopy-visible folding and collapsing of the cell walls. The results obtained and observations made are in agreement for microwave-blanched carrot slices. The rapid elimination of water from the cells may also boost the mechanical strength of the tissue by enhancing the crystallinity of cellulose and hemicellulose in the cell wall. It appears that microwave blanching produces some dehydration of the tissue. It can be assumed that changes in cell structure can be the cause of the tendency for enhanced firmness in microwave blanching, unlike blanching in water and steam. The variations in cell structure are reflected upon blanching, which clearly distinguishes between microwave, steam and hot water blanching. The large fissures that develop in water and steam-blanched samples give rise to a less firm product.

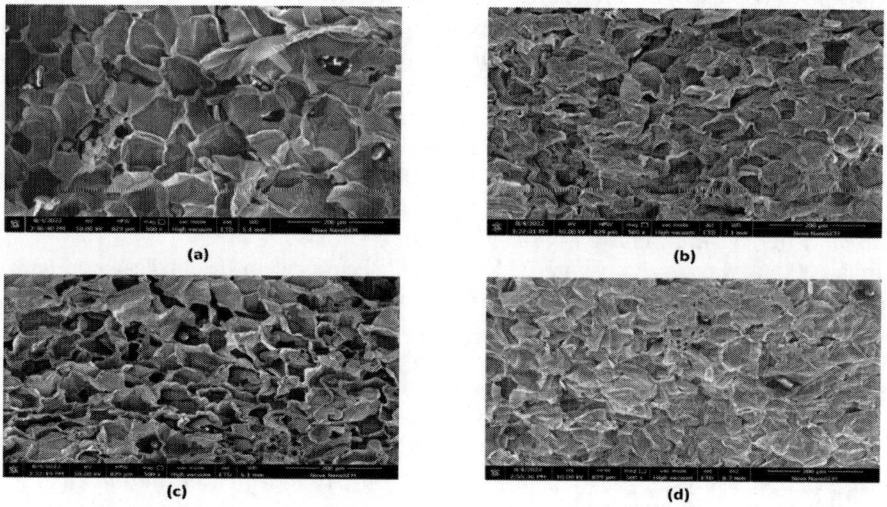

Figure 7. (a) Micrograph for control (b) Micrograph for steam blanched sample (c) Micrograph for hot water (90°C) blanched lotus rhizome (d): Micrograph for microwave blanched sample, 900 W.

Conclusion

Lotus rhizome is rich in nutrients but is highly susceptible to enzymatic degradation. Hence, it is not properly utilized in the food industry despite having health-promoting qualities, a good taste and texture. This study successfully tried to solve this problem by utilizing the green chemistry method of the microwave technique. It was done by optimizing the time of blanching methods and then finding the best method out of these after proper assessment of methods by quality analysis. Three methods were chosen: steam blanching, hot water blanching at 90°C and 50°C and microwave blanching at 360 W, 540 W and 900 W, respectively. After optimization, a quality test was performed on the blanched slices. Several properties, such as ascorbic acid loss, texture using hardness analysis, residual activity of POD and PPO and browning index, were determined. Losses in ascorbic acid and hardness were observed in the case of hot water blanching and steam blanching. The loss in ascorbic acid at 90°C was due to thermal degradation, while that at 50°C was due to leaching loss and mass transfer in the blanching medium. In contrast, in microwave blanching, values increased due to increased chemical extractability owing to heat and intact structure. Microstructural analysis also proved the efficacy of the microwave blanching method with little moisture loss. All the tests performed proved that microwave blanching at 900 W is the best method for the enzymatic inactivation of enzymes along with quality retention among all the methods used in this study. In fact, microwave blanching at all powers is far better than conventional blanching methods because it is environmentally friendly and green.

Hence, this method can be successfully utilized at optimized time and temperature combinations for processing lotus rhizome in the food industry and even at home.

References

[1] Ravichandran, S., and E. Karthikeyan. "Microwave synthesis-a potential tool for green chemistry." *International Journal of ChemTech Research* (2011): 3, 466-470.

[2] Grewal, Ajmer & Kumar, Karunesh & Redhu, Sonika & Bhardwaj, Shashikant. (2013). Microwave assisted synthesis: a green chemistry approach. *International Research Journal of Pharmaceutical and Applied Sciences*. 3, 278-285.

[3] Sruthi, A., Panjikkaran, S. T., Aneena, E. R., Pathrose, B., & Mathew, D. (2019).Insights into the composition of lotus rhizome. *Journal of Pharmacognosy and Phytochemistry*, 8(3), 350-355.

[4] Zhao, X., Shen, J., Chang, K. J., & Kim, S. H. (2014). Comparative analysis of antioxidant activity and functional components of the ethanol extract of lotus (*Nelumbo nucifera*) from various growing regions. *Journal of Agricultural and Food Chemistry*, 62, 6227–6235.

[5] Arroqui, C., A. Lopez, A. Esnoz, P. Virseda, Mathematical model of heat transfer and enzyme inactivation in an integrated blancher cooler, *Journal of Food Engineering,* Volume 58, Issue 3, 2003, 215-225.

[6] Wenlin Zhao, Wei Xie, Shenglan Du, Shoulei Yan, Jie Li, Qingzhang Wang, Changes in physicochemical properties related to the texture of lotus rhizomes subjected to heat blanching and calcium immersion, *Food Chemistry*, 2016, 211, 409-414.

[7] Behera, Gitanjali, Kalpana Rayaguru and Prakash Kumar Nayak. " Effect of microwave blanching on slice thickness and quality analysis of star fruit." *Current Research in Nutrition and Food Science Journal* 5.3(2017):274-281.

[8] Iwase, T., Tajima, A., Sugimoto, S. A Simple Assay for Measuring Catalase Activity: A Visual Approach. *Sci Rep* 3, 3081 (2013).

[9] Chisari M., Barbagallo R. N., Spagna G. (2007) Characterization of polyphenol oxidase and peroxidase and influence on browning of cold stored strawberry fruit. *J Agr Food Chem* 55(9):3469–3476.

[10] Severini C., Giuliani R., De Filippis A., Derossi A., De Pilli T. Influence of different blanching methods on color, ascorbic acid and phenolics content of broccoli. *J Food Sci Technol*. 2016 Jan;53(1):501-10.

[11] Brogan, A., Sharma, K., Perriman, A. Enzyme activity in liquid lipase melts as a step toward solvent-free biology at 150°C. *Nat Commun* 5, 5058 (2014).

[12] Almeida, Samara Lopes. Polyphenol oxidase and peroxidase enzyme assays in sweet potato cultivars harvested at different times. *Revista Caatinga*. 2019, 32, 537-542.

[13] Sanjuán, Neus & Hernando, Isabel & Lluch, M & Mulet, Antonio. (2005). Effects of low temperature blanching on texture, microstructure and rehydration capacity of carrots. *Journal of the Science of Food and Agriculture*. 85. 2071-2076. 10.1002/jsfa.2224. *Science* B, 10(8), 580–588.

[14] Podsedek A., Sosnowska D., Redzynia M., Anders B. (2006) Antioxidant capacity and content of *Brassica oleracea* dietary antioxidants. *Int J Food Sci Technol* 41(1):49–58.

[15] Sharma, B. R., Gautam, L. N., Adhikari, D., & Karki, R. (2016). A comprehensive review on chemical profiling of *Nelumbo nucifera*: Potential for drug development. *Phytotherapy Research*, 31(1), 3–26.

[16] Yuan, G. F., Sun, B., Yuan, J. and Wang, Q. M. (2009). Effect of different cooking methods on health-promoting compounds of broccoli. *Journal of Zhejiang University Science* B, 10(8), 580–588.

[17] Igwemmar, N., Kolawole, S. and Imran, I. (2013). Effect of heating on vitamin c content of some selected vegetables. *Scientific and Technology Research*, 2(11), 209–212.

[18] Oerlemans K., Barrett D. M., Suades C. B., Verkerk R., Dekker M. (2006) Thermal degradation of glucosinolates in red cabbage. *Food Chem* 95:19–29.

[19] Aamir M., Boonsupthip W. Effect of microwave drying on quality kinetics of okra. *J Food Sci Technol.* 2017 Apr;54(5):1239-1247.

[20] Xu, Y. Q., Yin, J. F., Yuan, H. B., Chen, J. X., Wang, F., 2007. Research advances of browning during processing of fruits and vegetables. *Storage Process* 7 (3), 11–14.

[21] Neri L., Hernando I. H., Pérez-Munuera I., Sacchetti G., Pittia P. Effect of blanching in water and sugar solutions on texture and microstructure of sliced carrots. *J Food Sci.* 2011 Jan-Feb;76(1):E23-30. doi: 10.1111/j.1750-3841.2010.01906.x. Epub 2010 Nov 29. PMID: 21535672.

[22] Fuchigami M., Hyakumoto N., Miyazaki K. 1995. Programmed freezing affects texture, pectin composition and electron microscopic structures of carrots. *J Food Sci* 60:137–41.

[23] Kidmose, U. and Martens, H. J. (1999), Changes in texture, microstructure and nutritional quality of carrot slices during blanching and freezing. *J. Sci. Food Agric.*, 79: 1747-1753.

Chapter 19

The Modeling of a Precision-Controlled Cooling Water System for Resonating the Cavity Structure of a Proton Accelerator

Satish Lukka[1,2,*]
Vimal K. Kotak[1,3]
Samiran Sengupta[1,3]
and Nilesh C. Gohel[3]
[1]Homi Bhabha National Institute, Anushakti Nagar, Mumbai, India
[2]Reactor Projects Division, BARC, Trombay, Mumbai, India
[3]Research Reactor Design and Projects Division, BARC, Trombay, Mumbai, India

Abstract

A proton accelerator is being developed at BARC as the front end of an accelerator-driven subcritical reactor system. Heat generated within the proton accelerator structure during its operation is removed by a low conductivity water (LCW) system with stringent control of the supply water temperature. One such typical accelerator structure in the Low Energy High Intensity Proton Accelerator (LEHIPA) is the Drift Tube Linac (DTL), which requires the LCW supply temperature to be controlled within ±0.2°C at 20°C. The LCW system for DTL consists of different components, such as a heat exchanger, control valves, holding tank, heater, pump, piping and instrumentation. In the present paper, the mathematical modeling of each of these components of the LCW system has been discussed. Most of these models are derived from first principles. For some components, test data obtained during inspection or testing at the manufacturer's facility were used to derive the model. Finally, all the mathematical models of the individual components were integrated to form a complete system model for the LCW system, and simulation was carried out to predict the system response against various possible load disturbances. The mathematical model will be quite useful in the process control design of cooling water systems with high-precision control requirements.

Keywords: mathematical modeling, LCW system, DTL-1, proton accelerator

---

* Corresponding Author's Email: satishl@barc.gov.in.

In: Sustainability in Chemical Processes through Digitalization and Green Chemistry Approaches
Editors: Susarla V. A. R. Sastry, Shravan Kumar and Ashish Kapoor
ISBN: 979-8-89113-257-3

1. Introduction

A proton beam is accelerated by setting up electric fields inside a resonating cavity structure made of OFHC copper in a low-energy, high-intensity proton accelerator. These electric fields penetrate through the copper surface and deposit nearly 60–65% of the electrical energy in the cavity structure. Due to energy dissipation in the cavity walls, any temperature fluctuation and the accompanying thermal expansion can make the cavity off-resonant or bring in higher-order resonant modes. Therefore, there is a stringent requirement to keep the walls of the cavity structures of the proton accelerator at a stable and uniform temperature. This is achieved by providing a number of flow channels through the cavity structure and supplying water at a controlled temperature. Since water flows through the copper accelerating structure, to minimize interferences of water with electric fields, low conductivity water (LCW) is used for cooling the accelerating structures [1].

Drift Tube Linac-1 (DTL-1) is one such resonating cavity structure that needs cooling water at a controlled temperature (20 ± 0.2°C). The DTL-1 cooling system is designed to meet this precision temperature-controlled cooling requirement, as shown in Figure 1. The cooling system removes approximately 560 kW of heat energy dissipated from drift tubes and transfers it to the chilled water system. The temperature at the inlet of the holding tank is controlled by manipulating the flow between the bypass and primary sides of the heat exchanger. Based on the signal from the temperature sensor measured before holding the tank, the PID controller generates a pneumatic signal, which actuates the three-way control valve to manipulate flow between the bypass and primary sides of the heat exchanger. The heat exchanger bypass valve is throttled such that the resistance of both the heat exchanger path and its bypass path is similar. This mixing scheme (process control loop P1) effectively controls T_{hm} at 19.5 ± 0.5°C. A holding tank is provided to increase the thermal inertia of the cooling system, i.e., reduce fluctuations in temperature at the inlet of DTL-1 caused by external parameters such as chilled water flow and temperature. The tank receives flow after mixing both streams, F_b and F_h, and there is a centrifugal recirculation pump downstream of the tank. A 2-way flow control valve is provided to maintain constant flow through the system by countering the hydraulic disturbances due to the operation of the 3-way flow control valve (process control loop P3) and variation in pump rpm. Finally, a heater is provided just before the inlet of DTL-1 for fine control of the LCW temperature to 20 ± 0.2°C (process control loop P2).

2. Mathematical Modeling of Components

2.1. Heat Exchanger

The heat exchanger acts as an interface between two process streams and facilitates heat transfer between hot and cold streams without direct contact. The heat exchanger used in the cooling system is a plate-type heat exchanger (PHE) to provide a high surface area to volume ratio compared to other types. Figure 2. The following assumptions were made for the modeling of energy transport in heat exchangers:

- Fluid volume is divided into a number of small control volumes, and a well-mixed approximation is assumed.
- Since the fluid temperature variation is small, constant properties (such as fluid densities, thermal conductivity and specific heat) are considered constant. All the properties were evaluated at the average temperature for each fluid stream.
- There is no phase change in either LCW or chilled water during energy transfer.
- Pure countercurrent and plug flow on both LCW and chilled water take place in the heat exchanger [2].
- There is complete mixing in the transverse direction of a fluid volume in any cell.
- There is no axial heat conduction in fluids along the direction of flow.
- There is no heat loss to the surroundings, i.e., adiabatic boundary conditions for all other surfaces except heat transfer plates [3].

Basic energy balance for the equation for the fluid:

$$\frac{\partial T_f}{\partial t} = -V_f \frac{\partial T_f}{\partial x} - \frac{UA_H}{H_f}\left(T_f - T_{fn}\right) \qquad \text{(Eq. 1)}$$

The energy balance equation is a nonlinear partial differential equation with space variable x, as velocity, heat transfer coefficient and temperature of fluid are variables. This equation is discretised and applied to both LCW and chilled water.

After discretization of Eq. 1 in space, the following equations are obtained.

For the LCW control volume:

$$\frac{dT_{N+i}}{dt} = V_h\left(\frac{T_{N+i-1} - T_{N+i}}{\Delta x}\right) - \frac{UA_H}{H_c}\left(T_{N+i} - T_i\right) \qquad \text{(Eq. 2)}$$

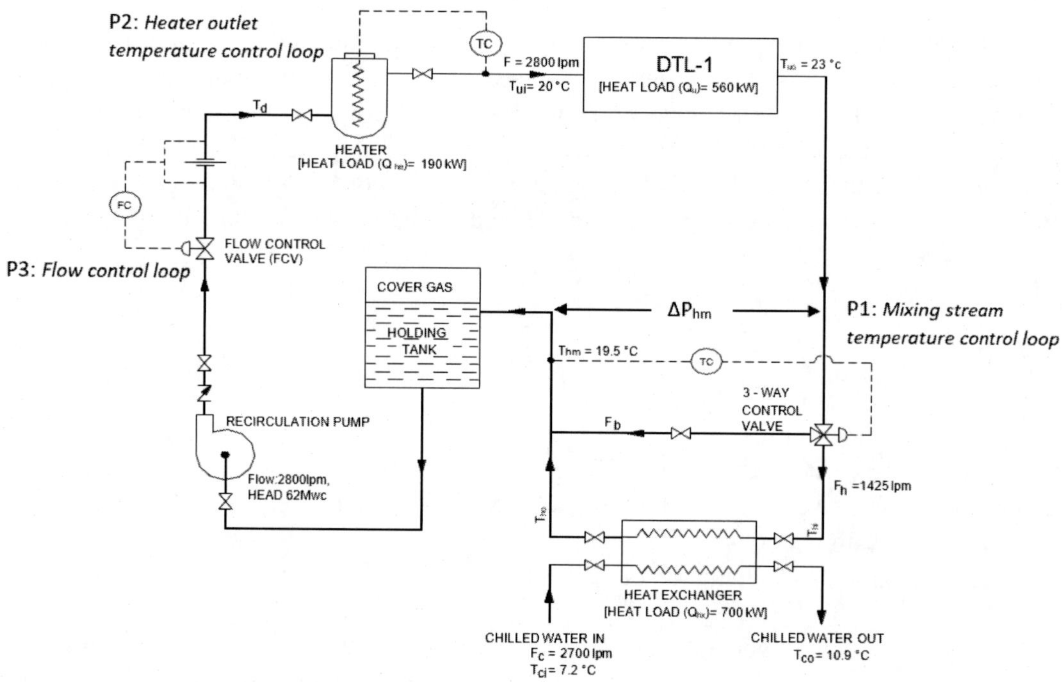

Figure 1. DTL-1 cooling system with steady-state operating parameters.

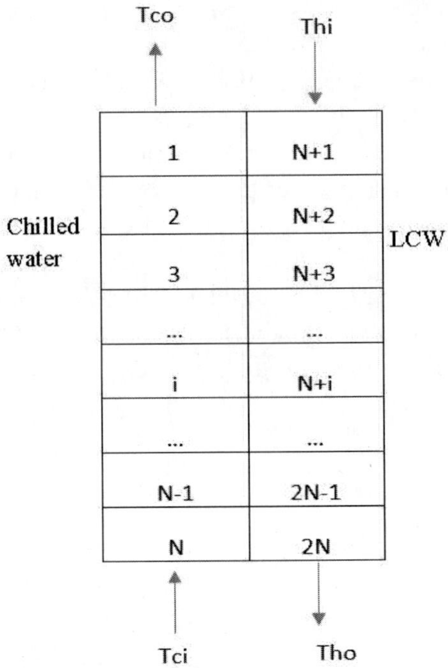

Figure 2. Nodalization scheme of the heat exchanger.

For chilled water control volume:

$$\frac{dT_i}{dt} = -V_c\left(\frac{T_{i+1}-T_i}{\Delta x}\right) + \frac{UA_H}{H_c}(T_{N+i} - T_i) \quad \text{(Eq. 3)}$$

Equations 2 and 3 combine to form a 2 N number of nonlinear ODEs. Subscript h refers to hot fluid (LCW), c refers to cold fluid (cold water), and i refers to the control volume number, which varies from 1 to N for chilled water and N+1 to 2 N for LCW, as shown in Figure 2. U is the overall heat transfer coefficient, which depends on the flow rates of the LCW and chilled water in the heat exchanger. Manufacturer data of U vs. flow of LCW with rated flow of chilled water is used in modeling [4]. Equations 2 and 3 are solved by the Runge–Kutta 4th-order method.

2.2. Heater

A heater is provided upstream of DTL-1 for fine control of the supply temperature within ±0.2°C around the set point. The total heater capacity is approximately 190 kW. The heating element consists of kanthal heating wire surrounded by MgO insulation, which is enclosed inside a tube sheath to avoid contact with water. The following assumptions were considered during the modeling and simulation of the heater:

- Volumetric heat generation is uniform along the length of the heating element.
- Heat conduction in the axial direction is neglected due to the low temperature difference.

- The arithmetic average of properties is considered for the interface between two materials.
- The basic conduction equation is discretized using the forward difference method.
- Coulson and Richardson's correlation for predicting the heat transfer coefficient of fluid in the shell side of the shell and tube heat exchanger is taken [5].

$$N_U = \frac{h_h d_e}{k_f} = j_h Re Pr^{\frac{1}{3}} \left(\frac{\mu}{\mu_w}\right)^{0.14} \quad \text{(Eq. 4)}$$

where $j_h = 0.002$ based on the flow rate.

Governing differential equations considered for the heat flow within the heating element

Heating wire

$$\left(\rho Cp_W \frac{dT_r}{dt}\right)_w = \left(K_W \frac{\partial}{\partial r}\left(\frac{\partial T_r}{\partial r}\right)\right)_w + \left(\frac{K_W}{r}\frac{\partial T_r}{\partial r}\right)_w + q_v \quad \text{(Eq. 5)}$$

Insulation

$$\left(\rho Cp_I \frac{dT_r}{dt}\right)_I = \left(K_I \frac{\partial}{\partial r}\left(\frac{\partial T_r}{\partial r}\right)\right)_I + \left(\frac{K_I}{r}\frac{\partial T_r}{\partial r}\right)_I \quad \text{(Eq. 6)}$$

Tube sheath

$$\left(\rho Cp_{ts} \frac{dT_r}{dt}\right)_{ts} = \left(K_{ts} \frac{\partial}{\partial r}\left(\frac{\partial T_r}{\partial r}\right)\right)_{ts} + \left(\frac{K_{ts}}{r}\frac{\partial T_r}{\partial r}\right)_{ts} \quad \text{(Eq. 7)}$$

where 't_s', 'I', 'w' represent the corresponding properties of the tube sheath, insulation and wire, respectively. The above equations were discretised both in the radial and axial directions by using equations 8 to 11 because the temperature varies in both directions. Equations 5 to 7 are solved by the Runge–Kutta 4th-order method.

$$\left(\frac{\partial}{\partial r}\left(\frac{\partial T_r}{\partial r}\right)\right)_i = \frac{T_{i+1} - 2T_i + T_{i-1}}{\Delta r^2}; \quad \text{(Eq. 8)}$$

$$\left(\frac{\partial T_r}{\partial r}\right)_i = \frac{T_{i+1} - T_i}{\Delta r}; \quad \text{(Eq. 9)}$$

$$\left(\frac{\partial T_f}{\partial x}\right)_j = \frac{T_{j+1} - T_j}{\Delta x}; \quad \text{(Eq. 10)}$$

$$\boldsymbol{r = i\Delta r} \quad \text{(Eq. 11)}$$

2.3. Flow Control Valve (FCV)

A 2-way flow control valve (FCV) maintains constant flow through the DTL-1 cooling system. FCV is provided at pump discharge but upstream of the heater. The linear transfer function model was predicted from the manufacturer's shop-tested frequency response data [6] by using the system identification technique. Various models have been tried to fit the shop-tested frequency response data. A model that better fits the shop-tested data with the lowest number of zeros and poles has been used to model the 2-way control valve. Finally, the transfer function (Eq. 12) with one zero and five poles showed a better fit (85%), and the phase difference with respect to frequency matches well with the experimental data, as shown in Figure 3. Nonlinear characteristics such as dead bands and hysteresis obtained from shop-tested data were also added to the above model. Figure 3 shows the frequency response data of the model (black) and experimental (green) FCV.

$$\frac{2200*s+40090}{s^5+18.26*s^4+570.4*s^3+4215*s^2+36630*s+45050} \quad \text{(Eq. 12)}$$

$$\frac{5.514*s^3+12770*s^2+244300*s+155000}{s^6+27.21*s^5+745.3*s^4+9090*s^3+72380*s^2+313000*s+137300} \quad \text{(Eq. 13)}$$

2.4. Three-Way Control Valve (TCV)

A three-way control valve is provided upstream of the heat exchanger. It regulates the flow rate required to be sent through the primary side of the heat exchanger. Thus, by varying the overall heat transfer coefficient, the mixing stream temperature (T_{hm}) is maintained at the desired set point. The linear transfer function model of TCV was obtained from the manufacturer's shop-tested frequency response data [7] by the system identification technique.

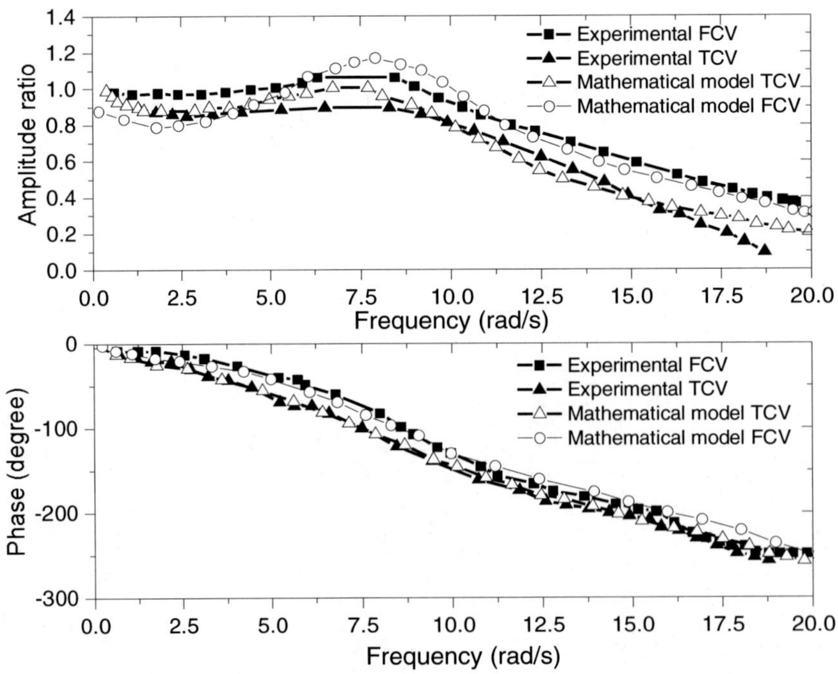

Figure 3. Model vs experimental frequency response comparison of control valves.

Various models have been tried to fit the shop-tested frequency response data. A model that better fits shop-tested data with the lowest number of zeros and poles has been used to model TCV. The transfer function (Eq. 13) with three zeros and six poles showed a better fit (94.22%), and the phase difference with respect to frequency matches well with the experimental data, as shown in Figure 3. Nonlinear characteristics such as dead bands and hysteresis obtained from shop-tested data were also added to the above model. Figure 3 shows the frequency response data for the model (blue) and experimental (red) TCV.

2.5. Holding Tank

A holding tank with a capacity of approximately 3300 liters is provided at the pump suction. It increases the thermal inertia of the system so that the effect of any change in temperature at the outlet of the heat exchanger due to any external disturbance is significantly dampened before it reaches the inlet of the DTL-1 component.

Since frequency-response data for the tank are not available, the system identification technique cannot be applied directly to the holding tank. Therefore, frequency response data are developed for the holding tank by using CFD simulation. Frequency response data obtained from the CFD model are shown in Figure 4. Then, a mathematical model was generated from these data using the system identification technique.

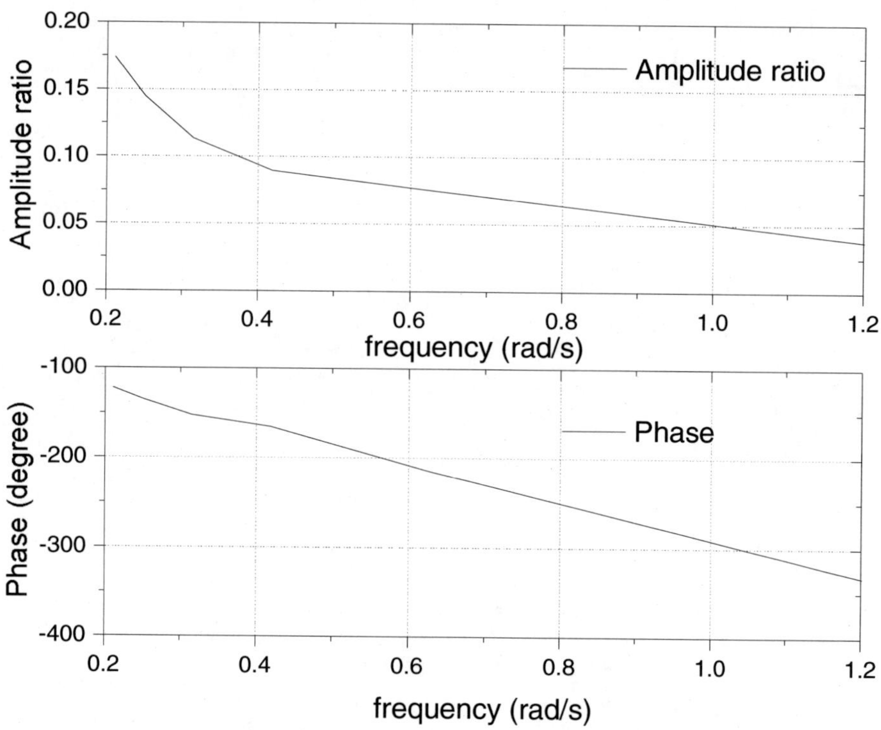

Figure 4. Frequency response data of the holding tank.

2.6. Mixing Process

A mixing process takes place between the bypass stream (F_b) and the LCW water outlet after the heat exchanger (F_h). The basic energy balance can be described as

$$\left(F_h \rho C_p T\right)_{ho} + \left(F_b \rho C_p T\right)_{hi} = \left(F \rho C_p T\right)_{hm} \qquad \text{(Eq. 14)}$$

T_{hm} can be calculated from the above equation. T_{hm} was measured considerably downstream after mixing both streams to minimize the effect of mixing disturbances.

2.7. Piping

As the fluid flows through the pipe in DTL-1, it takes a finite time to move from downstream of one piece of equipment to upstream of another piece of equipment. This time depends on the fluid flow velocity in the pipe and the length of the pipe. By taking this into consideration, transportation lags were calculated based on the actual layout.

2.8. Sensors

For temperature measurement, RTDs with a four-second time constant are used, whereas flow is measured using an orifice element with a differential transmitter.

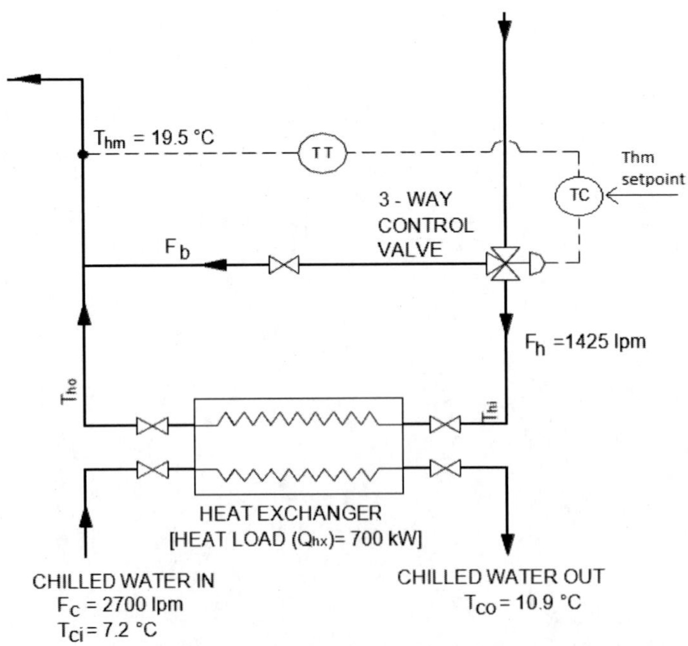

Figure 5. Heat exchanger temperature control loop.

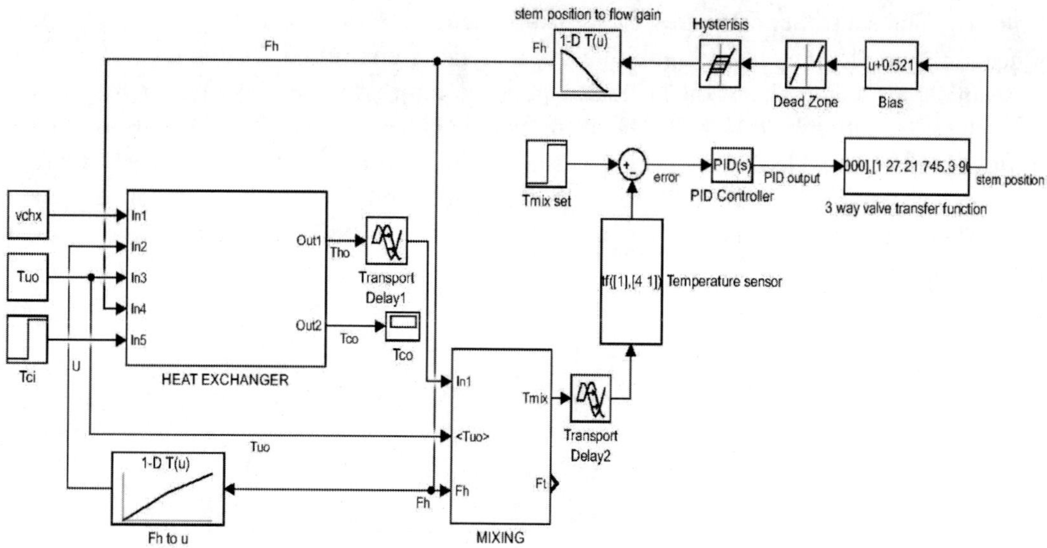

Figure 6. Heat exchanger temperature control loop model.

3. Integrated Modeling of the Cooling System

3.1. Heat Exchanger Temperature Control Loop (P1)

A mixing loop with a three-way control valve and heat exchanger is shown in Figure 5. Here, the controlled variable is the temperature of the mixing stream before the holding tank, and the manipulated variable is the LCW flow through the heat exchanger. The Zeigler-Nichols-Nichols loop tuning technique (Ultimate Method) was used to tune the control loop [7].

A mathematical model of the heat exchanger temperature control loop (Figure 5) is shown in Figure 6. The heat exchanger block contains the fundamental mathematical equations Eq 1 to Eq *3*. The overall heat transfer coefficient varies with the flowrate of LCW through the primary side of the heat exchanger, and the same was given to the model based on the flowrate through the primary side [8].

Transportation lags were added based on the actual piping layout of the installed system. Nonlinearities such as dead bands and hysteresis were also added to the TCV model. Bias was added to the controller output so that at zero error, there must be a certain flow through the heat exchanger. Transport delay 1 was added because fluid takes some finite time to reach the mixing point after the outlet of the heat exchanger, and transport delay 2 was added because fluid has to move to the sensor location before measuring T_{hm}.

3.2. Heater Outlet Temperature Control Loop (P2)

A heater is provided in the DTL-1 cooling system just before the inlet of the accelerating component (DTL-1) for fine control of temperature, as shown in Figure 1. The control loop of this system is shown in Figure 7. The process variable T_{ui} is measured and transmitted to the controller. The controller compares T_{ui} measured with the set point, and the necessary amount of heating is provided by the heater unit using a thyristor-controlled system.

A mathematical model of the heater temperature control loop is shown in Figure 8. Heater fundamental equations (Eq. 4 to Eq. 11) were modeled in the heater block. A transportation lag of 14 seconds was added to the model based on the actual piping layout. A PID controller regulates the heat input to the LCW based on the measured heater outlet temperature and set point. The Zeigler-Nichols-Nichols loop tuning technique (Ultimate Method) was used to tune the control loop [7].

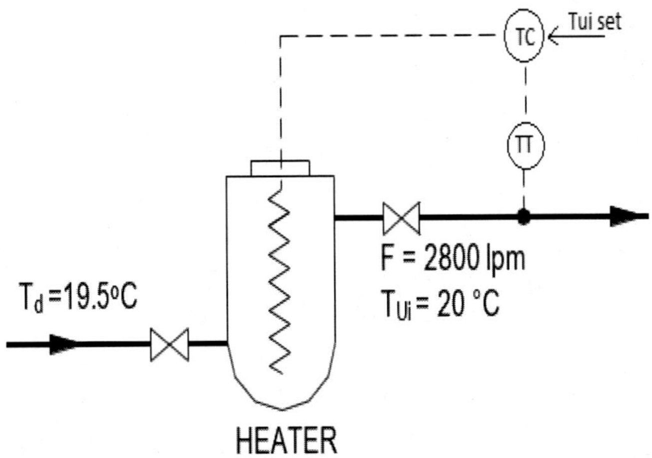

Figure 7. Heater outlet temperature control loop.

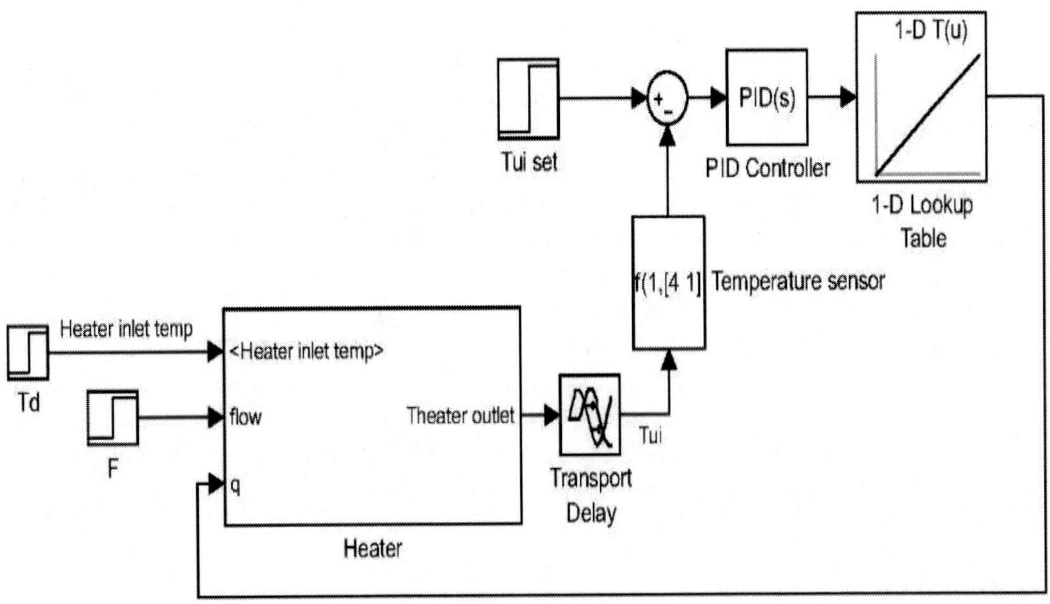

Figure 8. Heater outlet temperature control loop model.

3.3. Integrated System

Integrated system modeling was obtained by combining control loops P1 and P2, as shown in Figure 9. A mathematical model of the holding tank was also added. Transport delays were added at necessary locations by considering the actual piping layout.

4. Results and Discussion

The process conditions considered in the simulation for each component of the DTL-1 cooling system are shown in Figure 1. A mathematical model of the plate-type heat exchanger was simulated, and the temperature profiles of the LCW and chilled water obtained are shown in Figure 10. LCW was cooled from 23°C to 16°C, whereas chilled water was heated from 7.2°C to nearly 11°C. The temperature drop in LCW is almost twice that of chilled water because the flow rate of LCW is approximately half that of chilled water.

The heater mathematical model was simulated with an inlet water temperature of 19.5°C at a rated flow of 2800 l/min. The heating element temperature profile (from center to surface) along the heat transfer direction at the midplane of the flow path is shown in Figure 11. A relatively large temperature drop was observed in insulation since its thermal conductivity was low compared to other materials.

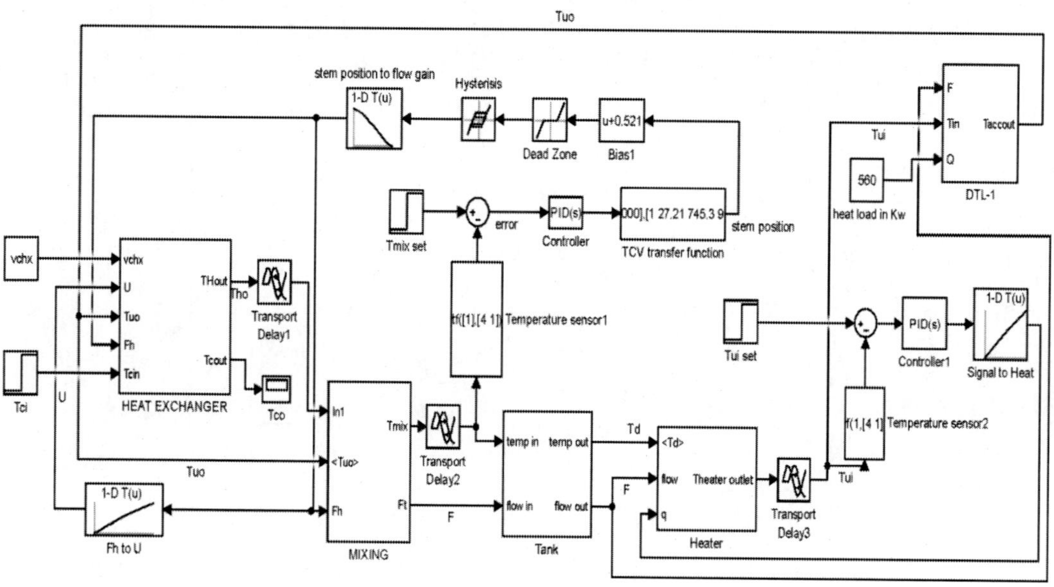

Figure 9. Integrated model of the DTL-1 cooling system.

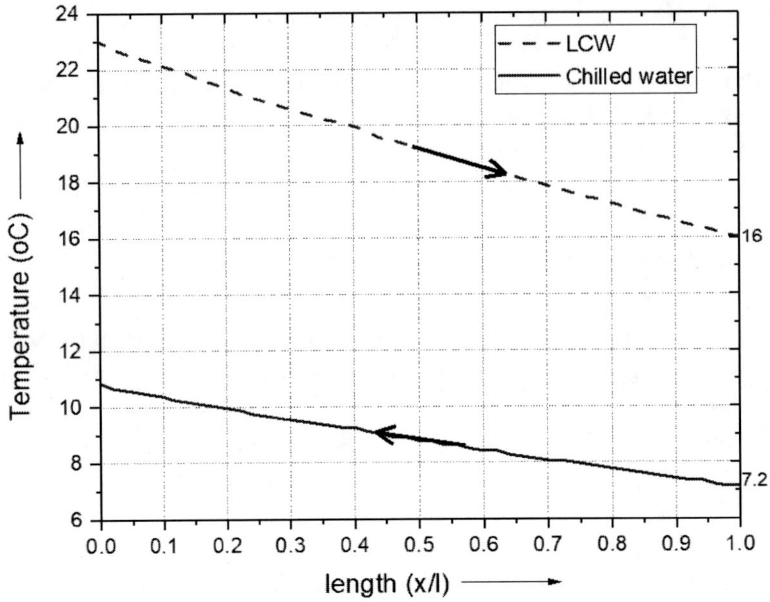

Figure 10. steady state temperature distribution of LCW and Chilled water along the flow in heat exchanger.

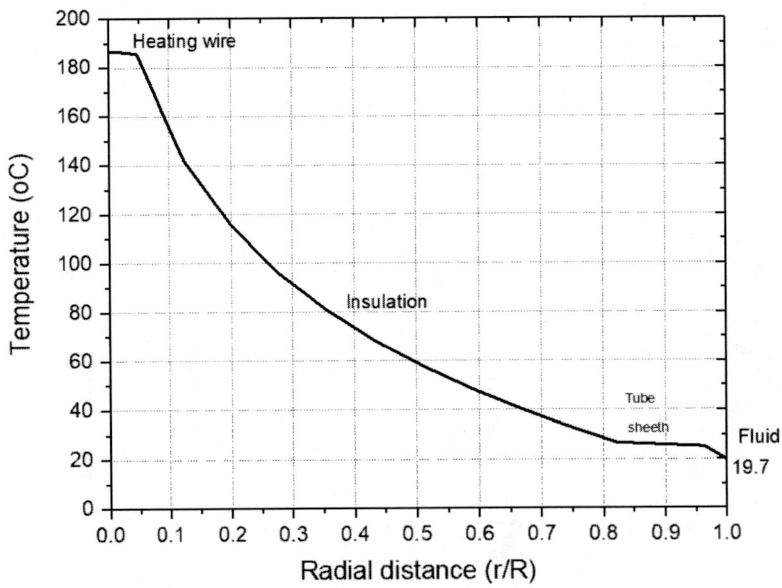

Figure 11. Temperature distribution of heating element.

4.1. Heat Exchanger Temperature Control Loop

Initially, the system was assumed to have steady-state operating parameters, as shown in Figure 1. The following disturbances were simulated, and the system response was predicted using the mathematical modeling of the cooling system as described in the previous sections.

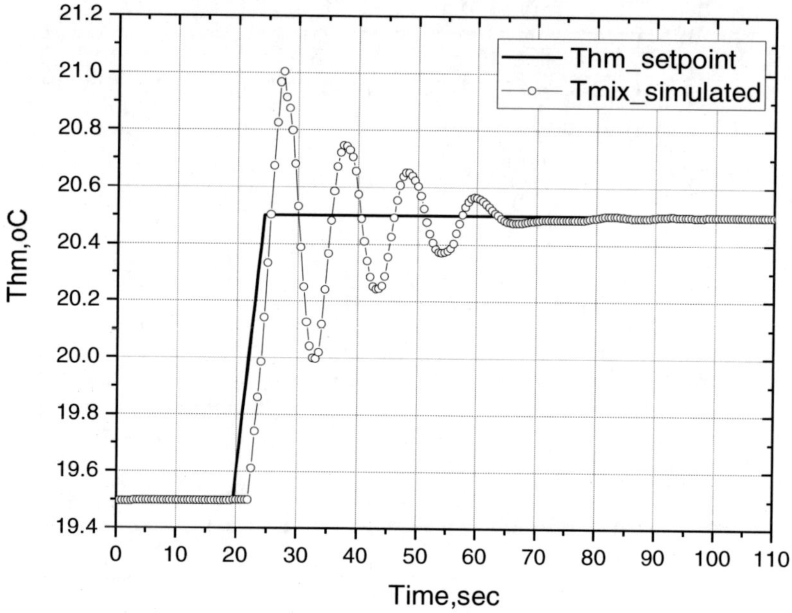

Figure 12. Change in T_{hm} from 19.5°C to 20.5°C.

Table 1. Response parameters

Parameter	value
Dead time	2.5 seconds
Rise time	3 seconds
Over shoot	0.54
Period of oscillations	10.3 seconds
Decay ratio	0.46

4.1.1. Increase in T_{hm} Set Point by 1°C

Initially, the system was maintained at steady state with Thm set point = 19.5°C, and as a disturbance, the set point changed (red) from 19.5°C to 20.5°C within 10 sec (ramp input with slope 0.1°C/sec). The response of the system is shown in Figure 12. Various parameters describing the system response are shown in Table 1. Initially, the system overshoots the final set point by 0.5°C, and the system oscillates around the new set point. Finally, oscillations die approximately 70 seconds after disturbing the system.

4.1.2. Step Change in Chilled Water Temperature

During the steady state of operation, when the chilled water temperature is increased by 1°C from 7.2°C to 8.2°C at t = 0 seconds (step input), the system response is shown in Figure 13. It can be observed that when the disturbance is given, for a moment, T_{hm} increases by 0.22°C after 6 seconds, and then the controller realizes the error and acts accordingly by increasing the flow through the primary side of the heat exchanger, thus bringing T_{hm} to the original steady state within 35 seconds after the original disturbance. The parameters describing the response are shown in Table 2.

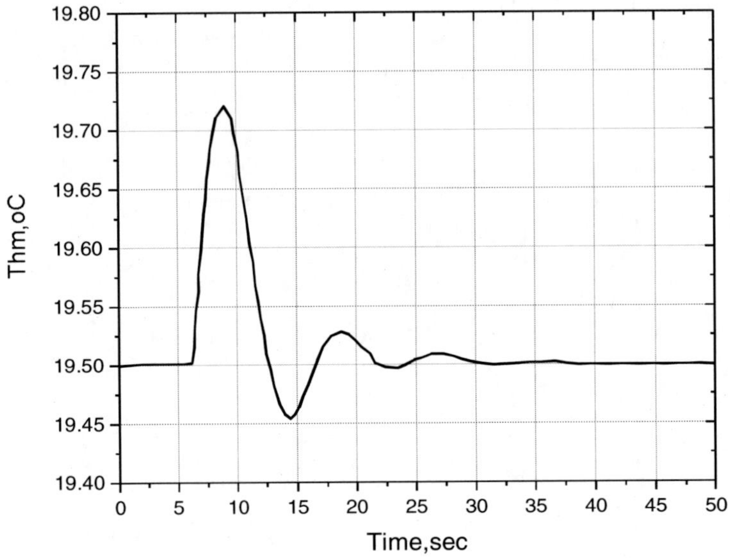

Figure 13. Mix mean temperature variation with a change in chilled water temperature.

Table 2. Response parameters

Parameter	**Value**
Decay ratio	0.136
Dead time	6 seconds
Period of oscillation	9.5 seconds

4.1.3. Step Change in Chilled Water Flow

During steady state, a step change in the chilled water flow rate was given to the system at t = 0 seconds. The chilled water flow decreases from 2700 lpm to 2000 lpm, and the effect of this on T_{hm} is given below in Figure 14. T_{hm} rises to approximately 0.13°C after a lag of 6 seconds. As the error increases, the controller increases the LCW flow through the primary side. Finally, T_{hm} was brought to the original steady state in 40 seconds. Parameters describing the response are shown in Table 3.

Table 3. Response parameters

Parameter	Value
Decay ratio	0.138
Dead time	6 seconds
Period of oscillation	9 seconds

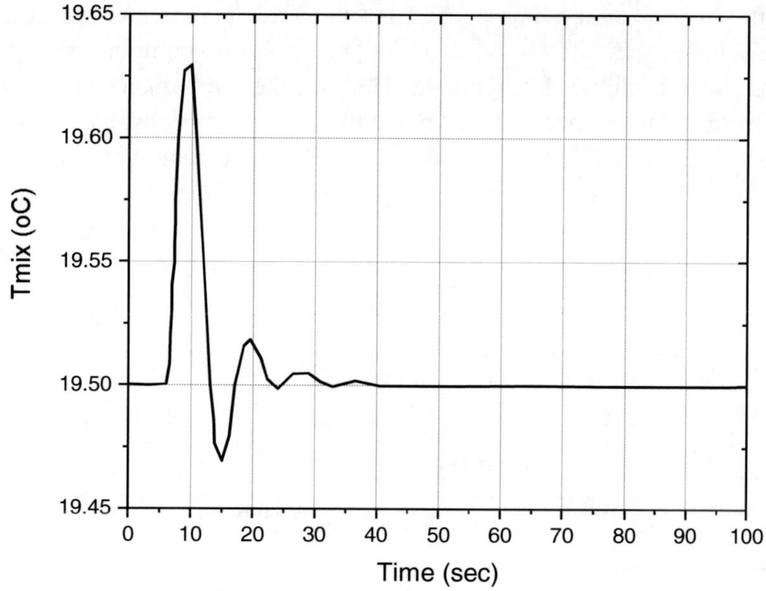

Figure 14. Mix mean temperature variation with a step change in chilled water temperature.

4.2. Heater Outlet Temperature Control Loop

4.2.1. Change in Heater Outlet Temperature Set Point

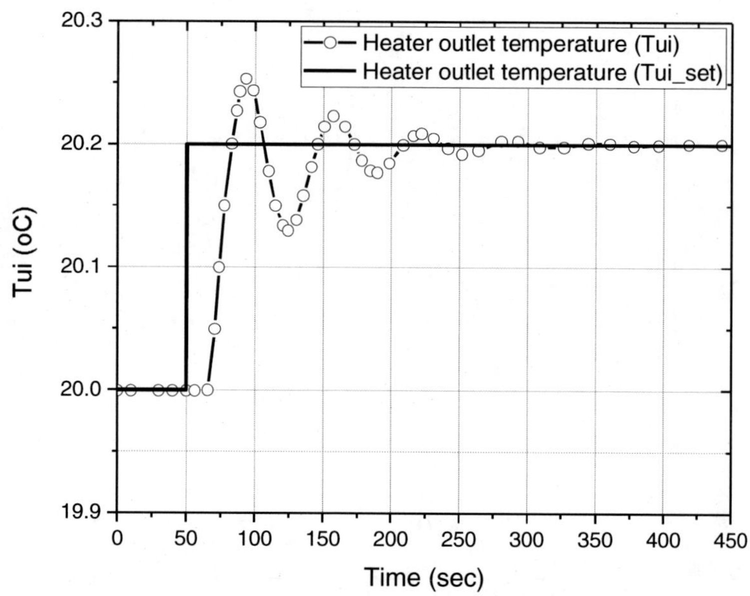

Figure 15. Heater outlet temperature variation to step change in set point.

Initially, the system operates in a steady state with a heater outlet temperature (T_{ui}) of 20°C. Now, a step change of 0.2°C (from 20°C to 20.2°C) in the set point is given at t=50 seconds. The system response is shown in Figure 15. Because the temperature sensor was physically located away from the heater outlet in the loop, the system started responding after 14 seconds, with a peak overshoot of 20.25°C observed at t = 90 seconds, and the temperature stabilized after 300 seconds of set point change. Parameters describing the response are shown in Table 4.

Table 4. Response parameters

Parameter	Value
Dead time	16 seconds
Rise time	17 seconds
Over shoot	0.25
Period of oscillations	63 seconds
Decay ratio	0.5

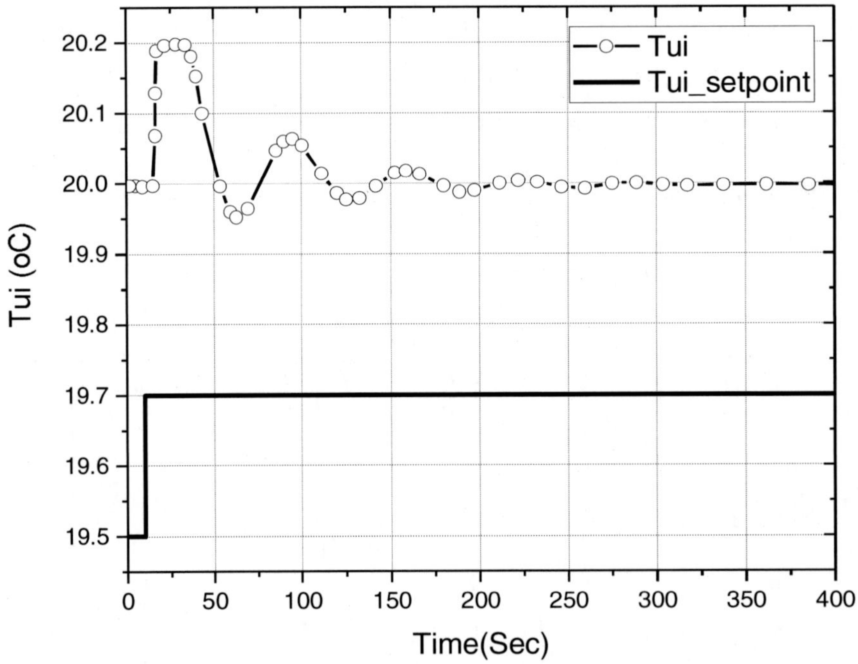

Figure 16. Heater outlet temperature variation to step change in heater inlet temperature.

Table 5. Response parameters

Parameter	Value
Decay ratio	0.25
Dead time	5 seconds
Period of oscillation	63 seconds

4.2.2. Change in Heater Inlet Temperature

The system was brought to steady state, and a step change in inlet (of heater) LCW water temperature from 19.5°C to 19.7°C was given. The system response is shown in Figure 16. The outlet temperature also changes by 0.2°C after 14 seconds because the temperature sensor is located away from the heater outlet in the actual piping system layout. Once the temperature change is realized by the controller, the power to the heating element is reduced, and the outlet temperature is brought back to its original steady state. Parameters describing the response are shown in Table 5.

4.3. Integrated System Response

4.3.1. Chilled Water Inlet Temperature Step Change

The system was brought to steady state operation with a total flow of 2800 lpm, T_{hm} = 19.5°C and heater outlet temperature of 20°C. A step change in chilled water inlet temperature from 7.2°C to 8.2°C was given at t = 50 seconds, as shown in Figure 17b. The system response to the above disturbance is shown in Figure 17a. The mix mean temperature (T_{hm}) rises by approximately 0.2°C after some lag before finally settling to the original steady-state temperature.

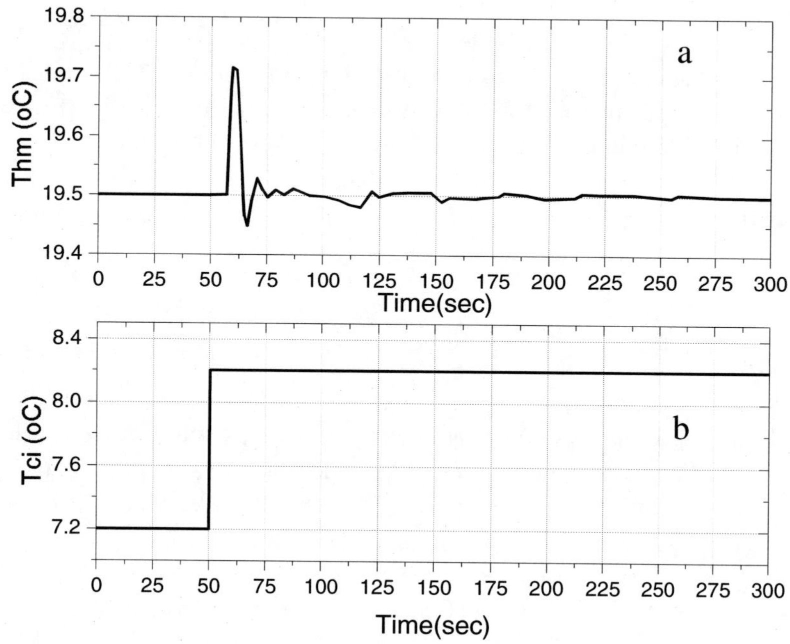

Figure 17. Integrated system response to a step change in chilled water temperature.

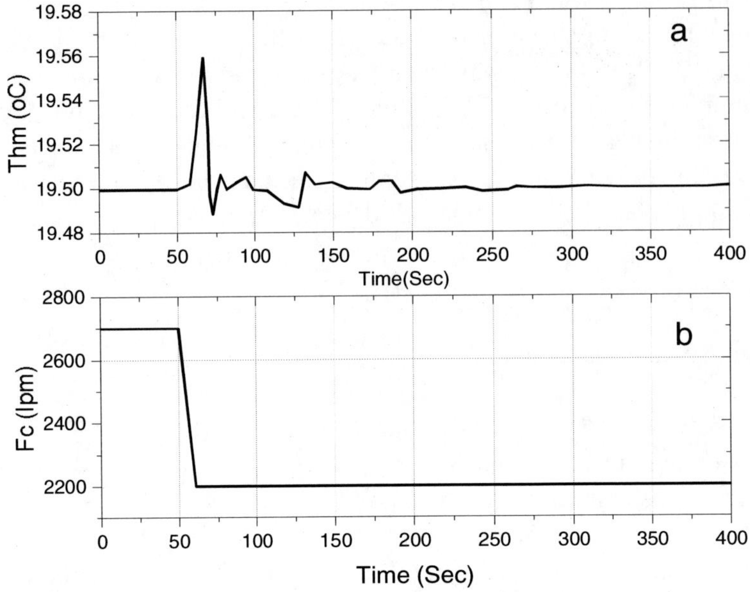

Figure 18. Integrated system response to ramp change in chilled water flow.

4.3.2. Chilled Water Flow Change

The system was brought to steady state operation with a total flow of 2800 lpm, a T_{hm} of 19.5°C and a heater outlet temperature of 20°C. Disturbance in chilled water inlet flow was given at t = 50 seconds from 2700 lpm to 2200 lpm, as shown in Figure 18b. The system response to the above disturbance is shown in Figure 18 a. The mix mean temperature (T_{hm}) rises by approximately 0.05°C after 5 seconds of lag before finally settling to the original steady-state temperature.

5. Conclusion

Mathematical models of various components in the DTL-1 cooling system, such as the heat exchanger, heater, FCV, TCV, holding tank piping and sensors, were presented. First principles were used for modeling most of the components, but models of FCV and TCV were developed using shop-tested frequency response data. An integrated model was developed from models of individual components, and tuning was performed by the Zeigler-Nichols method. The performance of individual and integrated control loops with various disturbances affecting the mix-mean temperature (T_{hm}) of the DTL-1 system were studied. It was observed that the temperature of chilled water has a stronger effect on the system than that of chilled water flow. The control scheme adopted in the DTL-1 cooling system is found to be very effective in meeting the requirement of keeping the supply temperature within $\pm$0.2°C around the set point.

Nomenclature

A_H	Heat transfer area for unit cell in heat exchanger (m^2)
Cp	Specific heat (J/kg.°C)
d_e	Equivalent diameter of shell side of heater (m)
F	Total flow in loop (lpm)
FC	Flow controller
F_h	LCW inlet flow to heat exchanger (lpm)
F_b	Chilled water flow to heat exchanger (lpm)
F_c	LCW bypass flow around heat exchanger (lpm)
H_c	Heat capacitance (J/°C)
h_h	Heat transfer coefficient of shell side of heater ($W/m^2.K$)
k	Thermal conductivity of LCW fluid in heater (W/m.K)
N_U	Nusselt number
P	Process control loop
Pr	Prandlt number
Q	Heat load of heat exchanger (W)
Qv	Volumetric heat generation (W/m^3)
Re	Reynolds number
T	Temperature(°C)
TC	Temperature controller
TT	Temperature transmitter
t	Time (s)
U	Overall heat transfer coefficient ($W/m^2.K$)
V	Velocity (m/s)
x	Length (m)

Greek Symbols

Δ	Temporal/spatial deviation
μ	Viscosity of fluid at bulk
μ_w	Viscosity of fluid near wall
ρ	density

Subscripts

c	Chilled water
ci	Chilled water inlet temperature from heat exchanger (°C)
co	Chilled water outlet temperature from heat exchanger (°C)
T_d	Inlet temperature heater (°C)
f	Fluid
h	Hot fluid (LCW)
he	Heater

hi	Heat exchanger inlet for LCW
ho	Heat exchanger outlet for LCW
i, N+i	Heat exchanger chilled water side, LCW side
u	DTL-1 heater
I	Insulation
ts	Tube sheath of heater rod
ui	DTL-1 inlet
uo	DTL-1 outlet
w	Heating wire

References

[1] RRDPD, BARC, (2013). "Design basis report of low conductivity water system of Low Energy High Intensity Proton Accelerator".

[2] Yimin, X., and Roetzel, W. (1999). Dynamic behavior of heat exchangers, Boston: WIT press.

[3] Khan, N. A. R. (1988). "The dynamic characteristics of a counter current plate heat exchanger," *International journal of heat and mass transfer*, pp. 1269-1278.

[4] ALFA LAVAL (INDIA) LIMITED, "Performance curves of plate type heat exchanger," 2012.

[5] Sinnot, R. (2005). Chemical Engineering Design:Chemical Engineering Volume 6, Elsevier.

[6] Fluid control research institute, *Certificate of testing on Three way flow control valve (FCV),* Kerala, 2013.

[7] Fluid control research institute, *Certificate of testing on Flow control valve (TCV),* Kerala, 2013.

[8] Stephanopoulos, G. (1984). Chemical process control: An introduction to theory and practice.

Chapter 20

The Process Simulation and Optimization of Dimethyl Ether Production from Biogas Using Aspen Plus

D. Venkata Padma[1]
S. V. A. R. Sastry[2]
S. Kishore Kumar[3,*]
and R. Hemalatha[1]

[1]Faculty, Department of Chemical Engineering, MVGRCE(A), Vizianagaram, A. P., India
[2]Faculty, School of Chemical Technology, HBTU, Kanpur, U. P., India
[3]PG Student, Department of Chemical Engineering, AUCE, Visakhapatanam, A. P., India

Abstract

Biogas mainly contains methane, carbon dioxide, water and nitrogen. Methane and carbon dioxide are the sources of carbon and hydrogen generated during syngas processing that can be used to convert them to Dimethyl Ether (DME). It has low particulate emissions, sulfur-free emissions and a high cetane number (55–60). The production of DME can be a two-step process or a one-step process. One-step DME synthesis is a highly exothermic process because it carries out the water gas shift reaction, methanol formation and methanol dehydration. In the present study, a single-step process is chosen because of its inherent capacity to replace several process units, namely, the dehydration reactor, preheater and separator, by a single reactor. One-step process technology is adopted in the present study for the processing of 5 kmol/hr or 125 kg/hr of biogas from four different sources as input to the process. The process was developed with the Aspen Plus simulator. It is also envisaged to know the sensitivity of the process for different operating parameters, namely, temperature, pressure, stream flow rates, stream composition, stream enthalpies, etc., and to identify optimum operating conditions for the lowest possible equipment sizes. A sensitivity analysis is carried out to evaluate the influence of different possible changes on the process. Therefore, the model can facilitate the development of small-scale DME production units at community poultry farms, dairy farms, community biogas units and even industrial and municipal waste treatment plants.

Keywords: biogas, Aspen Plus, simulation, dimethyl ether, methanol, alternative fuel

[*] Corresponding Author's Email: padmakishore9@gmail.com.

In: Sustainability in Chemical Processes through Digitalization and Green Chemistry Approaches
Editors: Susarla V. A. R. Sastry, Shravan Kumar and Ashish Kapoor
ISBN: 979-8-89113-257-3

1. Introduction

Due to the depletion of fossil fuel resources, considerable research has been conducted on alternative fuels to comply with domestic and transportation needs while minimizing environmental effects [1]. Even to reduce fuel consumption and greenhouse gas emissions, the thermodynamic principles of diesel engines reveal that only with fuel change can such a decrease be achieved [2]. The concept of a carbon neutral cycle brings to the forefront a new approach to converting biomass to alternative clean fuel [3].

For this conversion, biomass is converted to biogas in an anaerobic digester because, among biofuels, biogas is an interesting alternative fuel for combustion and to generate electricity [4]. Utilizing biogas results in the immediate generation of carbon dioxide and less energy. Untreated biogas reduces the performance of any engine [5].

There are six basic types of reforming processes available for the production of syngas from gases containing methane [6]. Auto thermal reforming (ATR) is one of the methods that suits small units well. Low operational temperatures and easy start-up are the advantages. It is an operation based on a combination of processes, namely, catalytic partial oxidation and adiabatic steam reforming [7].

Dimethyl ether (DME) has several advantages over diesel. It has low particulate emissions, sulfur-free emissions and a high cetane number (55–60) [8]. A high cetane number makes DME an attractive fuel for compression as well as ignition (CI) engines. However, DME has significantly different physical properties than diesel fuel, including low viscosity, negligible lubricant requirements and high vapor pressure [9].

From the literature, it is found that there are two processes available for the production of DME, i.e., two-step processes and one-step processes [10]. In a two-step process, the syngas is converted into methanol, followed by catalytic dehydration. In a single-step process [11], a bifunctional catalyst ($Cu/ZnO/Al_2O_3$+γ-Al_2O_3) is used in a single reactor as a mixture. A bifunctional catalyst is a catalytic mixture, one for methanol synthesis and the other for dehydration. One-step DME synthesis [12] is a highly exothermic process because it carries out the water gas shift reaction, methanol formation and methanol dehydration.

In the present study, a single-step process was chosen because of its inherent capacity to replace several process units, namely, the dehydration reactor, preheater and separator, by a single reactor. One-step process technology is adopted in the present study for the processing of 5 kmol/hr or 125 kg/hr of biogas from four different sources as input to the process.

The study is aimed at the design of a process with the Aspen Plus simulator. It is also envisaged to know the sensitivity of the process for different operating parameters, namely, temperature, pressure, stream flow rates, stream composition, stream enthalpies, etc., and to identify optimum operating conditions for the lowest possible equipment sizes. A sensitivity analysis is carried out to evaluate the influence of different possible changes on the process. Therefore, the model can facilitate the development of small-scale DME production units at community poultry farms, dairy farms, community biogas units, food industry wastes and even industrial and municipal waste treatment plants.

Table 1, Table 2 and Table 3 represent the design and optimum operating conditions developed and are given below.

Table 1. Optimum operating conditions

Blocks	Without recycle			With recycle		
	# K	* atm	kmol/hr	#K	* atm	kmol/hr
ATR	600	1.5	2	600	1.5	3
DMER	470	30	-	550	30	-
Recycle fraction of Off-gas	-	-	-	0.4	-	-

Table 2. Equipment size # Temperature, *Pressure and Oxygen flow rate

Equipment	Length, m	Dia, m
Auto thermal reactor	2	0.3
DME synthesis reactor	2.5	0.15

Table 3. Range of variables covered in the study

Variable	Range	
	ATR	DMER
Temperature, K	400-1100	470-600
Pressure, atm	1-3	5-70
Oxygen flow rate, kmol/hr	1-5	-
Methane, kmol/hr	2.25- 4.7	-

The process developed in the present study can be adopted for the utilization of various organic wastes generated during different activities and makes the environment clean by reducing greenhouse gases from different human activities.

2. Process Design Development

2.1. Inception for Alternative Fuel

The idea of designing a small-scale system that can convert the biogas produced from municipal wastewater, landfills, agricultural residues and poultry waste will lead to a carbon-neutral cycle. The development of a process design for DME manufacturing from biogas is performed to establish a new process that adapts various technologies from existing processes and is more profitable, better controlled, operationally safe and environmentally friendly. The following steps are selected for the process development:

1. Conversion of methane-containing sources into syngas
2. Direct conversion of syngas to DME.

Typical steps followed for designing the chemical process are as follows:

- Recognizing product importance

- Preliminary process synthesis involves determining reactions, separations and possible operation conditions from the literature and experimental data from research publications.
- Preparation of a flow sheet.
- Integrate and optimize the process.

The present work satisfies the preliminary design that determines the suitability of the process for further detailed estimation. In the preliminary design, the process is developed by applying various configurations of processing operations to find the path and mechanism that can convert biogas to DME. A steady-state model has been developed that can be used for semibatch and continuous operations.

3. Raw Materials and Product Specifications

3.1. Raw Materials

Biogas is obtained from the anaerobic digestion of organic products [13]. Reactions that take place in this process are grouped into stages of hydrolysis, acetogenesis (the process through which acetate is produced by anaerobic bacteria from a variety of carbon sources) and methanogenesis (the formation of methane from carbon dioxide and acetic acid by microbes) [14]. The composition and characteristics of the biogas depend on the raw material (agricultural wastes, stockbreeders and the organic part of urban and industrial solid waste and wastewater) as well as on the technology used. However, the important components of biogas are approximately 60% methane, 30% carbon dioxide, 6–4% water and 4–6% nitrogen [15–16]. The study is done on municipal wastewater biogas composition for developing the process, and then other sources are investigated. In Table 4, the composition of biogas obtained from municipal wastewater, landfills and agricultural residue is presented [17].

Table 4. Typical biogas compositions obtained from various sources in volume fraction

Biogas constituents	Municipal wastewater	Landfill	Agricultural residues
CH_4	60	45	68
CO_2	33	32	26
N_2	1	17	1
O_2	0	2	0
H_2O	6	4	5

Oxygen from a pressure swing air separation (PSA) unit is supplied to the process; two kinds of purities with their exergy values (exergy is the energy that is available to be used) provide the choice of selection based on the biogas composition. Table 5 shows two types of standard oxygen that can be produced by PSA.

Table 5. Purity and exergy values of two kinds of oxygen available from a standard air separation unit [18]

Component	Oxygen	Nitrogen	Argon	Exergy value(kJ/mol)
1	0.943	0.012	0.045	3.6406
2	0.995	0.001	0.004	3.8910

3.2. Product

DME has an oxygen concentration of 34.78% and can be burned without soot emission, while for traditional diesel fuels, simultaneous NO_x and soot emissions occur. It has a boiling point of -25°C, which is 20°C higher than that of LPG, and can be liquified at 0.54 MPa (20°C) [18–19]. It is noncarcinogenic, nonteratogenic (it does not cause birth defects), nonmutagenic (it does not change the genetic material) and nontoxic [20]. The calorific value of DME is 28.9 $\times 10^6$ J/kg as liquid and 59.44$\times 10^6$ J/Nm^3 as gas, which means that it is 1.37 times higher than methanol in the liquid phase and 1.65 times higher than methane in the gas phase. Finally, when we look at the liquid densities, DME has a liquid density 1.37 times higher than propane, which means that 85% of the energy of propane can be obtained from an equal-sized DME storage tank.

3.3. Basis to Develop Preliminary Process Design

A small-scale system is developed with a capacity of 5 kmol/hr of biogas. Upon developing the process, the DME produced would be approximately 0.9 to 1 kmol/hr (41–46 kg/hr).

The overall process is divided into two sections based on key conversion operations.

1. Auto thermal reformer section
2. DME synthesis section

Catalytic ATR has received much attention in research in recent years. It offers the advantages of small units, lower operational temperatures, easier start-up and a wider choice of materials. Moreover, it has higher energy efficiency compared to SMR or POX. ATR has major advantages over other reforming alternatives, mainly for low energy requirements and a lower process temperature than POX, a higher H_2/CO ratio, and an easily regulated H_2/CO ratio by the inlet gas composition. Although ATR has interesting potential in industrial applications, there has been only a limited amount of work reported in the field of reactor design and simulation.

3.4. Rate Equations

The rate of the reforming reactions was described by applying a Langmuir-Hinshelwood (LHHW) rate equation [21].

$CH_4 + H_2O \leftrightarrow CO + 3H_2$ 206.2 kJ/mol (R1)

$CH_4 + 2H_2O \leftrightarrow CO_2 + 4H_2$ 164.9 kJ/mol (R2)

$CO + H_2O \leftrightarrow CO_2 + H_2$ −41.1 kJ/mol (R3)

$CH_4 + 2O_2 \leftrightarrow CO_2 + 2H_2O$ −802.7 kJ/mol (R4)

The frequency factor, adsorption constant and desorption constant data for the ATR model [22] are given in Table 6. Equilibrium constants for R_1, R_2, and R_3 are given below.

Syngas from the ATR section is fed to the DME synthesis section. The process for producing DME from biogas came from one of the two commercial production methods.

1. A two-step process (the indirect method) consisting of methanol formation from synthesis gas followed by the dehydration of the methanol
2. A single-step process (the direct method), which is the direct formation of DME from synthesis gas [23].

Table 6. Rate constants and adsorption coefficients for the ATR model

Parameters $A\exp^{(-B/RT)}$	A mol/($kg_{cat}S$)	B J/mol
k_1	$1.17\times10^{15} bar^{0.5}$	240,100
k_2	$2.83\times10^{14} bar^{0.5}$	243,900
k_3	$5.43\times10^{5} bar^{-1}$	67,130
k_{4a}	$8.11\times10^{5} bar^{-2}$	86,000
k_{4b}	$6.82\times10^{5} bar^{-2}$	86,000
K_{CH4}	6.65×10^{-4} bar^{-1}	-38,280
K_{CO}	8.23×10^{-5} bar^{-1}	-70,650
K_{H2}	6.12×10^{-9} bar^{-1}	-82,900
K_{H2O}	$1.77\times10^{5} bar$	88,680
K_{CH4}^{c}	1.26×10^{-1} bar^{-1}	-27,300
K_{O2}^{c}	7.78×10^{-7} bar^{-1}	-92,800

ATR Reaction rate equation:

$$R_1 = \frac{k_1}{p_{H_2}^{2.5}}\left(p_{CH_4}p_{H_2O} - \frac{p_{H_2}^3 p_{CO}}{K_1}\right)\frac{1}{\theta^2} \quad (1)$$

$$R_2 = \frac{k_2}{p_{H_2}^{3.5}}\left(p_{CH_4}p_{H_2O}^2 - \frac{p_{H_2}^4 p_{CO_2}}{K_2}\right)\frac{1}{\theta^2} \quad (2)$$

$$R_3 = \frac{k_3}{p_{H_2}}\left(p_{CO}\ p_{H_2O} - \frac{p_{H_2}p_{CO_2}}{K_3}\right)\frac{1}{\theta^2} \tag{3}$$

$$R_4 = \frac{k_{4a}\ p_{CH_4}p_{O_2}}{\left(1+K^C_{CH_4}p_{CH_4}+K^C_{O_2}p_{O_2}\right)^2} + \frac{k_{4b}\ p_{CH_4}p_{O_2}}{\left(1+K^C_{CH_4}p_{CH_4}+K^C_{O_2}p_{O_2}\right)} \tag{4}$$

$$\theta = 1 + K_{CO}p_{CO} + K_{H_2}p_{H_2} + K_{CH_4}p_{CH_4} + K_{H_2O}\frac{p_{H_2O}}{p_{H_2}} \tag{5}$$

Equilibrium constants:

$$K_1 = e^{\left(\frac{-26830}{T}+30.114\right)}(bar^2)$$
$$K_2 = K_1K_3(bar^2)$$
$$K_3 = e^{\left(\frac{4400}{T}-4.036\right)}$$

In comparison with the two-step method, the single-step procedure is attracting more attention for its economic value and theoretical significance. A reaction model of DME synthesis is considered that is applied with the LHHW (Langmuir-Hinshelwood Hougen Watson) equation as an adsorption-desorption model on the surface of the DME catalyst. Overall DME synthesis reactions consist of three steps: methanol synthesis, dehydration and the water gas shift (WGS) reaction, as follows:

$CO + 2H_2 \leftrightarrow CH_3OH$ −43.4 kcal/mol-DME (R5)

$CO_2 + 3H_2 \leftrightarrow CH_3OH + H_2O$ −13.5 kcal/mol-DME (R6)

The formation of methanol is considered by two reaction mechanisms, as illustrated in both reactions R_5 and R_6. Reaction R_5 is the formation of methanol with CO and H_2 as major reactants. This reaction emits a high reaction enthalpy, while reaction R_6 is led by participating CO_2 and H_2 and has a lower reaction enthalpy than reaction R_5. It is assumed that both reactions can proceed to produce methanol simultaneously.

Methanol passes through the following dehydration reaction with the DME synthesis catalyst, as in reaction R_7. Reaction R_8 is shown as a WGS reaction, and it supplies CO_2 and H_2 for DME synthesis.

$2CH_3OH \leftrightarrow CH_3OCH_3 + H_2O$ −5.6 kcal/mol-DME (R7)

$CO + H_2O \leftrightarrow CO_2 + H_2$ −9.8 kcal/mol-DME (R8)

By integrating reactions R_5 to R_8, the overall DME reaction shows two mechanisms: reaction R_9 and reaction R_{10}. The combination of reactions R_5 and R_7 gives the overall reaction R_{10}; the combination of reactions R_5, R_7 and R_8 gives the overall reaction R_9; and if reaction R_8 proceeds slowly, the overall DME reaction leads to reaction R_{10}.

$3CO + 3H_2 \leftrightarrow CH_3OCH_3 + CO_2$ −58.8 kcal/mol-DME (R9)

$2CO + 4H_2 \leftrightarrow CH_3OCH_3 + H_2O$ −49.0 kcal/mol-DME (R10)

According to DME overall reactions R_9 and R_{10}, the DME reaction is an exothermal reaction that generates high amounts of heat. If the DME reactor temperature is increased by the reaction heat from DME synthesis, the thermodynamic equilibrium conversion is decreased.

3.5. Reaction Kinetics of DME Synthesis

The reaction kinetics of methanol synthesis as a major part of DME synthesis have been studied by many researchers [24–26]. Methanol synthesis is governed by reaction R_5, which is accelerated at low concentrations of CO_2 on $Cu/ZnO/Al_2O_3$+ γ-Al_2O_3 catalysts under industrial conditions. After all, with an increase in CO_2 concentration, the reaction rate of methanol decreases. Because DME synthesis is governed by catalytic reactions of syngas on the surface of the catalyst, it is applied to the LHHW model to simulate the DME synthesis reaction. It is assumed that each CO and CO_2 is comparatively adsorbed at an active site on the catalyst surface and that each H_2 and H_2O is also comparatively adsorbed at the other. Therefore, it is suggested that methanol production is led by both CO and CO_2 components throughout catalytic reactions under the LHHW model. To simplify the DME synthesis reaction, the overall reaction is divided into three sections: methanol synthesis, methanol dehydration and the WGS reaction [27].

The frequency factor, adsorption constant and desorption constant data for the DME synthesis reactor model are given in Table 7.

Table 7. Rate constants and adsorption coefficients for the DMER model [28]

Parameter Aexp($^{-B/RT}$)	A	B
k_5	7380	-54307
k_6	5059	-67515
k_7	1062	-43473
k_8	73976	-20436
K_{CO}	3.934×10^{-6}	37373
K_{CO2}	1.858×10^{-6}	53795
K_{H2}	0.6716	-6476
K_{CH3OH}	3.480×10^{-6}	54689

Reaction rate equations:

$$R_5 = \frac{k_5 f_{CO} f_{H_2}^2 (1-\beta_1)}{(1+K_{CO}f_{CO}+K_{CO_2}f_{CO_2}+K_{H_2}f_{H_2})^3} \quad \beta_1 = \frac{f_{CH_3OH}}{K_{f_1} f_{CO} f_{H_2}^2} \qquad (6)$$

$$R_6 = \frac{k_6 f_{CO_2} f_{H_2}^3 (1-\beta_2)}{(1+K_{CO}f_{CO}+K_{CO_2}f_{CO_2}+K_{H_2}f_{H_2})^4} \quad \beta_2 = \frac{f_{CH_3OH} f_{H_2O}}{K_{f_2} f_{CO} f_{H_2}^3} \qquad (7)$$

$$R_7 = \frac{k_7 f_{CH_3OH}(1-\beta_3)}{(1+\sqrt{K_{CH_3OH} f_{CH_3OH}})^2} \quad \beta_3 = \frac{f_{DME}\ f_{H_2O}}{K_{f_3} f_{CH_3OH}^3} \tag{8}$$

$$R_8 = \frac{k_8 f_{H_2O}(1-\beta_4)}{1+K_{CO} f_{CO}+K_{CO_2} f_{CO_2}+\sqrt{K_{H_2} f_{H_2}}} \quad \beta_4 = \frac{f_{CO_2} f_{H_2}}{K_{f_4} f_{CO} f_{H_2O}} \tag{9}$$

Equilibrium constants:

ln K_{f1}, K_{f2}: $4213/T - 5.752 \ln T - 1.707\times10^{-3}T + 2.682\times10^{-6}T^2 - 7.232\times10^{-10}T^3 + 17.6$

ln K_{f3}: $4019/T + 3.707 \ln T - 2.783\times10^{-3}T + 3.8\times10^{-7}T^2 - 6.561\times10^{-4}/T^3 - 26.64$

ln K_{f4}: $2167/T - 0.2258 \ln T - 1.037\times10^{-3}T - 2.331\times10^{-7}T^2 - 1.277$

The reaction rates of methanol synthesis are individually expressed in Eq. (6), led by CO, and Eq. (7), led by CO_2. where β is defined as the fugacity for equilibrium reactions of each component. The R_7 of DME synthesis through methanol dehydration is shown in Eq. (8), and the R_8 of the WGS reaction is shown in Eq. (9). The WGS reaction takes place on the Cu cluster among the DME catalyst, and it proceeds through two steps: supplying H_2 by adsorption of oxygen in the H_2O molecule and generating CO_2 between CO and adsorptive oxygen.

Assumption made during model development:

1. The overall process is operated at steady state with a constant input flow rate of syngas.
2. The DME synthesis catalyst is composed of homogeneous catalyst and charged with a constant void fraction of the catalyst bed in the DME synthesis reactor.
3. Catalytic poisoning effected by H_2S is neglected.

3.6. Overall Process

The sections (ATR and DME) generate heat during the reaction operations. Adiabatic reactor models used for the study increase the temperature of product streams much more than reactant streams. Heat energy is recovered by heat exchangers in product streams. Condensed water is separated from product streams by separators. The product stream of ATR is flashed in the flash separators to remove water; similarly, in the DME synthesis section, the product stream is sent to a series of flash tanks to remove water, methanol and DME from the off-gas [28–29].

Off-gas streams contain unreacted methane, carbon monoxide, hydrogen, nitrogen and carbon dioxide. A split of off-gas is recycled to the ATR section at the suction side of the compressor. A steady-state preliminary design model for DME synthesis is developed based on the flowsheet development method and contains all the required equipment connected with the process and recycling streams [30]. In the established model, various sources of biogas are introduced as feed, and the DME production rate is investigated.

3.7. Flowsheet Development

Based on the process selected, a flowsheet is established. For preliminary process synthesis, the flowsheet is prepared, including the reactors, separators, heat exchangers and compressors. Material and energy balances are performed using the Aspen Plus simulator. Attempts are made to improve the design of the process units and to achieve more efficient process integrations with the help of separation units, heat and power integration and sensitivity analysis.

3.71. Flow Sheet Description

The flowsheet developed is studied and simulated in two sections.

1. ATR section
2. DME synthesis section

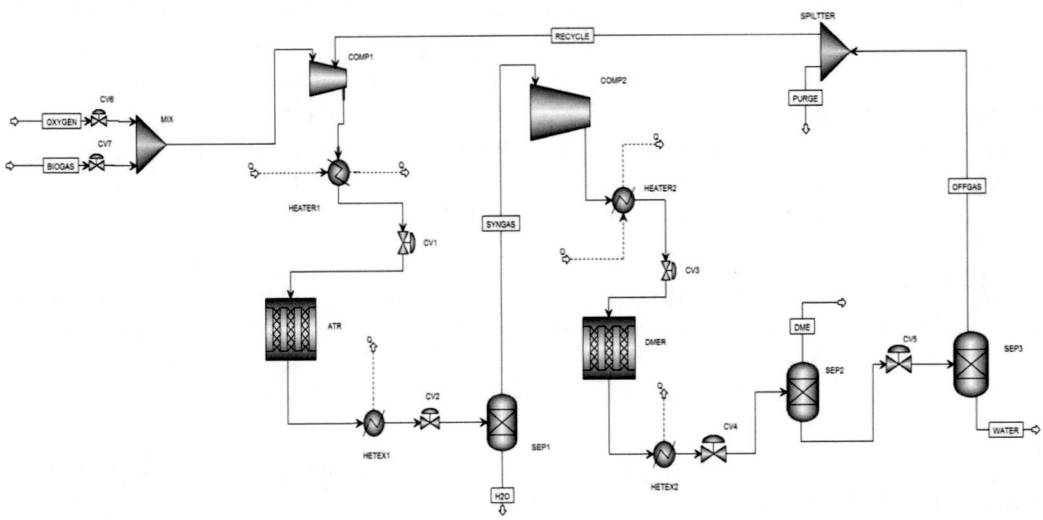

Figure 1. Final flowsheet with recycle.

Biogas is fed from a storage tank at the anaerobic station to the mixer, where oxygen is mixed in the mixing block. Oxygen is supplied from a pressure-swing air separation plant. By a control system, the feed and oxygen ratio are maintained and supplied to the compressor (COMP1) and heat exchanger (HEATER1). A compressor builds up the required pressure and is heated to the required temperature by a heater. The adiabatic fixed bed reactor model is applied with an RPLUG block as an Auto thermal reactor (ATR). Due to the exothermic reaction, the temperature of the product stream increases, and the heat energy is recovered by the heat exchanger. Water formed in the Auto thermal reactor is to be separated by a separator (SEP1), and syngas is sent to the second compressor (COMP2) without water. The DME synthesis reactor temperature and pressure are adjusted by the compressor and heat exchanger (HEATER 2). The reactions in the DME reactor (DMER) are exothermic, so the heat is recovered as in the previous section. Separation of methanol, water and DME was performed in a flash separator (SEP2). A split off-gas from the separator is sent to the flare for recycling. The simulation of process flow conditions is carried out as per the flow diagram shown in Figure 1.

The recycle stream from the off-gas separator (SEP3) is connected to the ATR feed compressor (COMP1). In every simulation run, a fraction of the off-gas is purged out of the splitter. Reactor sections were given priority over the separation systems because the separation techniques are well developed and can be applied to the present process. Process simulation is conducted for biogas of various compositions obtained from various sources.

4. Process Simulation

4.1. Objective of the Study

It is proposed to develop a plant for processing biogas and converting it into DME, which is a prospective fuel that can replace petroleum and LPG. It is proposed to design a small-scale plant that can process 5 kmol/hr of raw biogas into DME and to find suitable operating conditions for the process. The data obtained from the simulator are useful for controlling the plant safely. An operative physical model can be designed. The proposed plant is small-scale and can be adopted in rural areas where biogas can be generated. The smaller plant has several advantages, as listed below.

At low operating conditions, coke formation can be avoided, reducing failed safe loops [31].

- A pilot plant can be constructed with low investment.
- Easy startup and shutdown of the plant are possible.
- Minimum control system utilization.

It is proposed to acquire data from the simulator for different possible alternative designs for the process plant. Various schemes worked out are presented here under [32].

- Steady state simulation without recycling
- Steady-state simulation with recycling
- Steady state simulation with recycling and different feed sources (compositions)

5. Results and Discussion

5.1. Simulation and Results of Auto Thermal Reforming Sections

For a 5 kmol/hr feed of biogas from municipal wastewater with the following composition [33–34], simulated data were obtained for the ATR (auto thermal reformer).

Component	Mole fraction	Kmol/hr
CH_4	0.6	3
CO_2	0.33	1.65
H_2O	0.06	0.30
N_2	0.01	049

A simulation is carried out with a set of operating parameters, and the outlet gas composition is analyzed. The objective of the present work is to operate the plant at the lowest possible temperature and pressure, yielding the highest conversion with the fewest possible side reactions. Simulation of process flow conditions is carried out as per the flow diagram shown in Figure 2, which represents the ATR section. The inlet and outlet concentrations of various components and temperatures were monitored.

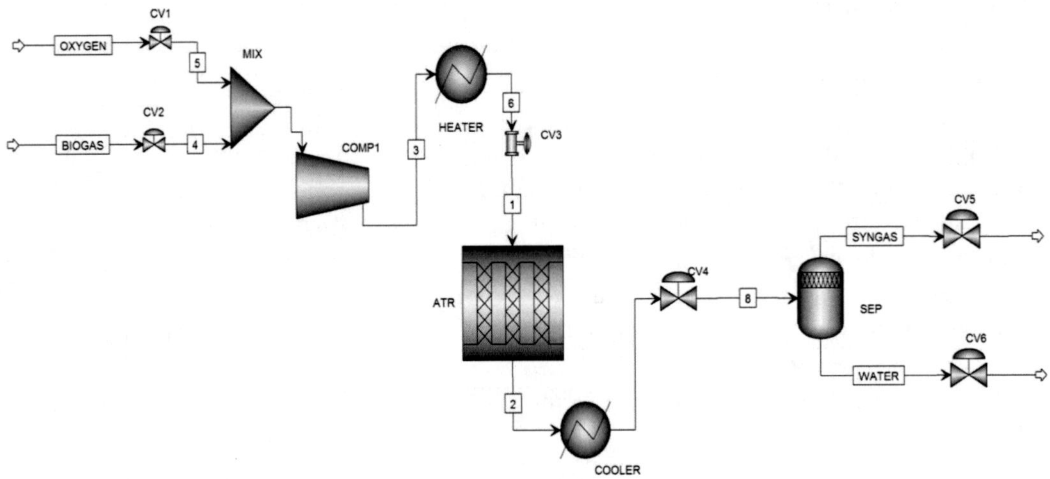

Figure 2. ATR section flowsheet.

5.2. Effect of Feed Temperature

In the present-day process, CO_2 separation from syngas is not performed. It is allowed in the DME section along with CO and H_2. Reaction R_6 will take place if the ratio H_2:CO is higher than 2 in the syngas, which produces additional methanol, favoring an increase in DME production.

A significant quantity of CO_2 can also be converted to methanol, and its concentration in purge gas can be decreased. However, to avoid the CH_4 slip from the reactor and the CH_4 conversion to CO_2 reaction, R_1 should take place instead of R_2. A graph is drawn as the reactor outlet temperature and H_2:CO ratio against feed temperatures and is shown in Figure 3. It is observed that at feed temperatures of 500 to 700 K, H_2:CO ratios are above 2.35, and upon an increase in temperature, the value decreases. The quality of syngas depends on the H_2:CO ratio and is approximately 2.5 for an ideal ATR model.

If the feed temperature is increased, the outlet stream temperature also increases. High temperatures reverse the water gas shift reaction R_3, which consumes hydrogen and carbon dioxide and produces CO and water. Thirty-three percent of the CO_2 in the feed will consume hydrogen produced from reaction R_1. The thermodynamic equilibrium of R_1 will be attained due to CO production by R_3 reducing the conversion of methane to CO and increasing methane conversion to CO_2. Thus, finally, syngas leaves the reactor with a lower syngas ratio and a high CO_2 and CH_4 content. Therefore, it is decided to conduct the reaction in the range of less sensitive feed temperatures (600 K and 700 K) to attain maximum concentrations of CO in the reactor outlet.

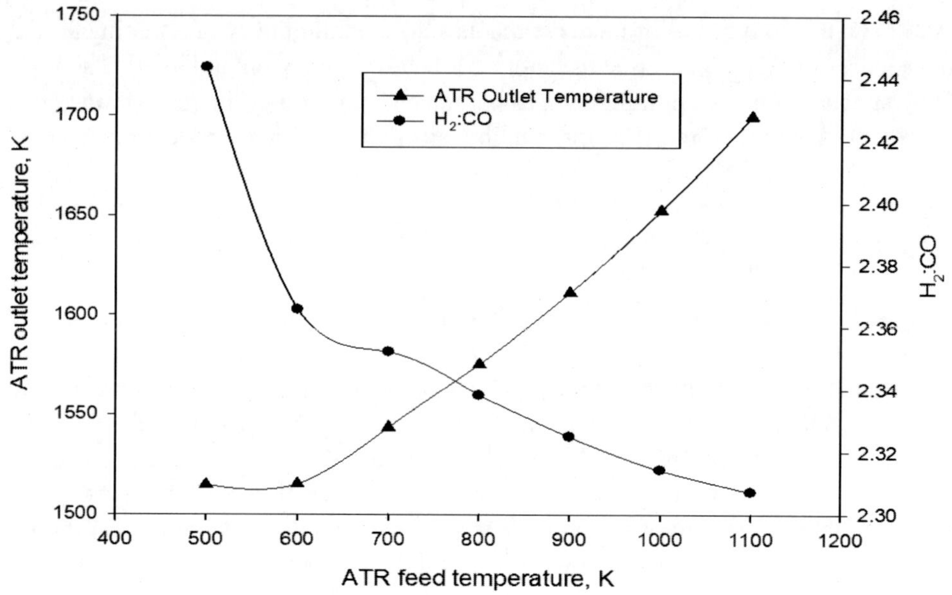

Figure 3. Effect of ATR feed temperature on outlet temperature and H_2:CO ratio in syngas.

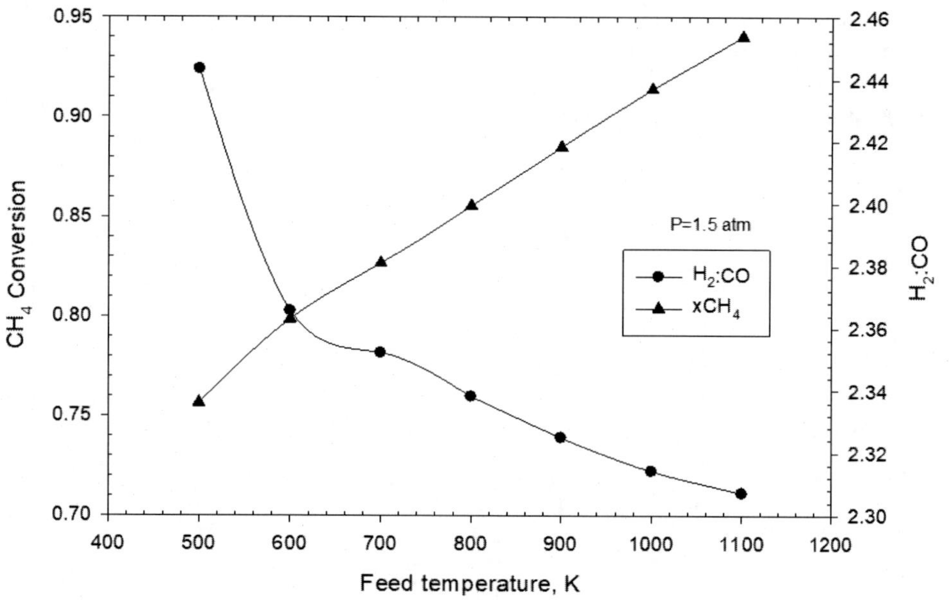

Figure 4. Effect of ATR feed temperature on CH_4 conversion and H_2:CO ratio in syngas.

Three interesting regions at 500 K, 600 K and 700 K in Figure 4 show how the feed temperature of the ATR reactor affects the CH_4 conversion and syngas H_2:CO ratio. At 500 K, the H_2:CO ratio is 2.44, with a low methane conversion of 75.6%. Under this condition, methane conversion takes place by reactions R_4 and R_2, which produce CO_2, H2O and H_2. Reaction R_1 is not initiated at this temperature, so less CO is produced. The 500 K feed temperature is not suitable for better DME conversion due to the lower CO composition.

At 600 K, the conversion of methane reaches 80%, while the syngas parameter (H_2 to CO ratio) decreases to 2.36, which is satisfactory for better conversion and yield. The third region of 700 K shows a small drop in the syngas parameter, but with an increase in the conversion, up to 82%, the syngas parameter drops, which leads to lower DME production.

5.3. Effect of Oxygen Flow Rate

The oxygen flow rate is varied from 1 to 5 kmol/hr while fixing the temperature at 700 K, and the pressure chosen is 1.5 atm, which is just above atmospheric pressure and is necessary to maintain the flow of gases through equipment. A graph of the composition of individual components in syngas has been plotted and is shown in Figure 5. According to reaction R10, it is necessary to maintain the maximum possible CO composition in syngas, as it is a limiting reactant in the production of DME. With increasing oxygen flowrate, the three components, namely, CO_2, CO and H_2, are observed in the product stream. CO is a limiting component, and its composition is to be maintained at a maximum value while maintaining the minimum possible CO_2 content, which is found at the 2 kmol/hr O_2 flowrate.

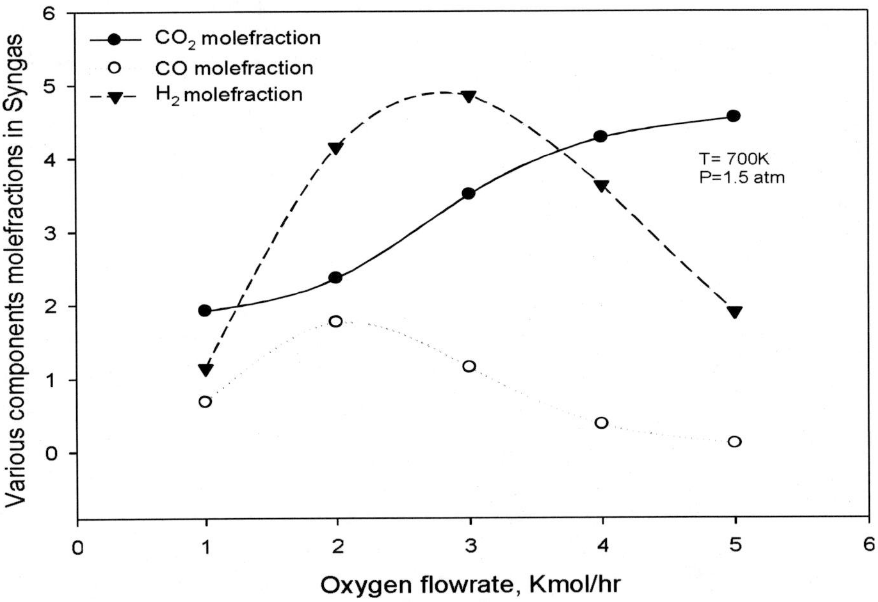

Figure 5. Effect of oxygen flow rate on syngas composition.

5.4. Effect of Reactor Pressure

Syngas processing shows an increase in the volume of the product stream as per reaction R_1 (4 moles of product are formed from 2 moles of reactant), and low pressures are favorable for the forward reaction (Le Chatelier's principle). Upon an increase in pressure from 1 to 2.5 atm, the conversion of methane decreased for the 500, 600 and 700 K reactor temperature conditions. At 1 atm pressure, the conversion is at its maximum, but the H_2:CO ratio is lower, which should be high enough for DME synthesis. The effect of feed temperatures on the H_2:CO ratio and conversion of methane with reactor pressure is shown in Figure 6. The pressure required for maximum conversion and the H_2:CO ratio are found for the present case. For a feed temperature of 500 K, the syngas ratio increased with pressure, while the conversion did not reach above 80% at 1 atm and decreased with an increase in pressure. For 600 K feed temperatures, above 80% conversion is obtained in the region of 1 to 1.5 atm. However, for 700 K and up to 2 atm pressure, more than 80% methane conversion is obtained. Practically, 1 atm pressure reactor operation is not reasonable in a packed bed. To maintain the flow and overcome the flow resistance, a pressure of 1.5 atm is selected as the optimum condition, where the conversion and H_2:CO ratio are 82.9% and 2.34, respectively, at a 2 kmol/hr oxygen flowrate for 700 K and 80% and 2.24 for 600 K. In the present case, no recycling is provided to the ATR section. The product composition obtained from ATR is given in Table 8.

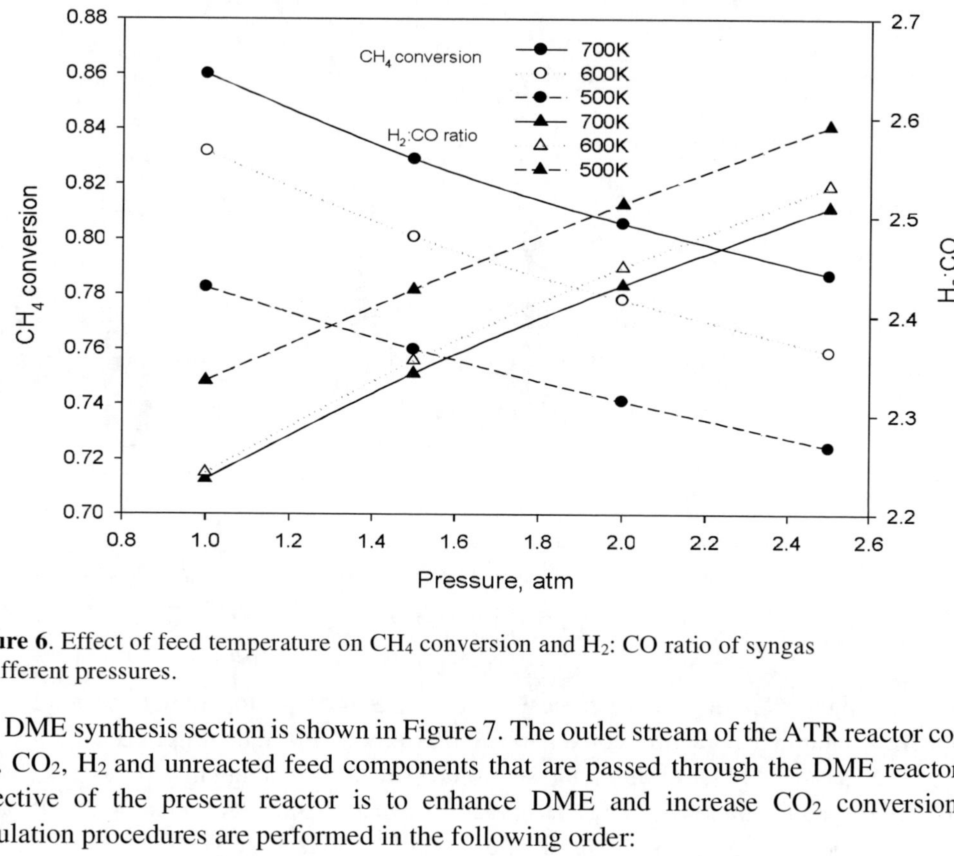

Figure 6. Effect of feed temperature on CH_4 conversion and H_2: CO ratio of syngas at different pressures.

DME synthesis section is shown in Figure 7. The outlet stream of the ATR reactor contains CO, CO_2, H_2 and unreacted feed components that are passed through the DME reactor. The objective of the present reactor is to enhance DME and increase CO_2 conversion. The simulation procedures are performed in the following order:

1. Apply the initial kinetic parameters in the DME synthesis reactor model as DMER.
2. Perform a sensitivity analysis for changing operation parameters.

Operating parameters such as conversion and yield are defined.

Table 8. The product composition obtained from ATR is shown below

Components	Kmol/hr
CH_4	0.5206497
CO_2	2.364507
CO	1.765343
H_2	4.152364
H_2O	1.106836
N_2	0.049

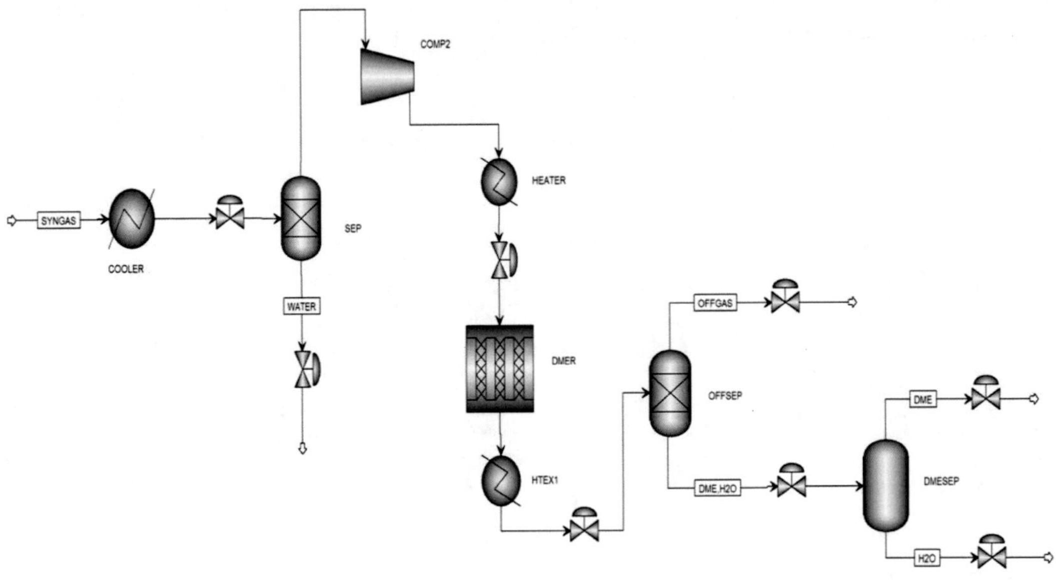

Figure 7. DME synthesis section flow sheet.

5.5. Effect of Pressure

Through simulation of the DME synthesis model, we are able to predict the tendency of performance parameters such as conversion, DME production rate (P_{DME}) and DME yield (Y_{DME}). The variables of sensitivity analysis (to determine the change in process variables by varying a process parameter) include temperature and pressure. It is observed that at low temperatures and high pressures, the production rate of DME is high. With increasing pressure in the DME reactor from 5 to 70 atm at the 500K DME reactor feed temperature, the production rate, conversion and yield profiles for each major product (DME, CO and CO_2) are shown in Figure 8. This indicates that the conversion of CO, PDME and Y_{DME} increases with pressure and reaches a nearly constant value at higher pressures, but the pressure beyond 30 atm has shown only a marginal effect. Therefore, 30 atm pressure is chosen, and the corresponding yield and rate are 89.6% and 44.87 kg/hr, respectively. The conversion rates of CO and CO_2 are 97.6% and 9.7%, respectively.

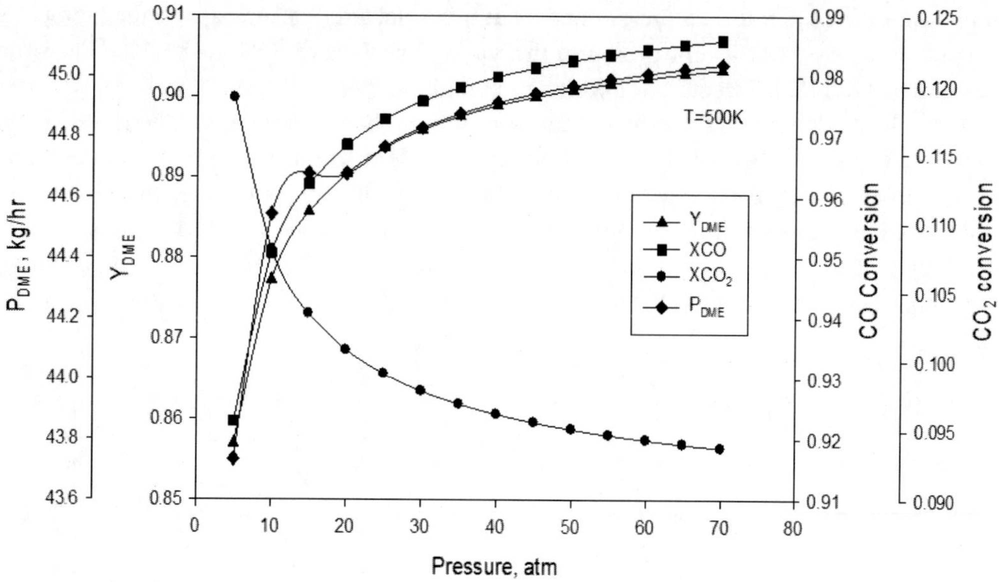

Figure 8. Effect of pressure on P_{DME}, Y_{DME}, conversion of CO and CO_2.

5.6. Effect of Temperature

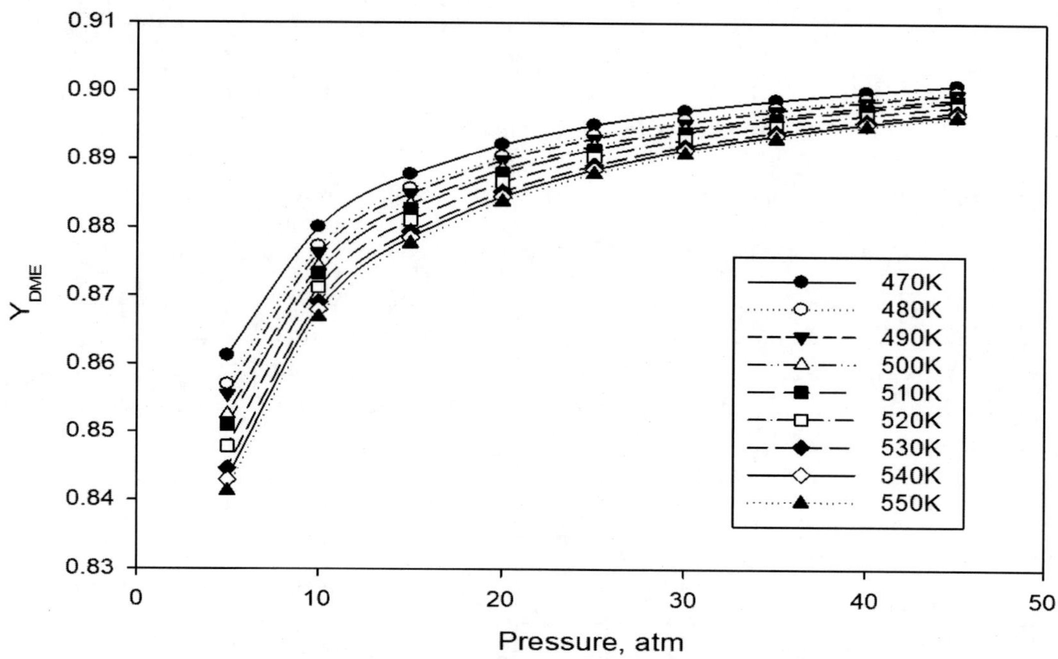

Figure 9. Increase in Y_{DME} with pressure: effect of feed temperature.

A graph is drawn as the yield of DME vs. pressure with feed temperature as a parameter and is shown in Figure 9. The graph reveals that an increase in pressure increases the yield of DME. The increase is substantial up to a pressure of 30 atm, beyond which a very marginal change is noticed. Therefore, a pressure of 30 atm is selected. An increase in feed temperature decreased the yield. At 30 atm pressure, the yield varied from 89% to 89.7%. Therefore, a temperature of 470 K is selected, for which the yield of DME is 89.5%.

Figure 10 demonstrates the effect of temperature on the conversion of CO, H_2 and CO_2 and the yield of DME. CO_2 conversion increases as the temperature increases, but CO and H_2 conversions are low, decreasing the yield of DME. In the present case, the reactor is not provided with recycling to design a smaller reactor; a higher yield of DME at 470 K is selected. When recycling is provided, reasonable feed temperatures and CO_2 conversion are preferred for optimum DME synthesis. Table 9. Represent the DMER product stream composition at 500 K and 30 atm.

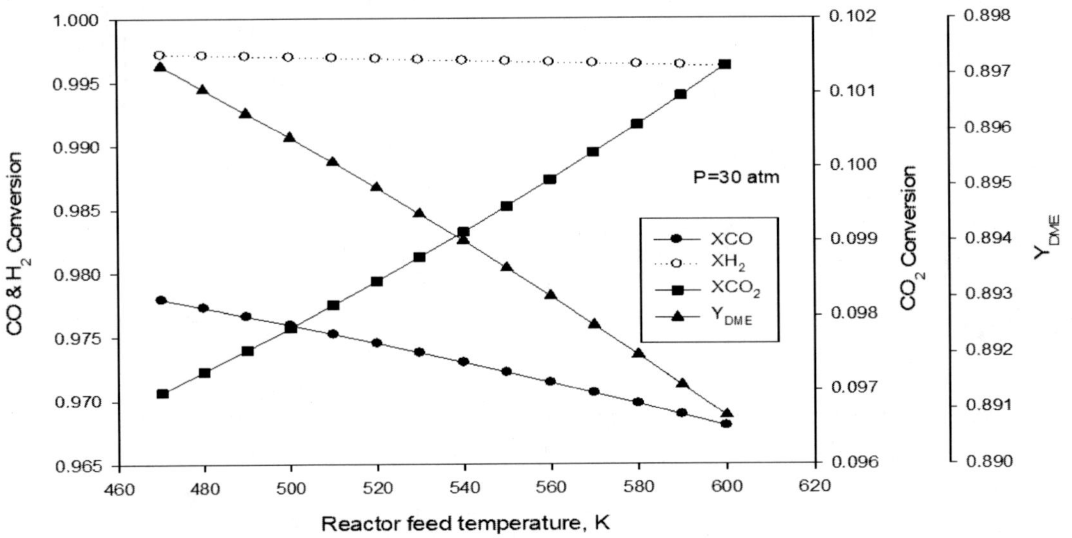

Figure 10. Effect of feed temperature on Y_{DME} and CO, H_2 and CO_2 conversion.

Table 9. DMER product stream composition at 500 K and 30 atm

Components	Kmol/hr
CH_4	0.52065
CO_2	2.133169
CO	0.042475
H_2	0.012615
H_2O	1.205884
O_2	0
N_2	0.049
METHANOL	0.005112
DME	0.974547
Total Flow kmol/hr	4.943452
Total Flow kg/hr	171.6054

5.7. Steady-State Simulation with Recycling

Considerable quantities of unreacted CH_4, CO and H_2 are generated by the present flow scheme. To reduce the content of these reactants in off-gasses, recycling is envisaged. The recycling scheme is shown in Figure 1. A recycle stream of off-gas from the DME section to the ATR section is connected to the suction side of the biogas compressor (COMPR1). A simulation is performed to investigate the effect of the recycle fraction on the reactor feed temperature, conversion, oxygen requirement and yield of DME. By keeping optimum operating conditions, the analysis of data arrived without recycling. The oxygen flow rate is varied from 1 to 5 at individual recycle fractions to find the maximum DME production, as shown in Figure 11. At the lower recycle fraction, the highest DME yielded a 2.5 kmol/hr flowrate of O_2. At an O_2 flow rate of 3 kmol/hr, it is observed that for 0.1 to 0.4 recycle fractions, DME production is 1.0 kmol/hr.

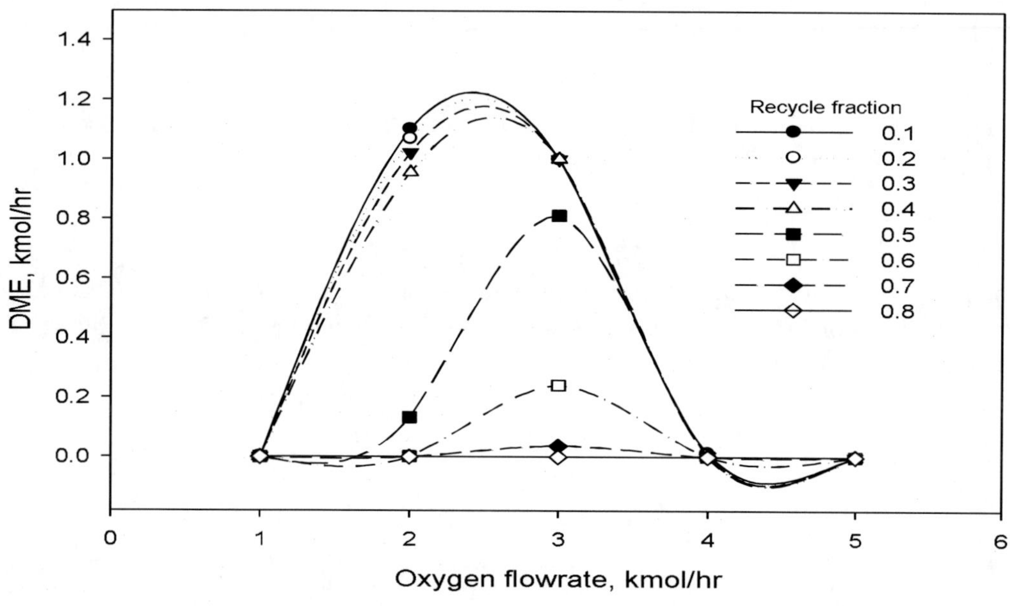

Figure 11. Effect of oxygen flow rate on DME production for various recycle fractions.

5.8. Auto Thermal Reactor

As a part of the product stream was recycled, optimum operating conditions could vary considerably. A study was performed to vary the reactor pressure from 1.5 to 2.5 atm, and the temperature of the reactor was obtained from a simulation study (Figure 12). Figure 12 is a plot showing DME production against feed temperature at different pressures in an auto thermal reactor. The graph reveals that the increase in temperature raised the DME production from 0.52 to 1.0 kmol/hr. From Figure 12, 600 K is selected as the optimum operating temperature, and the difference in DME production is very marginal for 1.5 atm and 2 atm operating pressures.

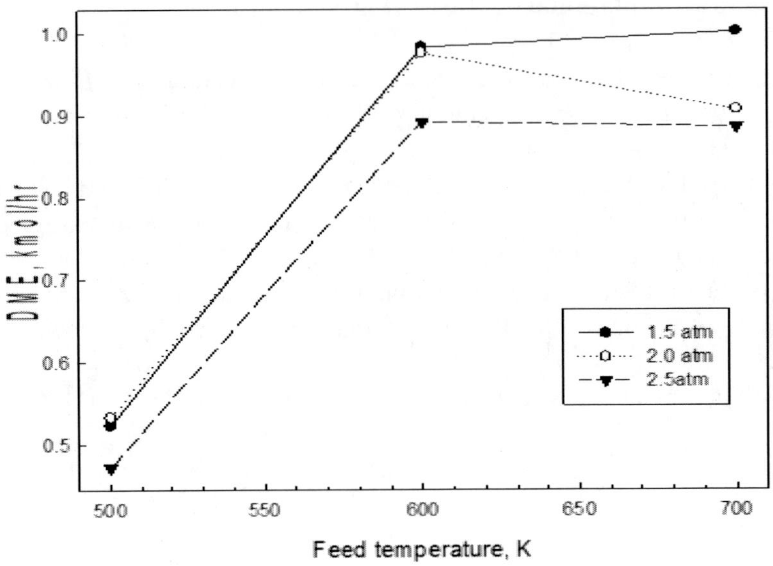

Figure 12. DME production for different feed temperatures with the recycling effect of ATR pressure.

5.9. DME Synthesis Reactor

A similar analysis is carried out for the DME reactor and shown in Figure 13. This indicates that a maximum of 1 kmol/hr of DME was realized at a temperature of 550 K, which has less sensitivity at 550 K, and a minimum pressure of 30 atm was recommended. However, strict control of temperature must be maintained to obtain maximum yield. DME production does not change largely with an increase in pressure from 30 atm.

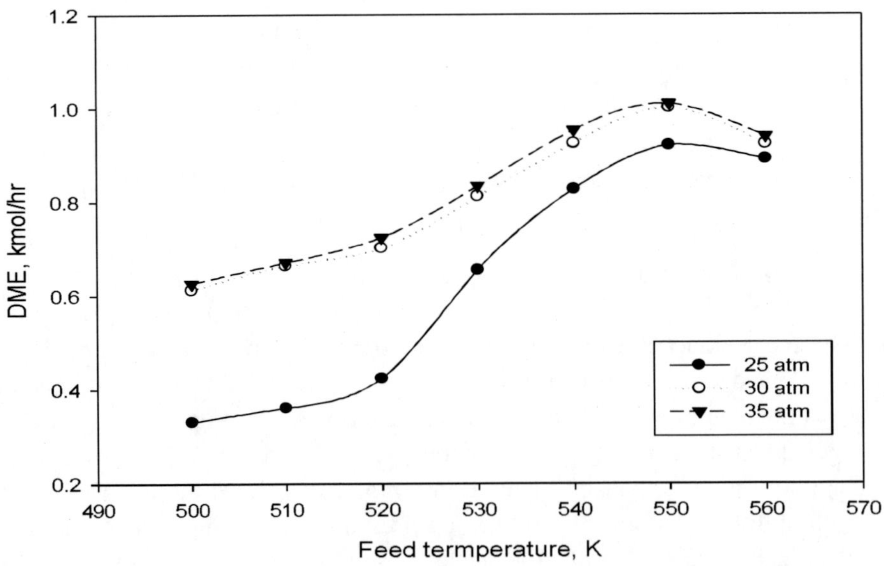

Figure 13. Effect of pressure on DME production at various feed temperatures of the DME synthesis reactor.

Reactions R_1, R_2, R_3 and R_4 progress along the length of the reactor. Simulation of the ATR reactor yield concentration profile along the length of reactor tubes. The concentration data of each reactant and product are plotted against the reactor length and shown in Figure 14.

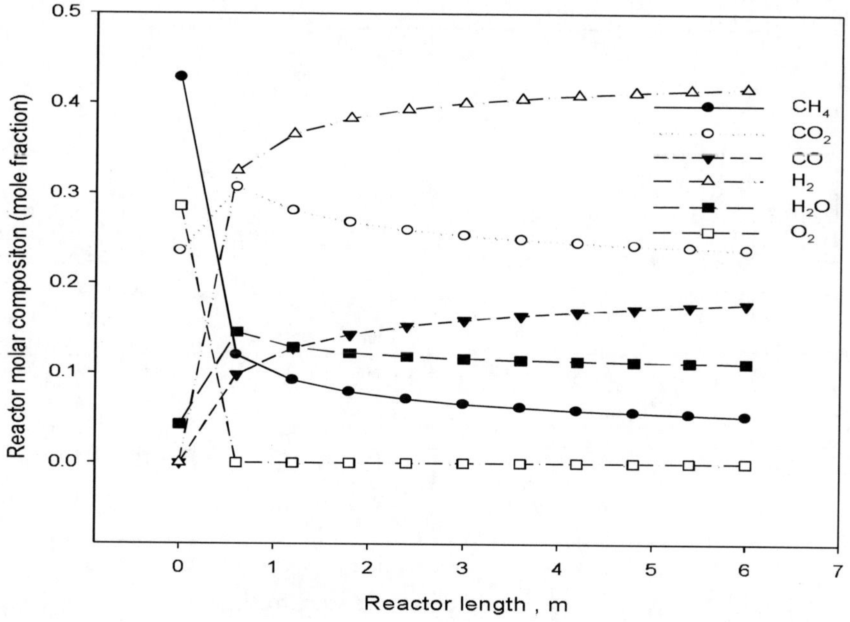

Figure 14. ATR reactor profile.

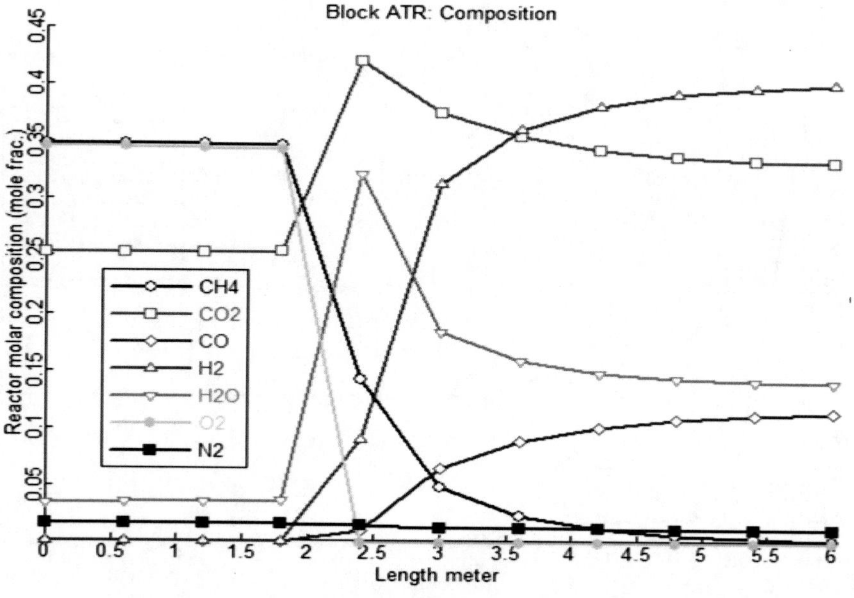

Figure 15. Component profiles in the ATR reactor with recycling.

It reveals that all the components stabilize after nearly 3 meters of reactor length. Therefore, it is recommended to use a reactor length between 2 and 4 m. ATR behavior with recycling is shown in Figure 15 and satisfies the determined size obtained for the simulation without recycling.

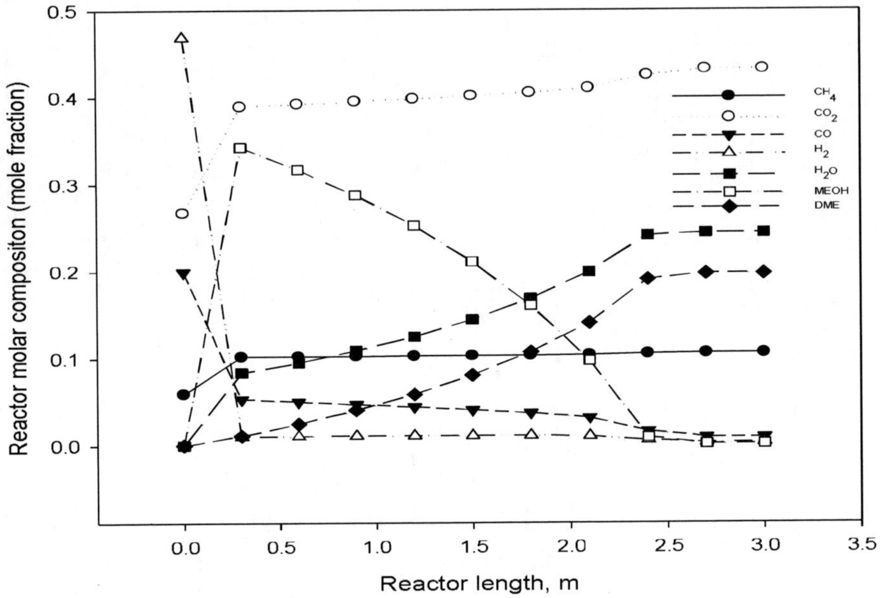

Figure 16. DME synthesis reactor profile.

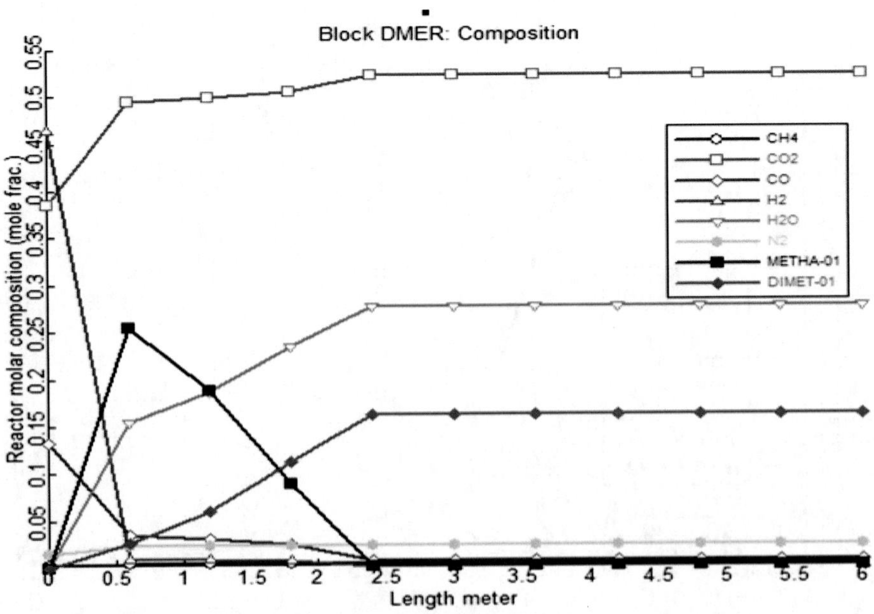

Figure 17. Component profiles in the recycled DME synthesis reactor.

Reaction R_4 occurs initially, followed by R_1 and R_2, as observed by the component profile. Reactions R_5, R_6, R_7 and R_8 progress along the length of the reactor. Simulation of the DME synthesis reactor yield concentration profile along the length of reactor tubes. The data for each reactant and product concentration are plotted against the length of the reactor and shown in Figure 16.

It shows that all the components are nearly stabilizing after 3 meters of length. Therefore, it is recommended to use a reactor length between 2.5 m and 3 m. Similarly, the DMER behavior is shown in Figure 17. Methanol formation occurs in reactions R_5 and R_6 and is converted into DME by reaction R_7. The recycling runs satisfy the 3 m optimum length of the reactor obtained during the simulation without recycling.

Table 8. Process optimum conditions obtained for biogas from municipal wastewater

Blocks	With recycle			Without recycle		
	Temperature, K	Pressure, atm	Oxygen flow rate, kmol/hr	Temperature, K	Pressure, atm	Oxygen flow rate, kmol/hr
ATR	600	1.5	3	600	1.5	2
DMER	550	30	-	470	30	-
Recycle fraction of Off-gas	0.4	-	-	-	-	-

5.10. Steady State Simulation Procedure with Recycle and with Different Feed Sources (Compositions)

Different feed sources are introduced to the process to investigate the production of DME and energy utilization at the optimum conditions obtained with the recycling study. The component content for 5 kmol of feed for each source is shown in Table 9.

Table 9. Composition of Biogas obtained from various sources

Biogas constituents	Municipal wastewater	Landfill	Agricultural residues	Domestic waste
	kmol/hr	kmol/hr	kmol/hr	kmol/hr
CH_4	3	2.25	3.4	3.2333
CO_2	1.65	1.6	1.3	1.2937
N_2	0.05	0.85	0.05	0.0249
O_2	0	0.1	0	0.0149
H_2O	0.3	0.2	0.25	0.2989
H_2S	0	0	0	0.1329
NH_3	0	0	0	0.001

Biogas generated from municipal wastewater and agricultural residues shows similar compositions. DME production and equivalent energy generated during processing of various feed sources are tabulated in Table 10. It is observed that landfill gas composition is not suitable for producing DME under optimum conditions.

Table 10. DME produced and energy generated for various biogas sources

Various Biogas sources	Feed		DME produced		Production efficiency*
	kmol/hr	kg/hr	kmol/hr	kg/hr	%
Municipal wastewater	5	127.5	1.0022	46.17	36.2
Landfill	5	137.12	-	-	-
Agricultural residues	5	117.662	1.1373	52.3966	44.5

* output/input = DME produce/feed (in mass).
Energy generation is in process.

The calorific values and flame temperature of biogas for each source are tabulated in Table 11. The conversion of biogas to DME facilitated improving the calorific value and flame temperature that meet the characteristics of a fuel.

Table 11. Calorific value and flame temperature of raw materials and product [33]

Various Biogas sources	MJ/kmol	MJ/m^3	Flame temperature, °C/mole
Municipal wastewater	481.575	21.4989	1897.3
Landfill	361.181	16.1242	1824.1
Agricultural residues	545.785	24.3654	1936.9
DME	1328.397	59.44×10^3	2130.8

Conclusion

This study comprises the simulation of process parameters to produce DME from biogas using ASPEN Plus software. Optimum operating parameters such as temperature, pressure, flowrate and dimensions of reactors were identified. A sensitivity analysis was carried out for the ATR and DME reactors. Optimal operating conditions for 5 kmol/hr of biogas from municipal waste are 600 K at 1.5 atm for the ATR reactor and 550 K at 30 atm for the DME synthesis reactor. An optimal oxygen flow rate of 3 kmol/hr and an off-gas recycle fraction of 0.4 were obtained. For various biogas feed sources, such as domestic gas, municipal wastewater, and agricultural residue, 45.12, 46.12 and 52.39 kg/hr of DME are produced at optimum operating conditions. However, landfill gas cannot be converted into DME under the defined process conditions. The developed process is able to handle up to 40% recycling of off-gas. The production efficiencies obtained in terms of mass for municipal wastewater and agricultural residue are 36.2% and 44.5%, respectively.

References

[1] Prasad Kaparaju, Maria Serrano, Irini Angelidaki. (2010). Optimization of biogas production from wheat straw stillage in UASB reactor. *Applied Energy*, *87*(12): 3779-3783.

[2] Lacour, S. Chinese, T. Alkadee, D. Perilhon, C. Descombes, G. (2012). Energy and environmental balance of biogas for dual-fuel mobile applications. *Renewable and Sustainable Energy Reviews*, *16*(3): 1745-1753.

[3] George A. Olah, Alain Goeppert, G. K. Surya Prakash. (2009). Chemical recycling of carbon dioxide to methanol and dimethyl ether: from greenhouse gas to renewable, environmentally carbon neutral fuels and synthetic hydrocarbons. *J. Org. Chem*, *74*(2):487-498.

[4] Xinyuan Jiang. Sommer, Sven G. Christensen, Knud V. (2011). A review of the biogas industry in China. *Energy Policy*, *39*(10): 6073–6081.

[5] Eric Favre, Roda Bounaceur, Denis Roizard. (2009). Biogas, membranes and carbon dioxide capture. *Journal of Membrane Science*, *328*(1-2):11–14.

[6] Preeti Gangadharan, Krishna C. Kanchi, Helen H. Lou. (2012). Evaluation of the economic and environmental impact of combining dry reforming with steam reforming of methane. *Chemical Engineering Research and Design*, *90*(11): 1956-1968.

[7] Halabi, M. H. Croon, M. H. J. M. de. Schaaf, J. van der. Cobden, P. D. Schouten, J. C. (2008). Modeling and analysis of Auto thermal reforming of methane to hydrogen in a fixed bed reformer. *Chemical Engineering Journal*, *137*(5): 568–578.

[8] Hyun Min Shim, Seung Jong Lee, Young Don Yoo, Yong Seung Yun, and Hyung Taek Kim. (2009). Simulation of DME synthesis from coal syngas by kinetics model. *Korean J. Chem. Eng.*, *26*(3): 641-648.

[9] Takashi Ogawa, Norio Inoue, Tutomu Shikada, Yotaro Ohno. (2003). Direct Dimethyl Ether Synthesis. *Journal of Natural Gas Chemistry*, *12*(4): 219-227.

[10] Takashi Ogawa, Norio Inoue, Tutomu Shikada, Yotaro Ohno. (2003). Direct Dimethyl Ether Synthesis. *Journal of Natural Gas Chemistry*, *12*(4): 219-227.

[11] Zhaoguang Nie, Hongwei Liu, Dianhua Liu, Weiyong ying, Dingye Fang. (2005). Intrinsic kinetics of dimethyl ether synthesis from syngas. *Journal of national gas chemistry*, *14*(1): 22-28.

[12] Farsi, M., R. Eslamloueyan, A. Jahanmiri. Modeling, simulation and control of dimethyl ether synthesis in an industrial fixed-bed reactor. *Chemical Engineering and Processing: Process Intensification*, *50*(1):85–94.

[13] Osorio, F. and Torres, J. C. (2008). Biogas purification from anaerobic digestion in a wastewater treatment plant for bio fuel production. *Renewable Energy*, *34*(10):2164–2171.

[14] Chiodo, V. Urbani, F. Galvagno, A. Modelo, N. Freni, S. (2012). Analysis of biogas reforming process for molten carbonate fuel cells. *Journal of Power Sources*, *206*: 215- 221.

[15] Bala, B. L. and Satter, M. A. (1991). System dynamics modeling and simulation of biogas production systems. *Renewable Energy*, *1*(5/6):723-728.

[16] Strevett, Keith A. Vieth, Robert F. and Domenic Grasso. (1995). Chemo-autotrophic biogas purification for methane enrichment: mechanism and kinetics. *The Chemical Engineering Journal*, *58*(1): 71-79.

[17] Kapdi, S. S. Vijay, V. K. Rajesh, S. K. and Rajendra Prasad. (2005). Biogas scrubbing, compression and storage: perspective and prospectus in Indian context. *Renewable Energy*, *30*(8): 1195–1202.

[18] Li Zhou, Shanying Hu, Yourun Li, Qihong Zhou. (2008). Study on co-feed and coproduction system based on coal and natural gas for producing DME and electricity. *Chemical Engineering Journal*, *136*(1): 31-40.

[19] Alonso-Vicario, A. Ochoa-Gomez, Jose R. Gil-Rio, S. O. Gomez-Jimenez-Aberasturi, Ramirez-Lopez, C. A. Torrecilla-Soria, J. Dominguez, A. (2010). Purification and upgrading of biogas by pressure swing adsorption on synthetic and natural zeolites. *Microporous and Mesoporous Materials*, *134*(1-3): 100-107.

[20] Junke Xu, Wei Zhou, Zhaojing Li, Jihui Wang, Jianxin Ma. (2010). Biogas reforming for hydrogen production over a Ni-Co bimetallic catalyst: Effect of operating conditions. *International journal of hydrogen energy*, *35*(23): 13013-13020.

[21] Jianguo Xu, Gilbert F. Froment. (1989). Methane steam reforming, methanation and water-gas shift: I. intrinsic kinetics. *AIChE Journal*, *35*(1):88-96.

[22] Zahedi nezhad, M. Rowshanzamir, S. Eikani, M.H. (2009). Auto thermal reforming of methane to synthesis gas: modeling and simulation. *International journal of hydrogen energy*, *34*(3): 1292-1300.

[23] Gorazd Bercic and Janez Levec. (1993). Catalytic dehydration of methanol to dimethyl ether, kinetic investigation and reactor simulation. *Ind. Eng. Chem. Res*, *32*(11): 2478-2484.

[24] Hadipour, A. and Sohrabi, M. (2006). Determination of the Kinetic Parameters and Dynamic Modeling of the Reactor for the Direct Conversion of Synthesis gas to Di-methyl ether. *Iranian Journal of Chemical Engineering*, *3*(4).

[25] Han Yuanyuan, Zhang Haitao, Ying Weiyong, Fang Dingye. (2009). Modeling and simulation of production process on dimethyl ether synthesized from coal-based syngas by one-step method. Process systems engineering. *Chinese journal of chemical engineering*, *17*(1): 108-112.

[26] Zhaoguang Nie, Hongwei Liu, Dianhua Liu, Weiyong ying, Dingye Fang. (2005). Intrinsic kinetics of dimethyl ether synthesis from syngas. *Journal of national gas chemistry*, *14*(1): 22-28.

[27] Yasari, E., M. Shahrokhi, and H. Abedini. (2010). Modeling, Simulation and control of a tubular fixed-bed dimethyl ether reactor. *Chem. Biochem. Eng*, *24* (4): 415–423.

[28] Tippayawong, N. and Thanompongchart, P. (2010). Biogas quality upgrade by simultaneous removal of CO_2 and H_2S in a packed column reactor. *Energy*, *35*(12): 4531-4535.

[29] Wei-Hsin Chen, Bo-Jhih Lin, How-Ming Lee, Men-Han Huang. (2012). One-step synthesis of dimethyl ether from the gas mixture containing CO2 with high space velocity. *Applied Energy*, *98*: 92–101.

[30] Hadipour, Ali, and Morteza Sohrabi. (2008). Synthesis of some bifunctional catalysts and determination of kinetic parameters for direct conversion of syngas to dimethyl ether. *Chemical Engineering Journal*, *137*(2): 294–301.

[31] *Optimization of Chemical Processes*, 2nd Edition, Edgar, Himmelblau, Lasdon.

[32] *Plant Design and Economics for Chemical Engineers*, 5th Edition, Max S. Peters, Klaus D. Timmerhaus, Ronald E. West.

[33] Caceres, C. X. Caceres, R. E. Hein, D. Molina, M. G. and Pia, J. M. (2012). Biogas production from grape pomace: Thermodynamic model of the process and dynamic model of the power generation system. *International journal of hydrogen energy*, *37*(13): 10111-10117.

[34] Ryckebosch, E. Drouillon, M. Vervaeren, H. (2011). Techniques for transformation of biogas to biomethane. *Biomass and Bioenergy*, *35*(5)1633-1645.

Chapter 21

The Modeling and Simulation of the Bio-Drying Performance of Municipal Solid Waste (MSW) Under Different Aeration Rates

Ravi Kumar Jatav
Ani Lawrance
and A. Arunagiri*
Department of Chemical Engineering, National Institute of Technology, Tiruchirappalli, Tamil Nadu, India

Abstract

The problem of efficient disposal of municipal solid waste has puzzled humans since the industrial revolution. Moisture-laden solid waste of approximately 60-70% (by wt.) has become one of the main hurdles in the field of waste-to-energy conversion technologies. It is necessary to eliminate excess moisture present in the solid waste using hybrid technological pathways, namely, biodrying, in which the heat generated by inherent microorganisms in the solid waste is used for drying. The performance of the biodrying process is estimated in terms of moisture removal by adjusting the air flow rate as the primary process control parameter. In the present simulation study, the aeration rate is varied from 115.2-172.8 m^3/d for the developed mass balance model, which integrates 18 first-order ordinary differential equations for a single layer by applying first principles based on process kinetics. The simulation is carried out using Simulink software with the MATLAB R2022a interface. The predicted final moisture content in the waste is found to change from 28.62% to 16.99% for the selected input range of aeration rates, which is fixed based on the experimental studies. From the predicted results, the optimum value of aeration flow rate is 172.8 m^3/d, which provides a final moisture content of 16.99%, since this is the minimum value required for the survival of microorganisms in municipal solid waste. The predicted output was compared with literature data and found to fit well.

Keywords: municipal solid waste, biodrying, aeration rate, modeling

---

* Corresponding Author's Email: aagiri@nitt.edu.

In: Sustainability in Chemical Processes through Digitalization and Green Chemistry Approaches
Editors: Susarla V. A. R. Sastry, Shravan Kumar and Ashish Kapoor
ISBN: 979-8-89113-257-3

1. Introduction

India, home to more than 1210 million people, is confronted with many hurdles in regard to essential facilities in general and waste management. The municipal population of the nation is higher than 377 million, which accounts for 31.16% of its entire population (Census of India 2011, GOI 2011). According to the findings of the central pollution control board (CPCB) of India, per capita waste production is escalating with exponential factors from 0.26 kg/day to 0.85 kg/day (CPCB India 2018 a).

According to several studies conducted in this field, approximately 80% to 90% of MSW is discarded in landfills without proper execution strategies. On the other hand, the sheer unawareness and unwillingness to follow proper procedures on the part of citizens is overburdening municipalities. Consequently, the potential hazards posed by MSW at landfills are tremendous, such as the release of greenhouse gases and contamination of groundwater resources due to the release of leachates.

The rapid rise of all sorts of wastes in general and municipal solid wastes in particular is posing much confusion and challenges for the administrative authorities to devise the most environmentally friendly yet cost-effective technologies capable of dealing with all angles of sustainable development and engineering in a broad sense. Amidst all the MSW that is being produced, approximately 70% of the waste is composed of compostable items, leading to major pollution. An approach for the proper disposal and improvement of this kind of waste is a main concern.

Conventionally, municipal solid waste is either incinerated or simply dumped in open places, leading to severe health and environmental issues. Incineration of high-moisture-laden MSW causes the production of toxic biproducts such as dioxins, which create several environmental problems. There is a need for appropriate pretreatment methodologies for the proper management of MSW and waste-to-energy conversion techniques. Microorganisms play a relevant role in MSW recycling, which appears to be cost-effective and causes less destruction to the environment. The biodrying process is set up to be the best option, where biological heat energy is successfully channeled for drying purposes so that the heavy moisture-laden solid waste is dried. The several end-use possibilities of the yield make the biodrying process adaptable.

Despite the fact that several experimental studies have been performed with various substrate compositions to determine the effect of biodrying, very few modeling attempts have been made in the literature. The biodrying of the organic component of MSW using structural materials was studied by Zawadzka et al. (2010) [1]. The relationship between inlet air characteristics and matrix temperature was examined. The inlet air was heated to 30°C and supplied at 86.35 m^3/h with an in-line blower, and 80 to 90% of the moisture was present at first. This led to the conclusion that the original moisture decreased by 50%. An increased airflow rate led to a greater proportion of water removal. Higher matrix temperatures were a result of the lower air flow rates. The heat of combustion of dried wastes and the calorific value of dried wastes ranged from 6750 kJ/kg to 12280 kJ/kg and 8050 kJ/kg to 10980 kJ/kg, respectively. Due to the energy balance, the energy production rates in dried waste ranged from 1.66 W/kg to 6.90 W/kg, and the efficiency of the process varied from 0.73 to 0.97. Sewage sludge biodrying was analyzed by Cai et al. (2012) [2], and the study parameter was moisture fluctuation, related to water input and evaporation. The beginning conditions are an initial moisture content of 66% to 82%, bulk density of 1040-800 kg/m^3, and a ratio of 3:2:1 for 20 days between the sewage sludge, biodried product, and sawdust. According to the experimental findings, there was a moisture drop in the bulk waste from 66 to 54%, which initially increased and then decreased. The daily mean water evaporation and water generation throughout the thermophilic phase of 2 to 9 days were 37.7±7.09 kg/d and 7.93±2.39 kg/d, respectively. Colomer-Mendoza et al. (2013) [3] experimented with biodrying garden waste in biodrying reactors. The primary study parameter was the aeration flow rate (AFR). The initial conditions for the experiment were an AFR of 0.88 L/(min. kg dry weight) to 6.42 L/(min. kg dry weight) at 25°C and 70% RH for 20 days. After 20 days, the result was a weight loss of 40-57% and a volume decrease of 40-60%. Greater weight loss was achieved by using a higher AFR and less bulking agent (BA), which was 15%. Sewage sludge biodrying was assessed by Yang et al. (2014) [4]. The main emphasis is on the significance of BA (bulking agent) and IMC (initial moisture content) on the weight of sludge, temperature, moisture content concentration, and volatile matter content. These settings served as the basis for this experiment's evaluation. The AFR was 0.035 m^3 per hour, and the IMC was 81.8% for 10 days. The results of the experiment showed that throughout 10 days of biodrying, 33.7% to 47.1% of the initial water was separated by consuming 12.3% to 21.2% of the initial volatile substrate of the sludge. Sludge that has been air-dried or biodried performed better than sawdust and shredded rubber in terms of weight reduction, temperature increase, and water removal among the bulking agents evaluated. IMC levels between 50 and 70% are necessary for effective biodrying.

Villegas et al. (2014) [5] and Cesar Huiliir et al. (2015) [6] examined the simultaneous effects of IMC and AFR during the biodrying of sewage sludge. The initial circumstances are an AFR of 3 L/min kg TS (total solid) and an IMC of 59-78%. The results of the experiment demonstrated that the IMC has a greater influence on biodrying efficiency than the AFR. The IMC water removal with the highest maximum temperature was 68%, and the IMC matrix removal with the lowest maximum temperature was 78%. Tom, A. P. et al. (2016) [7] concentrated on moisture and weight reduction to analyze the impact of matrix height on MSW biodrying. The reactor had a capacity of 0.565 m^3, and the aeration flow rate was 40 L/min. This experiment was carried out for two cases: (1) IMC 62.45%, 1.65 m of matrix height; and (2) IMC 66.4%, 2 m of matrix height. Based on this, the results showed that cases 1 and 2 each had an overall weight decrease of 26.32% and 19.01%, as well as a normal moisture reduction of 24.26% and 15.98%. Tom A.P. et al. (2016) [8] studied the parameters of moisture, volume, and weight decrease to examine the MSW biodrying process. Mixed MSW at AFR – 40 LPM and IMC - 61.25% for 33 days are the conditions. The weight of the mixed MSW substrate decreased by 33.94% in 20 days, the volume decreased by 56.5%, and the bulk density improved by 52% as a result of condensation of evaporated moisture. A final moisture content of 48.5% was achieved. A model for batch biodrying of sewage sludge was examined by César Huiliir et al. (2017) [9, 10] and concentrated on reducing volatile materials by balancing the temperature and moisture of the solid phase with those of the gas phase. At two predetermined initial moisture concentrations (58% and 68%), samples were obtained at three different aeration flow rates (1 L/min kg TS, 2 L/min kg TS and 3 L/min kg TS). It was also noticed that a faster AFR results in a reduced temperature. Convective energy transmission decreases at lower AFR, causing the solid phase to accumulate energy and reach a higher temperature.

According to previous studies conducted in this area, variations in temperature, microbial concentration, moisture, and concentration of degradable organic matter are important data required to obtain better performance from the system and for the design of biodryers. Mathematical modeling can serve as a useful tool to study heat transfer, biokinetic, and mass transfer phenomena in the biodrying process and support the optimization of these variables. A mathematical model that represents the abovementioned variables and their changes with time can be used to analyze the performance of the biodryer. The present study is aimed at devising a mathematical model for municipal solid waste. The combined equations were simulated numerically by using the interface of Simulink software in MATLAB R2022a. The simulated values were compared with the reported literature values of Ani Lawrance et al. (2022) [11] to verify the predictive ability of the model.

2. Methodology

2.1. Materials and Methods

The current mathematical model was developed using experimental data obtained from pilot-scale batch experiments carried out at CSIR-NIIST, Thiruvananthapuram, Kerala, for defined compositions of MSW with varying proportions of food waste, paper waste, and inert (plastic) material under different operating conditions. The development of a model based on molecular diffusion, biochemical kinetics, and convective transfer of moisture vapor across boundary surfaces of wet solids and gases is objective. For this analysis, the equations modeled for easily degradable food components in MSW were used. The horizontal layers of material, as idealized in the waste pile, are numbered from the bottom layer up. Leachate moves downward due to gravity as air enters the system from the bottom layer. By treating the entire biodrying pile as a single layer with a height of 2.5 m, an area of 0.16 m^2, and a volume of 0.40 m^3, the mass balance equations were developed, and the projected output was confirmed by the outcomes of the experimental biodryer. Solid waste biodrying is a complicated process involving several interconnected physical, chemical, and biological actions. The entire process of biodrying is thought to be a mixture of the following five processes, as shown in Figure 1: (i) hydrolysis (conversion of food COD to soluble COD); (ii) oxidation (aerobic oxidation of both food and soluble COD); (iii) evaporation (removal of moisture content of solid matrix via bioheat and aeration); (iv) biomass degradation (endogenous respiration of microorganisms); and (v) condensation.

These processes, which are shown in Figure 1, are considered for the generation of the mass balance equation for the model. Although the solid waste in the column has produced a porous matrix, which has filled with water depending on the original moisture level, this water includes soluble substrate COD, formed when food COD is hydrolyzed. Microorganisms degrade soluble COD and food waste based on the availability of oxygen. Aerobic decomposition leads to the formation of carbon dioxide, water, and new microbes, as well as heat. The biodrying pile is idealized as horizontal layers of material. The model is one-dimensional, i.e., a variation in time with respect to the depth of the pile is modeled.

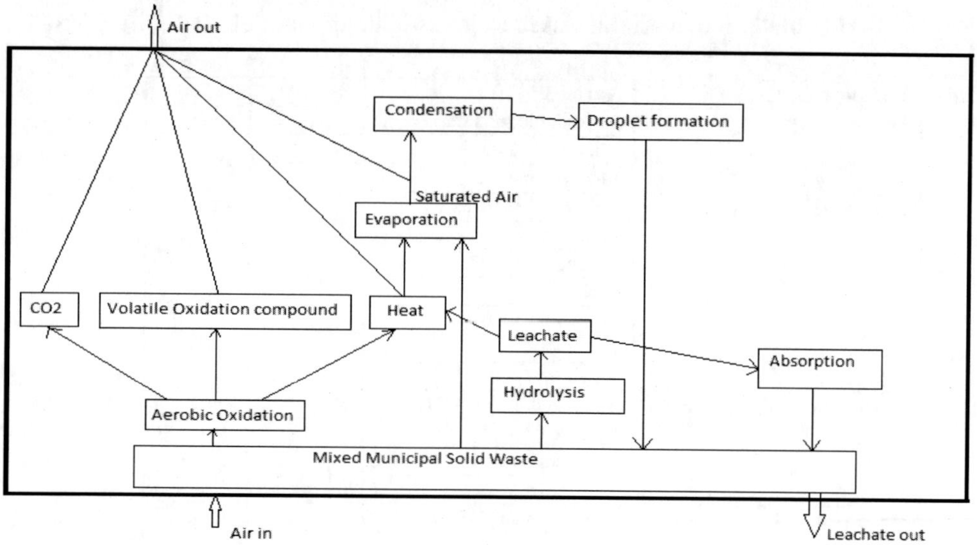

Figure 1. Mechanism of the biodrying process.

2.2. Assumptions

The formulation of equations is based on the assumptions listed below:

1. Throughout the operation, the system temperature and pressure are constant.
2. The aeration flow rate remains constant during the process.
3. The relative humidity (75%) of the system remains unchanged.
4. The biodrying pile of material is idealized as having horizontal layers.
5. The model is 1-dimensional, and the variation occurs along the height of the column with time.
6. The solid materials are homogeneous in each layer.
7. The total volume of the system remains constant.
8. No solid downward velocity or movement.
9. There is no heat loss to the surrounding area.

Table 1. Key variables used to investigate the model

Component	Variable	Symbol	Units
Food waste	Degradable food COD	x_f	kg/m^3
	Microbial COD	x_b	
	Moisture in food	m_f	
Paper waste	Degradable paper COD	x_p	kg/m^3
	Fungal biomass COD	x_{bfu}	
	Leachate COD	x_{ab}	
	Moisture in paper	m_p	
	Bacterial COD	x_{bp}	
inert	Inert COD	x_i	kg/m^3
	Moisture in inert	m_i	
Leachate as biodried product	Stabilized organics COD	x_s	kg/m^3
	Soluble COD	s	
	Bacterial biomass	x_{bs}	
	Free moisture	m_L	
Gas phase	Oxygen in air	O_2	mol/m^3
	Carbon dioxide in air	CO_2	
	Nitrogen in air	N_2	
	Water vapor in air	av	

2.3. Development of Model

Kinetics of the Process

Using stoichiometry, mass transfer, and reaction kinetics principles, the mass balance equations were obtained. Table 2 lists the rate terms used in the model. Anoxic hydrolysis, which depends on moisture, food COD, and the COD of the microbial biomass, is considered the first stage of organic matter decomposition. Food hydrolysis was based on the Contios relationship. Food waste must have at least 20% moisture for hydrolysis to occur, and the waste matrix must also support microbial activity [12]. In the rate equation, the dependence between moisture content and food waste is given by a Monod equation. To determine the effects of temperature on biological responses, the Arrhenius equation was utilized [12]. In our earlier study, [11] the description of the mass balances and kinetic equations were provided, and the numerical values of the empirical constants in the rate equations, the maximum rate constant, and the half-saturation constants were obtained from Ani Lawrance et al., 2022 [11]. The rate terms shown in Table 2 are defined using ~~all~~ these values. According to Table 2, the aerobic oxidation rate depends on the COD, oxygen, microbial biomass, and moisture content of the food. x_t in aerobic oxidation is the total of the COD of the food plus the bacteria.

Table 2. Rate equations

Rate term	Rate equation	Reference
Hydrolysis of food	$r_{hy} = k_{hy}.x_b \dfrac{\left(\frac{x_f}{x_b}\right)}{K_{s,1}+\left(\frac{x_f}{x_b}\right)} * \dfrac{\left(\frac{m_f}{x_f}\right)}{K_{s,2}+\left(\frac{m_f}{x_f}\right)}$	Wang and Witarsa (2016), [13]
Oxidation of food COD (r_{of})	$r_{of} = k_{of}.x_b \left\{ \dfrac{\left(\frac{x_f}{x_t}\right)}{K_s+\left(\frac{x_f}{x_t}\right)} * \dfrac{\left(\frac{m_f}{x_f}\right)}{K_{s,mf}+\left(\frac{m_f}{x_f}\right)} * \dfrac{O_2}{K_{s,pO}+O_2} \right\}$	Haug (1993) [12]
Oxidation of paper (r_{op})	$r_{op} = k_{op}.x_{bfu} \left\{ \dfrac{\left(\frac{x_p}{x_{tp}}\right)}{K_{sp}+\left(\frac{x_p}{x_{tp}}\right)} * \dfrac{\left(\frac{m_p}{x_p}\right)}{K_{smp}+\left(\frac{m_p}{x_p}\right)} \right\}$	
Paper absorbed organic matter COD (r_{oab})	$r_{oab} = k_{oab}.x_{bp} \left\{ \dfrac{x_{ab}}{K_s+x_{ab}} * \dfrac{O_2}{K_{s,pO}+O_2} \right\}$	
Oxidation of soluble COD (r_{os})	$r_{os} = k_{os}.x_{bs} \left\{ \dfrac{\frac{S}{m_L}}{K_{sb}+\frac{S}{m_L}} * \dfrac{O_2}{K_{s,pO}+O_2} \right\}$	
Decay of bacterial biomass (r_d)	$r_{dxb} = k_d.x_b$ $r_{dxbfu} = k_d.x_{bfu}$ $r_{dxbp} = k_d.x_{bp}$ $r_{dxbs} = k_d.x_{bs}$	Haug (1993) [12]
Evaporation of moisture from food(r_{ef})	$r_{ef} = (A+Bv).\left(\dfrac{x_f+m_f}{(x_f+m_f)_0}\right)^{\frac{2}{3}}.\left[1-\dfrac{P_{va}}{P_{fsat}}\right].\left[\dfrac{\frac{m_f}{x_f}}{K_f+\frac{m_f}{x_f}}\right].m_f$	Ani Lawrance et al. (2022) [11]
Evaporation of moisture from Paper(r_{ep})	$r_{ep} = (A+Bv).\left[1-\dfrac{P_{va}}{P_{psat}}\right].\left[\dfrac{\frac{m_p}{x_p}}{K_p+\frac{m_p}{x_p}}\right].m_p$	
Evaporation of moisture from inert material(r_{ei})	$r_{ei} = (A+Bv).\left[1-\dfrac{P_{va}}{P_{isat}}\right].\left[\dfrac{\frac{m_i}{x_i}}{K_i+\frac{m_i}{x_i}}\right].m_i$	
Evaporation of moisture from free water(r_{eL})	$r_{eL} = (A+Bv).\left[1-\dfrac{P_{va}}{P_{Lsat}}\right].\left(\dfrac{m_L}{(m_L)_0}\right)^{\frac{2}{3}}.m_L$	
Absorption of free water by paper(r_{abp})	$r_{abp} = A_p.m_L.\left(\dfrac{B_p}{B_p+\frac{m_p}{x_p}}\right)$	Ani Lawrance et al. (2022) [11]
Absorption of free water by soluble COD (r_{abps})	$r_{abp} = A_p.m_L.\left(\dfrac{B_p}{B_p+\frac{m_p}{x_p}}\right).\dfrac{S}{m_L}$	
Absorption of free water by inert COD (r_{abi})	$r_{abi} = A_i.m_L.\left(\dfrac{\frac{x_i}{x_p}}{B_i+\frac{x_i}{x_p}}\right).\left(\dfrac{C_i}{C_i+\frac{m_i}{x_i}}\right)$	

The bacterial biomass in food COD absorbed by paper and soluble leachate COD have a role in the expression and degradation of microbial biomass. The microbes are transformed into compost that is stable and organic. The first-order decay coefficient is referred to as k_d [13]. The expression for the evaporation rate was created using the two-film theory [14]. Utilizing the fundamental equation of mass transfer, the rate of evaporation is computed. The rate of evaporation depends on the amount of moisture present in the solid. The first term (A) in the evaporation rate equation represents water molecule diffusion, and the second term (Bv) represents convective evaporation. The third term illustrates the difference between the vapor pressure of the solid surface and bulk air. The fourth term demonstrates the movement of moisture internally, due to which a limited amount of moisture diffuses on the surface of solid particles. Leachate evaporates at a rate that is similar to the rate at which a liquid evaporates from a free surface without taking internal moisture movement into consideration.

Mass Balance

Mass balance equations have been developed for the important variables stated in Table 1 for the solid, liquid, and gas phases. Additionally, all mass balance equations were developed using the rate term shown in Table 2. Similar to a batch reactor, where the rate of accumulation inside the system must equal the rate of transformation, with a positive sign signifying the generation of components and a negative sign denoting the removal of components, the mass balance equations for the solid and liquid phases were determined. Gas phase balancing equations were constructed from the continuous flow assumption made for a continuous stirred tank reactor.

The empirical formulas for food COD and microbial COD in the model were expressed as $C_1H_2O_1N_{0.1}$ and $C_5H_7O_2N_{0.1}$, respectively, by assuming carbohydrate content [12]. The rate of composition change in each layer depends on the solid phase state as well as the situation of the liquid and gas flow streams entering.

Therefore, to simulate the complete waste pile, the mass balance of each variable in each layer must be represented. In addition, whereas there is no flow of solid and liquid phases in a reactor because these phases are considered batch reactors, in the case of the gas phase, it is considered a continuous stirred tank reactor (CSTR), so the flow of the gas phase is continuous in the reactor. The outcomes, shown in Table 3, are ascertained using this mass balance equation. The rate of oxidation and the rate of hydrolysis control the development and decay of microbial biomass in food materials at certain concentrations [11]. The variation in moisture content is caused by evaporation and hydrolysis. The stable biodried product is made up of the mineralized byproduct of bacterial endogenous respiration and the nondegradable byproduct of other degradable materials. The aeration flow rate represents the whole gas volume flux in the gas phase (Q). Furthermore, Vp is an indication of pore volume, which makes up approximately 60% of the total volume.

Table 3. Mass balance equations

Phase	**Expression**	**Reference**
Solid	$V\left(\frac{dx_f}{dt}\right) = V(-r_{of} - r_{hy})$ $V\left(\frac{dx_b}{dt}\right) = V\left(Y_{\frac{b}{x_f}}.r_{of} + Y_{\frac{b}{x_f}}.r_{hy} - r_{dxb}\right)$ $V\left(\frac{dm_f}{dt}\right) = V\left(-r_{ef} - Y_{\frac{L}{x_f}}.r_{hy}\right)$ $V\left(\frac{dx_p}{dt}\right) = -V.r_{op}$ $V\left(\frac{dx_{bfu}}{dt}\right) = V\left(Y_{\frac{bfu}{x_p}}.r_{op} - r_{dxbfu}\right)$ $V\left(\frac{dx_{ab}}{dt}\right) = V(-r_{oab} + r_{abps})$ $V\left(\frac{dx_{bp}}{dt}\right) = V\left(Y_{\frac{xbp}{xab}}.r_{oap} - r_{dxbp}\right)$ $V\left(\frac{dm_p}{dt}\right) = V(-r_{ep} + r_{abp})$ $V\left(\frac{dx_i}{dt}\right) = 0$ $V\left(\frac{dm_i}{dt}\right) = V(-r_{ei} + r_{abi})$ $V\left(\frac{dx_s}{dt}\right) = V\Big[Y_{\frac{xs}{xf}}.r_{of} + Y_{\frac{xs}{xp}}.r_{op} + Y_{\frac{xs}{s}}.r_{os} + Y_{\frac{xs}{xab}}.r_{oab}$ $+ Y_{\frac{xs}{b}}(r_{dxb} + r_{dbp} + r_{dbfu} + r_{dbs})\Big]$	Ani Lawrance et al., (2022) [11]
Liquid	$V\left(\frac{dm_L}{dt}\right) = V\left(Y_{\frac{L}{xf}}.r_{hy} - r_{abp} - r_{eL}\right)$ $V\left(\frac{ds}{dt}\right) = V(r_{hy} - r_{abps} - r_{os})$ $V\left(\frac{dx_{bs}}{dt}\right) = V\left(Y_{\frac{xbs}{s}}.r_{os} - r_{dxbs}\right)$	
Gas	$V_p\frac{dO_2}{dt} = Q.O_{2in} - Q.O_{2out}$ $- V.Y_{\frac{O_2}{x}}(r_{of} + r_{op} + r_{oab} + r_{os} + r_{dxb} + r_{dbfu} + r_{dbp} + r_{dbs})$ $V_p\frac{dCO_2}{dt} = Q.CO_{2in} - Q.CO_{2out}$ $+ V.Y_{\frac{CO_2}{O_2}}(r_{of} + r_{op} + r_{oab} + r_{os} + r_{dxb} + r_{dbfu} + r_{dbp} + r_{dbs})$ $V_p\frac{dN_2}{dt} = Q.N_{2in} - Q.N_{2out}$ $V_p\frac{da_V}{dt} = Q.a_{Vin} - Q.a_{Vout} + V.Y_V(r_{ef} + r_{ep} + r_{ei} + r_{eL})$	

Table 4. Input parameter

Input parameter	Initial value of input parameter	units
x_f	25	kg/m^3
x_b	1	kg/m^3
m_f	100	kg/m^3
x_p	25	kg/m^3
x_{bfu}	0.01	kg/m^3
x_{ab}	0.01	kg/m^3
m_p	25	kg/m^3
x_{bp}	0.01	kg/m^3
x_i	25	kg/m^3
m_i	3	kg/m^3
x_s	1	kg/m^3
s	0.0001	kg/m^3
x_{bs}	0.1	kg/m^3
m_L	0.0001	kg/m^3
O_2	8.64	mol/m^3
CO_2	0.02	mol/m^3
N_2	30.64	mol/m^3
a_v	1.16	mol/m^3

Numerical Simulation

The Simulink program in MATLAB 2022a was used to solve each of the 18 N ordinary differential equations shown in Table 3 that make up the model. It has been attempted to simulate all three phases, namely, gas, liquid, and solid. Most of the equations fell within the mass balance category. We also must deal with the rate equations (given in Table 2) of the system, which is another important aspect. Most of the simulation code is composed of constants, mass balance equations, rate equations, and outputs. The behavior of the code is affected by the initial values of the input parameters (shown in Table 4) as well as the constant parameter values, which are taken from Ani Lawrance et al. (2022) [11]. The mass balance equations were represented using MATLAB functions in Simulink. To produce simultaneous results for the whole system, all the variables, parameters, and constants are connected in a circular manner that can be run concurrently using any appropriate solver settings, such as ODE 15 in Simulink. In this analysis, a waste system with a weight density of 206 kg/m^3 is considered. Knowing the moisture content in each of the individual systems can help to determine the starting values of x_f, m_f, x_p, m_p, x_i, and m_i. For the sake of this investigation, the MSW composition is considered to be in the order of food, paper, and inert (61:24:15). The amount of moisture present in the food, paper, and inert material is taken as 80%, 50%, and 10%, respectively. The studies were conducted for airflow rates (Q) of 115.2 m^3/day, 129.6 m^3/day, 144 m^3/day, 158 m^3/day and 172.8 m^3/day for a time span of 15 days. Additionally, the reactor specifications are listed in Table 5.

Table 5. Reactor parameter

Parameter	Symbol	Value
Diameter of reactor	d	0.45 m
Height of reactor	h	2.5 m
Volume of reactor	V	0.4m^3
Pore volume	V_p	0.24m^3
Temperature during the process	T	48°C

3. Results and Discussion

3.1. Change in Moisture in Waste Solid

According to reports by Huilinir and Villegas, 2015 [6], the initial moisture content, AFR (aeration flow rate), relative humidity, extent of biological reaction, and inlet air temperature are the main factors affecting the rate of evaporation from the waste pile. At the same time, microbial activity will be inhibited when the waste's moisture content reaches 20% (w/w) [12].

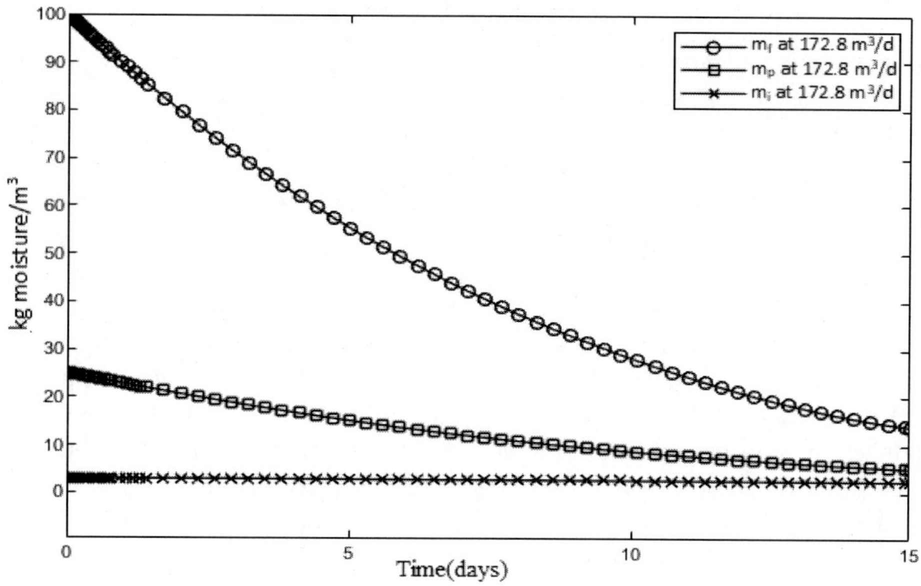

Figure 2. Simulation profile of moisture reduction in food waste (m_f), paper waste (m_p), and inert materials (m_i).

In the simulation model, municipal solid waste with an initial moisture value of 62% was simulated, and the aeration rate was set at 172.8 m^3/d. According to the simulation findings, the moisture content of food material reduces to 14.00 kg/m^3 from a starting value of 100 kg/m^3 with a removal of 86% as a response to hydrolysis and evaporation. Paper has a 79% decrease in moisture content. It has been demonstrated that the varied absorption qualities cause differences in moisture removal between paper and food. On the other hand, 16.67% of moisture is removed from inert materials. As shown in Table 1, the waste pile's ultimate moisture content was calculated to be 16.99%. Therefore, given the selected simulation settings, the simulation model predicted a loss of moisture equal to 83.01% of the starting moisture content. The expected outcomes were in line with the biodrying waste simulation result, which had a 16.99% final moisture content and an initial moisture content of 62% at an aeration flow rate of 172.8 m^3/d. A water removal percentage of 65.5–79 for MSW with 72% initial moisture content was observed [15] for 16 days of operation, which is comparable with the projected moisture loss [16], which demonstrated a loss of moisture from 85.7% to 89.8% after 10 days to 13 days for an organic portion of municipal solid waste combined with paper and pulp trash with an initial moisture content of 53.9% to 60%. The experimental analysis under greenhouse conditions for municipal solid waste at an air flow rate of 6 L/min kg showed a reduction in moisture content from 69% to 40% [17]. The experimental analysis of municipal solid waste conditions (MSW) under conditions of an aeration rate of 172.8 m^3/d revealed a reduction in water content from 62% to 21% [11].

As a result, the model enabled an accurate prediction of the waste matrix's moisture removal. Figure 2 illustrates the simulated profile of the moisture content in food, paper, and inert material for these three components of MSW.

3.2. Optimization Studies for Different Aeration Flow Rate (Q) Values

Behavior of Aeration Flow Rate

Figure 3 illustrates the optimization of the aeration flow rate (Q) for a constant composition of MSW. Overall moisture reduction was used as one of the optimization factors. For the sake of this analysis, the MSW is assumed to be composed of 61% food waste, 24% paper trash, and 15% inert waste, with the relative humidity and temperature set at 75% and 28%, respectively [11].

It is important to optimize the moisture composition in MSW. The MSW composition was fixed for paper, food, and inert moisture content. We had to verify the moisture level in MSW for various Q values, which ranged from 115.2 m^3/day to 172.8 m^3/day. Higher flow rates provide superior outcomes for a specific biodrying operation, as was clear from prior optimization research [1]. We used a bulk density of 206 kg/m^3 for MSW. The test cases that were utilized to carry out the optimization study are shown in Figure 3. The impacts of the aeration flow rate values as well as the MSW and IMC composition played the main role. Higher values of the aeration flow rate (Q = 172.8 m^3/day) and the compositions of food, paper, and inert material in municipal solid waste produce better outcomes for biodrying processes.

Behavior of Weighted Average Moisture

To compare the impacts of all the parameters collectively, the idea of a weighted average (as indicated in Table 6) is utilized. The weighted average moisture shows a considerable reduction in its value as time progresses. Based on the simulation result, the aeration flow rate is 115.2 m^3/d, and we found that the weighted average moisture is 21.19 kg/m^3. When we increased the aeration flow rate from 115.2 m^3/d to 172.8 m^3/d, we found that the weighted average moisture was reduced to 12.02 kg/m^3. This is the optimum value of moisture in MSW to promote the survival of microorganisms in MSW. This is verified by the research of Ani Lawarance et al. (2022) [11]. Higher airflow rates result in a greater reduction in overall moisture. This trend is likely due to the increase in biodrying processes taking place in them.

Table 6. Result based on the simulation

Case	AFR (m^3/d)	Total moisture left in MSW (%)	Weighted average (kg/m^3)
1	115.2	28.62	21.19
2	129.6	25.01	18.36
3	144	21.93	15.93
4	158.4	19.28	13.83
5	172.8	16.77	12.02

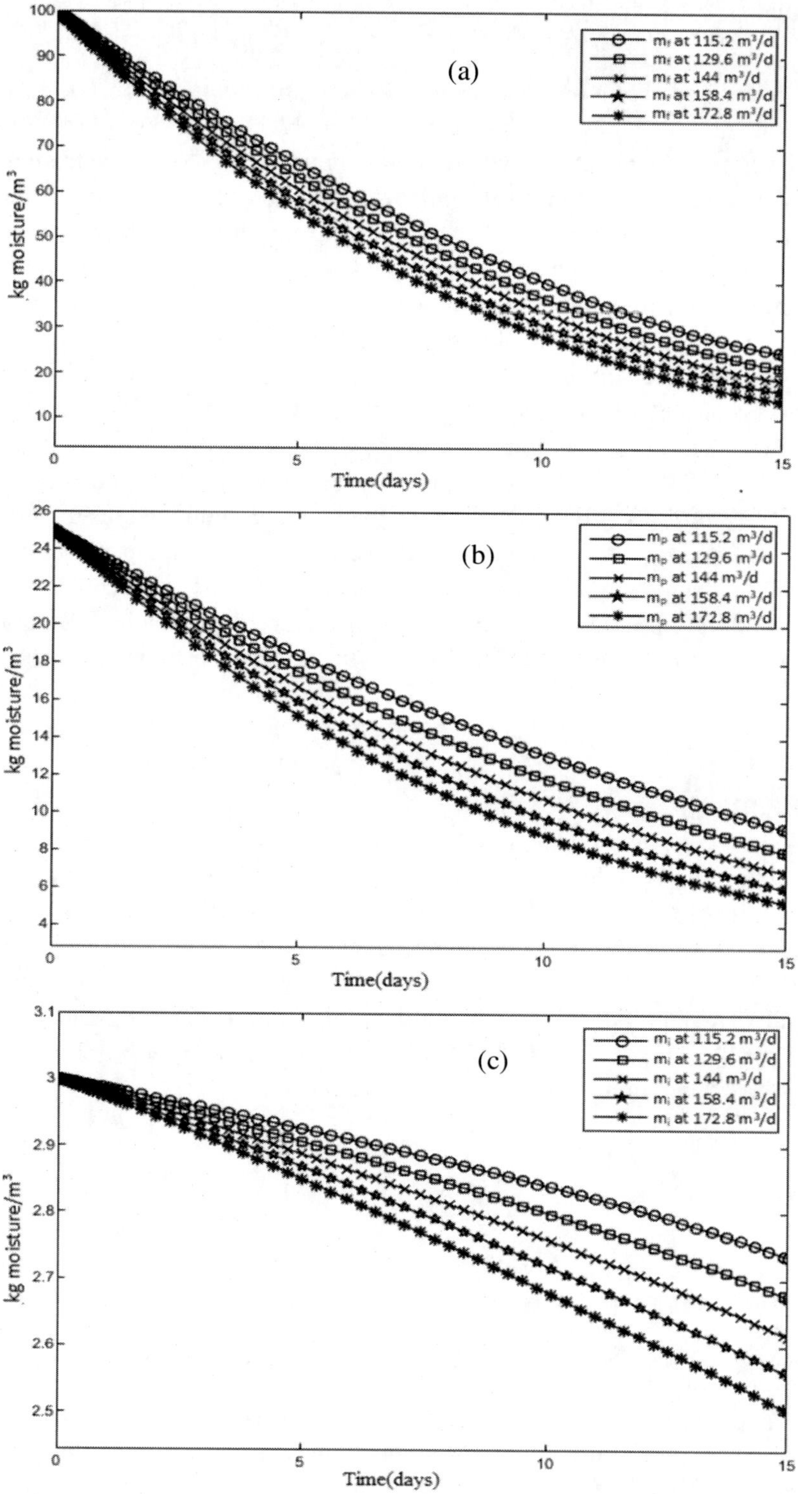

Figure 3. (a). Moisture reduction in food (m_f), (b). Moisture reduction in paper (m_p), and (c). Moisture reduction in inerts (m_i).

Conclusion

Modeling is performed over a municipal solid waste matrix with appropriate assumptions and rate expressions. It converted the process into 18 N simultaneous ordinary differential equations. The simulation results of these modeling equations are obtained from Simulink. The behavior of all compounds in municipal solid waste was observed. An optimization analysis of the biological system was performed. It was observed that higher airflow rates promoted the overall biodrying process. As the air flow rate was increased, an overall COD reduction and moisture reduction were observed. On the other hand, higher flow rates also showed low overall temperatures because a higher air flow rate promoted higher heat transfer rates, causing increased drying effects. Further optimization analysis was performed to find the optimum moisture reduction in MSW composition for a better drying process. Considering the effect of the aeration flow rate, the initial composition of MSW is fixed. We found that the optimum aeration flow rate was 172.8 m^3/d. The simulated result showed that the total moisture reduction is 83.02% at 172.8 m^3/d. We found consistency between the simulated result and the results of the experimental study. With the minimum of input data, such as operating parameters and initial parameter values, the mathematical model described in this study can be used to simulate the practical biodrying process. The current model's scope also includes combining the mass balance equation with the energy balance equation to examine how temperature affects the initial moisture content.

Acknowledgments

This research work was conducted with the permission of the All-India Council for Technical Education (AICTE), New Delhi. The experimental part of the study was conducted at the National Institute for Interdisciplinary Science and Technology (NIIST), Thiruvananthapuram. The simulation part of the study is performed at the National Institute of Technology, Tiruchirappalli. I would like to express my deep gratitude to Dr. A. Arunagiri and Dr. Ani Lawrance. I would also like to thank MoE, GoI, for providing the stipend during post-graduation at NIT Tiruchirappalli.

References

[1] Zawadzka, A., Krzystek, L., Stolarek, P., & Ledakowicz, S. 2010. Biodrying of organic fraction of municipal solid wastes. *Drying Technology*, *28*(10), 1220-1226.

[2] Cai, L., Gao, D., Chen, T. B., Liu, H. T., Zheng, G. D., & Yang, Q. W. 2012. Moisture variation associated with water input and evaporation during sewage sludge biodrying. *Bioresource Technology*, 117, 13-19.

[3] Colomer-Mendoza, F. J., Herrera-Prats, L., Robles-Martinez, F., Gallardo-Izquierdo, A., & Piña-Guzman, A. B. 2013. Effect of airflow on biodrying of gardening wastes in reactors. *Journal of Environmental Sciences*, *25*(5), 865-872.

[4] Yang, B., Zhang, L., & Jahng, D. 2014. Importance of initial moisture content and bulking agent for biodrying sewage sludge. *Drying Technology*, *32*(2), 135-144.

[5] Villegas, M., & Huiliñir, C. 2014. Biodrying of sewage sludge: Kinetics of volatile solids degradation under different initial moisture contents and air-flow rates. *Bioresource technology*, *174*, 33-41

[6] Huiliñir, C., & Villegas, M. 2015. Simultaneous effect of initial moisture content and airflow rate on biodrying of sewage sludge. *Water research*, *82*, 118-128.

[7] Tom, A. P., Haridas, A., & Pawels, R. 2016. Biodrying process efficiency: Significance of reactor matrix height. *Procedia Technology*, *25*, 130-137.

[8] Tom, A. P., Pawels, R., & Haridas, A. 2016. Biodrying process: A sustainable technology for treatment of municipal solid waste with high moisture content. *Waste management*, *49*, 64-72.

[9] Huiliñir, C., Pérez, J., & Olivares, D. 2017. A new model of batch biodrying of sewage sludge, Part 1: Model development and simulations. *Drying Technology*, *35*(6), 651-665.

[10] Huiliñir, C., & Pérez, J. 2017. A new model of batch biodrying of sewage sludge, Part 2: Model calibration and validation. *Drying Technology*, *35*(6), 666-679.

[11] Lawrance, A., Haridas, A., Savithri, S., & Arunagiri, A. 2022. Development of mathematical model and experimental Validation for batch biodrying of municipal solid waste: Mass balances. *Chemosphere*, *287*, 132272.

[12] Haug, R. T. 2018. *The practical handbook of compost engineering.* Routledge.

[13] Wang, Y., & Witarsa, F. 2016. Application of Contois, Tessier, and first-order kinetics for modeling and simulation of a composting decomposition process. *Bioresource technology*, *220*, 384-393.

[14] Treybal, R. E. 1981. *Mass-Transfer Operations.* McGraw-Hill Education, Singapore.

[15] Zhang, D. Q., He, P. J., Jin, T. F., & Shao, L. M. 2008. Biodrying of municipal solid waste with high water content by aeration procedures regulation and inoculation. *Bioresource Technology*, *99*(18), 8796-8802.

[16] Dominczyk, A., Krzystek, L., & Ledakowicz, S. 2014. Biodrying of organic municipal wastes and residues from the pulp and paper industry. *Drying Technology, 32*(11), 1297-1303.

[17] Zaman, B., Oktiawan, W., Hadiwidodo, M., Sutrisno, E., & Purwono, P. 2021. *Calorific and greenhouse gas emission in municipal solid waste treatment using biodrying. Global Journal of Environmental Science and Management*, *7*(1), 33-46.

Chapter 22

The Modeling of CO_2 Conversion and Capture Techniques Using Aspen Plus

S. K. Raja[1]
Munnangi Ananya[1]
C. Karthikeyan[2,*]
and P. Anand Kumar[1]
[1]Department of Chemical Engineering, St. Joseph's College of Engineering, Chennai, India
[2]Department of Chemical Engineering, Indian Institute of Technology Madras, Chennai, India

Abstract

Progressively, in day-to-day life, the amount of carbon dioxide emitted into the atmosphere is increasing, thereby increasing the carbon footprint. It has become a serious issue that requires our attention. Carbon dioxide capture and conversion can be a solution to this issue. This is a process where large amounts of carbon dioxide are captured from various sources using different capture techniques and later converted into useful products. This work focuses on analyzing and comparing different carbon dioxide capture techniques and finding the most effective capture technique. This work also focuses on converting the captured CO_2 into methanol, analyzing the methanol production plant by making changes in the physical, thermodynamic, and kinetic properties of the process, and finding the most efficient process conditions. Sensitivity analysis was carried out on different CO_2 capture techniques, such as post-combustion capture, precombustion capture, oxy-fuel combustion technique, direct air capture method, chemical looping method, and downstream to the conversion of methanol from carbon dioxide capture, to find the effect of different parameters, such as pressure, temperature, energy requirement, and other properties, on the amount of CO_2 captured and methanol conversion. The most effective CO_2 capture and the methanol plant were then fitted into a thermal power generation plant, which is run by coal, in an attempt to achieve zero discharge. ASPEN PLUS software was used to simulate the problem.

Keywords: carbon footprint; carbon dioxide; ASPEN PLUS; methanol; coal; zero discharge

---

* Corresponding Author's Email: karthikeyanchemz@gmail.com.

In: Sustainability in Chemical Processes through Digitalization and Green Chemistry Approaches
Editors: Susarla V. A. R. Sastry, Shravan Kumar and Ashish Kapoor
ISBN: 979-8-89113-257-3

1. Introduction

Considering the present or past years, it can be noticed that many chemical industries are emitting increasing amounts of CO_2 into the atmosphere. Compared with countries such as the United States, Canada, and China, India has been emitting lower amounts of CO_2. Simultaneously, CO_2 emissions in India are greater than those in South Korea, Germany, and some other countries. Comparing worldwide levels of intermediate, the chemical industries based in India are emitting up to seven percent (7%) of the overall carbon dioxide into the atmosphere this year. Emissions from chemical industries, the burning of fossil fuels, vehicle exhaust, and thermal power plants are increasing the amount of CO_2 in the atmosphere. The increased amount of CO_2 in the atmosphere is leading to global warming, which is a matter of serious concern. The use of the method of capture and conversion of CO_2 can be a possible solution to this enormous problem [1, 2]. There are different methods of CO_2 capture, and different techniques and technologies are incorporated into them to capture CO_2 efficiently. The different methods are the postcombustion method, precombustion capture method, oxy-fuel combustion technique, and direct air technique method. In this work, an attempt is made to reduce CO_2 emissions and attain net zero CO_2 discharge in a coal-fired thermal power plant.

2. Thermal Power Plant

In India, CO_2 emissions were approximately 1.93 tons last year. The burning of fossil fuels, road traffic, and waste gases from chemical industries are the main sources of CO_2. The need for electricity is not going to end, but the demand is increasing day by day. Nearly 75% of the electricity is generated by steam-electric power stations, which use nonrenewable resources such as coal and fossil fuel to produce steam, as it is the cheapest way to generate electricity [3, 4]. The emissions caused by this account for one-half of three (1.5/3) a part of CO_2 emissions to the atmosphere. There are nearly 8500 coal-fired power stations across the globe, of which 40 are in India, indicating a large amount of CO_2 emissions into the atmosphere every year. For this very reason, a coal-fired thermal power plant was chosen for this study, which may possibly reduce the intensity of global warming with time if incorporated. Figure 1 shows the outline of a coal-fired thermal power plant.

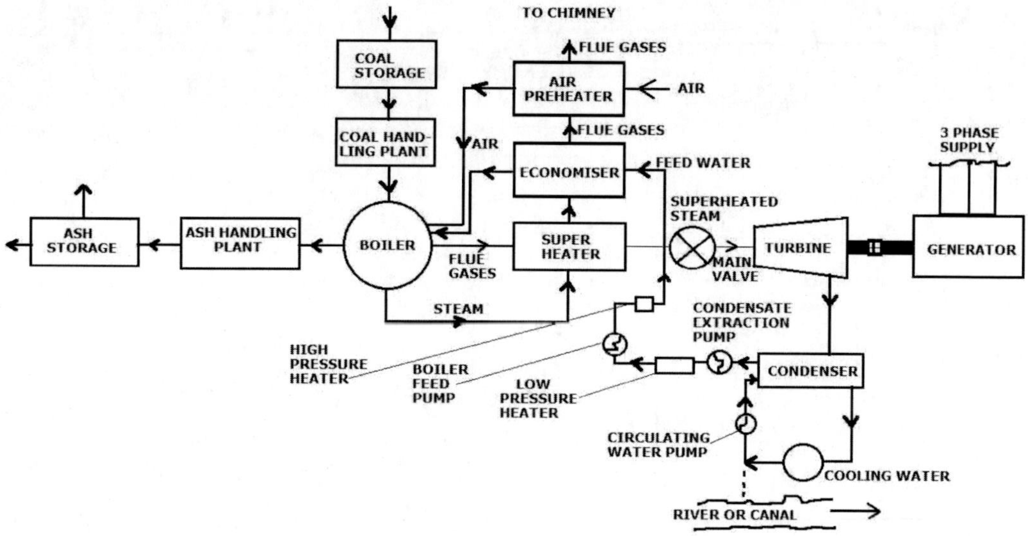

Figure 1. Thermal Power Plant.

3. Carbon Capture

Carbon capture involves the capture of carbon dioxide (CO2) emissions from industrial processes, such as steel and cement production, or from the burning of fossil fuels in power generation. The different carbon capture methods are as follows: 1. Direct air capture, 2. Oxyfuel combustion capture, 3. Precombustion carbon capture, 4. Postcombustion carbon capture. These carbon capture methods use different technologies such as physical and chemical absorption, physical adsorption, membrane separation, chemical looping, cryogenic distillation, and hydrate-based separation [5, 6].

3.1. Direct Air Capture

Sources

This method mostly uses absorption and adsorption techniques to capture CO_2 by using alkaline solvents and sorbents, as shown in Figure 2. The capturing efficiency of the direct air capture method roughly ranges from 85% - 95%. This method requires considerable energy and is expensive. Moreover, this method is not preferred in this case because the direct air capture method is mainly employed to remove the remaining CO_2 released into the atmosphere after removing most of the CO_2 at the source itself [7,8].

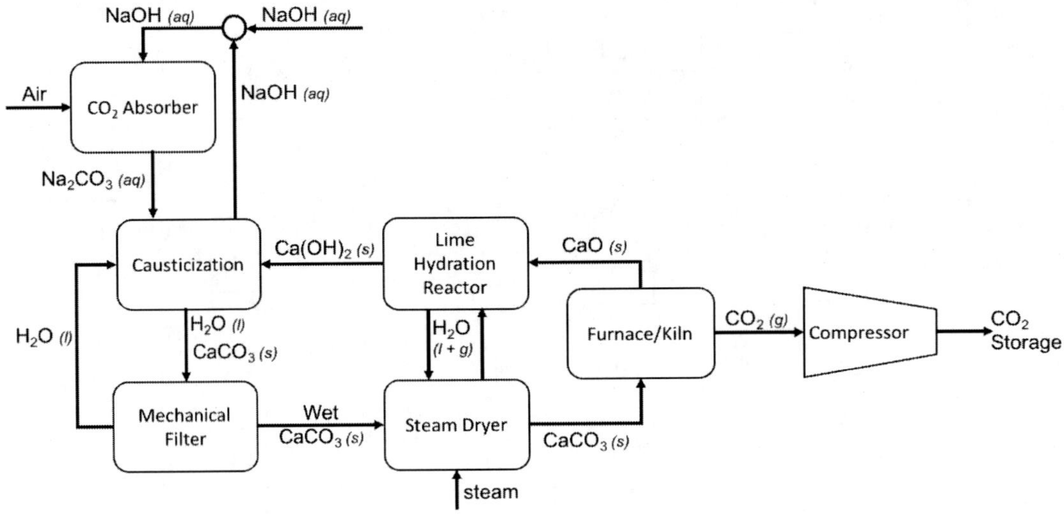

Figure 2. Direct Air Capture.

3.2. Oxyfuel Combustion Capture

The oxyfuel combustion capture method involves the process of burning the fuel in nearly pure oxygen instead of air. This will ensure that the combustion process takes place completely, which means that all the fuel combusted will be completely converted into CO_2. A brief flowchart of this process is shown in Figure 3. This method uses different techniques, such as chemical looping and cryogenic distillation techniques, to separate CO_2 from the flue gas [9,10]. Oxyfuel combustion capture has the highest efficiency among all the carbon capture methods, i.e., 100% CO_2 capture. This method is very energy intensive and hence is the most expensive method to capture carbon dioxide.

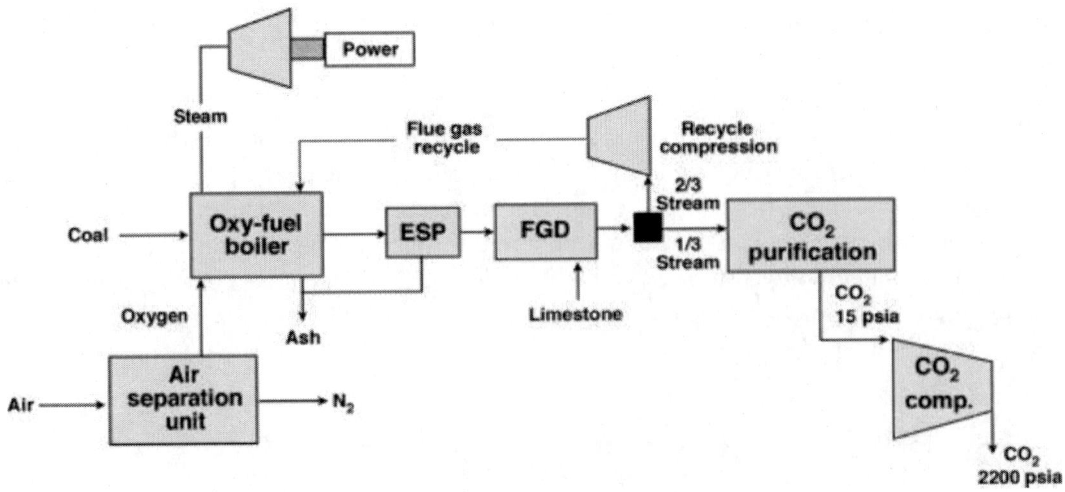

Figure 3. Oxyfuel combustion capture.

3.3. Precombustion Carbon Capture Method

Precombustion carbon capture refers to removing CO_2 from the fuel before combustion is completed. This method uses absorption, adsorption, and hydrate-based separation techniques to separate CO_2 from the syngas, which is briefly summarized in Figure 4. The high CO_2 concentration in the syngas enhances the adsorption efficiency, leading to the formation of a fuel that can be less harmful to the environment. The precombustion carbon capture method has high efficiency and facilitates relatively easier carbon removal from the fuel, but the capital costs of the base gasification process are more expensive [11].

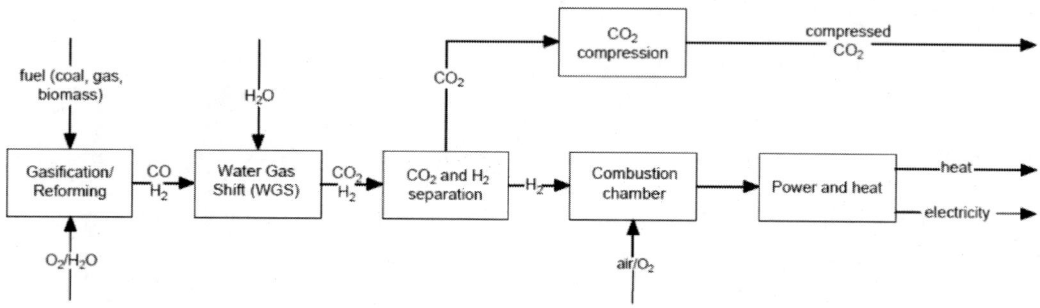

Figure 4. Pre-combustion Carbon Capture Method.

3.4. Post Combustion Carbon Capture

Post-combustion carbon capture involves separating CO_2 from exhaust gases created by burning fuel. This method uses various techniques, such as absorption, adsorption, membrane separation, chemical looping, and cryogenic distillation. This method is capable of removing up to 90% of CO_2 contained in the plant's exhaust gases. Postcombustion carbon capture is easier to retrofit into existing power plants than any other capture method. It reduces costs in CO_2 separation and lowers regeneration energy requirements. A brief flowchart of the postcombustion carbon capture method via adsorption is shown in Figure 5.

For this simulation, postcombustion carbon capture was chosen after considering all the pros and cons of all the carbon capture methods. Direct air capture, oxyfuel combustion capture, and precombustion capture were not chosen despite their high efficiencies of carbon capture because they cannot be easily retrofitted into existing power plants, are highly energy intensive and hence are more expensive.

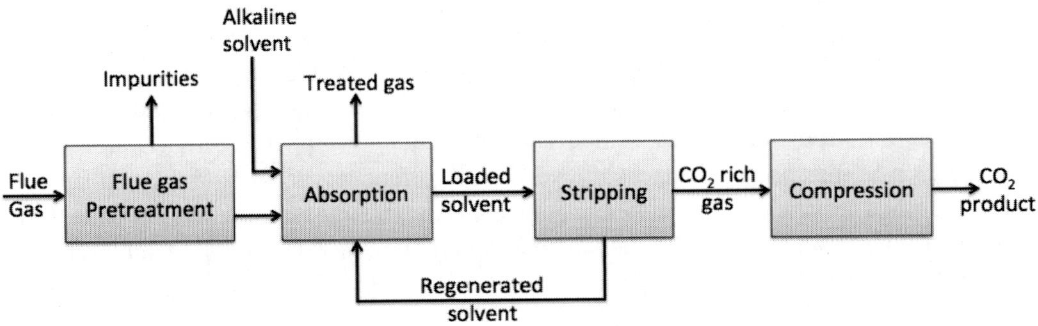

Figure 5. Post-combustion carbon capture.

4. Result

The CO_2 in the exhaust gas of the thermal power plant is captured by postcombustion carbon capture, as stated above. The CO_2 captured from the exhaust gas can either be sequestered or converted into useful products. The recently found method to store CO_2 is geologic sequestration, which has a larger storage capacity of CO_2 than any other sequestration method. However, like any other thing, it has its own disadvantages [12,13]. Although geologic sequestration seems to be a good option, in later years, the CO_2 stored in the geologic sites may slowly seep into the soil and groundwater, thereby contaminating them. Considering all this, the conversion of CO_2 into useful products is a far better option than sequestration. In this work, CO_2 was converted into methanol, which is a good fuel. Hydrogenation of CO_2 produces methanol. A flowchart of the methanol production plant is shown in Figure 6.

$$CO_2 + 3H_2 \longrightarrow CH_3OH + H_2O$$

Methanol is a very good fuel. In this work, the methanol obtained from the methanol plant is fed to the thermal power plant as fuel to produce steam. This will significantly reduce the need for coal, and it can possibly prove to be a good step toward self-sufficiency in coal-fired power stations [14].

In an attempt to achieve net zero CO_2 discharge in the thermal power plant, a postcombustion carbon capture method using absorption techniques was integrated with the thermal power plant. CO_2 in the exhaust gas of the power plant will be separated and captured from the exhaust stream, releasing the treated gas into the atmosphere, which will have a very minimal amount of CO_2. The captured CO_2 can either be sequestered or converted into useful products. In this work, the CO_2 captured is converted into methanol, which can be used as a fuel. Therefore, the methanol plant is also integrated with the postcombustion carbon capture in the power plant.

In this work, the flue gas from the power plant is directly fed to the postcombustion carbon capture plant, where CO_2 is captured. The captured CO_2 is then sent as a feed to the methanol production plant, where the CO_2 is mixed with H_2. In the methanol plant, CO_2 undergoes hydrogenation to produce methanol. Methanol, which has fuel characteristics, can be used as a fuel in the same thermal power plant for electricity generation.

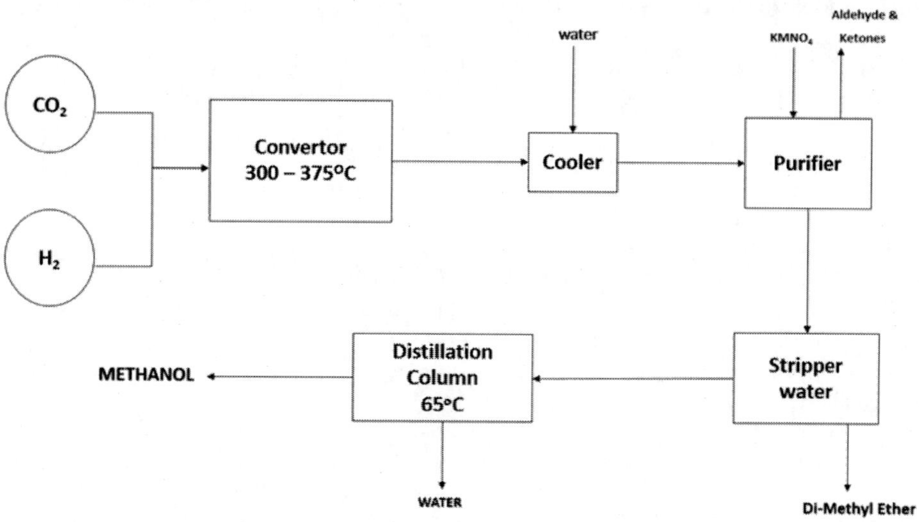

Figure 6. Flowchart of Methanol Synthesis.

The flowchart in the above figure was simulated in ASPEN PLUS software, and a sensitivity analysis was carried out to understand the effect of different parameters, such as temperature, pressure, flow rate, energy requirements, and other properties, on the final yield.

Conclusion

Although net zero discharge was not achieved, the CO_2 emitted into the atmosphere by the thermal power plant every year was greatly reduced. This can possibly be a step toward reducing CO_2 in the atmosphere, thereby also reducing the intensity of the effect of global warming with time. The conversion of CO_2 into methanol and its use as fuel in the same thermal power plant also reduces the amount of coal needed each day for power generation. Coal, as a nonrenewable resource, is exhausted with time due to overexploitation. The reduced use of coal indicates that over exploitation of the coal is also reduced, and it can be used for longer than before. Moreover, the amount of coal mined every year can also be reduced, which will in turn reduce the damage caused to the environment while mining the coal.

References

[1] Lucquiand Fran Mathieu, Jon Gibbins, On the integration of CO2 capture with coal-fired power plants: *A methodology to assess and optimize solvent-based postcombustion capture systems – 2021.*

[2] Xiaohong Huang, Junjun Guo, Zhaohui Liu, Chuguang Zheng, *Opportunities and challenges of Oxy-fuel Combustion, Oxy-fuel combustion, fundamentals, theory and practice, science direct – 2018.*

[3] Cesco Dalena*, Alessendro Senatore*, Alessia Marino*, Amalia Gordano, Marco Basile and Angelo Basile, Chapter - 1 *Methanol Production and Applications: An overview, Methanol (science and Engineering),* Elsevier – 2018.

[4] Kanniche Mohamed, Rene Gros-Bonnivard, Philippe Jaud, Jose Valle-Marcos, Jean-Marc Amann Chakib Bouallou, *Precombustion, postcombustion and oxy-combustion in thermal power plant for CO2 capture, applied therm engineering*, Elsevier – 2016.

[5] Office of Fossil Energy and Carbon Management, Precombustion carbon capture Research, energy.gov. – 2015.

[6] Rodriguez Jose A, ping Liu, Dario J. Stacchiola, Sanjaya D. Senanayake, Michael G. White, and Jingguang G. Chen, Hydrogenation of CO2 to methanol: Importance of Metal-oxide and Metal-Carbide Interfaces in the Activation of CO2, *ACS publications – 2015.*

[7] Wilcox Jennifer, *Carbon capture* – 2013.

[8] Rhode Emily, *Direct Air Capture pros and Cons, Treehugger* – 2013.

[9] Merkel Tim C., Haiqing Lin, Xiaotong Wei, Richard Baker, *Power plant postcombustion carbon capture: an opportunity for membranes* – 2012.

[10] Paul H. M. Feron, *Absorption-based PostCombustion capture of Carbon Dioxide, Woodhead publishing series in energy* – 2011.

[11] Yu Olexander. Perevertaylenko, Andriy O. Gariev, Theodoros Damartzis, eonid L. Tovazhnyanskyy, petro O. Kapustenko, Olga P.Arsenyeva, *Searches of cost effective ways for amine absorption unit design in CO2 postcombustion capture process, energy*, Elsevier – 2011.

[12] Chelsea A. Huff and Melanie S. Sanford, Cascade Catalysis for the homogeneous Hydrogenation of CO2 to methanol, ACS Publications – 2011.

[13] Mathieu Lucquiand, Jon Gibbins, On the integration of CO2 capture with coal-fired power plants: *A methodology to assess and optimize solvent-based postcombustion capture systems* – 2010.

[14] Ikhhlas Ghait and Tareq Al-Ansari, A review of carbon capture and utilization as a CO2 abatement opputunity within the EWF Nexus, *Journal of CO2 utilization 45 (2021) 101432*, Elsevier – 2010.

Chapter 23

An Experimental, Modeling and Optimization Study on the Formation of Solidified Natural Gas

Dhaval Patel*
Bhvikkumar Mahant
Omkar Singh Kushwaha
and Rajnish Kumar
Department of Chemical Engineering, Indian Institute of Technology Madras, Chennai, Tamil Nadu, India

Abstract

The world is moving toward the utilization of gaseous fuel for controlling CO_2 emissions. Currently, natural gas is the primary gaseous fuel that is being used. This natural gas is transported mainly as liquefied natural gas (LNG) and compressed natural gas (CNG). LNG has a vast network across the globe for transportation but faces challenges while distributing on a small scale. At the same time, CNG has serious safety concerns, which can be observed from the accidents in the last two decades. The solidified natural gas (SNG) formed when natural gas is solidified in an ice-like crystalline structure in the presence of water (also known as natural gas hydrate) may be a better alternative for this bottleneck.

This work investigates thermodynamic and kinetic studies on SNG formation with pure water and chemical promoters. In these experiments, synthetic natural gas containing the required proportion of methane, ethane, propane, butane, carbon dioxide, and hydrogen, similar to natural gas, replicates the behavior of commercial natural gas. The modeling study was performed using fundamental and artificial neural network (ANN)-based models, which were developed using MATLAB R2022. Both models took specific input parameters, such as temperature, pressure, and time, to predict the natural gas mole consumption as an output parameter. To find an efficient combination of parameters for the scale-up of the study, an optimization study was performed, and preliminary results are reported here.

Keywords: gas hydrate, modeling, optimization, artificial neural network, solidified natural gas

---

* Corresponding Author's Email: dpatel18999@gmail.com.

In: Sustainability in Chemical Processes through Digitalization and Green Chemistry Approaches
Editors: Susarla V. A. R. Sastry, Shravan Kumar and Ashish Kapoor
ISBN: 979-8-89113-257-3

1. Introduction

The world has been facing climate issues for several decades, and one of the prime reasons identified is increased carbon dioxide (CO_2) concentrations in the environment [1]. There have been many efforts to find and develop ways to tackle this problem. One of the most straightforward solutions is not to use fossil fuels for energy [1]. Considering the current advancements and technology, it is not feasible at the time, but we can move from coal to gaseous fuel. This approach may also considerably reduce CO_2 emissions. Although hydrogen is the cleanest-burning fuel, no sustainable technology is currently available. At present, natural gas is a primary gaseous fuel that can be seen as an intermediate fuel while transiting toward the aim of zero emissions. Natural gas is currently transported mainly in two forms: compressed natural gas (CNG) and liquefied natural gas (LNG) [2, 3]. Natural gas is liquefied under cryogenic conditions and transported as LNG. There is an extensive chain of transportation network LNG from one part of the world to another. However, it is uneconomical when transporting at small distances [4-6]. On the other hand, CNG suffers from severe safety issues, which can be observed from the many accidents that have occurred during the last two decades. Hence, there is a need for the development of a sustainable and safer medium for transporting natural gas [7].

A gas hydrate is an ice-like, crystalline, nonstoichiometric solid that forms when water and gas come in contact at moderate pressure and low temperature [8, 9]. There have been several applications of gas hydrates, such as in energy storage, desalination, gas separation, and carbon dioxide sequestration. It can be seen as a potential medium for transporting natural gas [5, 7]. Although methane hydrates have been studied widely, there is little work available on natural gas as a mixture of gases in the literature.

In the current work, synthetic natural gas (SNG) is used for the study, which replicates the composition of natural gas. The formation of gas hydrates involves gas–liquid contact, which involves interphase mass transfer and suffers from interphase mass transfer resistance. The experiments were performed with pure water and the surfactant sodium dodecyl sulfate (SDS) to study the effect on mass transfer limitation. A model requirement arises to study the hydrate formation process, and several models available for this process have different complexity levels [10-12]. There have been many attempts to make models that are easy to understand. The Peng-Robinson model (PR-model) was developed, which includes various parameters and is a relatively simple model [12]. The PR model is one of the fundamental models used for modeling gases thermodynamically. By incorporating different parameters related to gas hydrate formation, the fundamental model for gas hydrate formation is developed and compared with the experimental data [8]. An artificial neural network (ANN)-based model was developed to validate the results. This ANN model may reduce the experimental dependency and predict the behavior of novel additives [13, 14].

2. Material and Methods

2.1. Materials

The gas hydrate formation process involves the transfer of gas to a liquid (water), causing mass transfer resistance and reducing the rate of hydrate formation. This mass transfer limitation can be reduced with the help of different additives. One such chemical class is KHPs (KHPs are kinetic hydrate promoters, which can reduce the mass transfer limitation and increase the rate of hydrate formation), under which many surfactant molecules are studied to accelerate gas hydrate formation. Surfactant sodium dodecyl sulfate (SDS) with 99% purity was purchased from Himedia Ltd. Indogas, Chennai, supplied the 99.5% pure synthetic natural gas; the composition for the same is mentioned in Table 1. All solutions were prepared using Milli-Q water. In this paper, the same is denoted by pure water.

2.2. Methods

2.2.1. Experimental Setup

The experimental setup shown in Figure 1 was used for the current study. The experimental setup consisted of a double-window jacketed reactor containing polycarbonate windows with a volume of 145 cm^3 [15, 16]. The reactor is designed to withstand operations at high pressure and low temperatures. The desired temperature was maintained in the reactor using an external chiller unit from Siskin Profichill, India. A magnetic stirrer from Remi India provided the stirring for the solution and gas. The temperature and pressure data were captured by the RTD temperature sensor and Baumer-made pressure transducer, respectively, at an interval of 5 seconds and transmitted to the data acquisition unit (PPI Scan log 8C). These captured data were sent to the computational unit, where further data processing took place.

Table 1. Composition different gases of synthetic natural gas

Gaseous components	% w/w
CH_4	87.71
C_2H_6	5.21
CO_2	1.84
C_3H_8	3.62
$n\text{-}C_4H_{10} + i\text{-}C_4H_{10}$	1.62

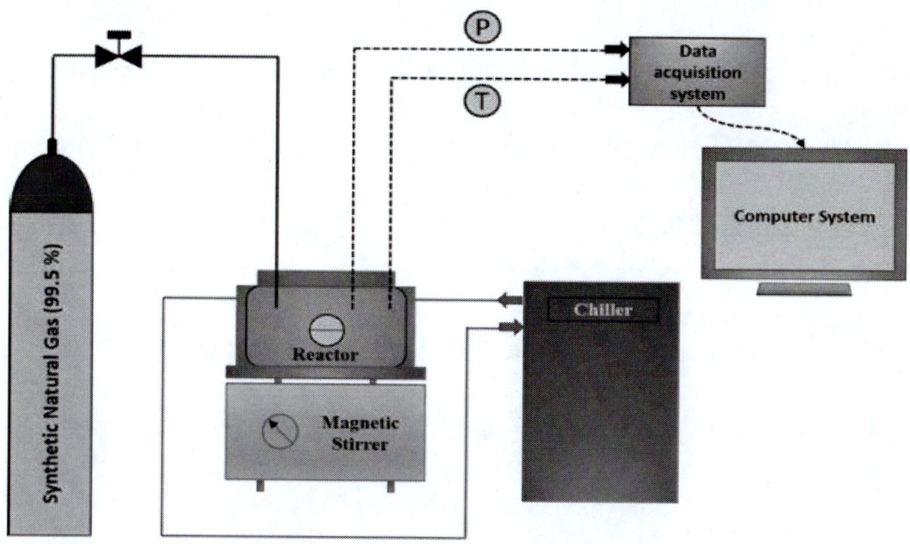

Figure 1. Schematic diagram of gas hydrate experimental setup.

2.2.2. Experimental Procedure

The reactor was cleaned and rinsed before every experiment with tap water, followed by Milli-Q water. The reactor was dried using a hot air dryer for 10 min. The required amount of additive was added to form the solution, which was stirred for 30 mins. After making the solution, it was supplied to the reactor. Then, the reactor was closed properly, and the required amount of gas was charged. The required temperature was set in the chiller, and a solution of ethylene glycol and water as a coolant circulated through the jacket. Once the temperature is reached, 5 MPa of synthetic natural gas is charged, which is sufficiently higher than the gas hydrate thermodynamic equilibrium pressure. The magnetic stirrer was started and set to 400 RPM. For the modeling study, experiments were performed for a specific amount of time.

2.2.3. Theory and Calculations

The process of gas hydrate formation is understood mainly in two parts: (1) the induction period and (2) the growth period [8, 9]. The time during which the first solid hydrate crystal nucleates is known as the induction time. It depends on several parameters and is a stochastic phenomenon that is difficult to reproduce accurately every time a new experiment is performed [8, 9]. On the other hand, the growth period depends on temperature and pressure conditions, supersaturation (driving force), and heat and mass transfer limitations.

1. The amount of gas encapsulated in the hydrate cages is calculated by taking the difference between the initial moles of gas charged in the reactor and the instantaneous moles of gas present at time "t" with the use of equation 1 [12].

$$(\Delta n_g)_t = \left(\frac{V_r P}{zRT}\right)_{t=0} - \left(\frac{V_r P}{zRT}\right)_{t=t} \quad (1)$$

where
V_r = volume of the gas in the reactor,
P = pressure in the reactor,
T = temperature in the reactor,
R = Universal gas constant
z = Compressibility factor, calculations based on Peng-Robinson's correlation

2. The forward difference formula calculates the hydrate formation rate in equation 2.

$$\left(\frac{d(\Delta n_g)_t}{dt}\right) = \frac{(\Delta n_g)_{t=0} - (\Delta n_g)_{t=t}}{\Delta t} \quad (2)$$

3. For the gas hydrate processes, the water-to-hydrate conversion gives an idea of the efficiency of the reaction, which can be obtained by using equation 3 [8].

$$\text{Conversion of water to hydrate (mole \%)} = \frac{\Delta n_g \times \text{Hydration Number}}{n_{water}} \times 100 \quad (3)$$

where
n_{water} = moles of water

2.2.4. Model

The process of hydrate formation, the physical reaction (Eq. 4) of natural gas with water, can be expressed as follows [8]: Figure 2 shows the algorithm for the model.

$$NG + MH_2O \leftrightarrow NG \cdot MH_2O \quad (4)$$

where M = hydration number

Considering that natural gas molecules form simple structure-II hydrates, the hydration number for the simple hydrates is related to the fraction filling small and large cavities [8]. Therefore, the hydration number for structure-II hydrate (Eq. 5) is calculated as follows [8]:

$$M = \frac{136}{(8\theta_L + 16\theta_S)} \quad (5)$$

where θ_S and θ_L are the fractional gas occupancies of small and large cavities, respectively. These are calculated from the basis of Langmuir adsorption theory (Eq. 6), as shown below [8]:

$$\theta_i = \frac{C_i f_{NG}}{1 + C_i f_{NG}} \quad (6)$$

where

C_i = Langmuir constant for natural gas in a type i cavity;

f_{NG} = fugacity of the natural gas in the gas phase

The Langmuir constant (C_i) for natural gas (Eq. 7) is obtained as follows [15, 16]:

$$C_i = \frac{A_i}{T} \exp\left(\frac{B_i}{T}\right) \quad (7)$$

where T = Temperature; A_i and B_i are the constants for calculating C_i, and the values for small and large cavities can be found in Table 2.

Table 2. Parameters A and B for Langmuir constant calculation

Structure-II	Small cavities		Large cavities	
Gas	A×1000 (K/atm)	B (K)	A×1000 (K/atm)	B (K)
methane				
ethane	0	0	240	2967
propane	0	0	5.455	4638
i-butane	0	0	189.3	3800
n-butane	0	0	30.51	3699

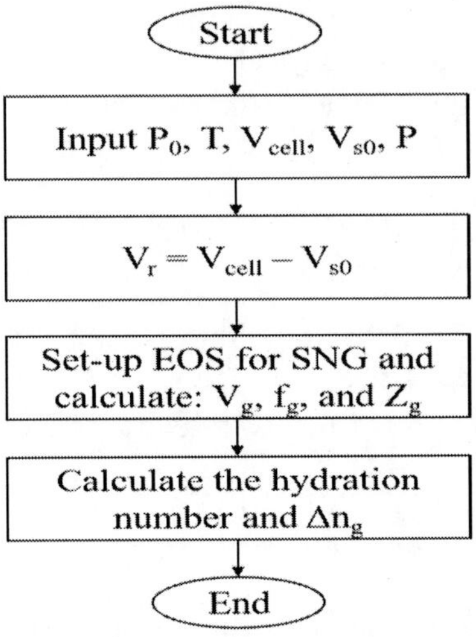

Figure 2. Algorithm to calculate the hydration number and gas mole consumption.

2.3. ANN Modeling

ANN was used to predict the natural gas mole consumption using three input parameters: temperature, pressure, and time, as illustrated in Figure 3. Utilizing the nnstart tool and the necessary number of data points, the ANN architecture was developed using MATLAB R2022 software. To forecast the model output, weights and biases are taken from the training model and put to use in mathematical computations [13, 14]. This model gives the molecular consumption as an output of the model after several hidden layers. To determine the model's effectiveness, the predicted and experimental values were compared. To improve the accuracy of the final model's prediction of the experimental value, the model was modified and trained using a variable number of hidden neurons.

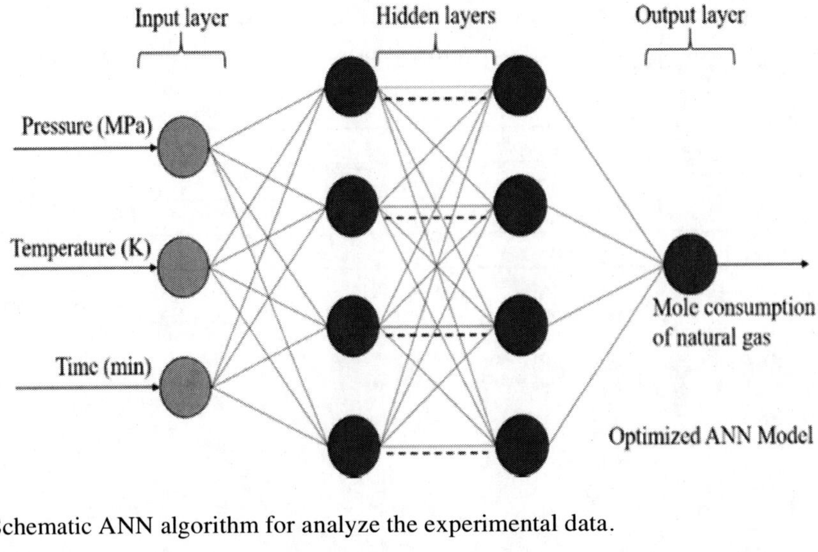

Figure 3. Schematic ANN algorithm for analyze the experimental data.

3. Results and Discussion

3.1. Mole Consumption of Natural Gas

The time-based natural gas mole consumption is represented in Figure 4. For the present study, pure water and sodium dodecyl sulfate (SDS), one of the most promising surfactants, were considered to understand the surfactant's effect on the hydrate formation rate. The molecular consumption of natural gas is greater when using 0.01 mol% SDS than when using pure water, which leads to a higher growth of hydrates. The model is well fitted to the experimental data, which can be observed in Figure 4. The low concentration of surfactants assures the overall economic process. The current work supports using sodium dodecyl sulfate (SDS) in hydrate research for the storage and transportation of natural gas, providing the basis for the scale-up of the process to the commercial level.

3.2. Artificial Neural Network (ANN) Model

The working principle of an ANN is similar to that of a human neural network. The input parameters were fed to the model and analyzed in several hidden neuron layers to achieve all possible permutations and combinations of the individual data to give the output [13, 14]. The model was then optimized to achieve accurate experimental results. Different layers of neurons are responsible for different accuracies and complexities in the models. In the current optimized model, a total of 10 different hidden neuron layers are incorporated to obtain a close output result with experiments. Figure 5 shows that the ANN is well fitted with the experimental data, which can be seen from the value of the linear regression of more than 0.95 between the data obtained from the model and experiments. ANN can predict the behavior of novel additives on hydrate growth and may reduce the experimental dependency.

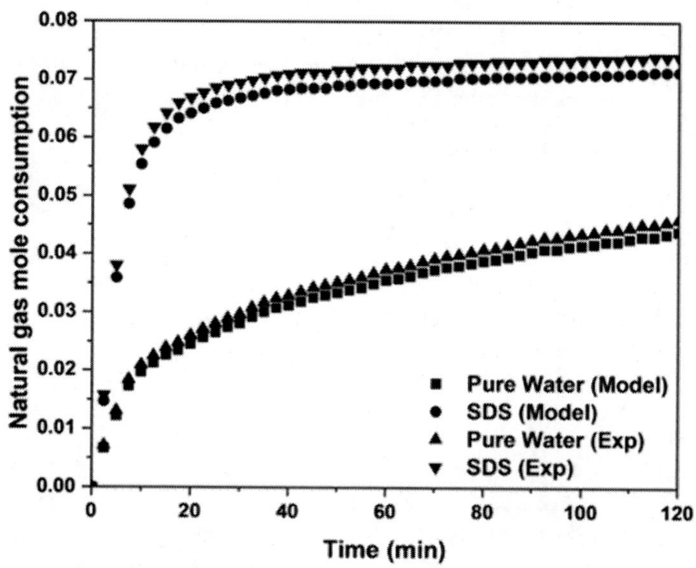

Figure 4. Comparison of model and experimental natural gas mole consumption.

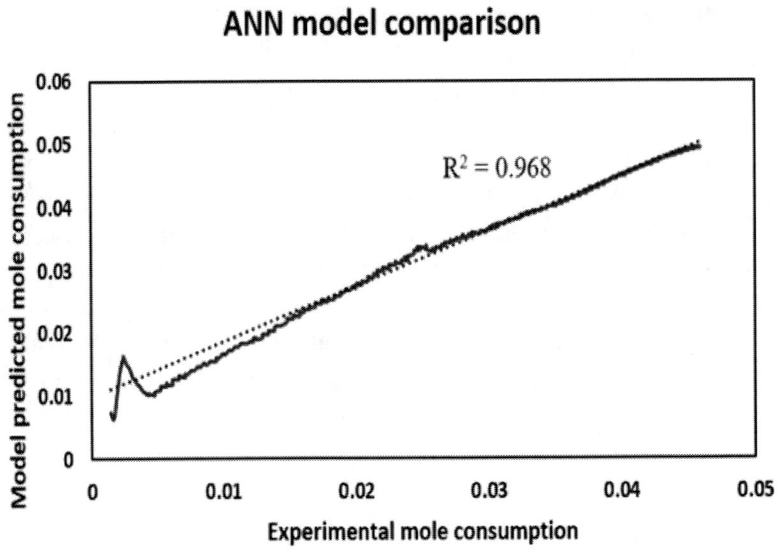

Figure 5. Comparison of ANN model predicted results and experimental data for molde consumption.

Conclusion

The novel approach for the transportation of natural gas has been investigated for the past several years, as the need to limit carbon dioxide emissions has been identified. Gas hydrate, being in a solid form, can accommodate more volume of the gas compared to other available technologies, and a medium to carry out this is water, which is freely available in nature. During this work, the fundamental Peng-Robinson model was developed to model the formation of gas hydrate, which is a good fit for the experimental data. An artificial neural network (ANN)-based model was developed, which is effective and may reduce the experimental dependency for future model development. Experiments and models have proven that solidified natural gas may be used as a future technology for transporting natural gas. As a future aspect, an ASPEN-based model can be developed for the feasibility study of the scale of the process.

Acknowledgments

The authors would like to thank the facilities provided by the Indian Institute of Madras for carrying out the research. Bhavikkumar acknowledges the prime minister's research fellowship for providing financial support.

References

[1] IEA, "*Global energy-related CO2 emissions*," 1990-2021, IEA, Paris.

[2] Ríos-Mercado, R. Z., & Borraz-Sánchez, C. (2015). Optimization problems in natural gas transportation systems: A state-of-the-art review. *Applied Energy,* 147, 536-555.

[3] Khan, M. I., Yasmin, T., & Shakoor, A. (2015). Technical overview of compressed natural gas (CNG) as a transportation fuel. *Renewable and Sustainable Energy Reviews,* 51, 785-797.

[4] Najibi, H. (2009). R. rezaei, J. Javanmardi, K. Nasrifar and M. Moshfeghian. *Appl. Therm. Eng,* 29, 2009-2015.

[5] Hao, W., Wang, J., Fan, S., & Hao, W. (2008). Evaluation and analysis method for natural gas hydrate storage and transportation processes. *Energy conversion and management*, 49(10), 2546-2553.

[6] Thomas, S., & Dawe, R. A. (2003). Review of ways to transport natural gas energy from countries which do not need the gas for domestic use. Energy, 28(14), 1461-1477.

[7] Zhang, C. S., Fan, S. S., Liang, D. Q., & Guo, K. H. (2004). Effect of additives on formation of natural gas hydrate. *Fuel,* 83(16), 2115-2121.

[8] Sloan, Jr, E. D., & Koh, C. A. (2007). *Clathrate hydrates of natural gases.* CRC press.

[9] Sloan, Jr, E. D. (2003). Fundamental principles and applications of natural gas hydrates. *Nature,* 426(6964), 353-359.

[10] Kushwaha, O. S., Meshram, S. B., Bhattacharjee, G., & Kumar, R. (2019). Molecular insights about gas hydrate formation. In *Advances in Spectroscopy: Molecules to Materials: Proceedings of NCASMM 2018* (pp. 311-322). Springer Singapore.

[11] Mahant, B., Kushwaha, O. S., & Kumar, R. *Application of pnictogen oxides in methane gas hydrate formation: experimental and modeling approach.*

[12] Smith, J. M. (1950). *Introduction to chemical engineering thermodynamics.*

[13] Zahedi, G., Karami, Z., & Yaghoobi, H. (2009). Prediction of hydrate formation temperature by both statistical models and artificial neural network approaches. *Energy Conversion and Management*, 50(8), 2052-2059.

[14] Foroozesh, J., Khosravani, A., Mohsenzadeh, A., & Mesbahi, A. H. (2014). Application of artificial intelligence (AI) in kinetic modeling of methane gas hydrate formation. *Journal of the Taiwan Institute of Chemical Engineers*, 45(5), 2258-2264.

[15] Patel, D., Mahant, B., Kumar, R., (2022). Investigation on synthetic natural gas storage in the form of gas hydrate. *4th Applied Energy Symposium: MIT A+ B (MITAB 2022).* Cambridge, USA.

[16] Sayani, J. K. S., Lal, B., & Pedapati, S. R. (2021). Comprehensive review on various gas hydrate modeling techniques: Prospects and challenges. *Archives of Computational Methods in Engineering,* 1-37.

Chapter 24

The Enhanced Temperature Control of a CSTR Using the Harris Hawks Optimization Algorithm

Siva Ramakrishna B[1,2,*]
Devendra Potnuru[2]
and M. S. N. Murty[1]
[1]Department of Chemical Engineering, Gayatri Vidya Parishad College of Engineering (A), Madhurawada, Visakhapatnam, India
[2]Department of EEE, Gayatri Vidya Parishad College of Engineering for Women (A), Madhurawada, Visakhapatnam, India

Abstract

Continuous stirred tank reactors (CSTRs) are the most important element in the process industry. Investigations into continuous hydrogen production using stirred tank reactors (CSTRs) are common. Since chemical processes are either exothermic (release energy) or endothermic (need energy input), energy must either be withdrawn from or added to the reactor to maintain a steady temperature. In this paper, the control problem of an ideal CSTR system is considered. Harris Hawks optimization (HHO) is implemented for temperature control. The controller gains are optimally tuned by minimizing the integral square error as an objective function. The validity of the proposed algorithm is verified using MATLAB simulations.

Keywords: CSTR, closed loop control, Harris hawks optimization, temperature control

1. Introduction

A Continuous Stirred Tank Reactor (CSTR) is a type of reaction vessel and is a useful tool for continuous chemical processing. The CSTR is a nonlinear system with lumped parameters from the perspective of system engineering. These reactors' mathematical models consist of a series of nonlinear ordinary differential equations. In process industries, temperature control is a

* Corresponding Author's e-mail: sivaramakrishna@gvpce.ac.in.

In: Sustainability in Chemical Processes through Digitalization and Green Chemistry Approaches
Editors: Susarla V. A. R. Sastry, Shravan Kumar and Ashish Kapoor
ISBN: 979-8-89113-257-3

crucial and frequent job. The temperature should be kept at a set point. There are numerous ways to regulate the temperature. For temperature management in the majority of applications, a PID controller is frequently employed. In nonlinear, complicated and ambiguous systems without a solid mathematical description, standard PID controllers have limitations.

On this topic, a number of researchers [1, 2] have worked. Similarly, several researchers [3, 4] have concentrated on optimizing PID gains for CSTRs. It is provided in [5, 6] how to regulate CSTR most effectively using various heuristic optimization methods. [7] outlines the use of the PSO and TLBO algorithms in PID controller tuning. Similarly, [8] improves CSTR control performance using the Jaya approach. In [9, 10], using firefly and hybrid firefly biogeography-based optimization algorithms, modified hybrid artificial bee colony algorithms and flower pollination methods were utilized for temperature control [9, 10].

2. Mathematical Modeling

The CSTR reactor is depicted graphically in Figure 1. A straight-forward exothermic reaction A→B occurs in the reactor, which is cooled by a coolant that circulates via a jacket surrounding the reactor. The primary dependent variables for the reactor are (a) the conversion of chemical An in the reacting mixture and (b) the temperature of the reacting mixture in the tank. They are mathematically described as

Material Balance:

$$\tau\frac{dX_A}{dt}+\left[1+\left(k_1+k_2\right)\tau\right]X_A=k_1\tau$$

Energy Balance:

$$\tau\frac{dT}{dt}+T-T_0+\frac{UA\left(T-T_c\right)}{\rho qC_p}=\frac{\left(-\Delta H_{rA}\right)C_{A0}X_A}{\rho C_p}$$

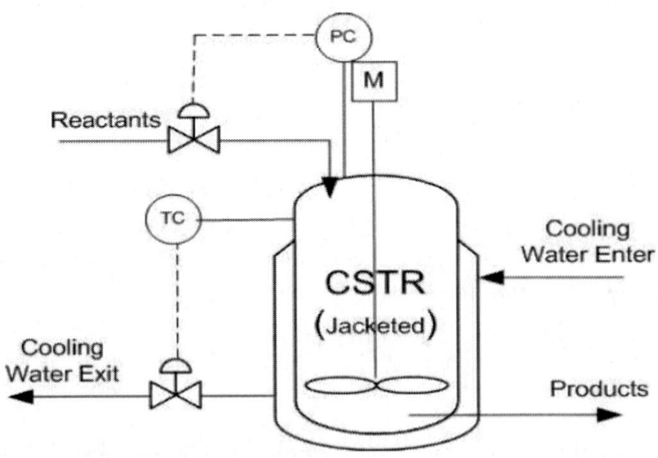

Figure 1. Continuous stirred-tank reactor (CSTR).

3. Harris Hawks Optimization (HHO)

This algorithm is based on how Harries hawks (Figure 2) acquire prey and the behaviors they exhibit while doing so. It includes phases of prey exploration and exploitation, such as surprise pounces and various Harris hawk attack strategies. Due to its primary characteristics of (i) a population-based approach and (ii) a gradient-free method, the algorithm is successful. For more details on the algorithm, one should refer to [11, 12].

Figure 2. Harris ʻhawk.

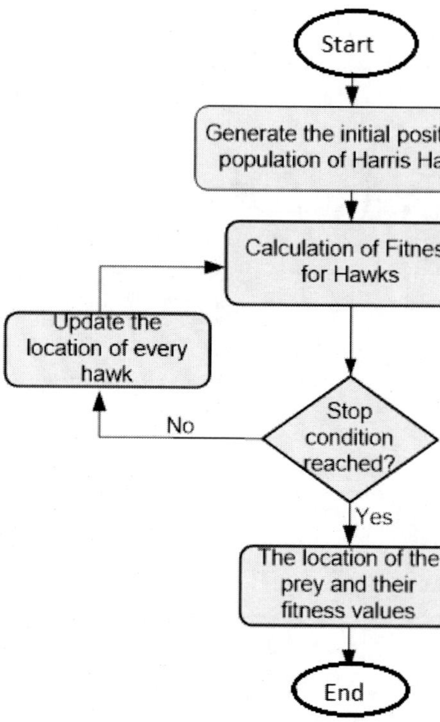

Figure 3. FFO Algorithm flowchart.

4. Result Analysis and Discussions

The MATLAB/Simulink model of the converter is developed for the parameters displayed in Table 1. In the simulation study, many examinations have been performed by considering different set points of temperatures and sudden variations. The final PID gains with the best fitness values are considered after running the FFO algorithm, as shown in Figure 3. Figure 4 shows the temperature control plot for a set point of 300 °C, as shown in Figure 4 (a). It is observed that the actual temperature tracks the reference value, and the error between the set point and the actual temperature is shown in Figure 4 (b). The error is negligibly low during steady-state operation, as shown in Figure 4(c). The PID gains obtained during algorithm implementation with the best position are kp = 25.8457; ki = 8.4864; and kd = 0.00032336.

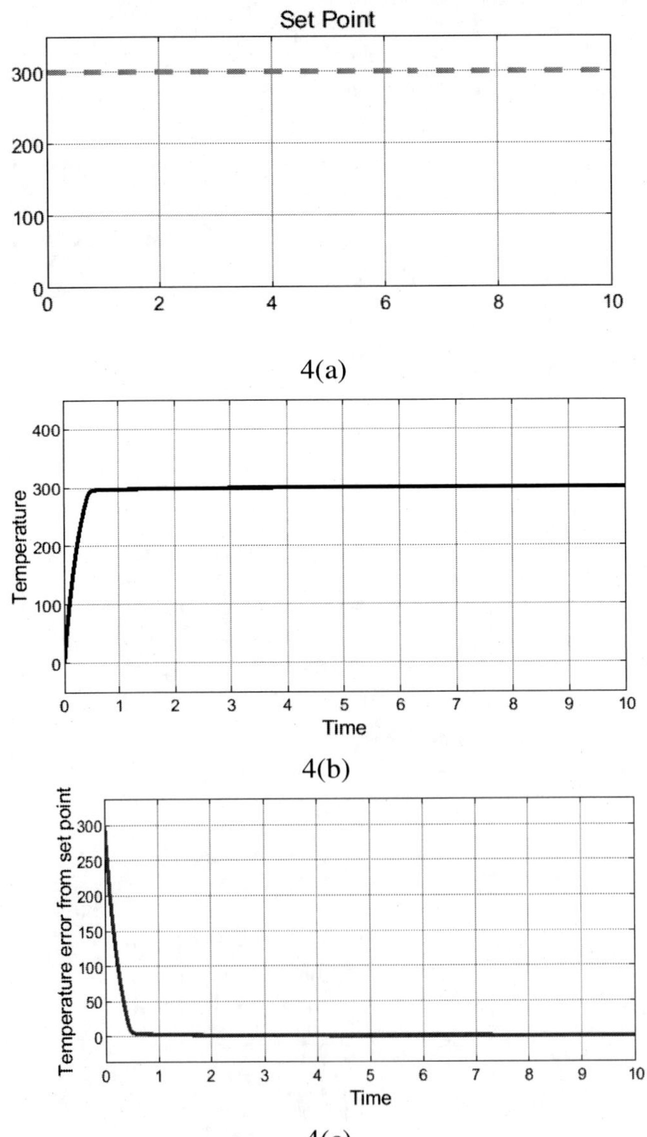

4(a) 4(b) 4(c)

Figure 4. Temperature control: (a) set point of 3000 °C (b) Actual temperature (c) Error.

Furthermore, a sudden increase in the temperature set point is used to validate the effectiveness of the control approach. As illustrated in Figure. 5, the temperature set point is first maintained at 300 °C for 5 seconds before being abruptly increased to 350 °C (i.e., 300 °C plus 50 °C) (a). As illustrated in Figure. 5, it is noted that the actual temperature tracks the temperature change for a shorter duration (b). While the thick line displays the actual temperature of the CSTR, the dotted line displays the temperature's set point. Figure 5 displays the relevant error plot (c).

Table 1. Parameters of CSTR

Parameter	Value
Volume (lit)	100
Density (kg/m3)	1000
Heat capacity of Mixture (J/kg K)	2390
Heat of reaction for A→B (J/mol)	$-5x10^4$
Universal Gas Constant (J/mol K)	8.314
ko	$7.2x10^{10}$
Overall Heat Transfer Coefficient (W/m^2 K)	5000
Equilibrium constant	1.5
Control parameters	$K_C = 25.8$, $\tau_I = 8.5$, $\tau_D = 0.003$

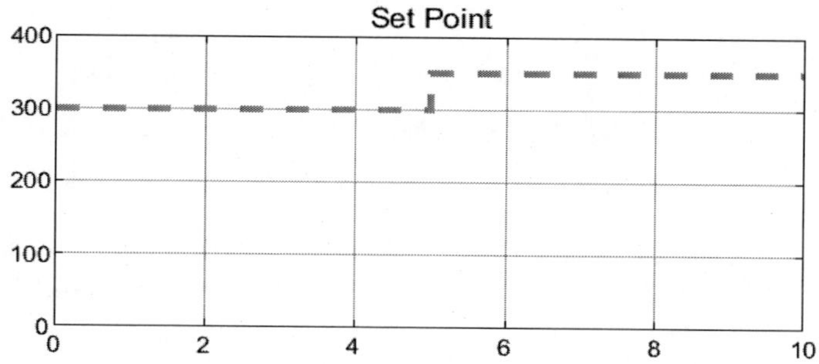

5(a)

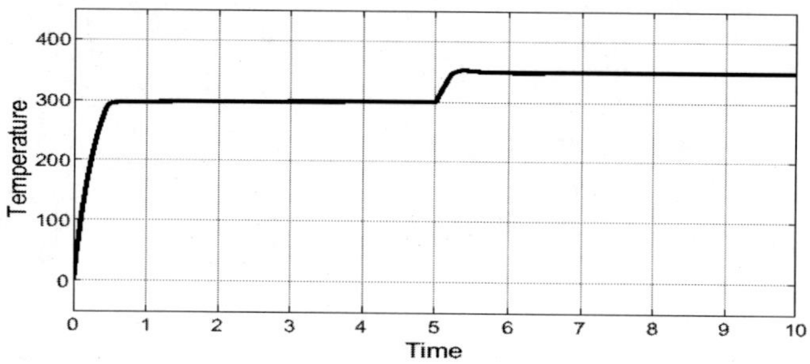

5(b)

Figure 5. (Continued)

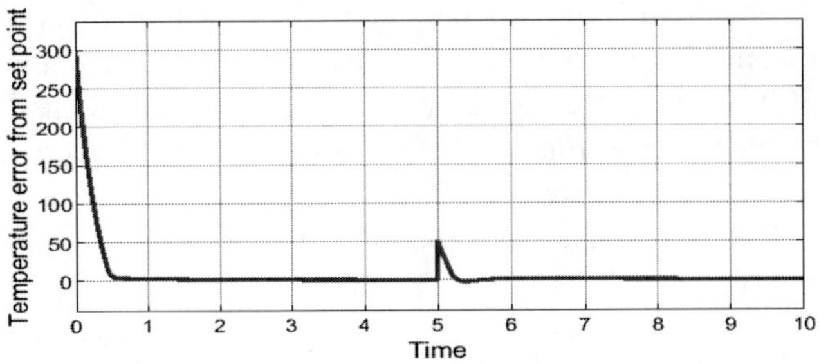

5(c)

Figure 5. Temperature control: (a) set point of 3000 °C with sudden increase of 500 °C, (b) Actual temperature (c) Error.

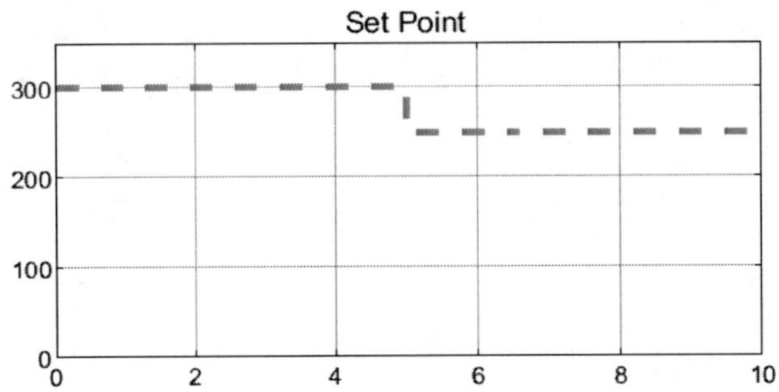

6(a)

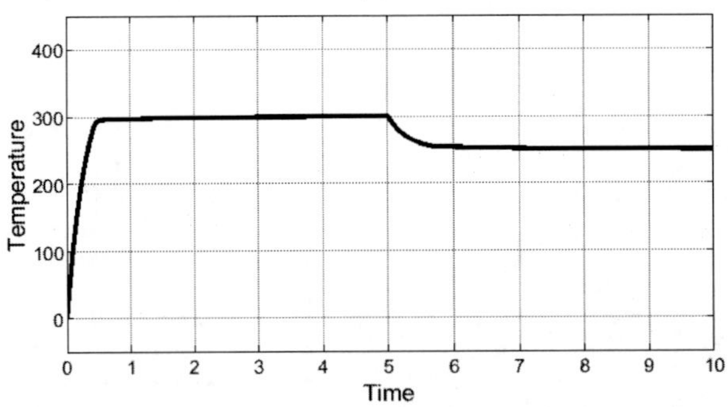

6(b)

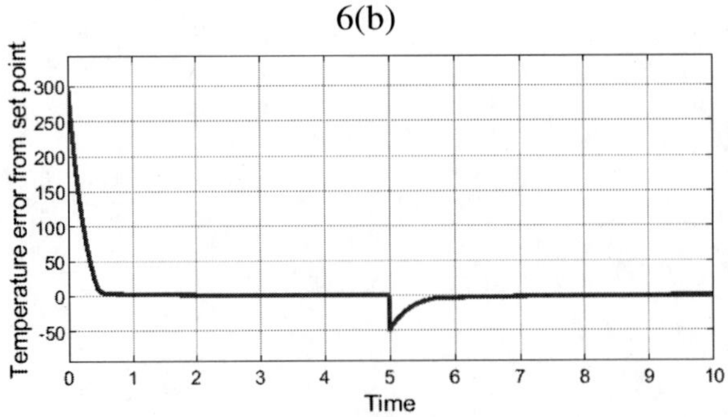

6(c)

Figure 6. Temperature control: (a) set point of 3000 °C with a sudden decrease of 500 °C (b) Actual temperature (c) Error.

For a sharp drop in temperature below the Set Point, the effectiveness of the control approach is also confirmed. The temperature set point is first maintained at 300 °C for 5 sec before being abruptly decreased to 250 °C (i.e., 300 °C – 50 °C) at t = 5 sec, as illustrated in Figure 6. (a). As shown in Figure 6, the real temperature is found to follow the temperature change in a shorter amount of time (b). The CSTR's actual temperature is shown by the thick line, while the dotted line displays the temperature's set point. During steady-state operation, the error is quite small, as seen in Figure 6. (c).

Conclusion

This work presents CSTR control using the HHO algorithm. The objective function for Harris-Hawkes optimization for temperature control is the integral square error between the set point and the actual temperature value. Consideration is given to the ideal CSTR system's control problem. By using the integral square error as an objective function, the controller gains are optimally tuned. Using MATLAB simulations, the validity of the proposed algorithm is confirmed.

References

[1] Pomberger, A., Mo, Y., Nandiwale, K. Y., Schultz, V. L., Duvadie, R., Robinson, R. I., & Jensen, K. F. (2019). A continuous stirred-tank reactor (CSTR) cascade for handling solid-containing photochemical reactions. *Organic process research & development*, *23*(12), 2699-2706.

[2] Wang, G., Jia, Q. S., Qiao, J., Bi, J., & Zhou, M. (2020). Deep learning-based model predictive control for continuous stirred-tank reactor system. *IEEE Transactions on Neural Networks and Learning Systems*, *32*(8), 3643-3652.

[3] Brindhadevi, K., Shanmuganathan, R., Pugazhendhi, A., Gunasekar, P., & Manigandan, S. (2021). Biohydrogen production using horizontal and vertical continuous stirred tank reactor-a numerical optimization. *International Journal of Hydrogen Energy*, *46*(20), 11305-11312.

[4] Zhou, W., Liu, H., He, H., Yi, J., & Li, T. (2018). Neuro-optimal tracking control for continuous stirred tank reactor with input constraints. *IEEE Transactions on Industrial Informatics*, *15*(8), 4516-4524.

[5] Goud, H., & Swarnkar, P. (2019). Investigations on metaheuristic algorithm for designing adaptive PID controller for continuous stirred tank reactor. *Mapan*, *34*(1), 113-119.

[6] Siddiqui, M. A., Anwar, M. N., & Laskar, S. H. (2021). Control of nonlinear jacketed continuous stirred tank reactor using different control structures. *Journal of Process Control*, *108*, 112-124.

[7] Matušů, R., Şenol, B., & Pekař, L. (2020). Robust PI control of interval plants with gain and phase margin specifications: Application to a continuous stirred tank reactor. *IEEE Access*, *8*, 145372-145380.

[8] Xin, L. P., Yu, B., Zhao, L., & Yu, J. (2020). Adaptive fuzzy backstepping control for a two continuous stirred tank reactors process based on dynamic surface control approach. *Applied Mathematics and Computation*, *377*, 125138.

[9] Bala, M. P., Priyanka, E. B., Prabhu, K., Priya, P. B., Bhuvana, T., & Raja, V. C. (2021, March). Real-time performance analysis of temperature process using continuous stirred tank reactor. In *2021 International Conference on Emerging Smart Computing and Informatics (ESCI)* (pp. 647-650). IEEE.

[10] Alshammari, O., Mahyuddin, M. N., & Jerbi, H. O. U. S. S. E. M. (2018). A survey on control techniques of a benchmarked continuous stirred tank reactor. J. Eng. Sci. Technol, 13(10), 3277-3296.

[11] Srivastava, V., & Srivastava, S. (2019). Control of continuous stirred tank reactor (CSTR) using nature inspired algorithms. Journal of information and optimization sciences, 40(2), 329-338.

[12] Sudha, T. (2021). Analysis of Proportional Integral Derivative Controller Parameters for CSTR. *WSEAS Transactions on Systems and Control*, *16*, 387-395.

Chapter 25

A Review on the Cavitation-Assisted Formulation of Nanocarriers for the Controlled Delivery of Poorly Soluble Drugs

Lakshay Vashishtha[1,*]
and Manisha V. Bagal[2]
Department of Chemical Engineering, BVCOE NM, Mumbai University, India

Abstract

Poorly water-soluble drugs have limited therapeutic effects, low bioavailability, and incomplete absorption by the gastrointestinal tract. Hence, these drugs become incapable of treating the affected organ or tissue. Many studies have been conducted to enhance solubility through nanoformulations. Particle size reduction is one of the tools of modern nanotechnology to overcome poor aqueous solubility. Recent studies have found that cavitation is a greener method for the production of nanoparticles. Cavitation is a phenomenon that involves the nucleation, growth, and collapse of bubbles in a very short time. The collapse of bubbles occurs at millions of places with the release of high temperatures and pressure, known as "hot spots." This released energy causes the formation of nanoparticles. In this review paper, an effort has been made to investigate the various reactors and the effects of varying the conditions for nanoparticle formation. This chapter also discusses the literature on the factors that contribute to the higher yield efficiency and small size of nanoparticles. There are several methods for the formation of nanoparticles, such as thermal decomposition, polyol method, microwave-assisted method, laser ablation, sputtering deposition, lithography, pulsed electrochemical etching, vapor deposition, and ol-gel. Of them, the use of the ultrasound-assisted method is considered good due to its high surface-to-volume ratio, high yield, environmental friendliness, no toxicity, etc. Therefore, the cavitation-assisted method is a step toward the development of green energy sources.

Keywords: cavitation, nanoparticles, drug delivery, green engineering, cavitation reactor

---

* Corresponding Author's Email: lakshayvashishtha11@gmail.com.

In: Sustainability in Chemical Processes through Digitalization and Green Chemistry Approaches
Editors: Susarla V. A. R. Sastry, Shravan Kumar and Ashish Kapoor
ISBN: 979-8-89113-257-3

Introduction

Many drugs are used for the treatment of many severe ailments. There are some drugs that are used for the treatment of cancer, rheumatoid arthritis, etc. Research says that 40% of drugs are not soluble in water, so they cannot transmit their analgesic and antipyretic effects properly. For example, nonsteroidal anti-inflammatory drugs (NSAIDs) are among the most commonly prescribed drugs around the world. They are heterogeneous organic compounds with anti-inflammatory properties that lack a steroid core. Unfortunately, the majority of routinely used NSAIDs are poorly soluble in water; therefore, these medications are categorized as biopharmaceutical class II pharmaceuticals (BCS). Due to the low solubility of drugs in water, incomplete absorption through the gastrointestinal tract occurs after oral administration. Additionally, many NSAIDs suffer from poor stability, a bitter taste, and a bad odor. These drawbacks can be overcome by proper formulation and particle size reduction using the tools of modern nanotechnology. These nanoformulation technologies not only improve solubility but also provide extra functionality, such as controlled and targeted medication release [1, 2]. Many similar nanoformulation methods have been used effectively to improve drug bioavailability by employing polymers, lipids, and other media. However, there is a great need to nanoformulate drugs to provide continuous and regulated release. There are several approaches, such as particle size reduction, crystal engineering, traditional colloidal carrier systems, nanoparticles, and nanoemulsions, that can be used for insoluble drug delivery. Many studies have been carried out, and the formulation of nanocarriers was found to be most effective [3, 4].

The science and engineering associated with the creation, formation, characterization, and application of materials and devices whose smallest functional unit in at least one dimension is on the nanometer scale is known as nanotechnology. Nanotechnology refers to the use of matter at the atomic and molecular levels, as well as the exploitation of its unique capabilities and properties at the atomic and molecular scales. Nanotechnology has a wide range of applications, and its advancement brings rapid changes to the research field and industries, such as disease detection and treatment, drug delivery, monitoring, environmental protection, the food sector, agriculture, and the construction of complex structures for electric circuits or airplanes. There has been much research done in the field of nanotechnology, but there has been less research done in the field of nanobiotechnology [5, 6].

Nanobiotechnology is the application of nanoscience to biosystems and the use of biological principles and materials to create a new device and system integrated at the nanoscale. Nanotechnology has made significant strides in biomedical and pharmacological applications. The material and device are designed with a high degree of functional specificity and allow for molecular interactions with cells and tissues. These nanomaterials are created in such a way that they interact with cells on a molecular level. These nanomaterials have been synthesized. Nanotechnology, in particular, has the potential to develop a delivery mechanism for encapsulated bioactive drug components by increasing their aqueous solubility, bioavailability, and absorption. Nanoparticles have the benefit of being stable against aggregation and gravitational separation. Furthermore, bioactive nanoparticles have greater bioavailability due to improved encapsulated bioactive adsorption and absorption in the small intestine [7, 8]. Most bioactive substances, such as drugs, are poorly soluble in water and are susceptible to degradation when exposed to environmental factors such as oxygen, light, and temperature. The encapsulation of bioactive substances with nanoparticles may improve their stability and lead to the development of novel functional drugs with the ultimate goal of improving human health [9, 10].

Nanocarriers such as SLN (solid lipid nanoparticles), polymer nanoparticles, metal nanoparticles, LNP (lignin nanoparticles), nanoemulsions, and carbon-based nanoparticles These nanocarriers are one of the tools of modern nanotechnology to overcome the poor aqueous solubility of drugs. Several methods have been developed for the synthesis of nanocarriers, such as solvent injection, droplet-phase aerosol synthesis, microemulsion, the membrane contactor method, high pressure homogenization, supercritical fluid technology, high pressure homogenization, and the sonication method. Ultrasound with an appropriate high frequency and ultrasonic power is transmitted to the liquid to create cavitation in the liquid. Ultrasonic cavitation has been widely used for laboratory-scale production of nanocarriers for nonaqueous drug delivery. In recent years, interest has grown in the use of ultrasound cavitation in the design and development of new dosage forms for nanocarrier formulations. Ultrasound irradiation has recently emerged as a novel source of energy for nanomaterial preparation. When compared to other energy sources, it has unusual reaction conditions that cannot be achieved by alternative methods. Because acoustic wavelengths are much larger than the dimensions of molecules, there is no direct interaction between ultrasound and molecular species. Ultrasonic energy deposition occurs via a process known as 'bubble implosion' or cavitation. Acoustic waves create bubbles, which are also known as cavities, when they pass through liquids. The cavities will grow as they oscillate and accumulate ultrasound energy. When these cavities collapse beyond certain critical size limits, highly concentrated energy is released in a very short period of time. The cavitation implosion is so localized and transient that the local temperature rises to 5000 K and the pressure exceeds 1000 bar. As a result, ultrasound is a major energy source for the preparation of nanomaterials. In biology, ultrasonication is commonly used to disrupt cell walls and obtain genes or proteins. However, ultrasound treatment has recently demonstrated promising properties, such as altering gene and protein expression. It was also reported that when human cells were exposed to ultrasound treatment, it was able to break the double strands of DNA. As a result, ultrasonication has been proposed for use in gene therapy. In this review, the power of ultrasound was harnessed to prepare drug-loaded nanocarriers. The goal of this chapter is to provide a broad overview and information on the benefits of ultrasound cavitation and its potential applications in pharmaceutical science. This chapter focuses on the principles of cavitation and the formation of nanoparticles. The principle and operation of ultrasound-assisted nanoparticle formulation, as well as major reviews, will be briefly described. This chapter also provides literature information on the factors that influence the rate and extent of an ultrasound-assisted nanoparticle formulation. The literature on nanoparticle formulation strategies with ultrasound cavitation and characterization techniques and various applications of nanoparticles in the solubilization of poorly water-soluble drugs and other active pharmaceutical ingredients (APIs) for controlled drug delivery and drug targeting has received special attention. Nanoparticle physicochemical properties are fundamentally different from conventional ones, particularly when nanoparticles travel through the human gastrointestinal tract (GI).

Small particles have a greater ability to pass through the mucus layer across the gut wall. This could lead to an increase in the absorption and bioavailability of nanoparticle-derived substances. As a result, consuming drugs containing nanocarriers may have specific health consequences, which may be beneficial for nonsoluble drug absorption to treat the affected organs.

Research and Development on Nanocarriers

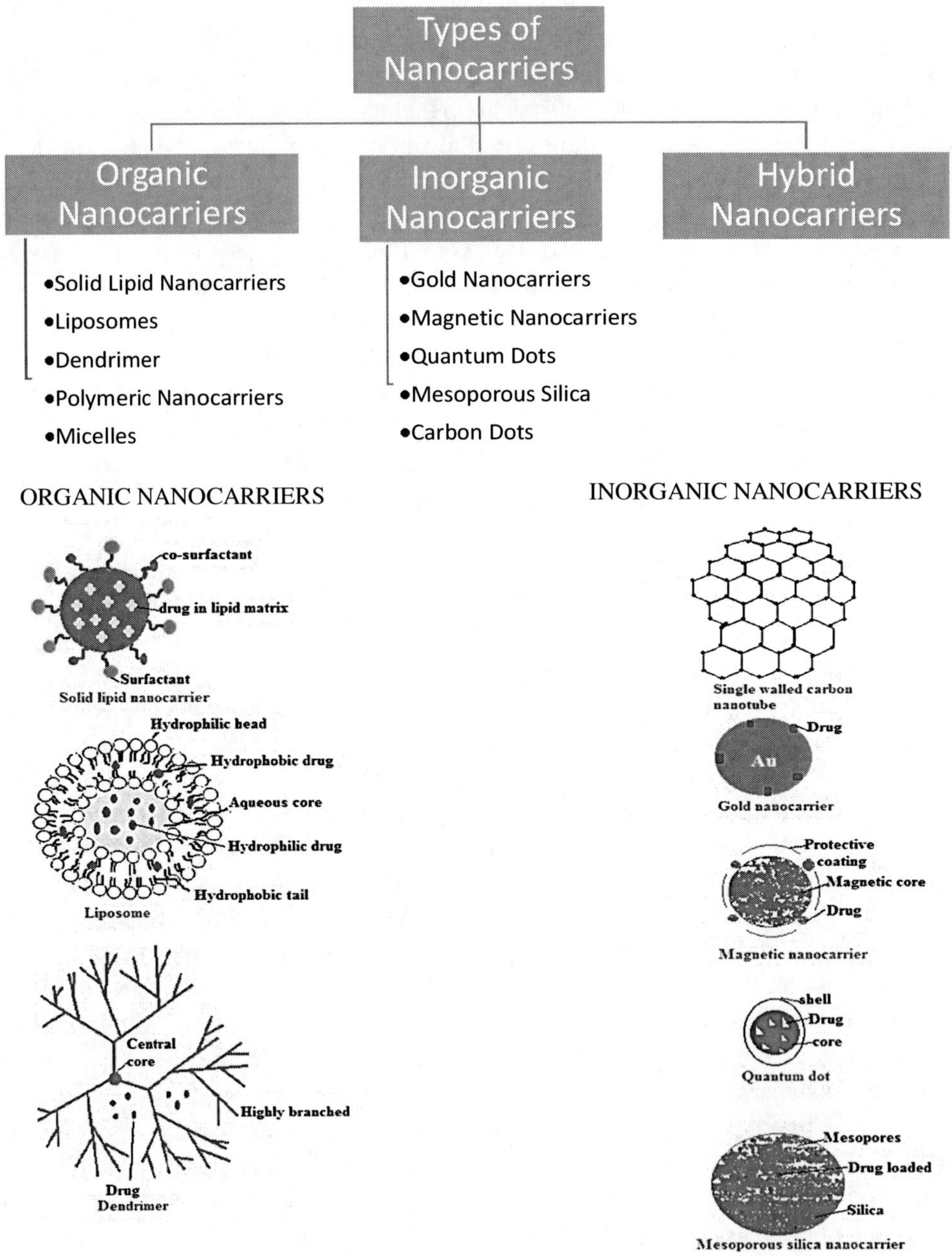

Figure 1. Types of Nanocarriers [7].

Nanocarriers are simply colloidal nanoparticles that are widely used to transport therapeutic agents or other substances to a target site. The diameter of the nanocarriers ranges between 1 and 100 nanometers (nm), with therapeutic nanocarriers having to be less than 200 nm due to the size of the body's microcapillaries. Because they are inactive and generally regarded as safe media, these nanocarriers have good biocompatibility. These nanocarriers will have a long circulation period with sustained drug release. A description of several types of nanocarriers is given in Figure 1. Modification of the physiochemical properties of nanocarriers, such as their surface, composition, and shape, can enhance their activity with decreased secondary effects. Thus, it creates a plethora of impacts in the field of drug delivery. Although a wide range of nanocarriers have been developed, only a few possess the remarkable ability to transport the drug to the targeted site. Some of the unique features of nanocarriers include the following:

- rapid dissolution velocity
- Enhanced stability
- Enhanced solubility
- Reduction in toxicity
- Sustained and targeted drug delivery

Depending on the materials and processes employed, nanoparticles can be created and built in a variety of structural shapes with varying physicochemical qualities.

Organic Nanocarriers

Solid Lipid Nanocarriers

There are a number of nanocarriers used for the delivery of less water-soluble drugs. Lipid-based drug carriers are gaining popularity among the different nanodrug carrier systems. Liposomes, self-emulsifying drug delivery systems (SEDDSs), nanostructured lipid carriers (NLCs), and solid lipid nanoparticles (SLNs) are examples of lipid-based drug delivery nanocarriers. SLNs offer several benefits over conventional drug delivery systems, including excellent biocompatibility, stability, nontoxicity, regulated drug release, and drug targeting [3]. Since then, much research has been conducted on SLNs, and certain SLN-based formulations are now commercially accessible. SLNs are made up of a biocompatible solid lipid, a surfactant, and the medication. At room and body temperatures, the medication is typically placed in a lipid matrix covered with a nontoxic surfactant and in the form of a solid dosage. Because of their ability to transport both lipophilic and hydrophilic medicines, SLNs have attracted much interest as drug carriers. Because of the exceptional biocompatibility of lipids, SLNs may be delivered by any current route. Solvent injection, supercritical fluid technology, droplet-phase aerosol synthesis, microemulsion, and membrane contactor techniques are some of the methods used to manufacture SLNs. Solid lipid nanoparticles are lipid particles consisting of high-melting-point triglycerides that can remain solid at room temperature but melt at higher temperatures. Thus, they are being used as lipid carriers to encapsulate more sensitive materials inside the solid lipid matrix.

Liposomes

Generally, for a given phospholipid formulation, longer irradiation periods and/or higher ultrasonic power resulted in a smaller mean size and a more homogeneous liposome population until most of the population reached the minimal size of a stable SUV (20–25 nm). The mechanism of ultrasound-induced liposome formation is related to transient cavitation, which induces "high energy" intervesicle collisions. During the collisions, short-lived phospholipid bilayer fragments are released. These fragments, due to the exposure of the hydrophobic regions to the aqueous phase, undergo rapid fusion and closure to rearrange into smaller vesicles. It was also suggested that extreme shear forces, induced by ultrasonic cavitation, narrow the size distribution of liposomal dispersions. Ultrasonication can be classified as a mechanical method for the production of LNPs. The phenomenon of US-induced cavitation (acoustic cavitation) consists of the formation, growth, and subsequent collapse of cavitation bubbles, which generate high heat and local pressure release. When ultrasound waves in the range of 20–150 kHz are applied to a liquid medium, the physical effects of cavitation, such as shockwaves and microjets, are predominant. On the other hand, ultrasound frequencies in the range of 150–2000 kHz mainly cause chemical effects due to the formation of hydroxyl radicals (HO·) in the local hotspots generated by cavitation. The release of local pressure and the generation of oxidizing radicals can be exploited to disintegrate and depolymerize the macromolecular structure of lignin to obtain particles on a nanoscale level. For this purpose, we sonicated an aqueous suspension of lignin (0.7%) with a US horn for 60 min at a frequency of 20 kHz (600 W) to obtain a homogeneous and stable nanodispersion of LNPs with an average diameter of 0.03 m. The water suspension of lignin (0.1% wt) was ultrasonicated at a frequency of 20 kHz, 130 W power, and 95% oscillation.

Dendrimer

Dendrimers are branched macromolecules that have a central core that gives rise to several arms. Dendrimers can be made from nucleotides, sugar molecules, and amino acids. They are multivalent, highly branched with diverse peripheral groups and have a distinct molecular weight, making them a one-of-a-kind source of drug delivery. Stepwise dendrimer synthesis can result in an irregular, well-arranged dendrimer branching pattern. The branch level added to the core is known as a generation, which, when extended, yields a large external group. With improved surface functionality, the drug molecules can be encapsulated within the core cavities via hydrophobic bonds, chemical interactions, or hydrogen bonds. Covalent bonds can also be used to attach drug molecules to terminal active groups. These dendrimers have a well-defined structure that can be fine-tuned to encapsulate a variety of drugs. Dendrimers with a single generation, on the other hand, can cause disassociation of attached molecules. The main mechanism of drug-dendrimer interaction is the formation of physical and chemical bonds. This dendrimer can be used in magnetic resonance imaging scanning, gene delivery, drug delivery, antiviral and vaccine delivery, and other applications. It is also useful for connecting with producers. Various anticancer drugs, such as cisplatin and doxorubicin, have been extensively linked with dendrimers to produce enhanced anticancer activity.

Polymeric Nanoparticles

Polymeric nanoparticles are colloidal, solid nanoparticles made from any biodegradable material made from polymers. It could be a reservoir type (nanocapsules) that dissolves or disperses drug molecules within the polymer matrix or a matrix type (nanospheres) that entraps drug molecules within the polymer matrix. Both types have the ability to chemically conjugate or adsorb the drug on its exterior. The biodegradation of polymeric nanocarriers inside the human body produces monomers that are easily degraded by metabolic pathways. Polymeric nanocarriers can be made from both natural (such as chitosan, gelatin, albumin, collagen, and alginate) and synthetic polymers (lactic-co-glycolic acid, polyethylene glycol, polyglutamic acid, and polycaprolactone). In terms of stability, drug payload, half-life in systemic circulation, and sustained drug release, this polymeric nanocarrier outperforms other nanocarriers.

Micelles

The colloidal aggregates formed by mixing detergent and water are referred to as'micelles' (liquid colloid). They are amphiphilic molecules with a nonpolar tail (hydrophobic) that faces the center and a polar head (hydrophilic) that contacts the external solvent. Similarly, an amphiphilic or detergent molecule in a nonpolar solvent can produce an inverse micelle with the head facing the center and the tail facing the outside. The size and shape of the micelle nanoparticles produced are determined by the solution conditions (temperature, ionic strength, and pH) and the nature of the amphiphilic molecule. The proper micelle formation is determined by the critical micellar concentration (surfactant concentration). Proper micelle formation will not occur below the critical micellar concentration. Polymeric micelles are formed in certain solvents alongside amphiphilic molecules by two copolymers.

Inorganic Nanocarriers

Gold Nanocarriers

Gold, a noble metal, has inspired nanotechnology to create gold nanoparticles that can be used in a variety of applications, such as photoacoustic imaging, chemotherapy, surface-enhanced resonance spectroscopy, gene therapy, and photothermal therapy. Gold nanoparticles can be synthesized using both top-down and bottom-up approaches. There are various anisotropes of gold nanoparticles, such as nanostars, nanorods, nanocages, nanoshells, and nanoprisms. Among the distinguishing properties of gold nanocarriers, the optical property is the most important one that draws them to the biomedical field. It allows various biomolecules, such as enzymes, carbohydrates, fluorophores, peptides, proteins, and genes, to bind to gold nanoparticles. This allows molecules to be transported effectively within the cell, overcoming the barriers that are present.

Magnetic Nanocarriers

Magnetic nanocarriers are typically composed of a magnetic core. Metal nanoparticles are more magnetic than metal oxide nanoparticles. This magnetic property, as well as its modified properties, allow it to be used for biosensing applications. Superparamagnetic nanoparticles are more susceptible to magnetic fields than paramagnetic nanoparticles. Because of their magnetic resonance, polymer-coated superparamagnetic iron oxide nanoparticles have been widely used in molecular imaging as contrast agents.

Quantum Dots

Atoms in the periodic table II-VI (Se, Zn, Te, and Cd) or III-V (In, As, and P) element groups are used to create quantum dots. They are colloidal nanocrystals that donate energy. The size of the quantum dot affects light emission in the UV-near IR region, with smaller quantum dots (2 nm) emitting blue fluorescence and larger quantum dots (5 nm) emitting red fluorescence. This optical property, combined with longer light emission and less photobleaching, distinguishes it from other organic dyes. As a result, it can be used for cell imaging. In mice, for example, a quantum dot–peptide conjugate is used for in vivo tumor vasculature targeting.

Mesoporous Silica Nanocarriers

Although mesoporous silica nanoparticles have been identified as promising drug carriers, their successful delivery to target tissues following systemic administration remains a challenge. Two different ultrasound frequencies, 0.5 and 1.6 MHz, with pressures ranging from 0.5 to 4 MPa, were used to drive real-time cavitation activity. The optimal ultrasound conditions were used to deliver dye-loaded nanoparticles as a model for drug-loaded nanocarriers, with fluorescence microscopy used to assess extravasation. The same nanoparticles were then coinjected with submicrometric polymeric cavitation nuclei to promote cavitation activity and lower the required in situ acoustic pressure for extravasation. The combination's overall cavitation energy and penetration were compared to mesoporous silica nanoparticles alone.

Carbon Nanotubes

The unique biological and physicochemical properties of CNTs make them an ideal and promising drug delivery vehicle. These are hollow tube-like structures made of graphene sheets rolled together at discrete and specific angles. Carbon nanotubes can be single-walled or multiwalled, depending on how many graphene sheets are rolled together. The cross section of these tubes can be 0.4–100 nm in diameter, with a length of a thousand times the diameter. Carbon nanotubes have a wide range of applications in drug delivery due to their unique properties, such as their high aspect ratio, ultralight weight with high surface area, nanosized needle structure, and distinguished chemical, thermal, mechanical, and electrical properties.

Hybrid Nanocarriers

Hybrid nanocarriers are nanocarriers that combine two or more organic and inorganic nanocarriers. It consists of organic–inorganic, inorganic–inorganic, and multicomponent systems. Examples of hybrid nanocarriers include lipid-polymer hybrids and ceramic-polymer hybrids. Combining two nanoparticles results in the dual nature of both nanoparticles, greatly enhancing their properties. Organic nanocarriers, such as liposomes, suffer from internal solution leakage as well as decreased stability. This allows it to be easily removed from the circulating blood. As a result, further stabilization makes it suitable for drug delivery. The hybrid nanocarrier system compensates for these drawbacks. The site of action, type of drug to be conjugated, physiological barriers during drug delivery, and nanocarrier stability and solubility all influence the choice of nanocarriers. The ultimate goal of selecting suitable nanocarriers for drug delivery is to increase therapeutic agent bioavailability with minimal or no side effects.

Mechanism of Cavitation-Assisted Nanocarrier Formulation

Ultrasound is made up of pressure waves that have a frequency that is higher than what humans can hear, usually above 20 kHz. Low-frequency ultrasound (frequencies less than 1 MHz), medium-frequency ultrasound (15 MHz), and high-frequency ultrasound (frequencies higher than 5 MHz) are the three types of ultrasound.

Ultrasonic waves propagate as mechanical longitudinal waves in soft tissue, causing alternating regions of compression and rarefaction in the medium. These density and pressure oscillations can have a number of biological consequences, which are discussed further below. Because of its cost-effectiveness, safety, noninvasiveness, and simplicity, ultrasound technology has been widely used in the medical field for both diagnostic and therapeutic purposes. Ultrasound parameters such as frequency and intensity are chosen for diagnostic applications in such a way that they produce minimal biological effects while allowing the production of images with high spatial and temporal resolution. Although ultrasound has many therapeutic applications, this review will concentrate on ultrasound applications for nanocarrier formulation. A description of the various stages of cavitation bubble formation is given in Figure 2.

There are two ultrasound parameters that are frequently manipulated to produce desired biological effects. Frequency and intensity of acoustics Reduced acoustic frequency increases penetration depth and the likelihood of cavitation, whereas increased intensity may control the amount of energy that can reach a target site. Other factors, such as exposure time and pulsing regime (i.e., duty cycle), can also be changed to produce a variety of beneficial biological effects.

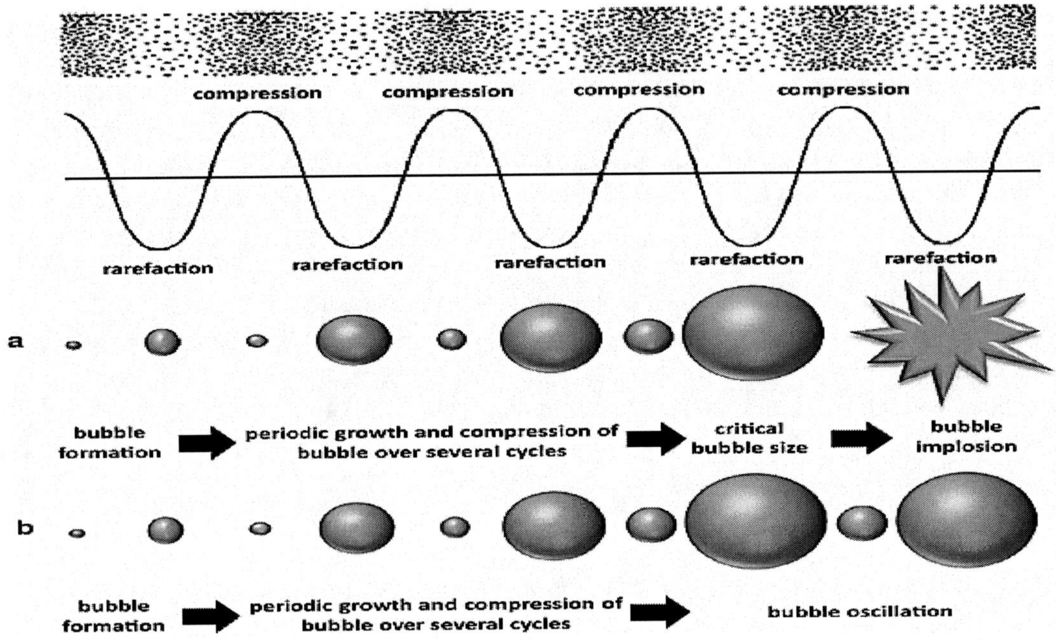

Figure 2. Stages of Cavitation Bubble Formation [7].

Methods of Nanocarrier Formation

High-Pressure Homogenization

In high-pressure homogenization, a coarse emulsion containing large oil droplets is initially prepared by blending the oil and water phases using a high-shear mixer before homogenizing them. During the subsequent homogenization process, the homogenizer forces the coarse emulsion to pass through narrow gaps under high pressures (3–20 MPa) to form a fine emulsion containing small droplets through a combination of shear, impact, and cavitation. The droplet size of the resulting fine emulsion depends on the homogenization pressure and the number of cycles used. The droplet size and size distribution of emulsions decrease with higher homogenization pressures and the number of cycles used.

Microfluidizer

To form nanocarriers, microfluidization employs extremely high pressures (up to 150 MPa). A coarse emulsion is prepared first, similar to high-pressure homogenization, and then fed into the inlet of a microfluidizer. However, in the interaction chamber of a microfluidizer, the emulsion is separated into two opposing streams. The emulsion's two streams can collide at high velocity (300 m/s) and very high pressures, generating large, intense disruptive forces that form small droplets. Several researchers have investigated the formation of nanocarriers via microfluidization and some of the major factors influencing the droplet size of microfluidized nanocarriers. The effect of emulsion composition and microfluidization conditions on the formation of nanocarriers was investigated. They discovered that increasing microfluidization pressures and cycles reduced the particle size of nanocarriers.

Ultrasonication

Ultrasound waves with a frequency greater than 20 kHz are used to generate imploding cavitation bubbles, which cause intense shock waves to break up large droplets into small droplets. The ultrasonic device includes a sonication probe that generates ultrasound waves to homogenize a preprepared coarse emulsion or a separate mixture of oil and water to reduce the size of the oil droplets. The droplet size is determined by the sonication time and energy used. The emulsion was first sonicated at 40 kHz for 8 minutes to form an emulsion. To form small droplets, the emulsion was sonicated at 40 kHz and then irradiated at 200 kHz and 1 MHz. The emulsion was first sonicated at 40 kHz for 8 minutes, resulting in an emulsion with a droplet size of 232 nm. When subjected to another cycle of sonication at 200 kHz, the droplet size decreased to 100 nm but did not decrease further at 1 MHz. This demonstrated that higher frequencies (> 200 kHz) aid in the breakdown of large droplets into smaller droplets. Figure 3 depicts a schematic illustration of the mechanical devices used to produce nanocarriers.

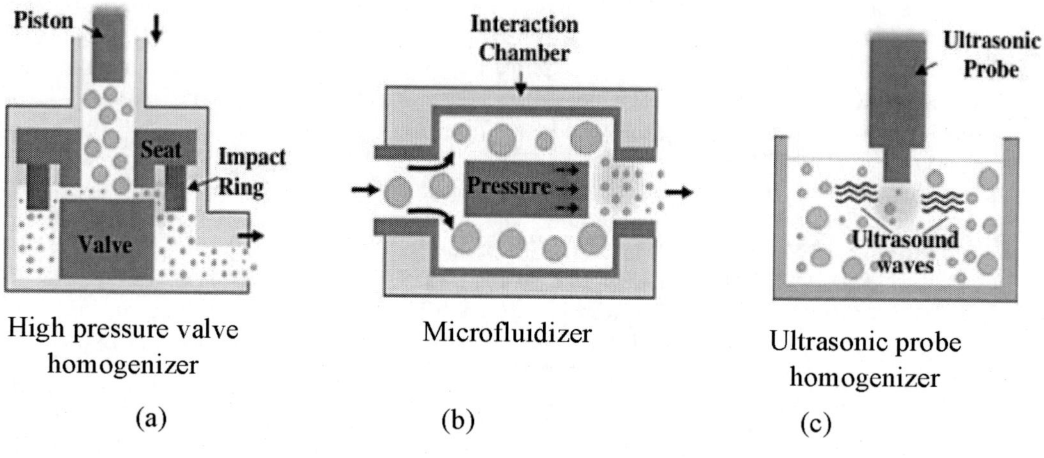

Figure 3. Schematic illustration of mechanical devices used to produce nanocarriers [10].

Conclusion

This paper reviews the rapidly evolving technology of combining ultrasound with various nanocarriers to treat affected organs with poorly soluble drugs. In vitro and in vivo studies on the efficacy of ultrasound-responsive nanocarriers have been reported. However, obstacles must be overcome for ultrasound-responsive nanocarriers to be successful in clinical settings. More research is needed to closely monitor the biodistribution and pharmacokinetics of ultrasound-responsive nanocarriers. Continuous technological advancements are required to create more biocompatible and effective ultrasound-responsive nanocarriers. Both the ultrasound parameters and the nanocarrier specifications must be optimized for successful use in future clinical trials.

References

[1] Kumar, R., Singh, A. K., & Garg, N. (2019). Acoustic cavitation assisted hot melt mixing technique for solid lipid nanoparticles formulation, characterization, and controlled delivery of poorly water soluble drugs. *Journal of Drug Delivery Science and Technology*, *54*, 101277. https://doi.org/10.1016/j.jddst.2019.101277.

[2] Kumar, R., Singh, A. K., Garg, N., & Siril, P. F. (2018). Solid lipid nanoparticles for the controlled delivery of poorly water soluble nonsteroidal anti-inflammatory drugs. *Ultrasonics Sonochemistry*, *40*, 686–696. https://doi.org/10.1016/j.ultsonch.2017.08.018.

[3] Tang, S. Y., Tan, K. W., & Manickam, S. (2012). Ultrasound Cavitation as a Green Processing Technique in the Design and Manufacture of Pharmaceutical Nanoemulsions in Drug Delivery System. *John Wiley & Sons, Inc. EBooks*, 153–208. https://doi.org/10.1002/9781118287705.ch7.

[4] Verdini, F., Gaudino, E. C., Canova, E., Tabasso, S., Behbahani, P. J., & Cravotto, G. (2022). Lignin as a Natural Carrier for the Efficient Delivery of Bioactive Compounds: From Waste to Health. *Molecules*, *27*(11), 3598. https://doi.org/10.3390/molecules27113598.

[5] Kumar, R., Singh, A. K., & Garg, N. (2019a). Acoustic Cavitation-Assisted Formulation of Solid Lipid Nanoparticles using Different Stabilizers. *ACS Omega*, *4*(8), 13360–13370. https://doi.org/10.1021/acsomega.9b01532.

[6] Chamundeeswari, M., Jeslin, J., & Verma, M. L. (2019). Nanocarriers for drug delivery applications. *Environmental Chemistry Letters*, *17*(2), 849–865. https://doi.org/10.1007/s10311-018-00841-1.

[7] Sharma, P. K., Dorlikar, S., Rawat, P., Malik, V., Vats, N., Sharma, M., Rhyee, J., & Kaushik, A. (2021). Nanotechnology and its application: a review. *Elsevier EBooks*, 1–33. https://doi.org/10.1016/b978-0-12-818154-6.00010-x.

[8] Paris, J. L., Mannaris, C., Cabañas, M. V., Carlisle, R., Manzano, M., Vallet-Regí, M., & Coussios, C. C. (2017). Ultrasound-mediated cavitation-enhanced extravasation of mesoporous silica nanoparticles for controlled-release drug delivery. *Chemical Engineering Journal*, *340*, 2–8. https://doi.org/10.1016/j.cej.2017.12.051.

[9] Narvekar, M., Xue, H., Eoh, J. Y., & Wong, H. Y. (2014). Nanocarrier for Poorly Water-Soluble Anticancer Drugs—Barriers of Translation and Solutions. *Aaps Pharmscitech*, *15*(4), 822–833. https://doi.org/10.1208/s12249-014-0107-x.

[10] Awad, N., Paul, V. K., AlSawaftah, N. M., Ter Haar, G., Allen, T. M., Pitt, W. G., & Husseini, G. A. (2021). Ultrasound-Responsive Nanocarriers in Cancer Treatment: A Review. *ACS Pharmacology & Translational Science*, *4*(2), 589–612. https://doi.org/10.1021/acsptsci.0c00212.

Chapter 26

The Synergistic Effect of Graphene Oxide in Polymer Flooding with Surfactant for Enhanced Oil Recovery

Bhuvanesh M.[1,3,4]
Ameelia Roseline A.[2]
and Kalaiselvam S.[1,3,4*]
[1]Department of Applied Science and Technology, AC Tech Campus, Anna University, Chennai, India
[2]Department of Electronics and Communication Engineering, Panimalar Engineering College, Chennai, India
[3]Centre for Industrial Safety, Anna University, Chennai, India
[4]Department of Mechanical Engineering, CEG Campus, Anna University, Chennai, India

Abstract

The recovery of oil in the existing reservoirs has to be increased to meet the rising energy demand, which is possible using enhanced oil recovery processes. The incorporation of nanoparticles in polymers has become trending in the development of nanotechnology for enhanced oil recovery processes. The viscosity of polymers and flooding process efficiency decrease because of polymer deformation. To prevent such deformation, nanoparticles are added to the injection fluid during the polymer flooding process. In this study, graphene oxide (GO) nanosheets were used, which are best known for their unique properties. They are first synthesized and then characterized. This study investigates the rheology, interfacial tension (IFT), and contact angle of GO sheets in the oil recovery process by enhancing the performance of the polymer hydrolyzed polyacrylamide (HPAM) and its synergistic effect with the presence of the cationic surfactant cetyltrimethylammonium bromide (CTAB). The performance of polymer flooding along with the surfactant using sand packs was investigated. The results showed significant improvements in the nanofluid viscosity, interfacial interactions, and wettability. The GO/HPAM/CTAB system exhibited excellent effects in decreasing IFT to a significant level and thereby attained an additional 10.62% of oil recovery attained using polymers in the core flooding experiment.

Keywords: nanoparticles, graphene oxide, polyacrylamide, surfactant, enhanced oil recovery

---

* Corresponding Author's Email: kalai@annauniv.edu.

In: Sustainability in Chemical Processes through Digitalization and Green Chemistry Approaches
Editors: Susarla V. A. R. Sastry, Shravan Kumar and Ashish Kapoor
ISBN: 979-8-89113-257-3

Nomenclature

SG	Specific gravity
MW	Molecular weight
XRD	X-ray Diffraction
FTIR	Fourier Transform Infrared Spectroscopy
EOR	Enhanced Oil Recovery
CEOR	Chemical Enhanced Oil Recovery
GO	graphene oxide
HPAM	partially hydrolyzed polyacrylamide
CTAB	Cetyl trimethyl ammonium bromide
NPs	nanoparticles
IFT	Interfacial tension in mN/m
CA	contact angle in degrees
$\Delta\rho$	density difference between the fluids in kg/m^3
ω	rotational velocity in rad/s

1. Introduction

Crude oil is considered to be the essential primary source of energy, considering the rising energy demand. Due to the unpredictable change in crude prices, oil companies often prefer the recovery of unrecovered oil from already exploited reservoirs instead of drilling new wells, which is not economical. There are several methods used for the recovery of oil, namely, primary, secondary, and tertiary methods. When the amount of oil recovered is substantially reduced with primary and secondary recovery methods, the EOR method is used to enhance production. Among the available EOR methods, the chemical EOR method has good potential for oil recovery because it uses polymers, surfactants, or combinations, although it has encountered challenges in achieving its objectives, including costly chemicals and poor performance in harsh reservoir conditions [1]. Nanoparticles help address these challenges.

Nanoparticles possess great potential for improving the rheological characteristics of polymers and the interfacial tension between the displacing and displaced fluids by being added along with polymer solutions or surfactants. The wettability is altered to be more water-wet using nanoparticles. It has the ability to reduce the adsorption of surfactant onto the rock surface.

It is evident from the literature that there are EOR studies carried out using surfactants and various kinds of nanoparticles in CEOR. Commonly used surfactants are SDS, CTAB, and TX-100 [2]. Some of the nanoparticles, such as ZnO, ZrO_2, and SiO_2, were used along with the abovementioned surfactants, and recovery studies were carried out.

Despite the attempts to show the effectiveness of various nanoparticles in EOR processes using surfactants, no experimental results have been reported previously in the literature with graphene oxide nanosheets [3, 4], which are said to possess unique surface properties as nanoparticles along with HPAM and surfactant CTAB. Here, the cationic surfactant CTAB, which possesses good stability [5], was selected for use along with the nanofluid containing

GO and HPAM. The IFT and contact angle measurements were carried out with the prepared nanofluid, and finally, with the flooding experiment, the ultimate oil recovery was obtained.

2. Experimental Section

2.1. Materials and Methods

Graphite flakes, sulfuric acid (H_2SO_4), citric acid monohydrate ($C_6H_8O_7$).H_2O, orthophosphoric acid (H_3PO_4), potassium permanganate ($KMnO_4$), hydrogen peroxide (H_2O_2), sodium chloride (NaCl), and cetyltrimethylammonium bromide ($C_{19}H_{42}BrN$) were purchased from Merck and SRL India. Crude oil (SG = 0.882) and HPAM (MW = 15 x 10^6 Da) powder were purchased from ONGC, Chennai. All the above mentioned materials were used as received without any further purification.

Graphene oxide sheets were synthesized using Hummer's method, as mentioned in the literature [6]. The morphology of the synthesized GO sheets was determined using a scanning electron microscope. The size of the synthesized particles was measured using an Anton Paar Litesizer 500 unit. The structural studies were carried out by X-ray diffraction (XRD, Bruker's D2 PHASER) using Cu K radiation in the range of 10° to 80°. Bruker Alpha Fourier transform infrared spectroscopy was used to measure the FTIR spectra of synthesized GO sheets. Raman spectra were obtained for the samples with 532 nm laser light under ambient conditions.

2.2. Preparation of Nanofluids

The synthesized GO nanosheets at different concentrations (0.02, 0.04, 0.06, 0.08, and 0.1 wt%) were dispersed in distilled water to form a stable dispersion using an ultrasonic bath for 20 min. Then, HPAM powders were introduced into the above aqueous GO suspensions with a designed concentration to prepare the GO/HPAM nanofluid. Another type of nanofluid, GO/HPAM/CTAB, was prepared in HPAM solution mixed with the surfactant CTABLE To allow enough time for the nanoparticles to come into contact with the polymer, the mixtures were gently stirred using a magnetic stirrer at room temperature for 8 hours. The concentration of surfactant, 0.04 wt%, corresponds to a CMC of 500 ppm HPAM solution. By varying surfactant concentrations in a 500 ppm aqueous HPAM solution, surface tension measurements were taken with a dynamic tensiometer, through which CMC was determined.[7].

2.3. Rheological Characterization

The rheological experiments [8] for the prepared nanofluids were performed using an Anton Paar Rheometer. The measurement system comprises cone-and-plate geometry, and the gap was preset to a standard 0.08 mm. Dynamic and steady measurements were observed to analyze different concentrated suspensions. The Peltier system was used to control the temperature of the system. The viscosity of all nanofluids was measured with shear rates ranging from 100 to 1000 s^{-1} at 60 °C.

Interfacial Tension Using Vonnegut's Formula

Interfacial tension between nanofluids and crude oil was measured by a spinning drop tensiometer using the spinning drop technique, as the IFT values obtained using surfactant are low. This technique uses Vonnegut's formula for the elongated drop, and the interfacial tension can be calculated by the following equation:

$$\sigma = \frac{\Delta\rho\omega^2 R^3}{4}$$

where

R – Radius of the drop at equator (m)
ω - Rotational velocity (rad/s)
Δρ - Density difference between the fluids (kg/m^3)
σ – Interfacial tension (N/m)

All the measurements of IFT were determined by analyzing the spinning drop through a digital camera and using designated processing software.

Wettability Alteration

The wettability alteration was studied through the measurement of contact angle. First, the sandstone substrate was soaked in a crude oil-toluene solution. After aging, the contact angle was measured for an oil-wet surface. The oil-wet surface was then treated with surfactant, nanofluid, and a surfactant-nanofluid hybrid solution. Finally, with these modifications, the water contact angle on the substrate was measured.

Core Flooding Experiment

Oil displacement efficiency is the primary measurement to conclude the effectiveness of nanoparticles and the application of nanofluids in EOR. The measurement was performed using core flooding equipment consisting of components such as a core holder, displacement pump, accumulators, pressure gauge, temperature controller, and collector. A sand pack was used to fill the core holder, and different nanofluid combinations were flooded into it to determine the recovery obtained. The specifications of the sand packs used in the experiments are given in Table 1.

Table 1. Sand pack properties

	Porosity (%)	*Permeability (mD)*	*Dimension (cm)*	
			Length	Diameter
GO/HPAM	26	305	30	5
GO/HPAM/CTAB	28	330	30	5

3. Results and Discussions

3.1. Characterization of GO

The surface morphology of the synthesized nanoparticles was studied using SEM, and Figure 1 shows that there is a thin sheet-like formation. The size of the particles was found to be 200 nm, which was also confirmed using a particle size analyzer, which gave a particle size of approximately 225 nm. The XRD pattern for the GO nanosheets is shown in Figure 2. Generally, a prominent diffraction peak at $2 = 27.4°$ is exhibited by graphite. The disappearance of the above peak and appearance of a new peak at 10.4° were due to oxidation, which confirms GO formation. The spectrum in Figure 3 shows the peak at 1066 cm^{-1}, which is attributed to C-O stretching. The peaks at 1288, 1587, and 1724 cm^{-1} show the C-O-C, C-OH, and C = O stretching vibrations, respectively. The broad peak at 3448 cm^{-1} is attributed to O-H stretching and the water content present in the material.

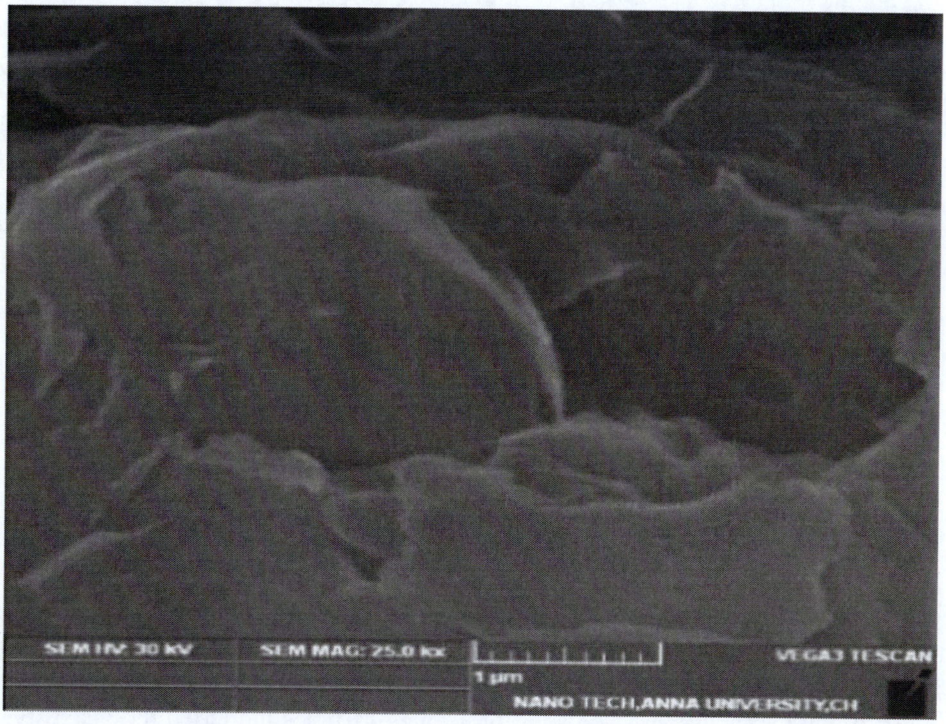

Figure 1. SEM image of GO nanosheets.

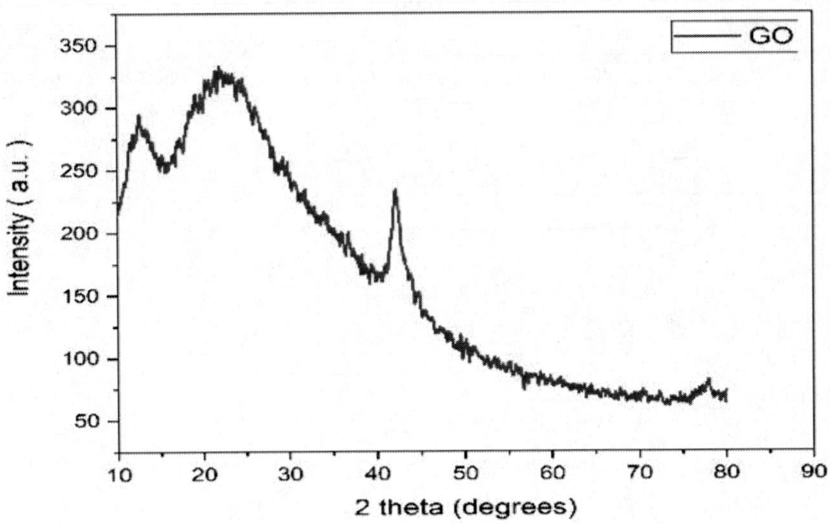

Figure 2. XRD pattern of synthesized GO (JCPDS Card No. 01-089-8491).

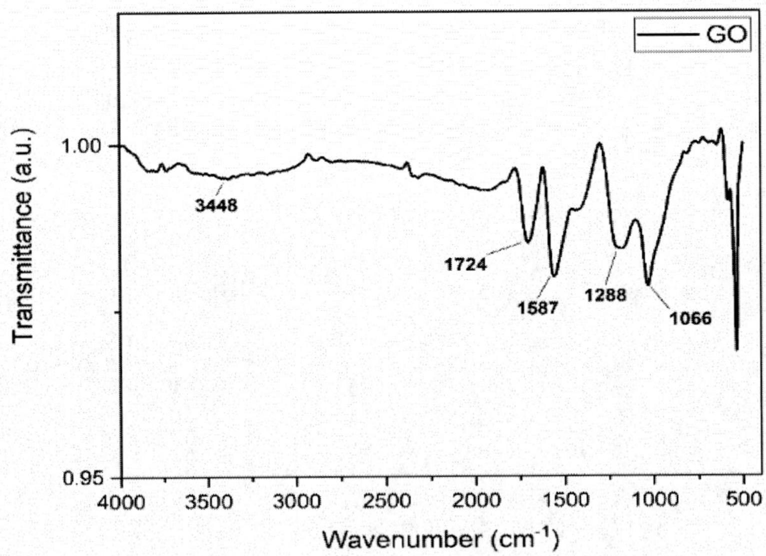

Figure 3. FTIR spectrum of GO.

3.2. Viscosity Measurements

The viscosity results of the nanofluids are shown in Figure 4. The fluids behave like non-Newtonian shear-thinning fluids over the shear rate range of 100 to 1000 s^{-1}. The viscosity of nanofluids with polymer and with polymer/surfactant is significantly higher than that of conventional polymer solutions. The addition of nanoparticles increased the viscosity, which enhances conventional polymer flooding and gives a good capillary number and mobility ratio [8, 9].

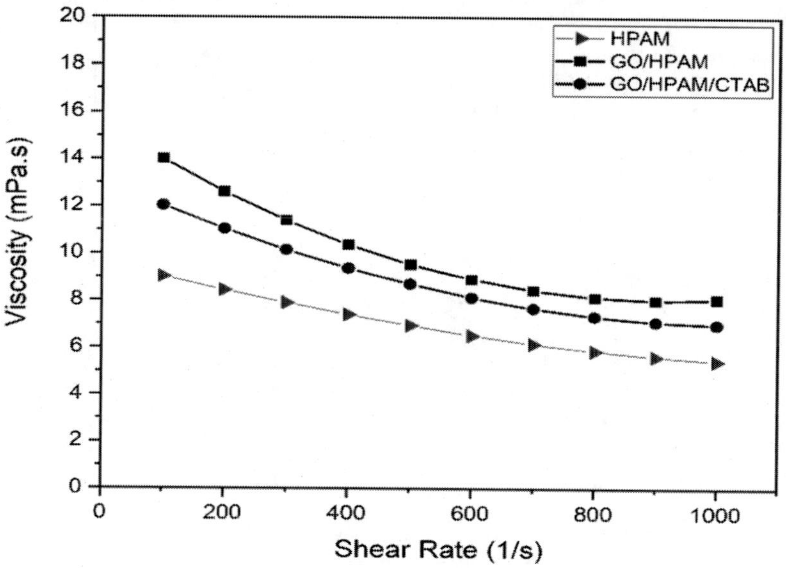

Figure 4. Viscosity of polymer nanofluids containing 0.1 wt% GO at 60 °C with and without CTAB for varying shear rates from 100 s^{-1} to 1000 s^{-1}.

3.3. IFT Measurements

The IFT measurements of the water/oil system for different nanofluids are illustrated in Figure 5. As shown, there is an increasing trend in IFT with an increase in nanoparticles after the critical concentration of nanoparticles with HPAM when the surfactant CTAB is not added. In the presence of the surfactant CTAB, it reaches a minimum value of 0.04%. A further increase in nanoparticle concentration increased the IFT values. The surface behavior of nanoparticles is changed by the adsorption of CTAB onto their surface. As the concentration of nanoparticles relative to CTAB is high, all CTAB molecules are adsorbed by the nanoparticles present. Therefore, some of the nanoparticles remain uncoated with CTAB, thus increasing the IFT.

3.4. Contact Angle Measurements

Figure 6 shows the contact angle variations with varying nanoparticle concentrations in the presence and absence of the surfactant CTAB. The contact angle of the water/rock surface/crude oil system was approximately 110°, which is said to be an oil-wet condition. As shown in the figure, the contact angle decreases up to an NP concentration of 0.04 wt% in both conditions, with and without the presence of surfactant. Beyond the optimum concentration, there is an increase in contact angle values. The main mechanism in the alteration of wettability is the development of a layer called the wedge layer at the crude oil, rock surface, or CTAB system.

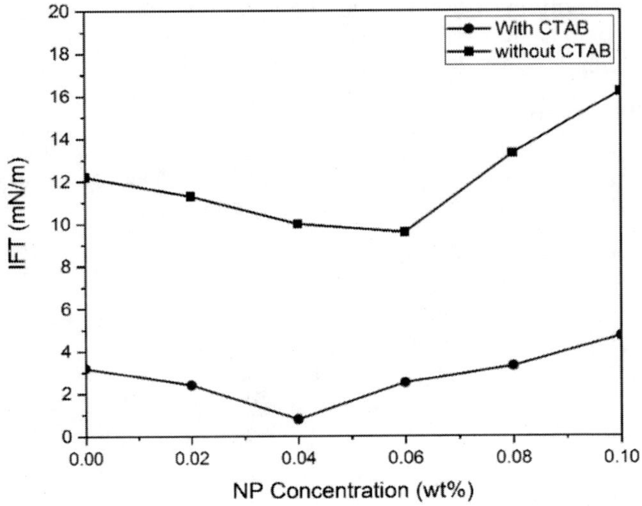

Figure 5. IFT measurements with varying NP concentrations of polymer nanofluid with CTAB (0.04 wt%) and without CTAB.

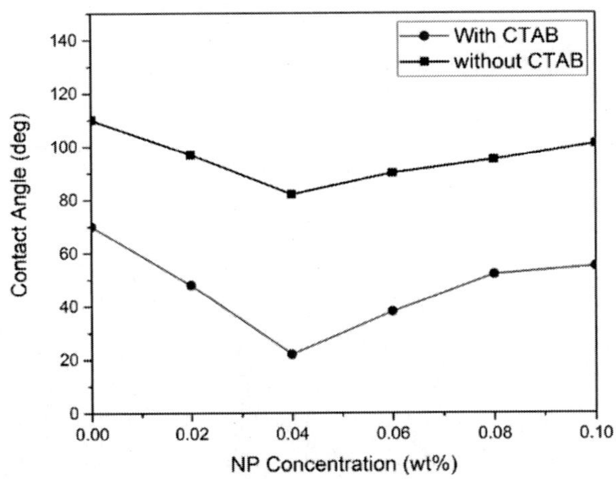

Figure 6. CA measurements with varying NP concentrations of polymer nanofluid with 0.04 wt% CTAB and without CTAB.

3.5. Oil Recovery

The mixing of GO nanosheets and the surfactant CTAB may minimize the adsorption of CTAB on rock surfaces, resulting in increased effectiveness of the surfactant in EOR projects. This indicates that nanofluid prepared using surfactant may result in better oil recovery and wettability alteration. To investigate the performance of the HPAM/GO and HPAM/GO/CTAB systems in improving oil recovery, both dispersions were tested for their oil displacement performance using the setup of core flooding. The concentrations of HPAM and GO used in this procedure are 500 ppm and 0.04 wt%, respectively. Both dispersions were injected into the sand pack to displace the crude oil. The flooding and recovery procedures are carried out as mentioned in the literature [10]. The recovery obtained is plotted with injection pore volume for both nanofluid dispersions, which is illustrated in Figure 7. As seen in the figure, the ultimate recovery obtained with the HPAM/GO/CTAB system is higher than the recovery obtained with a nanofluid containing only polymers, HPAM/GO. This might be due to the change in wettability and IFT caused by the addition of the surfactant CTAB.

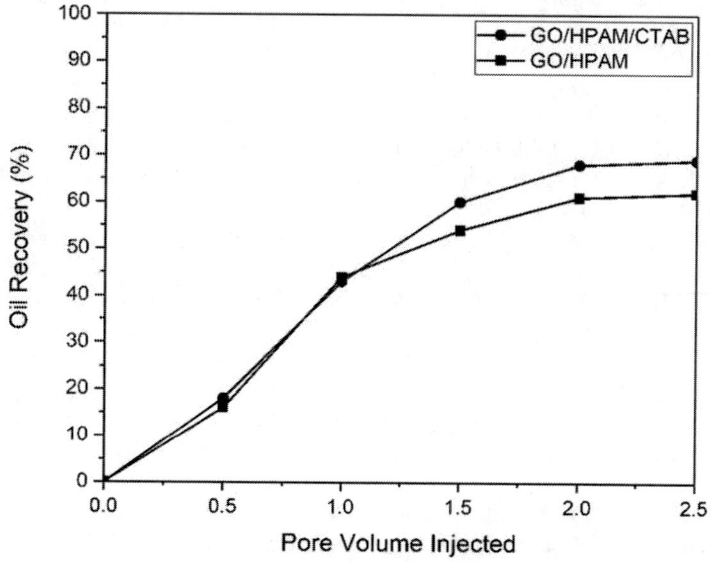

Figure 7. Production profile of polymer nanofluid flooding with and without CTAB as a function of injected PV.

Conclusion

In this study, the effect of the surfactant CTAB on GO nanosheets present in a polymer fluid was studied. The characterization of GO nanosheets has been reported. IFT and contact angle measurements were carried out for nanofluids with and without surfactant. It was observed that CTAB-assisted GO nanosheets were capable of significantly improving the wettability condition, which helped in achieving higher oil recovery. EOR core flooding experiments were performed, and the ultimate oil recovery was compared for both nanofluid dispersions, from which it was found that an additional 10.62% recovery was obtained for the HPAM/GO/CTAB system, more than that of the HPAM/GO dispersion.

Funding Details

The author, Bhuvanesh M., likes to thank Anna University, Chennai, for supporting this research work through the "Anna Centenary Research Fellowship-ACRF" (CFR/ACRF-2020/20145291143/Ph.D./AR15) provided by the Centre for Research (CFR). He also likes to thank the Department of Applied Science and Technology, Anna University, for providing research facilities established through DST-FIST (Order No. SR/FST/ET-1/2017/44).

References

[1] Aliabadian, E., Sadeghi, S., Rezvani Moghaddam, A., Maini, B., Chen, Z., & Sundararaj, U. (2020). Application of graphene oxide nanosheets and HPAM aqueous dispersion for improving heavy oil recovery: Effect of localized functionalization. *Fuel*, *265*. https://doi.org/10.1016/j.fuel.2019.116918.

[2] Dinesh, R., Karuppasamy, P., & Kalaiselvam, S. (2022). An experimental investigation and comprehensive modeling of thermal and rheological behavior of graphene oxide nano platelets suspended mineral oil nano lubricant. *Journal of Molecular Liquids*, *347*, 118357. https://doi.org/10.1016/j.molliq.2021.118357.

[3] Haruna, M. A., Pervaiz, S., Hu, Z., Nourafkan, E., & Wen, D. (2019). Improved rheology and high-temperature stability of hydrolyzed polyacrylamide using graphene oxide nanosheet. *Journal of Applied Polymer Science*, *136*(22). https://doi.org/10.1002/app.47582.

[4] Hirasaki, G. J., Miller, C. A., & Puerto, M. (2011). Recent Advances in Surfactant EOR. *SPE Journal*, *16*(04), 889–907. https://doi.org/10.2118/115386-PA.

[5] Imran Hussain, S., Ameelia Roseline, A., & Kalaiselvam, S. (2019). Enhancement of thermal conductivity and thermal stability of capric-lauric acid eutectic phase change material using carbonaceous materials. *Materials Research Express*, *6*(11). https://doi.org/10.1088/2053-1591/ab4768.

[6] Khoramian, R., Ramazani S. A., A., Hekmatzadeh, M., Kharrat, R., & Asadian, E. (2019). Graphene Oxide Nanosheets for Oil Recovery. *ACS Applied Nano Materials*, *2*(9), 5730–5742. https://doi.org/10.1021/acsanm.9b01215.

[7] Maghzi, A., Kharrat, R., Mohebbi, A., & Ghazanfari, M. H. (2014). The impact of silica nanoparticles on the performance of polymer solution in presence of salts in polymer flooding for heavy oil recovery. *Fuel*, *123*, 123–132. https://doi.org/10.1016/j.fuel.2014.01.017.

[8] Mohajeri, M., Hemmati, M., & Shekarabi, A. S. (2015). An experimental study on using a nanosurfactant in an EOR process of heavy oil in a fractured micromodel. *Journal of Petroleum Science and Engineering*, *126*, 162–173. https://doi.org/10.1016/j.petrol.2014.11.012.

[9] Panahpoori, D., Rezvani, H., Parsaei, R., & Riazi, M. (2019). A pore-scale study on improving CTAB foam stability in heavy crude oil−water system using TiO2 nanoparticles. *Journal of Petroleum Science and Engineering*, *183*. https://doi.org/10.1016/j.petrol.2019.106411.

[10] Pei, H., Shu, Z., Zhang, G., Ge, J., Jiang, P., Qin, Y., & Cao, X. (2018). Experimental study of nanoparticle and surfactant stabilized emulsion flooding to enhance heavy oil recovery. *Journal of Petroleum Science and Engineering*, *163*, 476–483. https://doi.org/10.1016/j.petrol.2018.01.025.

Chapter 27

The Green Synthesis of Silver Nanoparticles from the Residual Biomass of *Spirulina Platensis* and Its Applications

Bharathidasan Kanniappan
Naveen M.
and Perarasu Thangavelu*
Thermal and Bio Analysis Laboratory, Department of Chemical Engineering, Alagappa College of Technology Campus, Anna University, Chennai, Tamil Nadu, India

Abstract

Spirulina platensis sp. is used for the extraction of many value-added compounds, such as proteins, lipids, and omega-3 fatty acids. After extraction, this species is still found to be rich in flavonoids, terpenoids, phenolics, and other biological compounds, which act as reducing agents for metal ions. These residues are used in this study as a potential source for nanostructures, resulting in nanoparticles with optical and magnetic properties. Synthesized nanoparticles have fluorescent properties, which are suitable for many biosensing applications. Biotiol groups are nonessential amino acids that are found at higher levels in the blood and can cause serious health issues, so they must be detected in human blood. Therefore, we have developed a simple technique for the detection of biothiol groups in blood by using a fluorescent spectrophotometer. The antimicrobial properties of the synthesized nanoparticles were also studied.

Keywords: *spirulina platensis*, extraction, metal ions, nanoparticles, biosensing, antimicrobial, blood

1. Introduction

Phycocyanin, a proteinaceous pigment derived from *Nostoc linckia*, was used as a reducing agent in the biological production of silver nanoparticles [1]. Cunlan Guo *et al.* (2011) focused

* Corresponding Author's Email: vtparasu@annauniv.edu.

In: Sustainability in Chemical Processes through Digitalization and Green Chemistry Approaches
Editors: Susarla V. A. R. Sastry, Shravan Kumar and Ashish Kapoor
ISBN: 979-8-89113-257-3

on the protein-directed synthesis of silver nanoparticles. He reported on the synthesis of new water-soluble, stable, fluorescent Ag clusters via a facile, green method using denatured bovine serum albumin as a stabilizing agent. This is the first study to examine the production of nanoparticles utilizing deoiled algal biomass from dairy effluent. *A. dimorphus is* a green microalga with the potential to produce biofuels as well as being thermotolerant. After lipid extraction, the leftover biomass of *A. dimorphus* was utilized to create microalgal water extract, which was then used to synthesize silver nanoparticles [2]. Chen *et al.* focused on the preparation of fluorescent bovine serum albumin-confined silver nanoparticles (BSA-AgNPs), which were demonstrated to be a novel and environmentally friendly probe for the rapid detection of biothiols such as cysteine (Cyst), homocysteine (Hcy), and glutathione (GSH). By virtue of this specific response, a new, simple, and rapid fluorescent method for detecting biothiols has been developed. The proposed approach was successfully used to determine thiols in human plasma, and the recovery was 94.83–105.24%. For the first time, fluorescent metal nanoparticles are presented as novel probes for the sensitive detection of proteases. The sensing mechanism is based on trypsin digestion of the protein template of BSA-stabilized Au nanoparticles. The decrease in fluorescence intensity of BSA-Au nanoparticles caused by trypsin allows the sensitive detection of trypsin in the range of 0.01–100 g/mL. The present nanosensor for trypsin detection possesses red emission, excellent biocompatibility, high selectivity, and good stability. In addition, they demonstrated the application of the present approach in real urine samples, which suggested its potential for diagnostic purposes [3]. This study establishes a simple spectrophotometric and colorimetric method for the detection of ascorbic acid based on the growth of silver nanoparticles in bovine serum albumin-protected silver nanoparticles (BSA-AgNPs) and Ag+ mixtures. Due to the catalysis of BSA-AgNPs, ascorbic acid could reduce Ag+ to silver nanoparticles (NPs) at room temperature [4]. This study addresses fluorescent noble metal (Au, Ag) nanoparticles that have been biolabeled with bovine serum albumin (BSA) by wet chemistry. Spectroscopic and fluorescence studies have associated the involvement of pH and the composition of the reductant with the size and oxidation state of nanoclusters. Small silver nanoparticles labeled to the protein with a fluorescence emission in the red region are synthesized in the presence of a strong reducing agent and present only Ag (0) [5]. Nanoparticles are made of several metal atoms and have a diameter of less than 1 to 100 nm [6]. Some of the reported works were based on nanoparticles stabilized with proteins, enzymes, DNA, peptides, and graphene oxide. Such stabilized nanoparticles were found to have unique applications in optical sensors, biomedical devices, drug delivery, and catalytic processes. The optical properties of these protein-stabilized nanocrystals have attracted many applicative perspectives [7]. Metal nanostructures are toxic to living cells. Therefore, the synthesis and use of metal nanostructures in powder form are considered hazardous. Recovery of precious metal nanostructures after catalysis or from solutions is difficult. For these reasons, encapsulation of these highly reactive metal nanostructures in host films can be considered a potential answer to this problem. In addition, the development of metal NSs with incorporated host films is a fast-growing area due to their wide applications recyclability in various fields, such as catalysis, sensors, optics, electronics, functional and protective coatings [8]. In general, metal nanostructures are stabilized by a variety of coating materials, such as oligonucleotides, deoxyribonucleic acid, polyethylene amine, glutathione, and bovine serum albumin [9]. Nevertheless, innovations have been carried out in the areas related to raw extracts decocted from microalgal species, which also include *Spirulina platensis*, but researchers have not yet discovered a novel-based approach to extracting the residual biomass from the corresponding source of microalgal species [10]. Hence, after scaling up all the areas related to the extraction of raw biomass, it was found that the synthesis of fluorescent silver nanoparticles with residual biomass as decorators proved to be a subject of particular attention. Thus, bioconstructed silver nanoparticles (Ag NPs), which are incorporated from a renewable protein, are in exponential demand, as reported [11]. Thiols, majorly prioritized as cysteine (Cyst), homocysteine (Hcy), and glutathione (GSH), are engaged in capricious oxidation–reduction reactions and possess vital cellular actions involving cleansing and metabolic activities [12]. A standard proportion of biothiols in the human body is significant and acts as a critical parameter for maintaining normal health. Thus, if the measure of these biothiols (specifically Cyst, termed oxidative stress) decreases, it corresponds to ponderous health, strands of hair becoming depigmented, arthritis, cardiovascular diseases, and deliberate weakness. Measures of other vital biothiols may result in diseases such as cancer, aging, and many other skin and body-related ailments [13]. Thus, sensitive and specific detection of these biothiols proves to be of great importance. To date, many new technologies have been developed for periodic detection of these biothiols by various chemical and electrochemical processes, such as capillary electrophoresis, high-pressure liquid chromatography, and other mass spectrometric methods, but they possess sophisticated instrumentation, time-consuming methods, and nonrenewable sources for sensitive fluorometric detection [14].

Hence, the aim of this study was the biofabrication of silver nanoparticles (AgNPs) using the residual biomass of lipid-extracted *Spirulina platensis* as a protein stabilizer and utilizing it to detect a specific biothiol named cysteine (Cyst) in human blood [15]. The fluorescence intensity of protein-stabilized AgNPs was found to be quenched effectively in the presence of the corresponding biothiol cysteine (Cyst). Cysteine, a biothiol present in human blood, can be detected using nanoparticles, which proved to be a novel approach. The protein-capped AgNPs were synthesized from a renewable and bountiful algal species rich in protein, which made the approach more environmentally friendly. Further characterizations (FTIR, UV–Vis, SEM, TEM, and XRD) performed on the samples also implied similar inferences and proved the sensitized detection of cysteine by silver nanoparticles. This method was then successfully implemented for the sensitive detection of biothiols.

2. Materials and Methods

2.1. Materials

2.2.1. Microalgae Strain and Culture Medium

Spirulina platensis, which was obtained from Gandhi Gramme University, Dindugal, Tamil Nadu, was maintained in Zarrouk medium that consisted of (gL^{-1}): 16.8 $NaHCO_3$, 0.5 K_2HPO4, 2.5 $NaNO_3$, 1 K_2SO_4, 1 NaCl, 0.2 $MgSO_4.7H_2O$, 0.04 $CaCl_2$, 0.01 $FeSO_4.7H_2O$, 0.08 Na_2EDTA, and 1 ml of A5 micronutrient solution. The A5 micronutrient solution consisted of ($g\,L^{-1}$): 2.8 H_3BO_3, 1.8 $MnCl_2.4H_2O$, 0.220 $ZnSO_4.7H_2O$, 0.080 $CuSO_4.5H_2O$, and 0.02 $(NH_4)_6Mo_7O_{24}.4H_2O$ [16]. The temperature was maintained at 20°C, and illumination was provided by 40 W cool white fluorescent lights with a light intensity of 2500 lux under a 16 h:8 h (light:dark) cycle. The cells were subcultured in a 3 L reactor with modified Zarrouk medium.

The production-scale culture was then grown in a 20-liter container at ambient temperature (29–32°C). Five percent (v/v) of exponentially growing cells with a biomass concentration of 0.7 g.L^{-1} were used as inoculum. The suspended cells were harvested every 48 hours by filtration and washed once with distilled water. The harvested biomass was lyophilized at -80°C and then powdered, which was stored at 4°C for further analysis. All the abovementioned materials were purchased from Sisco Research Laboratory Pvt. Ltd., Chennai, Tamil Nadu, India.

2.2. Methodology

2.2.1. Preparation of Residual Biomass Extract

Dry *Spirulina platensis* biomass (powdered form) (5.0 g), standard chloroform ($CHCl_3$), standard methanol (CH_3OH), distilled water, and ethanol (C_2H_5OH) were used to obtain the extract. The residual biomass extract of the lipid-extracted biomass was obtained. Lipid extraction was performed according to a modified protocol of Bligh and Dyer, as reported by Ramanathan Ranjith Kumar *et al.* Dried residual biomass (5 g) was placed in a 200-ml beaker and mixed with 50 ml of distilled water. It was stirred for 10 minutes in a magnetic stirrer at 500 rpm. The mixture was sonicated in a probe-type ultrasonic generator for 15 minutes at 100 amplitude in an ON:OFF cycle, followed by heating in a hot plate stirrer at 100°C for 5 minutes at 1000 rpm. Then, the solution was centrifuged at 8000 rpm for 20 minutes under ambient conditions. The supernatant (residual biomass extract) was collected and stored at 4°C for further use.

2.2.2. Synthesis of Silver Nanoparticles (AgNps)

Residual biomass extract (50 ml), freshly prepared $NaBH_4$ (5 mM), and silver nitrate freshly prepared at a 1 mM concentration were used. For protein-coated Ag nanocluster preparation, 10 ml of microalgae residual biomass extract was added dropwise to 90 ml of 1 mM $AgNO_3$ in a 100 ml Erlenmeyer flask. After 5 minutes, 40 L of 5 mM $NaBH_4$ was added to the flask. The mixture was incubated for 24 hours under dark conditions at room temperature for the formation of protein-coated silver nanoparticles. The reduction of silver nanoparticles was monitored by visual color changes from green to reddish brown and UV–visible spectra [17]. Upon reaction with silver nitrate, the protein structure in BSA forms a linked chain structure with silver ions attached to it. With $NaBH_4$, these silver ions are reduced, and the formed silver nanoparticles agglomerate, leading to cluster formation with protein covering the outside.

2.2.3. Characterization Techniques

After the successful addition of various volumes of silver nitrate ($AgNO_3$) to the deoiled microalgal biomass extract, silver nanoparticles (AgNPs) were formed at different concentrations, keeping the pH at 10. Furthermore, the formed silver nanoparticles were incubated on a 12 to 24 hour cycle for an increase in the concentration of AgNPs in the solution; subsequently, we inferred through various characterization techniques that the formed nanoparticles depicted the properties of nanostructured materials (generally optical, magnetic, crystalline, lustrous in nature, and many more). UV–Vis spectroscopy, fluorescence spectroscopy, transmission electron microscopy, scanning electron microscopy, and X-ray diffraction are some examples of characterization methods.

2.2.4. UV–Visible Spectroscopy Analysis

The extract was prepared and stored for further preparation of AgNPs. Ultraviolet spectroscopy was performed, and it was inferred that the graphs had a large peak at the beginning and a slight shoulder curve, indicating the protein-rich species and its potential to cap silver nanoparticles, respectively. At pH 10, we carried out the experiment in a UV–visible spectrophotometer (Shimadzu, UV1800), in which values for wavelengths (ranging from 190 nm to 900 nm), absorbance, and wavelengths were tabulated for further reference and inference. The formed silver nanoparticles were confirmed by using a UV–visible spectrophotometer (Shimadzu, UV1800). The UV results showed that there were peaks at 424 nm, and after a certain range of wavelengths, the peaks started to lessen, which indicated that there was a formation of nanoparticles in the solution, and further characterization would lead us to a pragmatic decision on their purity.

2.2.5. FTIR Analysis

Fourier transform infrared spectroscopy (FTIR) is an analytical method for obtaining an infrared spectrum of a substance's absorption or emission. The absorbance and wavelengths were plotted in the first coordinates of a Cartesian plane, by which the curves were peaked, depicting the types and percentages of groups present in the dried biomass. The groups that were present in dried, deoiled microalgal biomass were carbohydrates, proteins, amino acids, O-H-bonded groups, and many other allied groups. FTIR analysis was also carried out for the deoiled microalgal biomass extract and the formed silver nanoparticles, and the plots (absorbance vs. wavenumber) depicted the presence of nanoparticles in the solution with various inverted peaks.

2.2.6. SEM Analysis

(SEM) is a microscopy technique that uses a focused beam of electrons to scan the surface of a sample to obtain pictures. Incident electrons with atoms in the sample create a variety of signals that carry information about the sample's surface topography and composition. The findings of a scanning electron microscopic investigation of a dried silver nanocluster sample revealed the topology and composition of the sample.

2.2.7. TEM Analysis

Transmission electron microscopy gave surface-induced high-resolution images of the silver nanoparticles at the start of the characterization and after a certain period of time. It also depicts the behavior of nanoparticles on a surface rich in carbon and shows how they agglomerate after a certain interval, indicating that silver is present in nanoparticles and that some properties of the metal are reflected in the process. The above-represented images show similar morphologies; the first one was performed initially at a time of 0 hrs, whereas the second image was taken after a time interval of 30 mins to view the behavior in a more enlarged way. Silver nanoparticles are randomly distributed and short-range aggregated due to limited diffusion on the protein surface, and the particle sizes are significantly uniform in comparison with the case of dried biomass.

2.2.8. X-ray Diffraction Analysis

The silver nanoparticles (Ag NPs) in the residual biomass extract were analyzed using XRD. Figure 1 shows the acquired XRD spectrum. X-ray is a crystallographic method of identifying the atomic arrangement within a crystal in which an X-ray beam impacts the crystal and causes the light beam to scatter in many different directions. A crystallographer may create a three-dimensional image of the abundance of particles within the crystal by analyzing the orientations and intensity of these diffracted beams. This method can be used for measuring the size of atoms, the lengths and types of chemical bonds, and the atomic-scale differences among various materials, especially minerals and alloys. This approach also revealed the structure and functioning of numerous biological substances, including vitamins, medicines, proteins, and nucleic acids such as DNA. It obeys Bragg's law, and its mathematical representation is given below:

$$n\lambda = 2d \sin\theta$$

where, '$n\lambda$' represents the integral number of wavelengths,'d' represents the interplanar distance, and 'Θ' is the scattering angle.

2.2.9. Detection of Cysteine Using Silver Nanoparticles (AgNPs)

Cysteine is a major type of biothiol that is present in considerable amounts in human blood serum and helps maintain the metabolism and high heme levels of plasma. This amino acid group plays a major role in preventing the negative effects of alcohol on the human bloodline. It helps in hair pigmentation, prohibition of edema, age-growth in children, etc. and its deficiency causes the diseases listed below: liver failure, cirrhosis, false hair growth, dwarfism, and many more related to the metabolism of the human body. Mixtures (20 mL) of the protein-capped AgNPs (30 mL, 0.01 M), sodium phosphate buffer (150 mL, 20 mM, pH 10), and biothiols (cysteine 0–40 mM) were equilibrated at ambient temperature for 30 min. Subsequently, the mixtures were subjected to fluorescence measurement with excitation at 375 nm. All experiments were performed in triplicate.

2.2.10. Fluorescence Emission Spectroscopy Analysis of Silver Nanoparticles (AgNPs) and Cysteine

Fluorescence spectroscopy was performed for different concentrations of cysteine, all maintained at a pH of 10. Initially, a blank zero was taken for the experiment, followed by spectroscopy of the formed silver nanoparticles. Different samples were prepared, each containing cysteine (1 M to 40 M in concentration) and diluted to 10 ml trials in 50 mL centrifuge tubes (Tarsons), where the remaining amount was the algal extract. Then, 1 ml of the silver nanoparticle solution was added to different concentrations of already prepared cysteine, a 10-minute incubation time was given for the nanoparticles to react with the thiol group, and further fluorescence emission spectroscopy was performed on those trials, stagnating the excitation potential to approximately 375 nm and the bandwidth to 2.5 accordingly. The fluorescent spectroscopy graphs depicted that there was an increase in quenching as the concentration rose in tenths of micromolars, clearly indicating that the protein-capped silver nanoparticles (AgNPs) detected and sensitized them to the presence of cysteine.

3. Results and Discussion

3.1. Characterization of Silver Nanoparticles (Ag NPs)

For the green synthesis of silver nanoparticles from residual biomass of *spirulina platensis,* the surface plasmon resonance band of the produced silver nanoparticles is centered at 425 nm. Many studies have reported a similar surface plasmon resonance band at 420 nm. The color of the mixture changed from colorless to yellow, and a strong absorption band near 420 nm could be found in their absorption spectra owing to the localized surface plasmon resonance (LSPR) of the produced silver in aqueous solution. Silver nanoparticles have a yellowish-brown color. Similar studies showed absorbance at 420 nm and 430 nm. Another study indicated that the UV–Vis peak at 417 nm may be correlated to the characteristic surface plasmon resonance of silver nanoparticles ranging between 420 and 480 nm. The absorbance changes at 420 nm had a good relationship with ascorbic acid concentration. This is the fundamental principle behind many color-based biosensor applications, different lab-on- a-chip sensors and diatom photosynthesis.

FTIR results depicted several peaks indicating different organic groups present in the dried and powdered biomass. By identifying specific functional groups resulting in effective capping of AgNPs, FTIR spectra revealed significant peaks of proteins that are involved in AgNP production [18-20]. FTIR analysis of microalgal extract and biosynthesized silver nanoparticles was performed to identify the possible biomolecules responsible for the reduction and stabilization of silver nanoparticles and is also represented in Figure 1.

The appearance of a signal in the FTIR spectra at 3203 cm1 indicates the presence of aromatic C-H groups. The peak at 3001 cm1 represents the stretching vibration of C-H bonds in rings. The peak at 2995 cm1 indicates the O-H group of either alcohols or phenols, indicating that phenols had a role in lowering the nanoparticle concentration. The appearance of a faint signal at 2395 cm1 indicates the existence of C-H stretched alkanes. The presence of C-O stretch in carboxyl connected to the amide linking in amide is shown by the peak at 1762 cm^{-1} by Lambert, who observed that the peak at 1643 cm^{-1} refers to N-H deformation, C = N, and C = O in "N-H primary amides II", oxime, and C=O in carbonyl compounds, respectively. According to Maisara, the peaks with a range of 531–535 cm1 indicate the vibrational peaks of PO43. Furthermore, the appearance of a faint band at 648 cm1 indicates the presence of -CC-H stretching vibrations in alkynes [21–23]. The bands at 514 and 407 cm1 indicate the presence of naphthalene and the S-CN bend in thiocyanates, respectively. According to Gole, proteins will be effective in binding nanomaterials by attracting negatively charged carboxylate groups in the enzymes present in the culture supernatant or by cysteines or free amine groups in the proteins. This idea is in line with our findings, which verified the remarkable stability of the biofabricated AgNPs contained in proteins. Functional groups such as -C-O-C-, -C = C-, and -C-O-, which are found in heterocyclic compounds such as proteins, are also present in the nanoparticles' stabilizing ligands. FTIR examination confirmed the existence of proteins on the surface of the synthesized nanoparticles, demonstrating that proteins operate as a stabilizing factor and inhibit the ingathering of the reduced AgNPs. Carbonyl functional groups from amino acid and protein peptide residues have a strong affinity for silver. It has been proposed that the stability of AgNPs formed in cell-free culture supernatants is due to the presence of a protein stabilizing factor that inhibits nanoparticle aggregation". Fourier transform infrared spectroscopy was performed for the formed silver nanoparticles, and the plots (absorbance vs. wavenumber) depicted the presence of nanoparticles in the solution with various inverted peaks as follows:

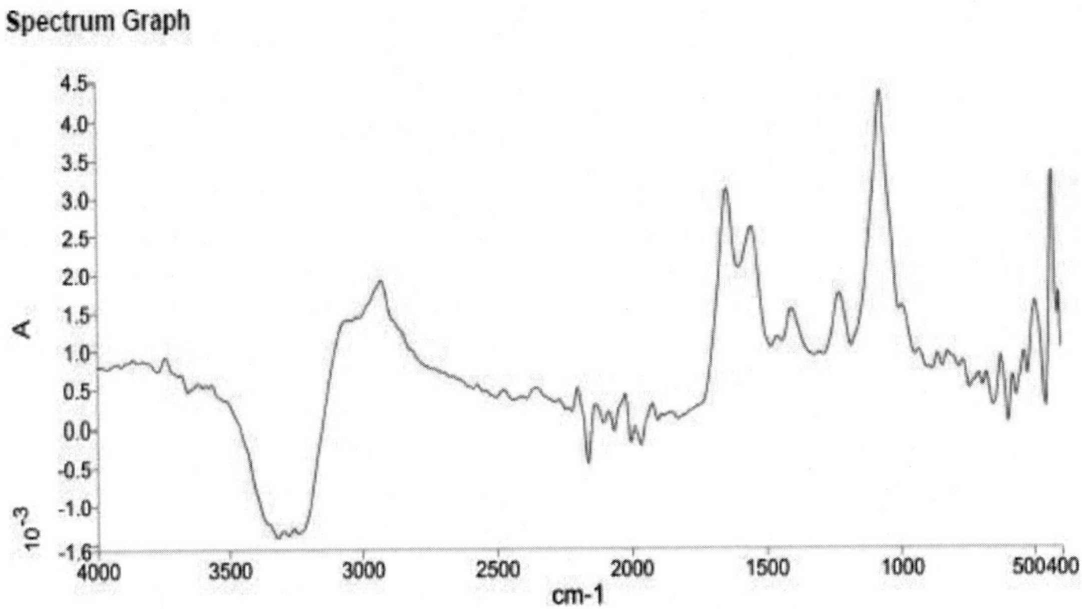

Figure 1. FTIR spectrum of residual biomass extract.

Figures 2(a) and 2(b) represent the SEM micrographs. Scanning electron microscopy (SEM) gave topographical images of the silver nanocluster samples obtained from *Spirulina platensis*, which are shown in the Figure. I agree with another study that the surface morphology of the nanoparticles was spherical or near spherical [23–25]. The results were in agreement, and electronic micrographs showed that they were uniform, smooth, and somewhat spherical in shape and were less than 100 nm in size.

FTIR results depicted several peaks indicating different organic groups present in the dried and powdered biomass. By identifying specific functional groups resulting in effective capping of AgNPs, FTIR spectra revealed significant peaks of proteins that are involved in AgNP production [18–20]. FTIR analysis of microalgal extract and biosynthesized silver nanoparticles was performed to identify the possible biomolecules responsible for the reduction and stabilization of silver nanoparticles [17, 20].

Figure 2(a). SEM micrographs of protein-induced AgNPs.

Figure 2(b). SEM micrographs of protein-induced AgNPs after 60 minutes.

Transmission electron microscopy gave surface-induced high-resolution images of the silver nanoparticles at the start of the characterization and after a certain period of time. Silver nanoparticles are randomly distributed and short-range aggregated due to limited diffusion on the protein surface, and the particle sizes are significantly uniform in comparison with the case of dried biomass. TEM examination of AgNPs revealed spherical silver nanoparticles with mean sizes ranging from 9.39 to 25.89 nm. [4, 21]. Chokshi et al. (2016) reported that they were polydispersed and predominantly spherical in shape, with a size range of 2–15 nm. A similar study clarifies that the frequency peak comes at approximately 10–15 nm [3] (Raj et al.,

2020). Most particles probably had an average size of 12 nm and 13 nm, respectively [22–23]. In agreement with another study, the crystalline particle size was 22.5 nm. Figure 3 depicts the TEM micrograph.

Figure 4 shows the XRD micrographs of Ag NPs. Depending on the FCC (face-centered-cubic) arrangement of SNCs, a dried sample of the Ag nanoparticle solution was collected for X-ray diffraction experiments, which revealed strong peaks corresponding to the (111), (200), and (220) Bragg reflections, with a lattice constant of a = 4.086 Å. An XRD graph for the obtained nanoparticles was recorded [23 - 25].

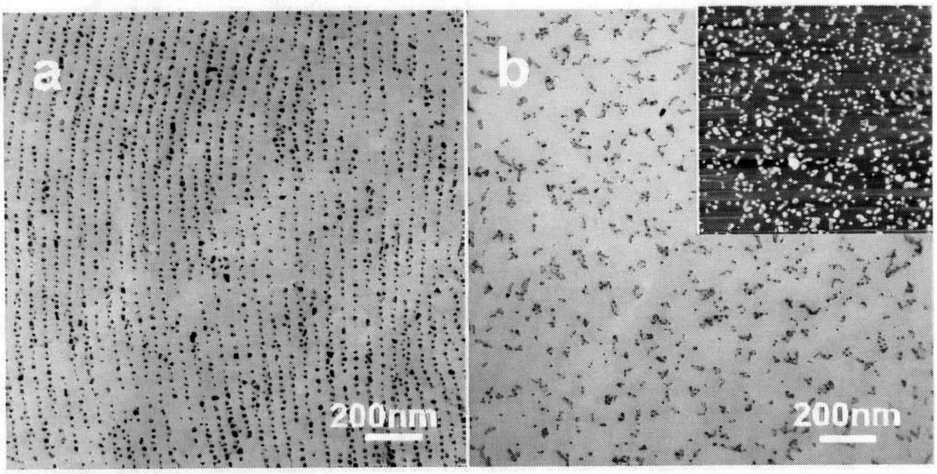

Figure 3. TEM micrographs of protein-induced Ag NPs.

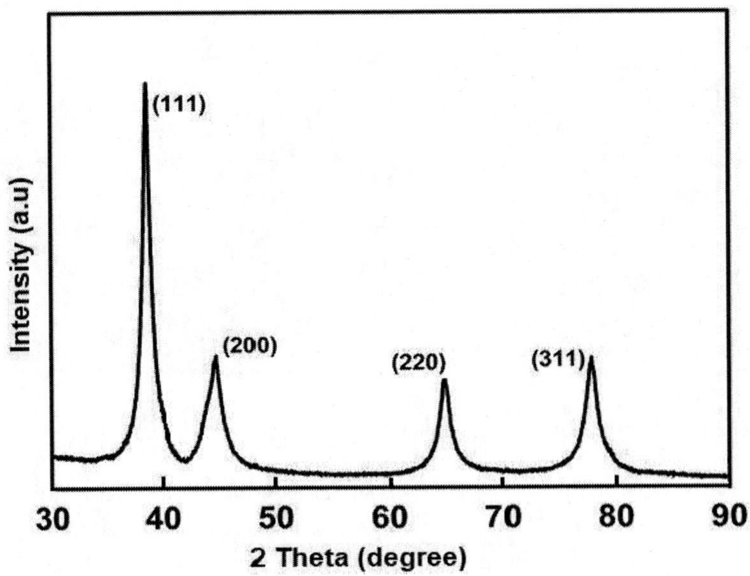

Figure 4. XRD micrographs of AgNPs.

3.2. Detection of Cysteine by Using Silver Nanoparticles (AgNPs)

Various trials of a well-known biothiol, "cysteine", were taken in amounts of 1 M, 10 M, 20 M, 30 M, and 40 M mixed with nanocluster solution, and the following characterizations were performed and compared with the results of silver nanoparticles to find the intensity at which cysteine was detected.

UV–visible spectra of AgNPs added at 1, 10, 20, 30, and 40 M concentrations with cysteine fluorescence spectroscopy were obtained for three different concentrations of cysteine, all maintained at a pH of 10. Initially, a blank-zero trial was taken for the experiment, followed by spectroscopy of the formed silver nanoparticles. Consequently, 1 ml of the silver nanoparticle solution was added to different concentrations of the previously prepared cysteine solution, and an incubation time of 10 min was given for the nanoparticles to react with the thiol group, eventually leaving it detectable. Furthermore, the fluorescence spectroscopy results proved that as the concentration of cysteine increased, the quenching decreased, and as it gradually decreased, it increased, portraying the sensitized detection of the thiol group in the presence of AgNPs. It also outlined many facts, including that the fluorescence graph for nanoparticles alone was lined up asymptotically (previously) to the x-axis, and the respective graphs for different concentrations of cysteine with AgNPs present in them rose exponentially. Chen *et al. found* that the fluorescence intensity of BSA-AgNPs was quenched efficiently by increasing the concentration of biothiol, corresponding to a red shift in the emission wavelength. However, the fluorescence of the silver nanoparticles was almost unchanged in the presence of other α-amino acids at 10-fold higher concentrations.

Fluorescence spectroscopy of AgNP-prepared Ag clusters showed high fluorescence emission at 637 nm and were stable even in 1 M NaCl. A similar study that focuses on the shift to higher absorbance wavelengths during the growth of AgNCs is accompanied by a 30 nm redshift in the excitation and emission wavelengths, indicating an increased cluster size. The red luminance of Ag-BSA was observed at 500 nm. First, the 30 μM solution is least quenched, depicting a larger concentration of thiol groups and less quenching. Second, human blood samples were collected from a local hospital (blood plasma was safe and cautiously handled) and incubated with 1 mL of AgNP solution. When the analysis was performed, it was inferred that cysteine, which is already present in considerable amounts in human blood, was detected (AgNPs quenched those) successfully. Primarily, the 20-M solution was more quenchable than the least concentrated one, which explained the sensitization behavior in a pragmatic manner. Conclusively, the highest quenching takes place in a 10 M solution that depicts the detection of the thiol group cysteine by the protein-capped silver nanoparticles proven to be biosensors that can be applied in real life with some repetitions that would scrutinize any errors and make the sensitization more precise.

4. Conclusion

The biofabrication (synthesis) of silver nanoparticles from an algal species, *Spirulina platensis,* was successfully carried out, and all characterizations were performed to inspect the inference. Primitively, the algae were cultivated on a large scale, and a portion of them was utilized to synthesize the nanoparticles. It was then determined that the algal species were rich in proteins, carbohydrates, and many types of amides, which gave a pragmatic idea that the nanoparticles that were to be produced would be protein-capped. A number of characterizations were performed on the procured nanoparticles, and it was found that the Ag NPs possessed properties specific to nanomaterials, such as optical, magnetic, and lustrous properties. Majorly, UV–visvisible spectroscopy, Fourier transform infrared spectroscopy, and fluorescence emission spectroscopy proved that the attained nanoparticles also possessed properties of silver and depicted phenomena such as surface plasmon resonance and quenching. Scanning electron microscopy and transmission electron microscopy portrayed the topographical images of the obtained silver nanoparticles (AgNPs) and explained their behavioral properties as enlargement increased and the surface to react was altered.

References

[1] Ameen, F., Abdullah, M. M., Al-Homaidan, A. A., Al-Lohedan, H. A., Al-Ghanayem, A. A., and Almansob, A. (2020). Fabrication of silver nanoparticles employing the cyanobacterium Spirulina platensis and its bactericidal effect against opportunistic nosocomial pathogens of the respiratory tract. *Journal of Molecular Structure*, 1217, 128392.

[2] Chokshi, K., Pancha, I., Ghosh, T., Paliwal, C., Maurya, R., Ghosh, A., and Mishra, S. (2016). Green synthesis, characterization and antioxidant potential of silver nanoparticles biosynthesized from deoiled biomass of thermotolerant oleaginous microalgae: Acutodesmus dimorphus. *RSC Advances*, 6(76), 72269–72274. https://doi.org/10.1039/c6ra15322d.

[3] El-Naggar, N. E. A., Hussein, M. H., and El-Sawah, A. A. (2017). Biofabrication of silver nanoparticles by phycocyanin, characterization, *in vitro* anticancer activity against breast cancer cell line and *in vivo* cytotxicity. *Scientific Reports*, 7(1), 1–20. https://doi.org/10.1038/s41598-017-11121-3.

[4] El-Naggar, N. E. A., Hussein, M. H., and El-Sawah, A. A. (2018). Phycobiliprotein-mediated synthesis of biogenic silver nanoparticles, characterization, *in vitro* and *in vivo* assessment of anticancer activities. *Scientific Reports*, 8(1), 1–20. https://doi.org/10.1038/s41598-018-27276-6.

[5] El-Sayed, A. E. K., and El-Sheekh, M. (2018). Outdoor Cultivation of Spirulina platensis for Mass Production. *Notulae Scientia Biologicae*, 10(1), 38–44. https://doi.org/10.15835/nsb10110177.

[6] Gaviria-Arroyave, M. I., Cano, J. B., and Peñuela, G. A. (2020). Nanomaterial-based fluorescent biosensors for monitoring environmental pollutants: A critical review. In *Talanta Open* (Vol. 2). https://doi.org/10.1016/j.talo.2020.100006.

[7] Hamouda, R. A., Hussein, M. H., Abo-elmagd, R. A., and Bawazir, S. S. (2019). Synthesis and biological characterization of silver nanoparticles derived from the cyanobacterium Oscillatoria limnetica. *Scientific Reports*, 9(1), 1–17. https://doi.org/10.1038/s41598-019-49444-y.

[8] [8] Hu, L., Han, S., Parveen, S., Yuan, Y., Zhang, L., and Xu, G. (2012). Highly sensitive fluorescent detection of trypsin based on BSA-stabilized gold nanoclusters. *Biosensors and Bioelectronics*, 32(1), 297–299. https://doi.org/10.1016/j.bios.2011.12.007.

[9] Hussain, I., Singh, N. B., Singh, A., Singh, H., and Singh, S. C. (2016). Green synthesis of nanoparticles and its potential application. *Biotechnology Letters*, 38(4), 545–560. https://doi.org/10.1007/s10529-015-2026-7.

[10] Khalkho, B. R., Kurrey, R., Deb, M. K., Shrivas, K., Thakur, S. S., Pervez, S., and Jain, V. K. (2020). L-cysteine modified silver nanoparticles for selective and sensitive colorimetric detection of vitamin B1 in food and water samples. In *Heliyon* (Vol. 6, Issue 2). https://doi.org/10.1016/j.heliyon.2020.e03423.

[11] Kulakova, I. I., and Lisichkin, G. v. (2022). Biosensors Based on Graphene Nanomaterials. *Moscow University Chemistry Bulletin*, 77(6), 307–321. https://doi.org/10.3103/S0027131422060049.

[12] le Guével, X., Hötzer, B., Jung, G., Hollemeyer, K., Trouillet, V., and Schneider, M. (2011). Formation of fluorescent metal (Au, Ag) nanoclusters capped in bovine serum albumin followed by fluorescence and spectroscopy. *Journal of Physical Chemistry C*, 115(22), 10955–10963. https://doi.org/10.1021/jp111820b.

[13] Oves, M., Rauf, M. A., Aslam, M., Qari, H. A., Sonbol, H., Ahmad, I., and Saeed, M. (2022). Green synthesis of silver nanoparticles by Conocarpus Lancifolius plant extract and their antimicrobial and anticancer activities. *Saudi Journal of Biological Sciences*, 29(1), 460-471.

[14] Pandit, C., Roy, A., Ghotekar, S., Khusro, A., Islam, M. N., Emran, T. bin, Lam, S. E., Khandaker, M. U., and Bradley, D. A. (2022). Biological agents for synthesis of nanoparticles and their applications. *Journal of King Saud University - Science* (Vol. 34, Issue 3). https://doi.org/10.1016/j.jksus.2022.101869.

[15] Rajkumar, G., and Sundar, R. (2022). Biogenic one-step synthesis of silver nanoparticles (AgNPs) using an aqueous extract of Persea americana seed: Characterization, phytochemical screening, antibacterial, antifungal and antioxidant activities. *Inorganic Chemistry Communications* (Vol. 143). https://doi.org/10.1016/j.inoche.2022.109817.

[16] Raza, S., Ansari, A., Siddiqui, N. N., Ibrahim, F., Abro, M. I., and Aman, A. (2021). Biosynthesis of silver nanoparticles for the fabrication of non cytotoxic and antibacterial metallic polymer based nanocomposite system. *Scientific Reports*, 11(1). https://doi.org/10.1038/s41598-021-90016-w.

[17] Ranjitha, S., Aroulmoji, V., Selvankumar, T., Sudhakar, C., and Hariharan, V. (2020). Synthesis and development of novel sensitizer from spirulina pigment with silver doped TiO2 nano particles for biosensitized solar cells. *Biomass and Bioenergy*, 141, 105733.

[18] Raj, S., Singh, H., Trivedi, R., and Soni, V. (2020). Biogenic synthesis of AgNPs employing Terminalia arjuna leaf extract and its efficacy toward catalytic degradation of organic dyes. *Scientific Reports*, 10(1), 9616. https://doi.org/10.1038/s41598-020-66851-8.

[19] Samanta, A., and Medintz, I. L. (2016). Nanoparticles and DNA-a powerful and growing functional combination in bionanotechnology. *Nanoscale*, 8(17), 9037–9095. https://doi.org/10.1039/c5nr08465b.

[20] Sargazi, S., Fatima, I., Hassan Kiani, M., Mohammadzadeh, V., Arshad, R., Bilal, M., Rahdar, A., Díez-Pascual, A. M., and Behzadmehr, R. (2022). Fluorescent-based nanosensors for selective detection of a wide range of biological macromolecules: A comprehensive review. In *International Journal of Biological Macromolecules* (Vol. 206, pp. 115–147). https://doi.org/10.1016/j.ijbiomac.2022.02.137.

[21] Shantanam, S., and Mueller. (2018). HHS Public Access. *Physiology & Behavior*, *176*(1), 139–148. https://doi.org/10.1016/j.freeradbiomed.2014.11.013.The

[22] Sharma, A. R., Lee, Y. H., Bat-Ulzii, A., Bhattacharya, M., Chakraborty, C., and Lee, S. S. (2022). Recent advances of metal-based nanoparticles in nucleic acid delivery for therapeutic applications. *Journal of Nanobiotechnology*, 20(1), 1–21. https://doi.org/10.1186/s12951-022-01650-z.

[23] Singh, J., Dutta, T., Kim, K. H., Rawat, M., Samddar, P., and Kumar, P. (2018). "Green" synthesis of metals and their oxide nanoparticles: Applications for environmental remediation. *Journal of Nanobiotechnology*, 16(1), 1–24. https://doi.org/10.1186/s12951-018-0408-4.

[24] Yang, X. H., Ling, J., Peng, J., Cao, Q. E., Wang, L., Ding, Z. T., and Xiong, J. (2013). Catalytic formation of silver nanoparticles by bovine serum albumin protected-silver nanoclusters and its application for colorimetric detection of ascorbic acid. *Spectrochimica Acta - Part A: Molecular and Biomolecular Spectroscopy*, 106, 224–230. https://doi.org/10.1016/j.saa.2012.12.097.

Chapter 28

The Green Synthesis and Characterization of Silver Nanoparticles for Cleaning and Microbial Applications

Vaishali Ajay Giri*
Chandrika Sengar
S. V. A. R. Sastry
and Ashish Kapoor
Department of Chemical Engineering, H.B.T.U. Kanpur, India

Abstract

Nanotechnology is a forefront technology with wide applications in the areas of biomedical fields, optics, electronics, chemical industries, defense, etc. NPs have an extremely small size and a large S:V ratio, which is why their physical and chemical properties are altered. This makes nanotechnology the center of attention for the scientific community. There are many methods for the synthesis of NPs, but several nanoparticles possess toxicity, which leads to the development of clean, nontoxic, biocompatible, and eco-friendly methods that bring the GS of nanoparticles into the picture. GS-zed NPs are eco-friendly, nontoxic, and cost-effective. Silver NPs are one of the most essential and captivating nanomaterials among many metallic NPs, comprising unique properties (e.g., size and shape depending on optical, electrical, and magnetic properties). The properties of silver NPs can be incorporated into wide applications in biomedical fields, such as their bactericidal ability, wound and burn healing properties, disinfection properties, and cleaning properties. The characterization techniques used for Ag NPs are FTIR, UV–vis, SEM, AFM, etc. In this paper, specific applications of silver NPs in self-cleaning and as an antimicrobial agent will be focused on.

Keywords: biomedical, cleaning agent, green synthesis, microbial applications, silver nanoparticles

* Corresponding Author's Email: 210202002@hbtu.ac.in.

In: Sustainability in Chemical Processes through Digitalization and Green Chemistry Approaches
Editors: Susarla V. A. R. Sastry, Shravan Kumar and Ashish Kapoor
ISBN: 979-8-89113-257-3

1. Introduction

Nanotechnology is the study and manipulation of matter in the nanoscale range (1-100 nm) [1]. The term nanotechnology was first presented by Dr. Richard P. Feynman in 1959. Nanoscale materials have extraordinary properties compared to bulk materials due to their extended S:V ratio and extremely small size. The abovementioned unique optical, electrical and magnetic properties make nanomaterials capable of endless applications in the biomedical [2], cosmetics [3], electronics [4], food [5], wastewater treatment [6], etc., fields.

The synthesis of nanomaterials is performed in two categories:

- Top-down approach: nanomaterials are generated by size reduction
- Bottom-up approach: nanomaterials are generated by assembling small entities such as atoms and molecules.

There are various methods for synthesizing NPs, including physical, chemical and biosynthesis techniques. The most commonly used physical methods are inert gas condensation, laser ablation, arc discharge, etc. In general, physical methods are time consuming and require plenty of time to attain thermal stability and consume high energy because the temperature rises near the source material. Chemical synthesis requires many hazardous chemicals that are costly and ultimately increase the pollution load in the environment [7], bringing biosynthetic techniques to the center of attention. Biosynthetic techniques are safer and nontoxic, and the raw materials are easily available, which makes them a better alternative than chemical and physical methods. Biological syn involve plants, bacteria, fungi, etc. Recently, silver Npts have attracted much interest because of their antibacterial, antimicrobial, antifungal, anticancer, antidiabetic, and anti-inflammatory characteristics [1].

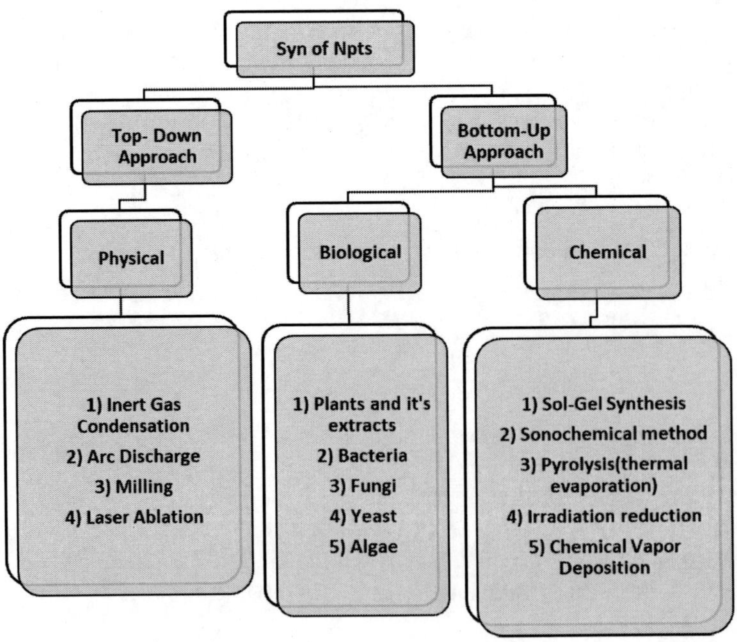

Figure 1. Synthesis of nanoparticles [8].

In GS, bioreduction is performed to obtain the elemental form of metal ions in the size range of 1-100 nm [7]. The GS process is considered to be simpler and more economical than conventional methods and can also be easily converted to large-scale operations. Many NPs have been synthesized through GS, such as silver, gold, iron, zinc, palladium, etc. [7].

2. Approaches and Methods for Syn of NPs

There are two main approaches for the syn of NPs, based on which there are several methods through which NP syn can be performed, including physical, chemical, and biological methods. Figure 1 shows the aforementioned.

Top-down approach: In this approach, NPs are formed by reducing the size of bulk particles to the desired size.

Bottom-up approach: In this approach, NPs are formed by assembling small entities atom-by-atom or molecule-by-molecule.

2.1. Physical Methods

The most commonly used methods in physical synthesis are high-energy ball milling and (HEBM), laser ablation, pyrolysis, evaporation-condensation, laser pyrolysis, arc discharge, atomization, annealing, and metal sputtering [7]. Earlier, NPs of numerous materials, such as gold (Au), silver (Ag), and fullerenes, were synthesized by evaporation and condensation techniques [8, 9]. Physical methods are much more preferred than chemical methods because of the lower usage of hazardous chemicals and the lesser solvent contamination in the thin nanofilms formed. In the inert gas condensation method, the metal precursor or metal oxide is placed in a boat at the tube furnace center and is vaporized into an inert atmosphere such as argon or nitrogen at atmospheric pressure and subsequently condensed to NPs [10]. Grinding and milling are performed to reduce the size of bulk materials and turn them into NPs with high energy input. In the laser ablation method, a laser beam is used as a primary excitation source for the vaporization of the target (metal precursor) from the surface, generating clusters of metal vapors in an inert gas atmosphere with high energy, which are further condensed to nanoparticles and thin films [11].

2.2. Chemical Methods

The most popular method in chemical synthesis is the sol-gel method. In this method, the formation of solvated metal precursors or stable solutions of alkoxide (the sol) is followed by the polycondensation reaction (gelation), aging, drying, dehydration, and finally densification of the gel. Various other methods used in chemical synthesis include irradiation reduction and chemical vapor deposition. The main drawback of chemical synergy is that it involves toxic chemical reagents that generate nonecofriendly byproducts [8].

2.3. Biological Methods

In this method, NPs are synthesized by plant extracts, algae, fungi, bacteria, and yeast. The prerequisite for GS is a metal ion solution and a reducing agent. The cell constituents act as stabilizing and capping agents, so the addition of reducing agents from outside is not needed in most cases [12].

3. GS of Silver NPs

GS of metal NPs is a cost-effective, uncomplicated, and environmentally benign substitute for conventional methods that are conducive to production in abundance and do not require complex reagents [13]. Living organisms are broadly divided into five kingdoms, namely, the Monera, Protista, Fungi, Plantae, and Animal kingdoms, among which silver NPs are synthesized using four of them: Monera, Protista, Fungi, and Plantae [7]. Figure 2 shows the contribution of each method of biosynthesis of nanoparticles (plant-derived, fungi-derived, bacteria-derived, algae-derived, and biomolecule-derived) to the total number of publications.

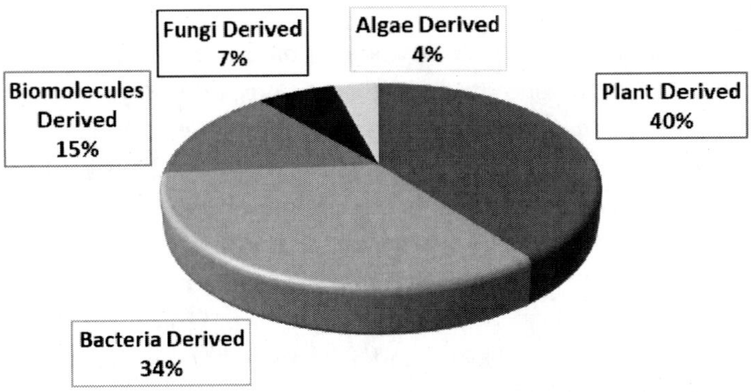

Figure 2. Number of publications of each keyword from 2010-2020 (based on the data from reference) [14].

3.1. Plant Derived

Plants contain large amounts of antioxidants and metabolites necessary for the biosynthesis of silver NPs (such as amino acids, proteins, alkaloids, terpenes, polysaccharides, phenols, and vitamins) that help reduce silver ions to NPs [14, 15]. The biological molecules obtained from plant extracts are found to be a better alternative to reagents in GS. Plant parts that are commonly used for the GS of metal NPs are leaves, flowers, stems, bark, roots, pulp, seeds, etc.

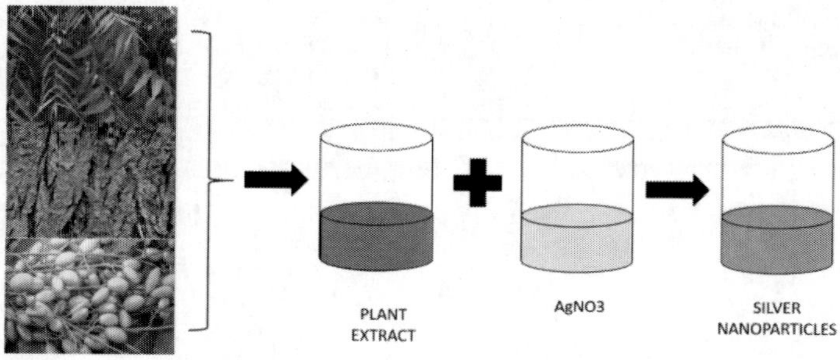

Figure 3. Silver nanoparticles through plants..

As shown in Figure 3, silver NPs were synthesized using the aqueous extract of Azadirachta indica (Neem) leaves. The neem leaf extract acted as a reducing and capping agent without using toxic chemicals, and after reducing Ag+ ions for 15 minutes, well-dispersed, spherical-shaped silver NPs with a diameter of 34 nm were formed. These Ag NPs were found to be active against both gram-negative and gram-positive bacteria [16].

3.2. Bacteria Derived

Microorganisms are armed with great potential for surviving and even adapting to extreme conditions such as pressure, temperature, pH, and nutrient availability. These qualities of the microorganisms make them potential biofactories in metal NP production [14]. Metal NPs synthesized from bacteria can be either intracellular or extracellular; generally, gold and silver NPs are made using this method [17]. Some bacteria are resistant to silver NPs since silver can accumulate on the cell wall, promoting their utilization in the recovery of Ag from ore materials [16, 17]. In the intracellular method, the molecular constituents and enzymes that are present in the cell reduce silver ions, and the synthesis of Ag NPs occurs inside the cell, whereas in the extracellular method, Ag NPs accumulate outside the cell. This method of syn for silver NPs is associated with difficulties in growth and culture maintenance; therefore, plant-derived syn is usually preferred [18]. For example, Ag-NPs were synthesized by using the culture supernatants of psychrophilic bacteria [19]. The bacteria continue to grow even after the syn of Ag-NPs, and the major pitfall of using bacteria for NP syn is that the rate of syn is very slow and only limited sizes and shapes could be synthesized [20].

3.3. Fungi Derived

Fungi-derived NPs are more beneficial than other microorganisms (bacteria) since fungi are easy to handle and fabricate in the lab, and fungi also grow more quickly [7]. Fungi, similar to bacteria, have high binding capacity, the ability to accumulate NPs, and intracellular intake ability, which is why fungi are preferred for the synthesis of metallic NPs [21].

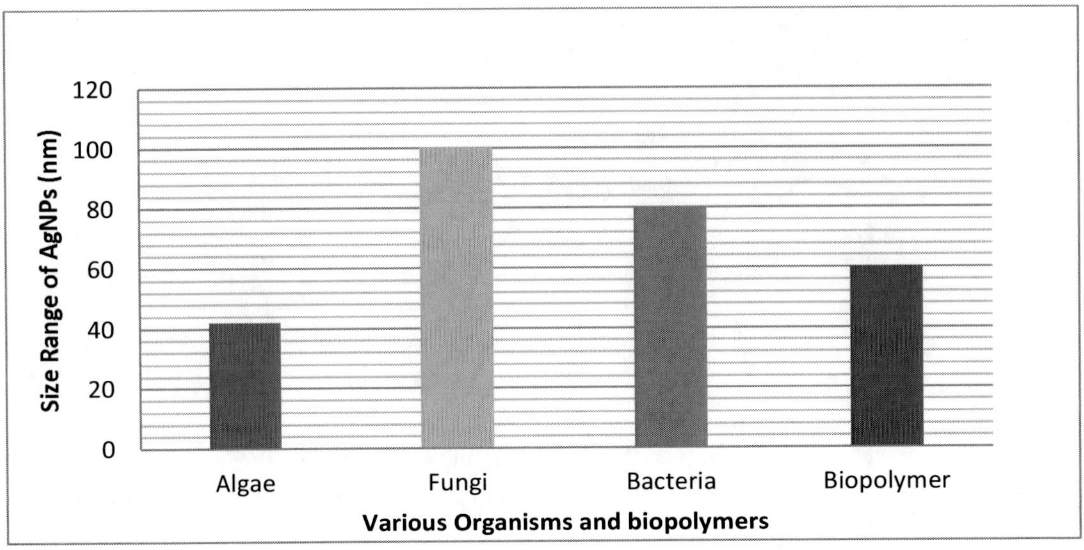

Figure 4. Size range of silver NPs formed at room temperature range (Based on the data of reference) [12]).

The initial step in the synthesis of silver NPs through fungi is the adsorption of Ag^+ ions on the surface of fungal cells. This is due to the electrostatic interaction between Ag^+ ions and carboxylate groups having a negative charge, which are present in enzymes. Then, the Ag^+ ions are reduced to their elemental form by the enzymes in the cell wall. The NPs formed are not in the solution but on the surface of mycelia in fungal-derived syn [22]. Figure 4 shows the size range of silver NPs formed at room temperature.

4. Mechanism of Ag NP Synthesis

The abovementioned biological methods for the syn of silver NPs occur due to the presence of proteins, phenols, carbohydrates, enzymes, flavonoids, alkaloids, terpenoids, etc., in either plant extracts or in bacteria, fungi, and yeast [12]. The reduction of Ag^+ ions to Ag^0 is due to the electron transfer from the OH groups present in the biomolecules of the extract or organism; therefore, reduction depends on the extract and organism used for syn. The syn process occurs in three phases: the first is the reduction of silver ions, followed by agglomeration and then stabilization of silver particles to form nanosized particles [18]. Figure 5 shows the stepwise transformation of silver ions.

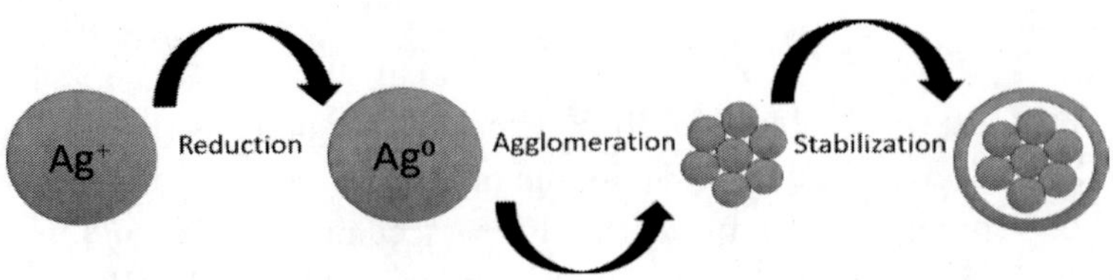

Figure 5. General mechanism for syn of silver Npts (based on the reference) [12]).

5. Factors Affecting Ag NP Synthesis

1. The pH of the reaction mixture is an important parameter that plays a prime role in the syn of silver Npts. The difference in hydrogen ion concentration leads to different shapes and sizes of NPs. It was observed from various studies that predominantly large-sized particles (above 25 nm) were formed at lower pH values of the reaction medium, whereas relatively small-sized particles were formed at higher pH values [7, 12].
2. The reaction temperature for the syn of NPs is another important parameter. The major three methods of syn (physical, chemical, and biological) require different temperatures for the syn. Researchers found that the temperature of the reaction medium determines the nature of the nanoparticles synthesized [23].
3. Techniques used for the synthesis of NPs are also of importance since the NPs formed from various methods have different properties. Biologically derived NPs are considered to be less toxic and eco-friendly, hence making them more acceptable than traditional methods.
4. Reaction time greatly influences the type and quality of NPs formed. When the time of reaction is varied, it may affect the NPs formed; either the NPs may grow or shrink during the process, which ultimately affects the potential of NPs [23].
5. Pressure applied in the reaction medium affects the structural changes, such as the shape and size of the NPs formed. The rate of reduction is fast when the reaction medium is at ambient temperature [24].

6. Characterization Techniques Used

Char is the key to the analysis of the generated silver NPs since it elucidates the structural components such as size, shape, and various other information related to the functional groups, and according to Char, the applications are defined. Currently, various techniques are used, a few of which are atomic force microscopy (AFM), scanning electron microscopy (SEM), UV–visible spectroscopy, Fourier transform infrared spectroscopy (FTIR), X-ray diffraction, etc.

- UV–visible spectroscopy: UV–visible spectroscopy is used to measure the ultraviolet extinction (absorbance and scattering) of light passing through each sample and the bioreduction of silver NPs. From the reference [25], the peaks studied in the UV–vis graph confirmed the syn of Ag NPs when the reaction mixture was scanned under 300 and 700 nm of a double-beam spectrophotometer [25].
- Fourier transformed infrared spectroscopy (FTIR): This method is used to determine the functional groups (biomolecules) present in the NPs that are responsible for the reduction and capping of silver NPs [26].
- Scanning electron microscopy (SEM): SEM was used to identify the morphology of the silver NPs formed, and a high-resolution image was obtained.
- X-ray diffraction (XRD): XRD is used to understand the chemical composition, structural crystallography, and physical properties of materials.

7. Applications

There are various applications of silver NPs that can be further subcategorized, such as biomedical applications, which can be subcategorized as drug delivery, cancer treatment, etc., and food industry applications, which could be further divided into applications of packaging, storage, etc. Therefore, there are a wide range of applications for silver NPs, some of which are shown in Figure 6.

In this paper, only antimicrobial and self-cleaning applications are the focus.

- **Antimicrobial properties:**

Silver has been recognized for its antimicrobial properties for centuries in various countries. Of the various metal NPs, silver has been mostly utilized against microbes because of its fatal nature toward bacterial cells and fast production [27]. Even though, until now, the accurate mechanism behind the antimicrobial properties of silver NPs remains obscure, certain theories have been expanded to suggest that the Ag NPs could accumulate on the surface of the cell walls, puncture the cell, and cause cell death [28].

- **Self-cleaning properties:**

Self-cleaning or disinfecting properties come from the bactericidal effects shown by silver NPs. We know that patients in health care zones are open to various kinds of infections, and this is a matter of concern worldwide, so for this purpose, silver, proving its self-cleaning property, is used as a thin layer of coating on the tiles of the floor, glass windows, buildings, bathrooms, hospitals, schools, etc., so that people could be saved from infections.

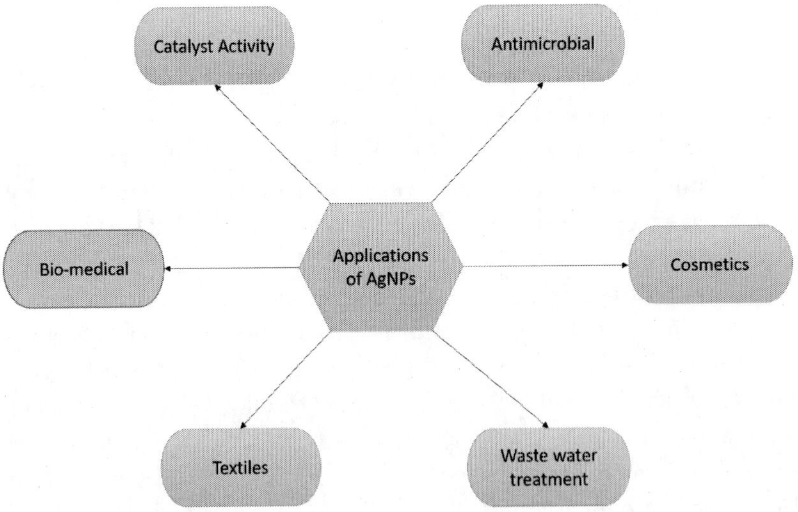

Figure 6. Applications of silver nanoparticles.

Conclusion

GS is a major focus for many researchers because of its less toxic chemical requirements and because of its less hazardous byproducts. NPs have unique properties compared to bulk materials because of their extremely small size and large S:V ratio. Silver NPs can be synthesized by physical, chemical, and biological methods, among which biological methods are preferred due to their eco-friendliness and cost effectiveness. The characterization of silver NPs is performed using various methods, such as FTIR, XRD, SEM, UV vis-spectrophotometers, etc., for crystallographic imaging, topography and morphology, chemistry, and several other aspects related to the functional groups attached to the silver NPs. Several factors that affect the syn of silver NPs are pH, temperature, time, techniques used, and pressure applied. In this paper, the applications of silver nanoparticles as antimicrobial agents and their self-cleaning properties are discussed. There are many more applications of silver NPs in the biomedical field, such as in specific drug delivery, cancer treatment, wound and burn healing, the food industry, and wastewater treatment.

The major challenge in the GS of silver NPs, or in general, metallic NPs, is the postuse leftover or waste-generated disposal risk. Handling and storing such small particles is also a tedious task.

References

[1] K. R, G. G, J. A, G. M, Rapid Green Synthesis of Silver Nanoparticles (AgNPs) Using (*Prunus persica*) Plants extract: Exploring its Antimicrobial and Catalytic Activities, *J Nanomed Nanotechnol.* 08 (2017). https://doi.org/10.4172/2157-7439.1000452.

[2] Chaichi, A., E. Sheikh, S. Mukhopadhyay, M. R. Gartia, Trends in Nanotechnology Development in Medical Applications, *Encyclopedia of Materials: Plastics and Polymers.* (2022) 913–919. https://doi.org/10.1016/B978-0-12-820352-1.00112-7.

[3] S. Raj, S. Jose, U. S. Sumod, M. Sabitha, Nanotechnology in cosmetics: Opportunities and challenges, *J Pharm Bioallied Sci.* 4 (2012) 186. https://doi.org/10.4103/0975-7406.99016.

[4] W. Zeng, L. Shu, Q. Li, S. Chen, F. Wang, X. M. Tao, Fiber-based wearable electronics: A review of materials, fabrication, devices, and applications, *Advanced Materials.* 26 (2014) 5310–5336. https://doi.org/10.1002/adma.201400633.

[5] Statement of Removal: Nanotechnology –New Lifeline For Food Industry (Critical Reviews in Food Science and Nutrition, (2015), 10.1080/10408398.2013.802662), *Crit Rev Food Sci Nutr.* (2020). https://doi.org/10.1080/10408398.2013.802662.

[6] S. Sharma, P. Dhingra, S. Jain, Advancements & challenges of nanotechnology in waste water treatment, *Mater Today, Proc.* (2022). https://doi.org/10.1016/J.MATPR.2022.09.481.

[7] G. Pal, P. Rai, A. Pandey, Green synthesis of nanoparticles: A greener approach for a cleaner future, in: *Green Synthesis, Characterization and Applications of Nanoparticles*, Elsevier, 2019: pp. 1–26. https://doi.org/10.1016/b978-0-08-102579-6.00001-0.

[8] Hasan, K. M. F., L. Xiaoyi, Z. Shaoqin, P. G. Horváth, M. Bak, L. Bejó, G. Sipos, T. Alpár, Functional silver nanoparticles synthesis from sustainable point of view: 2000 to 2023 – A review on game changing materials, *Heliyon.* (2022) e12322. https://doi.org/10.1016/j.heliyon.2022.e12322.

[9] E. Abbasi, M. Milani, S. F. Aval, M. Kouhi, A. Akbarzadeh, H. T. Nasrabadi, P. Nikasa, S. W. Joo, Y. Hanifehpour, K. Nejati-Koshki, M. Samiei, Silver nanoparticles: Synthesis methods, bioapplications and properties, *Crit Rev Microbiol.* 42 (2016) 173–180. https://doi.org/10.3109/1040841X.2014.912200.

[10] Abou El-Nour, K. M. M., A. Eftaiha, A. Al-Warthan, R. A. A. Ammar, Synthesis and applications of silver nanoparticles, *Arabian Journal of Chemistry.* 3 (2010) 135–140. https://doi.org/10.1016/j.arabjc.2010.04.008.

[11] Semaltianos, N. G. Nanoparticles by laser ablation, *Critical Reviews in Solid State and Materials Sciences.* 35 (2010) 105–124. https://doi.org/10.1080/10408431003788233.

[12] Srikar, S. K., D. D. Giri, D. B. Pal, P. K. Mishra, S. N. Upadhyay, Green Synthesis of Silver Nanoparticles: A Review, *Green and Sustainable Chemistry*. 06 (2016) 34–56. https://doi.org/10.4236/gsc.2016.61004.

[13] Malik, P., R. Shankar, V. Malik, N. Sharma, T. K. Mukherjee, Green Chemistry Based Benign Routes for Nanoparticle Synthesis, *Journal of Nanoparticles*. 2014 (2014) 1–14. https://doi.org/10.1155/2014/302429.

[14] Rana, A., K. Yadav, S. Jagadevan, A comprehensive review on green synthesis of nature-inspired metal nanoparticles: Mechanism, application and toxicity, *J Clean Prod*. 272 (2020). https://doi.org/10.1016/j.jclepro.2020.122880.

[15] Mustapha, T., N. Misni, N. R. Ithnin, A. M. Daskum, N. Z. Unyah, A Review on Plants and Microorganisms Mediated Synthesis of Silver Nanoparticles, Role of Plants Metabolites and Applications, *Int J Environ Res Public Health*. 19 (2022). https://doi.org/10.3390/ijerph19020674.

[16] Ahmed, S., Saifullah, M. Ahmad, B.L. Swami, S. Ikram, Green synthesis of silver nanoparticles using Azadirachta indica aqueous leaf extract, *J Radiat Res Appl Sci*. 9 (2016) 1–7. https://doi.org/10.1016/j.jrras.2015.06.006.

[17] Rafique, M., I. Sadaf, M. S. Rafique, M. B. Tahir, *A review on green synthesis of silver nanoparticles and their applications,* http://Dx.Doi.Org/10.1080/21691401.2016.1241792. 45 (2016) 1272–1291. https://doi.org/10.1080/21691401.2016.1241792.

[18] Alharbi, N. S., N. S. Alsubhi, A. I. Felimban, Green synthesis of silver nanoparticles using medicinal plants: Characterization and application, *J Radiat Res Appl Sci*. 15 (2022) 109–124. https://doi.org/10.1016/j.jrras.2022.06.012.

[19] Raheman, F., S. Deshmukh, A. Ingle, A. Gade, M. Rai, Silver nanoparticles: Novel antimicrobial agent synthesized from an endophytic fungus Pestalotia sp. Isolated from leaves of Syzygium cumini (L), *Nano Biomed Eng*. 3 (2011). https://doi.org/10.5101/nbe.v3i3.p174-178.

[20] Rauwel, P., S. Küünal, S. Ferdov, E. Rauwel, A review on the green synthesis of silver nanoparticles and their morphologies studied via TEM, *Advances in Materials Science and Engineering*. 2015 (2015). https://doi.org/10.1155/2015/682749.

[21] Sastry, M., A. Ahmad, M. Islam Khan, R. Kumar, *Biosynthesis of metal nanoparticles using fungi and actinomycete*, 2003.

[22] Vahabi, K., G. A. Mansoori, S. Karimi, Biosynthesis of Silver Nanoparticles by Fungus Trichoderma Reesei (A Route for Large-Scale Production of AgNPs), *Insciences Journal*. (2011) 65–79. https://doi.org/10.5640/insc.010165 .

[23] Patra, J. K., K. H. Baek, Green Nanobiotechnology: Factors Affecting Synthesis and Characterization Techniques, *J Nanomater*. 2014 (2014). https://doi.org/10.1155/2014/417305.

[24] Tran, Q. H., V. Q. Nguyen, A. T. Le, Corrigendum: Silver nanoparticles: Synthesis, properties, toxicology, applications and perspectives (Advances in Natural Sciences: Nanoscience and Nanotechnology 4 (033001) doi: 10.1088/2043-6262/4/3/033001), *Advances in Natural Sciences: Nanoscience and Nanotechnology*. 9 (2018). https://doi.org/10.1088/2043-6254/aad12b.

[25] Pallavi, S. S., H. A. Rudayni, A. Bepari, S. K. Niazi, S. Nayaka, Green synthesis of Silver nanoparticles using Streptomyces hirsutus strain SNPGA-8 and their characterization, antimicrobial activity, and anticancer activity against human lung carcinoma cell line A549, *Saudi J Biol Sci*. 29 (2022) 228–238. https://doi.org/10.1016/j.sjbs.2021.08.084.

[26] Tufail, S., M. Zeeshan, S. Hanif, A. Sajjad, M. Zia, Synthesis and morphological & biological characterization of Campsis radicans and Cascabela thevetia petals derived silver nanoparticles, *Biochem Syst Ecol*. 105 (2022). https://doi.org/10.1016/j.bse.2022.104526.

[27] Nisar, P., N. Ali, L. Rahman, M. Ali, Z.K. Shinwari, Antimicrobial activities of biologically synthesized metal nanoparticles: an insight into the mechanism of action, *Journal of Biological Inorganic Chemistry*. 24 (2019) 929–941. https://doi.org/10.1007/S00775-019-01717-7/FIGURES/7.

[28] Venkata Padma, D., S. V. A. R. Sastry, Kinetic Studies on Simultaneous Biosorption of Divalent Copper and Hexavalent Chromium using Mallet Flower Leaf Powder, *Journal of Water and Environmental Nanotechnology*. 5 (2020) 204-217. https://doi.org/10.22090/JWENT.2020.03.002.

Chapter 29

The Synthesis of Nanocellulose Using Plants and Agro-Wastes and Its Applications in Water Purification

Chandrika Sengar*
Vaishali Ajay Giri
Ashish Kapoor
and S. V. A. R. Sastry
Department of Chemical Engineering, HBTU Kanpur, U. P., India

Abstract

Growing populations, industrial development, and improperly managed domestic and agricultural practices contribute to water pollution and drinking shortages. Hence, efficient water purification technologies are needed. Nanocellulose-based nanomaterials contribute a key function in the purification of effluents from different sources. The nontoxic, biodegradable, biocompatible, and renewable nature of nanocellulose (NC) drives more utilisation of it. Lower hydrophilicity and poor physical and chemical stability of cellulose, which reduce its absorption capacity, have prompted the researchers to look at ways of synthesising nano-sized cellulose using plants, bacteria, and agro-waste as they upgrade the adsorption capabilities. Nanocellulose consists of hydroxyl groups that make it a potent alternative for the scavenging of different contaminants such as pharmaceuticals, heavy metals, and dyes. That is why synthesis techniques become crucial for the effective removal of pollutants. Several surface-modified nanocelluloses are utilised in water purification. In this work, we examine the potential of nanocellulose in diverse water purification technologies.

Keywords: synthesis, adsorption, hydroxyl groups, nanocellulose, water purification

1. Introduction

Recent advances in industrialization and urbanisation may have given us a better way of living, but their aftereffects cannot be overlooked. Therefore, availability and quality of potable water

* Corresponding Author's Email: 210202001@hbtu.ac.in.

In: Sustainability in Chemical Processes through Digitalization and Green Chemistry Approaches
Editors: Susarla V. A. R. Sastry, Shravan Kumar and Ashish Kapoor
ISBN: 979-8-89113-257-3

are two of the most frequently encountered issues that have emerged from these evolutions. Nanotechnology has opened new ways of purifying water using nanomaterials. Different and many nano-metals, nano-metals oxides, polymeric nanomaterials, etc. are being used to purify water, but some of the nanomaterials are costly, such as gold and silver nanoparticles, and some are non-biodegradable nanomaterials. This has shifted the attention of scientists to using nanocellulose in water purification because of its natural availability, high aspect ratio, biodegradability and accessibility of plenty of OH groups for binding dyes, heavy metals, and other pollutants [1].

Various outstanding properties of nanocellulose are used as an alternative to petrochemical products. However, the synthesis methods of nanocellulose also determine the potential of nanocellulose, energy consumption, operating costs, and many others, which motivates the adoption of eco-friendly, cost-effective, sustainable, and efficient methods for synthesising nanocellulose.

2. Nanocellulose: Structural Arrangement and Functional Group

Structure: Cellulose, the most abundant biopolymer, was first discovered by French chemist Anselme Payen in the year 1838 [2]. Polysaccharide cellulose, having a conformation like a flat ribbon, consists of repeating units of cellobiose. This homopolymer is made up of a long linear chain of molecules of glucose that repeats two anhydro glucose units of the chemical formula (C6H10O5)n. Here, the degree of polymerization is represented by the letter n, and n varies according to the cellulose source material. These units are linked together through an oxygen atom that is outside the aromatic ring. This oxygen is covalently bonded to one carbon atom of one glucose ring and another carbon atom of the adjoining ring (1-4 linkage) and so called the 1–4 glycosidic bond [3, 4], as shown in Figure 1. Parameters that influence the lignocellulose structure depend on lignin amount, how well the cellulose fibres are protected by lignin, how well the cellulose fibres are covered with hemicellulose, cellulose surface area, cellulose strength, and the degree of crystallinity. Therefore, the strength and viable characteristics of lignocellulosic structures can be altered by changing the above attributes [5].

Functional Group: The glucose unit in cellulose consists of the OH group that gives the reactive surface to nanocellulose. However, it also results in the agglomeration of several stacked units of cellulose because of van der Waals forces of attraction and intermolecular hydrogen bonding between the oxygen of glucose molecules and the -OH group of adjacent glucose molecules. Resulting parallel and multiple-stacked cellulose chains give poorly dispersed nanocellulose in polymer matrices and hydrophobic or nonpolar solvents. This issue can be diminished by surface modification by converting the hydroxyl group into a desirable functional group.

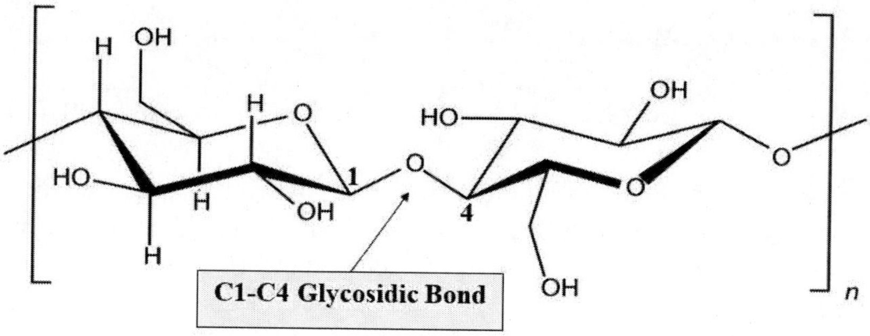

Figure 1. Structure of Cellulose.

3. Sources of Nanocellulose

Sources of nanocellulose are lignocellulosic biomass, which consists of three main components: cellulose, hemicellulose, and lignin, as shown in the pie charts in Figure 2. Plants, agricultural residues, animals, and bacteria are pre-treated and isolated to synthesise nanocellulose. Here, we illustrate some sources of nanocellulose, which is mainly extracted from plants and agricultural residues (Table 1).

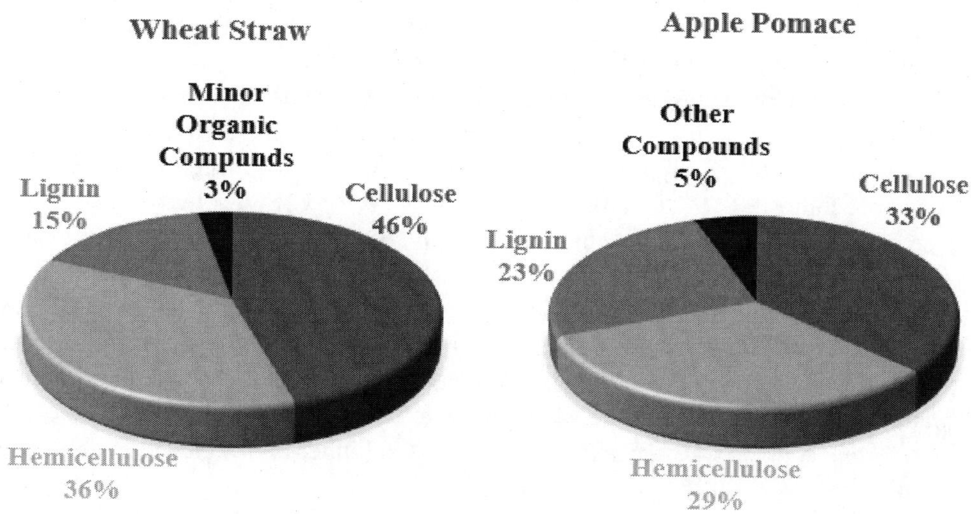

Figure 2. Chemical Compositions of Lignocellulosic Feedstock (% Dry Basis) based on data of [6].

Table 1. Different Parts of Plants and Agro-Wastes Used in Synthesis of Nanocellulose

Source	Category	Examples
Plants	Seed	Cotton
	Bast (Skin)	Flax, Rattan, Jute, Kenaf, Ramie, Hemp and Vine
	Leaf	Sisal, Agave, Fique, Pineapple, Lotus
	Fruit	Coconut (Coir), Kapok
	Algae	Cladophorales, Ulva Lactua, Cystoseria Myricaas
Agro-wastes	Stalk	Rice (straw and husk), Barley husk, Wheat straw, Bamboo, Grass, Maize straw, Lemon peel, Sunflower stalks, Bagasse etc.

4. Synthesis of Nanocellulose

Nanocellulose is formed by breaking down at least one dimension of the cellulose into nanoscales. Top-down and bottom-down, these two approaches are adopted to prepare nanocellulose. Pre-treatment and isolation of cellulose, followed by surface modifications, are the main steps to developing multifunctional nanocellulose. Physical treatments are mainly done after the surface modification. To avoid impurities, most of the raw materials are rinsed with deionized water and heated in an oven. After that, all the chemical and physical methods are applied.

5. Pre-Treatments

A high rate of yield and recovery of nanocellulose, minimal energy consumption, and low operating costs must be considered while choosing the pre-treatment method. Among several methods, hydrolysis is the most efficient and easiest method to deconstruct the lignocellulosic structure by cleaving the hydronium ion and other chemical bonds. Pre-treatment methods are classified primarily into three categories: 1.) chemical pre-treatment (treatment); 2.) physical pre-treatment (treatment); and 3.) biological pre-treatment (treatment) [6, 7]. One thing to notice while considering the synthesis procedures of different types of nanocellulose is that pre-treatments and treatments are interchangeable according to feasibility, type of availability of raw material, desired product, cost, and energy requirement.

Chemical Pre-treatment Method for Nanocellulose: Many acids, such as HCl, HBr, H_3PO_4, H_2SO_4, HNO3, etc., are utilised to break down the hard structure of cellulose. However, concentration, value of pH, hydrolysis time, and stability of acid affect the morphology and yield of nanocellulose [8, 9]. For example, if a 64% (w/w) H_2SO_4 solution with wheat bran cellulose is constantly stirred for 90 min, 60 min, and 30 min at 45°C, then centrifugation and dialysis are also applied to acid hydrolyzed cellulose to get cellulose nanocrystals. Results have shown that longer hydrolyzed materials have a higher surface area [10]. Although sulfuric acid isolates nanocellulose (CNC-cellulose nanocrystals) well, some problems, including corrosion of equipment, environmental incompatibility, long processing time, low thermal stability, and high water usage [6], are its limitations. Therefore, ionic liquids, organic acids [11], oxalic acid [12], and formic acid [13] are replacing sulfuric acid and other toxic acids as better and lower-toxic acid alternatives.

Sodium hydroxide is the most commonly used alkali to remove the hemicellulose and the lignin from the biomass matrix by targeting the esters and glycosidic side chains where the saponification reaction takes place. NaOH with a 10:1 ratio (NaOH to the biomass) delignifies the rice husk, treated for 2 hours at 80°C [14]. 5% (w/v) NaOH solution is added to treated wheat bran residue for 4 hours at 70°C under mechanical stirring [10], which helps in delignification. It is a greener process as compared to acid pre-treatment. KOH, CaOH2, and NH_3 are also used to prepare nanocellulose. Alkali swells the cellulose by increasing its surface area, and solubilization of lignin occurs, opening the chance to degrade the cellulose into sugar. To overcome this over-degradation, diluted acid treatment followed by alkali pre-treatment is done. For a high rate of yield of nanocellulose, bleaching treatment by $NaClO_2$ along with the alkali procedure is done to form cellulose nanofibers [15]. The bleaching process may be repeated many times until the desired lignin elimination is obtained. 7% NaOH over dry matter at 100°C for 150 min pre-treats the wheat straw afterwards for lignin elimination, and for each g of fibre, 0.3g of $NaClO_2$ (acidified via 2% acetic acid) is added to the 3% (w/v) fibre suspension at 75°C for 1 h for 3 cycles [16]. H_2O_2 is replacing $NaClO_2$ [17, 18] as a favourable chemical compound in order to protect the ecological system.

6. Physical Pre-Treatment Method of Nanocellulose

Chipping, grinding, and milling decrease the size of cellulose biomass by shear forces. Milling and grinding reduce the size more than chipping and do not produce any toxic chemicals [19]. Ball size, grinding time, moisture content, and cellulose weight to ball weight affect the production of nanocellulose. 500 g of lotus leaves were sent to the ball mill with 80 ceramic balls of diameter 3 cm for different time intervals such as 120, 96, 72, 48, and 24 h at 25 rpm, which gave average particle sizes of each 30g of lotus leaf powder as 19.5, 31, 52.3, 65.8, and 71.9, respectively [20]. This experiment concludes that higher adsorption efficiency is obtained by the smallest particle because of the higher available surface area. The only flaw of this method is that it is a high-energy-consuming process, whereas the grinding method consumes less energy than the high-pressure homogenization method but damages the structure of CNFs due to the presence of strong shear forces [21]. Potent ball milling techniques can be wet or dry. To limit de-crystallisation in dry milling, wet ball mills are often used, where liquid is added to recrystallize the biomass present in the ball mill [22]. It is based on a top-down approach.

7. Biological Pre-Treatment Method of Nanocellulose

This pre-treatment uses microorganisms such as fungi and bacteria. Brown-rot fungi, soft-rot fungi, white-rot fungi, and different genera of bacteria are used to disintegrate the cellulosic biomass, wherein a biological reaction takes place to hydrolyze the hard structure. White-rot fungi are the most efficient among all other microorganisms. Microbial pre-treatment has been witnessed to be an economic and efficient pre-treatment.

8. Treatments for Isolation of Nanocellulose

Oxidation, enzyme hydrolysis and utilization of deep eutectic solvents and ionic liquids are the main treatment methods to synthesize the nanocellulose. Two main oxidation routes are being operated to deconstruct the lignin and hemicellulose matrix from aromatic ring are 2,2,6,6-tetramethylpiperidine-N-oxyl (TEMPO)-oxidation and ammonium persulfate (APS) oxidation other than $NaClO_2$ and H_2O_2. The main objective of oxidation of lignocellulosic biomass is to minimise the consumption of energy during mechanical treatment and helping the alkaline pre-treatment. TEMPO-oxidation reduces the surface charges of fiber and provides colloidal suspension stability to form (cellulose nanocrystals) CNCs. TEMPO-oxidation is mostly used to synthesize CNCs and CNFs with the combination of catalyst NaBr and NaClO as a bleaching agent in basic medium [22, 23] 3g of Bleached bagasse were suspended in distilled water with

TEMPO (0.3 mmol) and NaBr (4.8 mmol). Then 15% NaClO was added with pH of 10 adjusted by NaOH with continuous stirring and at the termination of reaction pH was set to 7 and product was firstly centrifuged at 10,000 rpm and then dialyzed for 1 week using deionised water. Prepared product was then sent to the mechanical defibrillation [24]. Some Bleaching processes are often used to have faster reduction of lignin making use of chlorite and sodium acetate. However, employing these toxic reagents are not safe for the environment, thus ozone, hydrogen peroxide and APS (ammonium persulfate) came into picture because of their less toxicity. APS is a bleaching agent to solubilize lignin and cellulose of amorphous portions. The major drawback of APS is the longer processing time that confines its utilization at large-scale for nanocellulose production [25].

Cations and anions of salts are combined to form ionic liquids having the liquid state at room temperature. 1-ethyl-3-methylimidazolium acetate (EmimOAc), ($EmimHSO_4$), 1-butyl-3-methylimidazolium chloride (BmimCl), (BmimOAc), ($BmimHSO_4$) and 1-hexyl-3-methylimidazolium hydrogen sulphate ($HmimHSO_4$) etc. are the different types of potential ILs [26]. Low vapour pressure, nonflammability, chemical stability and recyclability make it a better and effective organic solvent to produce nanocellulose. The working principle of ILs is to solubilize the lignin and hemicellulose content by replacing the prior hydrogen bonds already present in the biomass matrix. New hydrogen bonds are formed between hydroxyl groups on cellulose and anions present on ILs [27]. But due to the presence of sulphate ion in some ILs causing a toxic environment and high cost are also raising the questions of the use of ILs. Here, Deep eutectic solvents (DES) being economical and lower toxic came to the surface in order to manufacture nanocellulose.

Recyclability, low toxicity and biodegradability of DES attract the researchers that advances the production of nanocellulose at full-scale [28]. DES is prepared by generating eutectic mixture of hydrogen bond acceptor (urea, quaternary ammonium, D-fructose and glycerol etc.) and hydrogen bond donor (amines, nicotinic acid, choline chloride and carboxylic acid etc.). DES treatment reduces the energy consumption, swells the cellulose fibre and assists the efficient cleaving of the hydrogen bond present in the cellulose fibre. Oxalic acid-choline chloride [29] and sulfamic acid-glycerol DES [30] have shown their potential in synthesising cellulose nanocrystals and cellulose nanofibril respectively.

Enzymatic hydrolysis is an eco-friendly approach, carried out at mild pressure and thermal conditions. It results in a lower energy consumption process. NC is produced by collaborating working of various enzymes on pre-treated cellulose biomass then post -treated mechanically or acid post-treatment is done [31]. Enzymatic cocktails consisting of (cellulases) exoglucanases, endoglucanases and β-glucosidases is added with the auxiliary enzymes such as xylanase to produce improved cellulose nanostructures [32]. Concentration ratio of cellulase to xylanase decides the morphology and structure of nanocellulose produced. Other than xylanase, α-amylase is also used to hydrolyse the wheat grain wherein 0.6% (w/v) of α-amylase at 70°C is mechanically stirred for 60 min to extract cellulose [10].

9. Surface Modification of Nanocellulose

Surface modification of nanocellulose is mostly done for the removal of contaminants present in water that are harmful and precarious for the ecological system. There are many types of surface modifications that have been done so far. The hydroxyl group of nanocellulose is targeted to get the desired modifications and further applications in water purification. Here, we will be discussing some techniques for surface functionalization. The general procedure to obtain CNC and CNF is illustrated in Figure 3.

Sulfonation passes the negative charges on the nanocellulose surface, improving its dispersion efficiency in water because of the formation of a negative electrostatic layer of sulphate groups on the surface [33], whereas thermal stability is gained by NaOH eutralization. H_2SO_4 and $NaHSO_3$ are common examples of this technique. In this method, hydroxyl groups are converted into sulphate alf-esters, which give the highly stable colloidal suspension of CNCs [34]. The negatively charged surface of nanocellulose attracts the cations (heavy metals) present in the water. In this series, there are many other surface functionalization techniques that impart negative charges on the nanocellulose surface. Phosphorylation changes the C2, C3, and C6 positions of cellulose into phosphate groups (PO_4^{-3}) [35, 36] by H_3PO_4, $(NH_4)HPO_4$, $POCl_3$, P2O5, and organophosphate compounds. The oxidation method is also used to modify the surface by using TEMPO-oxidation, in which the primary hydroxyl group ($-CH_2-OH$) is converted into aldehydes or carboxylic acid and the secondary hydroxyl group (-OH) does not get oxidised, and then finally into carboxylic acid by stable nitrosyl radicals [37], inducing a negatively charged surface that provides good dispersibility in water and a hydrophobic surface. Periodate and chlorite oxidation are also done to remove contaminants from water. Sialylation [38] converts the hydroxyl groups into sialons, and acetylation [39] increases the contact angle, indicating the increase in hydrophobicity of the surface. Both are well-known routes. Positively charged surfaces can also be induced by amination, which is used for the removal of negatively charged dyes and anions. Examples are ethylenediamine, diethylenetriamine, 3-aminopropyl-triethoxysilane, etc. [40].

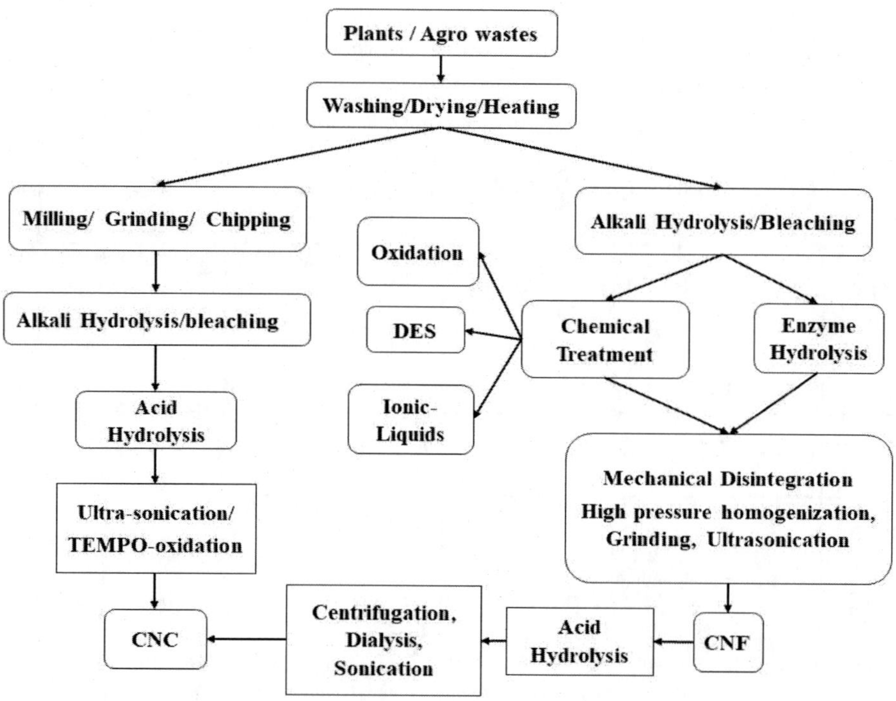

Figure 3. Isolation Techniques of Nanocellulose.

All these pre-treatment, treatment, and post-treatment steps may change the morphology and properties of the original nanocellulose, making it less valuable if done in the wrong way. The main purpose of grafting is to improve the thermal, physical, and mechanical properties of nanocellulose, but no significant adsorption abilities of grafted nanocellulose have been reported yet because of the steric hindrance induced by the density of the grafting polymer. However, grafting doesn't affect the functional groups of nanocellulose associated with adsorption. [41] Polyethyleneimine, polylactic acid, polymethacrylic acid, polyitaconic acid, polyacrylic acid, polyvinyl alcohol, polyvinyl acetate, etc. are some polymers used for grafting. Free-radical and ring-opening polymerization: these two reactions take place while modifications by grafting [42].

10. Mechanical Defibrillation

Milling, ultrafine grinding, high-pressure homogenization, ultrasonication, etc. are the physical methods to obtain defibrillated nanocellulose. High-pressure homogenization (HPH) uses high pressure and velocity. Shear forces influence the fibre structure and diminish its size to the nanometer range, where the hydrogen bond collapses [43]. 10 passes are required to get a 1 wt% cellulose nanofiber suspension of wheat straw, wherein 3 passes were at 600 bars and 900 bars and 4 passes were at 300 bars [16]. Structural damage to CNF, high operating costs, and the clogging issue of fibre at the orifice are the major drawbacks of HPH [7]. The clogging problem can be minimised by mechanical pre-treatment that reduces the size of the fibre before treating the cellulose with HPH [44]. HPH technique can be employed at large scale at the cost of high energy consumption that can be reduced with development of proper pre-treatment methods [45]. Besides, the HPH technique does not need any solvent and generates no wastewater [7]. HPH is an effective and quick continuous mechanical process with lots of homogenization cycles. The necessary degree of cell can be obtained by varying the pressure [46].

Table 2. Different Routes to Synthesize Nanocellulose

Raw Material	NC Type		Pre-Treatment	Surface Modification	Physical Treatment	Reference
Wheat Straw	CNF	CNF Type-1	NaOH+Acidic $NaClO_2$		High-Pressure Homogenization	[16]
		CNF Type-2		TEMPO-oxidation		
Lotus Leaf	CNC		-		Ball Milling	[20]
Bagasse	CNC		Bleaching	TEMPO-oxidation	Ultrafine Grinding	[24]
Rice Husk	CNF		NaOH+Acidic $NaClO_2$	TEMPO-oxidation	High-Pressure Homogenization	[14]
Wheat Bran	CNC		α-Amylase (Enzyme)+ acidic NaOH Hydrolysis	Acid Hydrolysis with sulphonation		[10]

Ultrasonication is another treatment method. It is a mechanical approach to fabricating nanocellulose. In this method, absorption of applied ultrasonic energy by the water molecules occurs, which induces cavitation, and small gas bubbles form and grow on the cellulose surface.

Defibrillation of the material is caused by the resulting hydrodynamic forces [47, 48]. Cellulose concentration, ultrasonication power, and processing time are independent parameters [49]. The heat generated during the process is controlled by the use of a water pool. Process optimisation, challenges in equipment design, and high energy consumption restrict the vast application of this method at the industrial scale. Recently, ultrasound-assisted methods are being adopted for minimising the toxic chemicals and improving the properties and yield of nanocellulose [50, 51]. Ultrasonication is also used at the pretreatment stage of cellulose feedstock. Several physical (enhanced surface area and structural disorientation) and chemical (reduced degree of polymerization of components) changes have been reported in cellulosic feedstock (Teo et al., 2020). A major drawback of ultrasonication is the high energy requirement. It is a cost-effective, highly efficient, and eco-friendly technique to produce nanocellulose. Different routes to synthesise nanocellulose is given in Table 2.

11. Application of Nanocellulose in Water Purification

Adsorption, photocatalytic degradation, coagulation/flocculation, and antimicrobial filtration are the chief processes that are used for water purification. Among these, adsorption is the most widely followed mechanism to purify the water because of its ease of operation, low cost, and high efficiency. Nano-sized cellulose has a high surface area, which increases its adsorption capacity. The aspect ratio of nanocellulose depends on sources and synthesis procedures; generally, CNF has a higher aspect ratio than that of CNC. The higher the aspect ratio, the higher the mechanical strength. Mechanical stiffness indicates cohesion and material stiffness. Adsorbents must have high mechanical stiffness to sustain other forces induced by separation.

Table 3. Different Surface-Modified Nanocellulosic Adsorbents

NC Type	Source	Surface Modification	Contaminants	Removal Efficiency/ Adsorption Capacity	Reference
CNF	Wheat Straw	TEMPO-oxidation	Methylene Blue	100% (almost)	[16]
		-	Methylene Blue	50-60%	
CNC	-	Polyethleniminne and β-cyclodextrin, crosslinked with polyacrylamide	Methyl Orange	155.93 mg/g	[51]
CNC	Lotus leaf	Non-selective Oxidation to -COOH	Methylene Blue	92.43% (27.96 mg/g)	[20]
			Formaldehyde	90.8% 1.18 (mg/g)	
CNF	Bagasse	TEMPO-Oxidation (TiO_2 - Nanocomposite)	Brilliant Blue	162 mg/g	[24]
CNC	Lotus leaf	Non-selective oxidation to - COOH	Ni^{+2}	39.7%	[20]
			Co^{+2}	44.9%	
			Fe^{+3}	50.2%	
			Mn^{+2}	54.3%	
			Cu^{+2}	56.3%	
CNF	Rice husk	Tempo-oxidation	La^{+3}	100.7 mg/g	[14]
			Pb^{+2}	193.2 mg/g	
CNC	Wheat bran	-	Cu^{+2}	5.82 mg/g	[10]
			Pb^{+2}	7.71 mg/g	
			Cd^{+2}	8.22 mg/g	

A high crystallinity degree decreases nanocellulose solubility and increases chemical resistance. Adsorption can be enhanced by surface functionalization, where the presence of a hydroxyl group on the nanocellulose surface is very important. Adsorbents should be stable in water to resist fouling during their application. For enhanced stability in water, nanocellulosic adsorbents are functionalized by anionic groups to improve dispersion in water. Another factor that determines the stability of an adsorbent is the surface tension. A favourable feature for anti-biofouling is high surface tension. Nanocellulosic adsorbents are used to remove harmful heavy metals, organic pollutants, oil-water separation, and suspended particles. Recently reported nanocellulosic adsorbents are listed in Table 3.

Conclusion

Sustainability is a salient feature that cannot be ignored in the current world. Greener approaches must be prioritised, even when choosing the raw material. However, the growth of industrialization and urbanisation opposes sustainability. So, it is the responsibility of everyone to lean towards a sustainable future. Cellulose is one of the best examples when it comes to eco-friendliness because of its biodegradability and non-toxicity. The applications of cellulose are in every field. Nanocellulose is an emerging nanomaterial because of its viable properties. We have discussed its application in water purification for removing heavy metals and dyes by adsorption mechanisms. Different types of treatment methods have been looked at. Among these, chemical pre-treatment followed by mechanical treatment came out as the most cost-effective route. Hydroxyl groups present on nanocellulose play a vital role in agglomeration that can be fixed by surface moderation. There are many moderation techniques, among which TEMPO-oxidation is considered the most sustainable approach. There are also other novel approaches to synthesising nanocellulose that can be an even better alternative to sustainability.

References

[1] Reshmy, R. "Potential of nanocellulose for wastewater treatment," *Chemosphere*, vol. 281, p. 130738, Oct. 2021, doi: 10.1016/J.CHEMOSPHERE.2021.130738.

[2] Ioelovich, M. "Nanocellulose-fabrication, structure, properties, and application in the area of care and cure," in *Fabrication and Self-Assembly of Nanobiomaterials: Applications of Nanobiomaterials*, Elsevier Inc., 2016, pp. 243–288. doi: 10.1016/B978-0-323-41533-0.00009-X.

[3] Moon, R. J., A. Martini, J. Nairn, J. Simonsen, and J. Youngblood, "Cellulose nanomaterials review: Structure, properties and nanocomposites," *Chem Soc Rev*, vol. 40, no. 7, pp. 3941–3994, Jun. 2011, doi: 10.1039/c0cs00108b.

[4] Azizi Samir, M. A. S., F. Alloin, and A. Dufresne, "Review of recent research into cellulosic whiskers, their properties and their application in nanocomposite field," *Biomacromolecules*, vol. 6, no. 2. pp. 612–626, Mar. 2005. doi: 10.1021/bm0493685.

[5] Abraham, E. "Extraction of nanocellulose fibrils from lignocellulosic fibres: A novel approach," *Carbohydr Polym*, vol. 86, no. 4, pp. 1468–1475, Oct. 2011, doi: 10.1016/j.carbpol.2011.06.034.

[6] Pradhan, D., A. K. Jaiswal, and S. Jaiswal, "Emerging technologies for the production of nanocellulose from lignocellulosic biomass," *Carbohydrate Polymers*, vol. 285. Elsevier Ltd, Jun. 01, 2022. doi: 10.1016/j.carbpol.2022.119258.

[7] Teo, H. L. and R. A. Wahab, "Towards an eco-friendly deconstruction of agro-industrial biomass and preparation of renewable cellulose nanomaterials: A review," *International Journal of Biological*

Macromolecules, vol. 161. Elsevier B. V., pp. 1414–1430, Oct. 15, 2020. doi: 10.1016/j.ijbiomac.2020.08.076.

[8] Pawcenis, D., M. Leśniak, M. Szumera, M. Sitarz, and J. Profic-Paczkowska, "Effect of hydrolysis time, pH and surfactant type on stability of hydrochloric acid hydrolyzed nanocellulose," *Int J Biol Macromol*, vol. 222, pp. 1996–2005, Dec. 2022, doi: 10.1016/J.IJBIOMAC.2022.09.289.

[9] Bacha, E. G. "Response Surface Methodology Modeling, Experimental Validation, and Optimization of Acid Hydrolysis Process Parameters for Nanocellulose Extraction," *S Afr J Chem Eng*, vol. 40, pp. 176–185, Apr. 2022, doi: 10.1016/J.SAJCE.2022.03.003.

[10] Xiao, Y., Y. Liu, X. Wang, M. Li, H. Lei, and H. Xu, "Cellulose nanocrystals prepared from wheat bran: Characterization and cytotoxicity assessment," *Int J Biol Macromol*, vol. 140, 2019, doi: 10.1016/j.ijbiomac.2019.08.160.

[11] Ramadhani, D. V. "Effect of lignocellulosic composition of Reutealis trisperma waste on nanocrystalline cellulose properties," *Biocatal Agric Biotechnol*, vol. 45, p. 102516, Oct. 2022, doi: 10.1016/j.bcab.2022.102516.

[12] Henschen, J., D. Li, and M. Ek, "Preparation of cellulose nanomaterials via cellulose oxalates," *Carbohydr Polym*, vol. 213, pp. 208–216, Jun. 2019, doi: 10.1016/J.CARBPOL.2019.02.056.

[13] Liu, C. "Properties of nanocellulose isolated from corncob residue using sulfuric acid, formic acid, oxidative and mechanical methods," *Carbohydr Polym*, vol. 151, pp. 716–724, Oct. 2016, doi: 10.1016/J.CARBPOL.2016.06.025.

[14] Zhan, C. "Rice husk based nanocellulose scaffolds for highly efficient removal of heavy metal ions from contaminated water," *Environ Sci (Camb)*, vol. 6, no. 11, pp. 3080–3090, Nov. 2020, doi: 10.1039/d0ew00545b.

[15] Pakutsah, K. and D. Aht-Ong, "Facile isolation of cellulose nanofibers from water hyacinth using water-based mechanical defibrillation: Insights into morphological, physical, and rheological properties," *Int J Biol Macromol*, vol. 145, pp. 64–76, Feb. 2020, doi: 10.1016/J.IJBIOMAC.2019.12.172.

[16] Morcillo-Martín, R., E. Espinosa, L. Rabasco-Vílchez, L. M. Sanchez, J. de Haro, and A. Rodríguez, "Cellulose Nanofiber-Based Aerogels from Wheat Straw: Influence of Surface Load and Lignin Content on Their Properties and Dye Removal Capacity," *Biomolecules*, vol. 12, no. 2, 2022, doi: 10.3390/biom12020232.

[17] Rosa, S. M. L., N. Rehman, M. I. G. de Miranda, S. M. B. Nachtigall, and C. I. D. Bica, "Chlorine-free extraction of cellulose from rice husk and whisker isolation," *Carbohydr Polym*, vol. 87, no. 2, pp. 1131–1138, Jan. 2012, doi: 10.1016/J.CARBPOL.2011.08.084.

[18] Hafemann, E., R. Battisti, D. Bresolin, C. Marangoni, and R. A. F. Machado, "Enhancing Chlorine-Free Purification Routes of Rice Husk Biomass Waste to Obtain Cellulose Nanocrystals," *Waste Biomass Valorization*, vol. 11, no. 12, pp. 6595–6611, Dec. 2020, doi: 10.1007/s12649-020-00937-2.

[19] Arora, A., P. Nandal, J. Singh, and M. L. Verma, "Nanobiotechnological advancements in lignocellulosic biomass pretreatment," *Mater Sci Energy Technol*, vol. 3, 2020, doi: 10.1016/j.mset.2019.12.003.

[20] Yang, H., "Facile and Effective Preparation of the Lotus Leaf-based Adsorbent by Exposing Cellulose Nanocrystal for Waste Water Treatment and Air Purification," *Fibers and Polymers*, vol. 21, no. 2, 2020, doi: 10.1007/s12221-020-9264-6.

[21] Krishna Shailaja, A. and B. Pranaya Ragini, "Nanocellulose: Preparation, Characterization and Applications," *Pharmaceutics and Pharmacology Research*, vol. 5, no. 1, 2022, doi: 10.31579/2693-7247/039.

[22] Rana, A. K., E. Frollini, and V. K. Thakur, "Cellulose nanocrystals: Pretreatments, preparation strategies, and surface functionalization," *Int J Biol Macromol*, vol. 182, pp. 1554–1581, Jul. 2021, doi: 10.1016/J.IJBIOMAC.2021.05.119.

[23] Zhang, H. "Extraction and comparison of cellulose nanocrystals from lemon (Citrus limon) seeds using sulfuric acid hydrolysis and oxidation methods," *Carbohydr Polym*, vol. 238, p. 116180, Jun. 2020, doi: 10.1016/J.CARBPOL.2020.116180.

[24] Al-Ahmed, Z. A., A. A. Hassan, S. M. El-Khouly, and S. E. El-Shafey, "TEMPO-oxidized cellulose nanofibers/TiO2 nanocomposite as new adsorbent for Brilliant Blue dye removal," *Polymer Bulletin*, vol. 77, no. 12, 2020, doi: 10.1007/s00289-019-03068-4.

[25] Mateo, S., S. Peinado, F. Morillas-Gutiérrez, M. D. la Rubia, and A. J. Moya, "Nanocellulose from agricultural wastes: Products and applications—a review," *Processes*, vol. 9, no. 9. MDPI, Sep. 01, 2021. doi: 10.3390/pr9091594.

[26] Haron, G. Abdalla Suliman, H. Mahmood, M. Hilmi Bin Noh, and M. Moniruzzaman, "Ionic liquid assisted nanocellulose production from microcrystalline cellulose: Correlation between cellulose solubility and nanocellulose yield via COSMO-RS prediction," *J Mol Liq*, vol. 368, p. 120591, Dec. 2022, doi: 10.1016/J.MOLLIQ.2022.120591.

[27] Rieland, J. M. and B. J. Love, "Ionic liquids: A milestone on the pathway to greener recycling of cellulose from biomass," *Resources, Conservation and Recycling*, vol. 155. 2020. doi: 10.1016/j.resconrec.2019.104678.

[28] Zargar, S., J. Jiang, F. Jiang, and Q. Tu, "Isolation of lignin-containing cellulose nanocrystals: life-cycle environmental impacts and opportunities for improvement," *Biofuels, Bioproducts and Biorefining*, vol. 16, no. 1, 2022, doi: 10.1002/bbb.2261.

[29] Lim, W. L., A. A. N. Gunny, F. H. Kasim, S. C. B. Gopinath, N. H. I. Kamaludin, and D. Arbain, "Cellulose nanocrystals from bleached rice straw pulp: acidic deep eutectic solvent versus sulphuric acid hydrolyses," *Cellulose*, vol. 28, no. 10, 2021, doi: 10.1007/s10570-021-03914-7.

[30] Li, W. "Facile preparation and characteristic analysis of sulfated cellulose nanofibril via the pretreatment of sulfamic acid-glycerol based deep eutectic solvents," *Nanomaterials*, vol. 11, no. 11, 2021, doi: 10.3390/nano11112778.

[31] Ribeiro, R. S. A., B. C. Pohlmann, V. Calado, N. Bojorge, and N. Pereira, "Production of nanocellulose by enzymatic hydrolysis: Trends and challenges," *Engineering in Life Sciences*, vol. 19, no. 4. Wiley-VCH Verlag, pp. 279–291, Apr. 01, 2019. doi: 10.1002/elsc.201800158.

[32] Bondancia, T. J., C. Florencio, G. S. Baccarin, and C. S. Farinas, "Cellulose nanostructures obtained using enzymatic cocktails with different compositions," *Int J Biol Macromol*, vol. 207, 2022, doi: 10.1016/j.ijbiomac.2022.03.007.

[33] Luo, J., N. Semenikhin, H. Chang, R. J. Moon, and S. Kumar, "Post-sulfonation of cellulose nanofibrils with a one-step reaction to improve dispersibility," *Carbohydr Polym*, vol. 181, 2018, doi: 10.1016/j.carbpol.2017.10.077.

[34] Thomas, B. "Nanocellulose, a Versatile Green Platform: From Biosources to Materials and Their Applications," *Chemical Reviews*, vol. 118, no. 24. American Chemical Society, pp. 11575–11625, Dec. 26, 2018. doi: 10.1021/acs.chemrev.7b00627.

[35] Ait Benhamou, A., "Extraction, characterization and chemical functionalization of phosphorylated cellulose derivatives from Giant Reed Plant," *Cellulose*, vol. 28, no. 8, 2021, doi: 10.1007/s10570-021-03842-6.

[36] Fiss, B. G., L. Hatherly, R. S. Stein, T. Friščić, and A. Moores, "Mechanochemical Phosphorylation of Polymers and Synthesis of Flame-Retardant Cellulose Nanocrystals," *ACS Sustain Chem Eng*, vol. 7, no. 8, pp. 7951–7959, Apr. 2019, doi: 10.1021/acssuschemeng.9b00764.

[37] Thakur, V., A. Guleria, S. Kumar, S. Sharma, and K. Singh, "Recent advances in nanocellulose processing, functionalization and applications: A review," *Materials Advances*, vol. 2, no. 6. 2021. doi: 10.1039/d1ma00049g.

[38] Indarti, E., Marwan, R. Rohaizu, and W. D. Wanrosli, "Silylation of TEMPO oxidized nanocellulose from oil palm empty fruit bunch by 3-aminopropyltriethoxysilane," *Int J Biol Macromol*, vol. 135, pp. 106–112, Aug. 2019, doi: 10.1016/j.ijbiomac.2019.05.161.

[39] Barbosa, R. F. S., A. G. Souza, F. F. Ferreira, and D. S. Rosa, "Isolation and acetylation of cellulose nanostructures with a homogeneous system," *Carbohydr Polym*, vol. 218, pp. 208–217, Aug. 2019, doi: 10.1016/j.carbpol.2019.04.072.

[40] Aoudi, B., Y. Boluk, and M. Gamal El-Din, "Recent advances and future perspective on nanocellulose-based materials in diverse water treatment applications," *Science of the Total Environment*, vol. 843. Elsevier B. V., Oct. 15, 2022. doi: 10.1016/j.scitotenv.2022.156903.

[41] Qiao, A. "Advances in nanocellulose-based materials as adsorbents of heavy metals and dyes," *Carbohydrate Polymers*, vol. 272. Elsevier Ltd, Nov. 15, 2021. doi: 10.1016/j.carbpol.2021.118471.

[42] Kedzior, S. A., J. O. Zoppe, R. M. Berry, and E. D. Cranston, "Recent advances and an industrial perspective of cellulose nanocrystal functionalization through polymer grafting," *Current Opinion in Solid State and Materials Science*, vol. 23, no. 2. Elsevier Ltd, pp. 74–91, Apr. 01, 2019. doi: 10.1016/j.cossms.2018.11.005.

[43] Nasir, M., R. Hashim, O. Sulaiman, and M. Asim, "Nanocellulose: Preparation methods and applications," in *Cellulose-Reinforced Nanofibre Composites: Production, Properties and Applications*, Elsevier Inc., 2017, pp. 261–276. doi: 10.1016/B978-0-08-100957-4.00011-5.

[44] Abdul Khalil, H. P. S. "Production and modification of nanofibrillated cellulose using various mechanical processes: A review," *Carbohydrate Polymers*, vol. 99. 2014. doi: 10.1016/j.carbpol.2013.08.069.

[45] Nechyporchuk, O., M. N. Belgacem, and J. Bras, "Production of cellulose nanofibrils: A review of recent advances," *Industrial Crops and Products*, vol. 93. 2016. doi: 10.1016/j.indcrop.2016.02.016.

[46] Kalia, S., S. Boufi, A. Celli, and S. Kango, "Nanofibrillated cellulose: Surface modification and potential applications," *Colloid and Polymer Science*, vol. 292, no. 1. Springer Verlag, pp. 5–31, Jan. 01, 2014. doi: 10.1007/s00396-013-3112-9.

[47] Phanthong, P., P. Reubroycharoen, X. Hao, G. Xu, A. Abudula, and G. Guan, "Nanocellulose: Extraction and application," *Carbon Resources Conversion*, vol. 1, no. 1, pp. 32–43, Apr. 2018, doi: 10.1016/J.CRCON.2018.05.004.

[48] Mazela, B., W. Perdoch, B. Peplińska, and M. Zieliński, "Influence of chemical pre-treatments and ultrasonication on the dimensions and appearance of cellulose fibers," *Materials*, vol. 13, no. 22, pp. 1–14, Nov. 2020, doi: 10.3390/ma13225274.

[49] Salimi, S., R. Sotudeh-Gharebagh, R. Zarghami, S. Y. Chan, and K. H. Yuen, "Production of Nanocellulose and Its Applications in Drug Delivery: A Critical Review," *ACS Sustainable Chemistry and Engineering*, vol. 7, no. 19. 2019. doi: 10.1021/acssuschemeng.9b02744.

[50] Hoo, D. Y. "Ultrasonic cavitation: An effective cleaner and greener intensification technology in the extraction and surface modification of nanocellulose," *Ultrason. Sonochem*, vol. 90, p. 106176, 2022, doi: 10.1016/j.ultsonch.2022.106176.

[51] Venkata Padma, D., S. V. A. R. Sastry, "Kinetic Studies on Simultaneous Biosorption of Divalent Copper and Hexavalent Chromium using Mallet Flower Leaf Powder," *Journal of Water and Environmental Nanotechnology*. vol. 5, no. 3, 2020 204-217. https://doi.org/10.22090/jwent.2020.03.002.

Chapter 30

An Assessment of Methane Generation from Municipal Solid Waste Using the LandGEM Model for Kanpur City, India

Shubham Chandra*
and Rajiv Ganguly
Department of Civil Engineering, Harcourt Butler Technical University, Kanpur, India

Abstract

Municipal solid waste disposal is an enormous challenge for developing countries such as India due to the increase in population, rapid urbanization, and waste generation. The practice of open dumping of municipal solid waste (MSW) in India leads to major global warming issues due to the presence of methane and carbon dioxide in landfill gases. The present study assesses the potential of methane generation from MSW determined in Kanpur city by using the LandGEM 3.02 version model developed by USEPA for the duration of 2015–2020. It was calculated that the average and total methane generation for the study period were 228.89 x 10^4 m^3/year and 1373.35 x 10^4 ^{m3,} respectively. The default input parameters for the values of k and L_o were modified to be more representative of the study conditions. The k and L_o values used for the study were 0.05 and 38 m3/Mg, respectively. It was further observed that the methane generation potential from MSW increased by a factor of 7.72 for 2020 compared to the base year of 2016. This suggests that enormous volumes of methane gases are emitted from the open dumpsite in Kanpur, which should be collected and harnessed for fuel and electricity production in Kanpur city.

Keywords: municipal solid waste, methane emissions, LandGEM 3.02

1. Introduction

Numerous climate changes are brought on by the substantial rise in municipal solid waste generation in India. Uncontrolled municipal waste disposal at open landfills and dump sites is

* Corresponding Author's Email: sc862278@gmail.com.

In: Sustainability in Chemical Processes through Digitalization and Green Chemistry Approaches
Editors: Susarla V. A. R. Sastry, Shravan Kumar and Ashish Kapoor
ISBN: 979-8-89113-257-3

currently a major challenge for municipal authorities. Methane, carbon dioxide, and a very small amount of other gases are produced as a result of these unmanaged waste disposal methods, which have serious negative impacts on the environment. In tropical nations such as India, a large portion of the generated solid waste contains organic content, which has a significant potential to emit greenhouse gases [1]. Methane gas produced during organic matter decomposition has a global warming potential approximately 20 times greater than that of carbon dioxide, which is the main greenhouse gas released as a result of open dumping.

2. Site Location

According to previous estimates, Kanpur City produces approximately 1500 metric tonnes of municipal solid waste every day, of which 80% is collected and dumped on site [2]. The only location used to dispose of MSW in the city of Kanpur is the Panki dump site. The dump site is located at 26° 46 36.72 N and 80° 34 51.46 E latitude and longitude, respectively. Approximately 4.6 km^2 of land is designated as a Panki dump site, which has operated for only 21 years. The recorded annual precipitation and temperature by the municipal corporation are 804 mm and 28 °C, respectively. There are no facilities for composting or covering landfills.

3. Methodology

3.1. MSW Generation in Kanpur City

Over the course of the study, the total amount of municipal solid waste produced is calclated as the product of Kanpur city's predicted population in a given year and the solid waste generation rate. Eighty percent of the total MSW produced in the city is dumped at the Panki open dumpsite [2], which is taken into account as the overall collection efficiency. The total MSW generation trend is shown in Table 1.

3.2. Methane Emission by LandGEM 3.02

The United States Environmental Protection Agency's (USEPA) LandGEM 3.02 modeling tool is used for quantifying greenhouse gas emissions and several other air pollutants found in municipal solid waste dumps. LandGEM 3.02 follows first-order decay decomposition and calculates the amount of landfill gas generation. Equation (1) shows the equation for the LandGEM 3.02 model.

$$QCH4 = \sum_{i=1}^{n} \sum_{j=0.1}^{1} kLo \frac{Mi}{10} e^{-ktij} \quad (1)$$

where Q_{CH4} = yearly methane generation (m^3/year), Lo = methane generation potential (m^3/Mg), I = annual time increase, J = the time increment, K = constant of methane generation, and Mi = mass of solid waste in the eighth year (Mg).

4. Results and Discussion

4.1. Total Municipal Solid Waste Generated

The total quantity of MSW generated in Gg/year in every year is calculated by using the anticipated population and the annual rate of waste generation for the period 2015–2020 [3]. The findings show that 350282 m3 of MSW was produced in Kanpur City in 2015. Furthermore, by taking into account the change in the MSW generation rate from 0.398 to 0.910 kg/capita/day for 2015 and 2020, respectively, as indicated in Table 1, the potential of MSW generation in the city increases by 2.37 times, from 350282 m3 in 2015 to 830160 m3 in 2020. The previous research carried out for Kanpur city (2010–2030) assumed a consistent growth rate of 0.6 kg/capita/day for the production of municipal solid waste. [2].

Table 1. Total MSW generation in Kanpur City

S.no	Year of forecasting	Estimated population (persons)**	MSW generation rate kg/capita/day	Total MSW (Mg/year)	Total MSW reaching dumping site (Mg/year)
1.	2015	3015060	0.398	437852	350282
2.	2016	3036069	0.6	664899	531919
3.	2017	3058239	0.679	757939	606351
4.	2018	3080571	0.76	854550	683640
5.	2019	3100285	0.840	950547	760438
6.	2020	3123990	0.910	1037633	830107

** Population source *https://worldpopulationreview.com/world-cities/kanpur-population)*

4.2. Estimation of CH_4 Emissions

Landfill gases generated from MSW generally consist of four components: approx. 50% methane, 45–50% CO_2, and a very small amount of NMOCs. In previous literature, it was found that there was no gas generation in the initial year of MSW deposition [3]. By using LandGEM 3.02, the total volumes of LFG and CH4 were 2746.70 x 104 m3 and 1373.35 x 104 m3, respectively. The Lo default value is 170 m3/Mg, which is site-specific for the USA, considering the value of Lo for Kanpur city as 38 m3/Mg [2]. Annual landfill gas emissions and trends for duration (2015–2020) are shown in Table 2 and Figure 1, respectively.

Table 2. Methane emission from MSW

S.no	Forecasted year	Estimated population (persons)**	Total waste reaching dumping site (Mg/year)	Total Landfill gas generation x 10^4 (m^3/year)	Methane emission x 10^4 (m^3/year)
1.	2015	3014060	350282	0	0
2.	2016	3036069	531919	130.16	65.08
3.	2017	3058239	606351	321.46	160.73
4.	2018	3080571	683640	531.10	265.55
5.	2019	3100285	760438	759.22	379.61
6.	2020	3123990	830107	1004.76	502.38
Total emission				2746.70	1373.35
Average emission				457.78	228.89

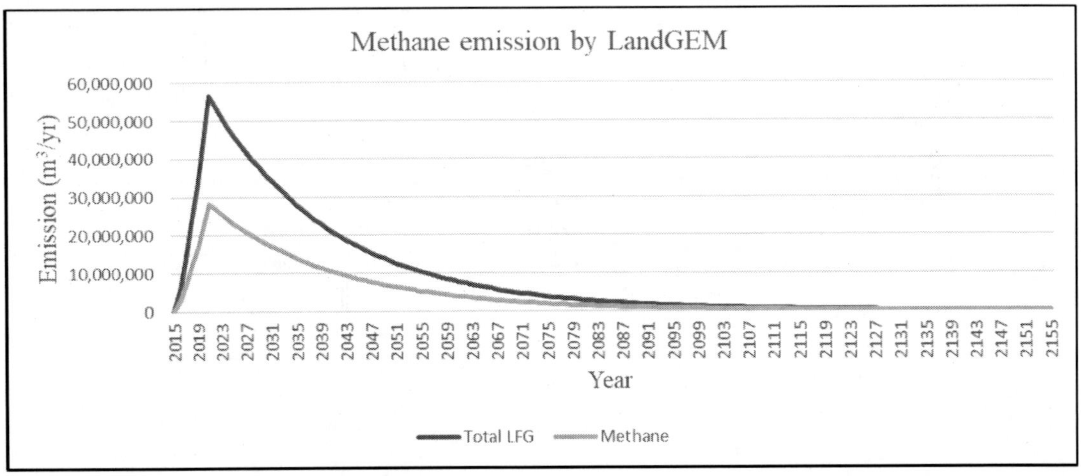

Figure 1. Methane emission trend from generated MSW.

Conclusion

In this study, an attempt was made to quantify the methane generation potential for the Panki open dumping site (Kanpur), India, by using the LandGEM 3.02 model. The total landfill gas and methane emissions estimated for 2015 to 2020 by the LandGEM 3.02 model result in 2746.70 x 10^4 m^3 and 1373.35 x 10^4 m3, respectively. The average rate of methane generation was 228.89 x 10^4 m^3/year. Due to the uncertainty associated with the value of K (between 0.04 and 0.06), the variation in methane emissions was 1098.68 x 10^4 m^3 to 1648.02 x 10^4 m^3. It can be concluded that the collection of landfill gases helps Kanpur Municipal Corporation and other management authorities plan for an energy recovery system and efficient utilization of the gases generated.

Acknowledgment

The author thanks the authorities of the Kanpur municipal corporation and development authority for providing pertinent data that were used in the study, as well as the prior research of other researchers to which the author has made references. The author is also appreciative of the editor's and reviewer's remarks for strengthening the manuscript.

References

[1] Srivastava, A. N. and Chakma, S., 2020. Quantification of landfill gas generation and energy recovery estimation from the municipal solid waste landfill sites of Delhi, India. *Energy Sources, Part A: Recovery, Utilization, and Environmental Effects*, 1-14. https://doi.org/10.1080/15567036.2020.1754970.

[2] Kaushal, A. and Sharma, M. P. 2016. Methane emission from Panki Open Dump Site of Kanpur, India. *Procedia Environmental Sciences*. 35:337-347. https://doi.org/10.1016/j.proenv.2016.07.014.

[3] Mehra, A. and Bhargava, A. 2020. Sustainable Municipal Solid Waste Management - A case study of Kanpur, India. *Journals of new Development in Chemistry*, Vol-3 Issue-1(12-22). https://doi.org/10.14302/issn.2377-2549.jndc-20-3516.

Chapter 31

The Numerical Simulation and Comparison of Flow through Y and Spherical Junction Ventilator Splitters

Aniruddh Mukunth
Raj Shree Rajagopalan
and Naren Rajan Parlikkad*
School of Chemical & Biotechnology, SASTRA Deemed to be University, Thanjavur, Tamil Nadu, India

Abstract

Mechanical ventilation is a technique in which air and oxygen mixtures are pumped into the lungs of patients who have low levels of oxygen and are unable to breathe on their own. During emergencies such as the outbreak of a pandemic such as COVID-19, hospitals run short of ventilator devices to meet the growing number of patients who require ventilator support. A single ventilator device should be used to ventilate multiple patients using a flow splitter as an emergency resort. This study aims to analyze the effect of different junction geometries and split angles on the hydrodynamics of flow through a typical ventilator splitter used for two patients having the same compliance and requiring equal flow rates. Three-dimensional spherical junctions and Y-shaped flow splitters were constructed, having an inlet diameter of 21.8 mm and splitting into two exit arms of 19 mm each. For the analysis, the Y-splitter was compared with the spherical junction splitter, which has a split angle over a range of 30° to 180°. Additionally, spherical junction splitters with diameters varying from 40 to 60 mm were analysed. A transient 3D model was developed in ANSYS 2019 R1 to simulate the flow through the ventilator splitter for a range of inlet mass flow rates. A splitter with an equal outlet mass flow rate was identified and recommended.

Keywords: CFD, mechanical ventilation, ventilator splitter, split angle

* Corresponding Author's Email: pnaren@scbt.sastra.ac.in.

In: Sustainability in Chemical Processes through Digitalization and Green Chemistry Approaches
Editors: Susarla V. A. R. Sastry, Shravan Kumar and Ashish Kapoor
ISBN: 979-8-89113-257-3

Nomenclature

A_i Area, (m^2)
$\overline{u}$ Average outlet velocity, (m/s)
u_i Outlet velocity at the i th arm, (m/s).

Greek Symbols

γ Uniformity index, (-).

1. Introduction

Acute respiratory distress syndrome (ARDS) is a type of medical condition of the lungs in which the maintenance of the required levels of oxygen and carbon dioxide in the body becomes difficult for the person suffering from it. A mechanical ventilator is a device required for patients with ARDS or internal blockages in the lungs. The tidal volume is the volume of air that is exhaled or inhaled during the respiratory cycle. The tidal volume of air required for the patients is supplied based on compliance and airway resistance in the lungs. The mechanical ventilator provides positive pressure to supply the required tidal volume of air to the patient. Based on the patient s condition, the ventilator can be kept in appropriate modes, such as assist-control, synchronized intermittent mechanical ventilation, and pressure support ventilation. The tidal volume, respiratory rate, and positive end-expiratory pressure are set to facilitate breathing.

SARS-CoV-2 is part of the coronavirus family, which uses healthy cells to multiply and evade other cells [1]. This type of coronavirus infects the lower and upper parts of the trachea and lowers the ability to exchange oxygen and carbon dioxide. Patients require external support for breathing to supply the oxygen needed to the body and for the lungs to recover as ARDS progresses. A shortage of medical equipment, such as mechanical ventilators, was observed during the surge in the second wave of coronavirus cases in India, and an increase in the mortality rate was also observed [2]. As an emergency resort, using a ventilator splitter, multiple patients were connected to a single ventilator to distribute the flow among two patients.

The ventilator splitter design must be carefully chosen, as even flow distribution among the patients is critical, as medically higher-level care is required for ARDS patients. Therefore, the ventilator splitter must provide equal and sufficient tidal volume to two or more patients with a minimum pressure drop. In the present work, a preliminary approach was used to study the hydrodynamics by understanding the behavior and distribution of the flow for patients with equal compliance. The analysis was carried out by varying the inlet air pressure of the patients and maintaining a constant inspiratory pressure. It is also assumed that the inspiratory flow of air is continuous and constant.

2. Literature Review and Objectives

2.1. Previous Study on Ventilator Splitters

Experimental and simulation reports on mechanical ventilation for multiple patients were consolidated as part of the current work. They used a ventilator splitter with a 180º angle of split at the junction for mechanical ventilation for multiple patients [3]. They reported that a 180º angle split was acceptable based on their experimental study with simulated lung models. They reported the acceptance of the design based on sensors in the mechanical ventilator that ensure set values of pressure and tidal volume are supplied without difficulty. Similarly, experiments with a custom two-port 3D-printed splitter with an approximately 90º split angle with healthy humans [4] also reported the acceptance of the design based on sensors in a mechanical ventilator. However, the actual flow distribution ratio at outlet arms was not reported [3, 4]. The use of a ventilator splitter would not guarantee adequate airflow to the patients. In all the studies, the designs used for ventilation were not conclusive enough in terms of equal flow distribution [5]. The authors suggested performing simulation studies through CFD to understand the presence of dead zones or any other flow parameters involved in the flow distribution. The transient state time marching method of solution was used to model the breathing cycle using Open FOAM for various restrictors to analyze the efficacy of using ventilator splitters for patients with different lung compliance [6]. The pressure drop between the inlet and outlet planes was analysed, but the equality of flow distribution was not discussed.

2.2. Previous Study on General Flow Splitters

The experimental and simulation results of various flow splitters used in other applications, such as heat exchangers, were also studied as part of the current work. Experiments with water for compact parallel-flow heat exchangers for a nine-number bundle tube with diameters of 2 mm and 3 mm and various inlet headers [7, 8] They reported flow maldistribution as nonuniformity (standard deviation) measured using i and β. Flow ratio, i, is measured as the ratio of volumetric flow rate in the i th tube to total volumetric flowrate and, the average flow ratio for the total tubes. In their experiments, they used a volumetric flowrate ranging from 0.5 LPM to 3.0 LPM. They reported nonuniformity ranging from 0.0013 to 0.0096 for various configurations of the inlet header. CFD studies with air for plate-fin heat exchangers on various numbers and configurations of splitter plates [9-11] The mass flow rate varied from 0.09 to 0.24 kg/s in their study. The flow maldistribution was measured as a standard deviation calculated using outlet channel mass flow rates. The flow maldistribution for different inlet headers was found to be from 0.05867 to 0.00498 kg/s. These studies indicate that for any configuration or design of the geometry, there is flow disturbance or uneven flow distribution at the outlet caused by the junction. Therefore, a detailed study of the flow splitter is required to understand the hydrodynamics of splitting and identify the design and geometry to obtain an equal flow distribution.

2.3. Medical Data for Mechanical Ventilation

The medical data to be used for a single patient were collected directly from the Meenakshi Hospital in Thanjavur. The medical data were collected from two critical care patients and a test lung ventilator. The commonly used range of pressure, tidal volume, and respiratory rate data for a single patient is presented in Table 1.

Table 1. Medical data for mechanical ventilation

Variable	Range	Unit
Pressure (gauge)	15-30	mbar
Tidal Volume	350-800	mL
Inspiration Time	1-1.2	s

3. Computational Model

A computational model was developed for simulating flow through the Y-splitter and the spherical junction-splitter. The computational domain consists of an inlet pipe and two outlet pipes joined smoothly at the junction. The 2-dimensional images of the Y-splitter and spherical junction-splitter CAD models with a split angle of 90° created using ANSYS 2019 R1-SpaceClaim are given in Figures 1 and 2. A standard k-epsilon model with a standard wall function was utilized to model the flow behavior, as the flow was transient and turbulent in character. The fluid was assumed to be incompressible. The effect of gravity was not considered in the simulation. In the simulation, air was used as a fluid material with a density of 1.225 kg/m^3 and a viscosity of 1.7894 x 10^{-5} kg/m2. The mechanical ventilator, on the other hand, supplies the patients with a mixture of air and oxygen.

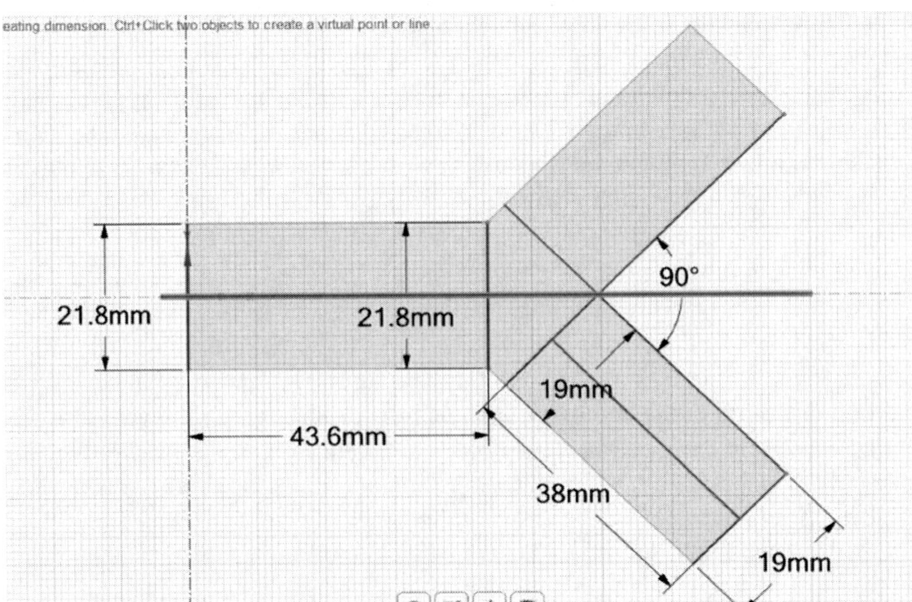

Figure 1. Y-splitter (90°) CAD model dimensions.

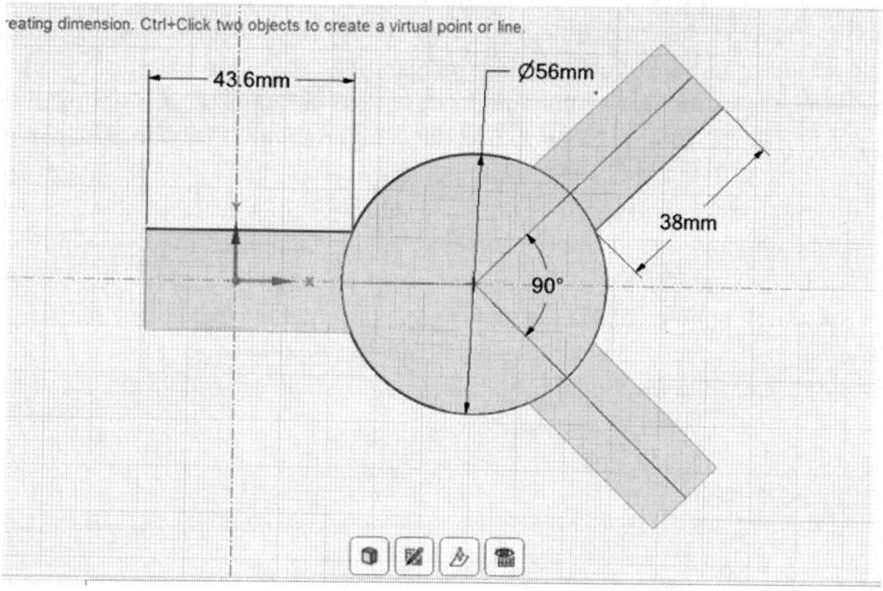

Figure 2. Spherical Junction-Splitter (90°) CAD model dimensions.

3.1. Boundary Conditions

The required tidal volume for two patients ranged from 700 mL to 1600 mL. The inspiration time of the breath is assumed to be constant at 1.2 s. The continuous inlet velocity based on the tidal volume and inspiration time was computed as 1.56–3.57 m/s. The Reynolds number is calculated based on an inlet diameter of 21.8 mm and is tabulated in Table 2.

The inlet and outlet boundary conditions were taken as a velocity inlet and pressure outlet with a gauge pressure of 2000 Pa, respectively. No-slip boundary conditions were used for the wall.

3.2. Mesh Independence Study and Sampling Studies

To establish that the simulated solution is independent of the mesh used in the numerical solution, a mesh-independent study was carried out. A polyhedral mesh with three inflation layers was used in the study, and the study indicated that 0.9 mm and 1.25 mm polyhedral meshes can be used for the Y-splitter and spherical junction splitter simulations, respectively.

Table 2. Superficial velocity vs Reynolds number

Superficial velocity (m/s)	Reynolds number
1.56	2315
2.07	3071
2.57	3813
3.07	4555
3.57	5297

Sampling times denote the time period of averaging performed in an unsteady state simulation, resulting in time-averaged flow profiles. The sampling time chosen must capture the values to represent the flow behavior for the entire duration. The outlet parameters with various sampling times were recorded to find the sampling time. The values were recorded for sample times of 1 s, 2 s, 4 s, and 6 s with a frequency of 1 time step for the Y-splitter and spherical junction splitter with a 90° split angle. On the outlet surface, 50 radial bands were created, and the circumferential average velocity in the axial direction at the outlet arms was measured for various sample times. There was no variation in the outlet mass velocity for the Y-splitter for sampling times of 1 s, 2 s, 4 s, and 6 s. Hence, a sample time of 1 s was used in further simulations for the Y-splitter. Similarly, no variation between the 4 s and 6 s samples was observed for the spherical junction splitter, and the sampling time was chosen as 4 s.

4. Results and Discussion

4.1. Simulation Results

A three-dimensional unsteady state model with standard k-ε turbulence closure was employed to simulate the flow. The Y-splitter and spherical junction splitter with split angles ranging from 30° to 120° was simulated over a range of air flow rates between 35 LPM and 80 LPM. The uniformity index, γ is calculated using the following expression.

$$\gamma = 1 - \frac{\sum_{i=1}^{n}[(|u_i - \overline{u}|)A_i]}{2\overline{u}\sum_{i=1}^{n} A_i} \quad (1)$$

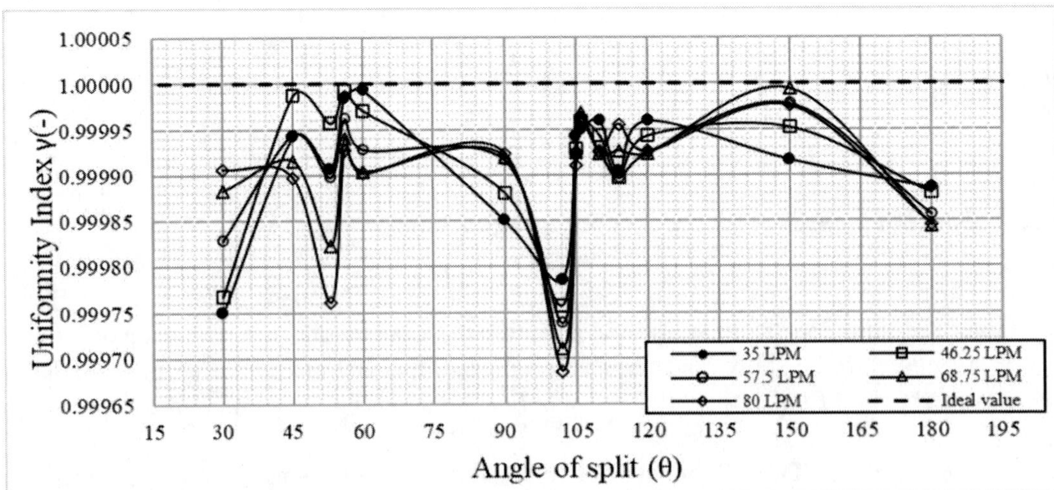

Figure 3. Simulation results of Y-splitter.

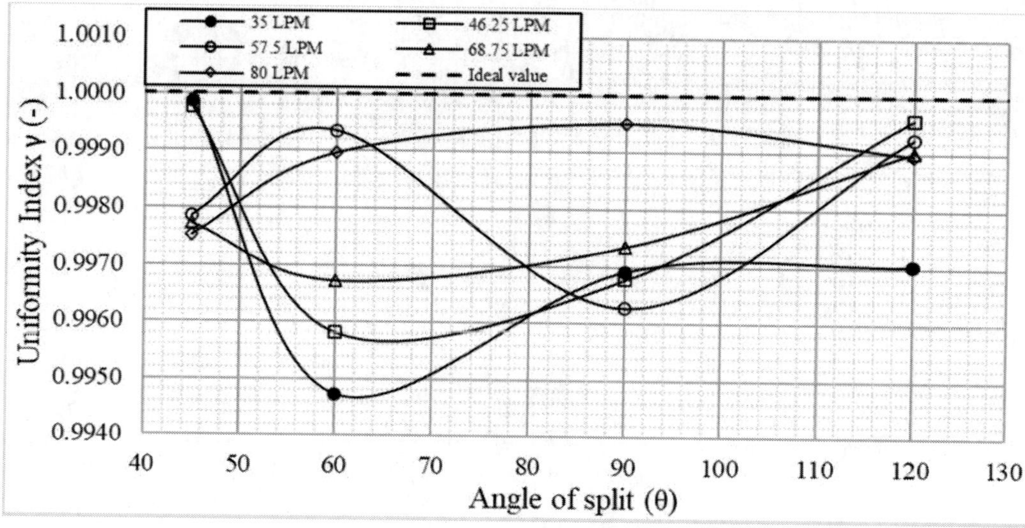

Figure 4. Simulation results of the spherical junction splitter.

The ideal flow distribution is represented by a uniformity index value of 1. Any value below 1 indicates nonuniformity in flow distribution, with 0 corresponding to the highest possible maldistribution in flow.

A summary of the simulation results is presented in Figures 3 and 4. As the expected differences in the flow uniformity index are known to be less, the number of significant digits was kept higher. Thus, it was necessary to maintain more precision to compare the results. It was observed from the preliminary analysis (45º, 60º, 90º, and 120º) that for the spherical junction splitter, the uniformity index is lower and the pressure drop is higher than for the Y-splitter. Hence, the spherical junction splitter cannot be recommended for use in mechanical ventilation and was not considered for further study.

4.2. Pressure and Velocity Contours

To comprehend the mal-distribution of flow, pressure and velocity contours were plotted and studied for the Y-splitter and the spherical junction splitter. The splitter designs with the least and highest deviation from the ideal uniformity index were recorded and compared. A low-pressure region at the sharp bends and an unequal pressure distribution at the junction were observed at an inlet velocity of 3.57 m/s. Recirculation of flow was also observed at certain sections of the low-pressure region in both splitters. The velocity contour and recirculation region for the Y-splitter with 150° and 102° angles of split at an inlet velocity of 3.57 m/s Comparing the recirculation of flow from the figures, it is observed that the 150° split angle is less than the 102° split angle for the Y-splitter. The velocity contour and recirculation regions for the spherical junction splitter with 120° and 45° angles of split at an inlet velocity of 3.57 m/s. In both spherical junction-splitter designs, dead zones were present in the junction just after the inlet pipe. The recirculation is less at the 120° split angle when compared to the 45° split angle. The mal-distribution of flow could be attributed to the recirculation zones. This observation from Y-splitter and spherical junction splitter contours suggests that minimizing recirculation is necessary to create an even flow distribution.

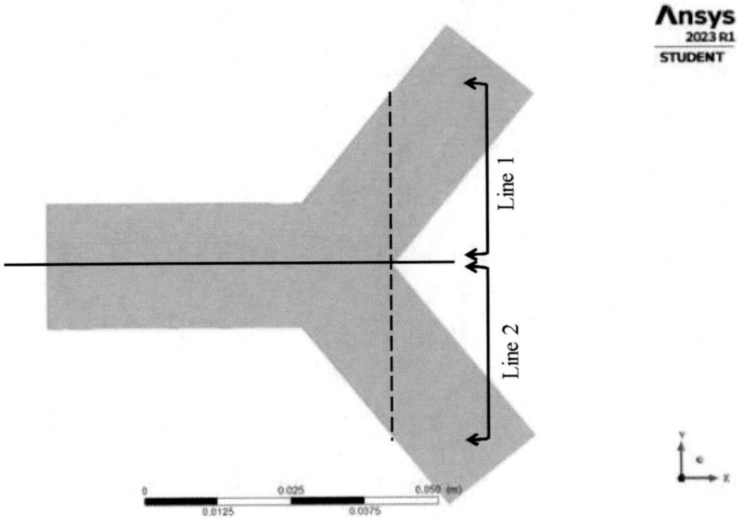

Figure 5. At the Z = 0 plane with a line at the Y-splitter junction of 102°.

The location of the maximum velocity is skewed away from the center at the outlet arms, as shown in the velocity contour plotted for the Y-splitter and spherical junction splitter.

As shown in Figure 5, two lines (Line 1 and Line 2) parallel to the Y-axis were plotted at the junction of the Y-splitter geometries to determine the pressure distribution between the points on the lines. The flow distribution is equal if the pressure distribution is symmetrical or the pressure difference between the points on Line 1 and Line 2 is zero. The point pressure values at 15 equidistant points from the X-axis (X = 0) on both lines were recorded. The average pressure difference at the junction is calculated for the best and worst splitting angles for each flow rate and is presented in Table 3. A low pressure difference at the junction tends to provide less flow maldistribution, as observed from Table 3.

Table 3. Average pressure difference at the junction of the best and worst Y-splitter geometries at various flow rates

Inlet Velocity (m/s)	Angle (θ)	Splitter Design -Deviation	Average junction pressure difference (Pa)	Uniformity index, γ (-)	Deviation from ideal ratio (%)
1.56	30°	Higher	0.00415	0.99975	0.025
	60°	Lower	0.00395	0.99999	0.0007
2.07	102°	Higher	0.01246	0.99976	0.0242
	55°	Lower	0.00836	0.99999	0.0007
2.57	102°	Higher	0.01851	0.99974	0.02617
	150°	Lower	0.00912	0.99998	0.0022
3.07	102°	Higher	0.02600	0.99971	0.0289
	150°	Lower	0.01283	0.99999	0.000649
3.57	102°	Higher	0.03108	0.99969	0.031410
	150°	Lower	0.023326	0.99998	0.002484

The Y-splitter performs better than the spherical junction splitter, as more recirculation is seen in the latter. The Y-splitter with a split angle of 1500 provides the least deviation from the ideal uniformity index (g = 1). The uniformity index is dependent on both the flow rate and the split angle. At certain ranges between 550–560 and 1050–1060, the split ratio becomes independent of the flow rate. From the velocity and pressure contours, the possible reasons for the maldistribution of flow are understood. The findings of the study indicate that to attain equal flow, the pressure distribution must be uniform at the junction, and low-pressure regions must be avoided. If there is recirculation present in the lower pressure region, then the recirculation must also be minimized to obtain a better flow distribution across the exit arms.

Conclusion

In the present study, the hydrodynamics of the ventilator splitter were studied by comparing the Y-splitter and the spherical junction splitter to identify the geometry for equal distribution of flow. By performing a preliminary analysis, it was observed that the Y-splitter distributed the flow more equally than the spherical junction splitter. The Y splitter with 14 different split angles was studied further. The uniformity index was dependent on the flow rate as well as the split angle. For the Y-splitter, the least and highest deviations from the ideal uniformity index were observed at a 150° split angle and a 102° split angle, respectively. The pressure and velocity contours were constructed to understand the flow mal-distribution.

The use of symmetric planes to mesh a section of symmetric geometry has been done in many studies. This is also a commonly suggested meshing technique in CFD studies [9]. This method of symmetric meshing was used by Longest and Vinchurkar [10] to study the effect of particle deposition in bifurcating airway models where only one-quarter of the geometry was meshed. Tran and Kim [11] meshed half of their geometry because of symmetry in the geometry for the Pico propeller hydro turbine applied in fish farms. In the current work, the complete domain of the geometry was simulated instead of simulating only a particular section to understand the flow mal-distribution of the outlet arms. It was evident from the results of the present study that there is a mal-distributed flow despite the symmetric geometry. In other words, the flow is asymmetrical in nature. Hence, simulating flow by assuming symmetric domains is not appropriate to capture the hydrodynamics in such systems. The method of reducing the computational domain to a particular section based on geometric similarity may not be appropriate.

Acknowledgments

The authors thankfully acknowledge the financial support provided by the Indian Institute of Chemical Engineers (IIChE) for carrying out B.E. or B.Tech. project work on this subject. We thank Meenakshi Hospital, Thanjavur, for helping us collect all the medically and technically relevant data. We extend our thanks to SAȘTRA Deemed University for providing remote server access to perform the simulation with ANSYS 2019 R1.

References

[1] *COVID-19 and the lungs.* (n.d.). Retrieved March 1, (2023) https://www.nhlbi.nih.gov/coronavirus/lungs.

[2] Ghosh, D. D., Sarkar, A., & Chouhan, D. P. (2021). COVID-19 second wave: District level study of concentration of confirmed cases and fatality in India. *Environmental Challenges*, 5, 100221. https://doi.org/10.1016/j.envc.2021.10022.

[3] Neyman, G., & Irvin, C. B. (2006). A single ventilator for multiple simulated patients to meet disaster surge. *Academic emergency medicine: official journal of the Society for Academic Emergency Medicine,* 13(11), 1246–1249. https://doi.org/10.1197/j.aem.2006.05.009.

[4] Ayyıldız, S., Dursun, A. M., Yıldırım, V., İnce, M. E., Gülçelik, M. A., & Erdöl, C. (2020). 3D-printed splitter for use of a single ventilator on multiple patients during COVID-19. *3D Printing and Additive Manufacturing,* 7(4), 181-185. https://doi.org/10.1089/3dp.2020.0102.

[5] Advincula, R. C., Dizon, J. R. C., Chen, Q., Niu, I., Chung, J., Kilpatrick, L., & Newman, R. (2020). Additive manufacturing for COVID-19: Devices, materials, prospects, and challenges. *MRS communications*, *10*(3), 413–427. https://doi.org/10.1557/mrc.2020.57.

[6] Duke, D. J., Clarke, A. L., Stephens, A. L., Djumas, L., & Gregory, S. D. (2022). A computational fluid dynamics assessment of 3D printed ventilator splitters and restrictors for differential multipatient ventilation. *3D printing in medicine*, *8*(1), 2. https://doi.org/10.1186/s41205-021-00129-1.

[7] Wang, C. C., Yang, K. S., Tsai, J. S., & Chen, I. Y. (2011). Characteristics of flow distribution in compact parallel flow heat exchangers, part I: Typical inlet header. *Applied Thermal Engineering*, 31(16), 3226–3234. https://doi.org/10.1016/j.applthermaleng.2011.06.004.

[8] *https://doi.org/10.3390/en13061323.*

[9] *Learning center resources.* (n.d.). Retrieved March 1, (2023), https://www.comsol.com/support/learning-center/article/Using-Symmetry-to-Reduce-Model-Size-35921.

[10] Longest, P. W., & Vinchurkar, S. (2007). Effects of mesh style and grid convergence on particle deposition in bifurcating airway models with comparisons to experimental data. *Medical Engineering & Physics, 29*(3), 350-366. https://doi.org/10.1016/j.medengphy.2006.05.012.

[11] Tran, B. N., & Kim, J. (2019). Design and analysis of a Pico Propeller Hydro turbine applied in fish farms using CFD and experimental method. *Journal of the Korean Society of Marine Environment and Safety*, 25(3), 373-380. https://doi.org/10.7837/kosomes.2019.25.3.373.

Chapter 32

The Treatment of Industrial Wastewater Containing Dyes Using Bio-Adsorbent Made from Different Leaves

Anjali Awasthi*
Khushi Rastogi
Vineet Maheshwari
Susarla Venkata Ananta Rama Sastry
and Ashish Kapoor
Department of Chemical Engineering, Harcourt Butler Technical University, Kanpur, Uttar Pradesh, India

Abstract

Dye effluents are non-biodegradable wastes that are produced by a multitude of industries, including pharmaceuticals, leather, paper, textiles, etc. Typically, dyes scatter or absorb sunlight that penetrates a water system, preventing light from absorption in the water. This, in turn, significantly slows photosynthesis and has a negative impact on the growth of microorganisms. Moreover, it reduces the amount of dissolved oxygen that aquatic organisms may absorb. Acute and chronic toxicity are caused by the presence of dye in water bodies. Therefore, the treatment of dyes is an important aspect. Among various techniques, the adsorption process is one of the simplest and most effective. The advantage of this process is its operational simplicity. Treatment of dye waste-water using bio-adsorbent made from different leaves is ecological and eco-friendly. Leaf-based bio-adsorbents are prepared by the following steps: washing, drying, crushing, acid treatment, and pyrolysis. These were found to be very effective in dye removal. In this work, an overview is given for comparison of various leaf-based bio-adsorbents used for the treatment of dyes.

Keywords: adsorption, bio-adsorbent, wastewater, dye

* Corresponding Author's Email: anjaliawasthi33@gmail.com.

In: Sustainability in Chemical Processes through Digitalization and Green Chemistry Approaches
Editors: Susarla V. A. R. Sastry, Shravan Kumar and Ashish Kapoor
ISBN: 979-8-89113-257-3

1. Introduction

One of the major challenges that our world faces today is the problem of water contamination or pollution. The category of dye-related water pollution is one of the most important ones. These contaminants may be carcinogenic, poisonous to ecosystems, plants and animals, or have other dangerous characteristics. Ozonation, oxidation processes, chemical coagulation, flocculation, ion exchange, chemical precipitation, ultrafiltration, reverse osmosis, etc., are some of the traditional methods used to remove dyes from wastewater. Currently, adsorption is rapidly gaining popularity as an alternative to conventional treatment methods. Commercially, activated carbon is the most widely used adsorbent. However, all these techniques have some significant drawbacks, such as high operating costs, adsorbent costs, reusability, separation costs, the formation of hazardous byproducts, or high energy demands.

Although the biosorption process has been around for a while it has only recently gained popularity as a low-cost, environmentally beneficial technique. Biosorption is the process by which chemical species from aqueous solutions are taken up by biological, physical, and chemical systems and deposited onto biological materials. Agricultural waste, crab shells, chitin, seed husks, nutshells, eggshells, fruit pods, and a wide variety of other materials have purportedly been used as bio-adsorbents. The utilization of plant waste products for color removal by biosorption from wastewater has increased during the last few decades.

Table 1. Summary of the characterization of leaf-based bio-adsorbents

S. No.	Bio-Adsorbent	FTIR Peaks Present Group	Present Peak (cm^{-1})	SEM/TEM Morphology	BET/Particle Size Analyzer Surface Area (m^2/g)	References
1.	Pineapple Leaf	CH group OH group C-O group C = O group C-N Group	3750 3050 1735 2910 1060 1400 1435 1255	Porous	5.24	[2]
2.	Posidonia Oceanica	C=O group OH group COOH group NH_2 group C-O group S-O group S = O group	1670-1780 3250-3700 2400-3300 3300-3500 1050-1300 550-650 1040-1200	–	–	[3]
3.	Lotus Leaf	C = O group OH group CH_3 group CH_2 group C-O-C group Aromatic group	1622 3424 2922, 1433 2852 1055 1530	Porous	7.171	[4]
4.	Posidonia Oceanica	OH group CH group C=C group Halo compound Phenolic group	3400 2925, 2854 1630 1080, 700, 1235 1390	Porous	-	[11]

S. No.	Bio-Adsorbent	FTIR Peaks		SEM/TEM	BET/Particle Size Analyzer	References
		Present Group	Present Peak (cm^{-1})	Morphology	Surface Area (m^{2}/g)	
5.	Tea Leaf Powder	Ether, Alkene, and aromatic carbon rings (carbon bond, i.e., C = C). C–H stretch from alkanes	1100 1400 and 1600 1400 1700–1600 2900–3000	Porous surface	26.35	[10]
6.	Oak Leaf	Carboxylic acid molecules (O-H vibration). Polyphenolic components (C = C tensile vibration). C-H vibration of rotation Epoxy rings C = C, C-C C-O-C stretching vibrations and O-H	1000-3400 1622.80 1253.95 1054.00	-	-	[6]
7.	Pineapple Leaf	-OH vibration C-H stretching of aliphatic—CH groups bending N-H of amines bending-OH C–O stretching	3325 and 3318.4 2913–2920 1595–1585.8 1365 and 1375 1034.3	Porous surface	9.1 m^2/g	[8]
8.	Lemongrass Leaf	OH stretching CH of alkane N = C stretching O = C group CH_3 group C = C aliphatic O-N group O-C group N-C group Bending H-C OH bending of carboxylic acids	3571–3554 2929–2958 2225 1736-1700 1386 655 1514 1088-1042 1126 1468–1462 1273–1175	Microporous surface	836.04 m^2/g	[12]
9.	Cassava Leaf Powder	C-O-C group C = C group C ≡ N group N-H group OH stretching	3325 and 3318.4 2913–2920 1595–1585.8 1365 and 1375 1034.3	-	-	[7]
10.	Mango Leaf	Stretching of C-H O-H C-O C = C	2925 3448 1057 1654	Porous surface	-	[9]

2. Preparation of Leaf-Based Bio-Adsorbent

Gulmohar leaf-based bio-adsorbent was prepared. Collected gulmohar leaves were thoroughly washed with raw water and then distilled water to remove impurities. Then, it was dried in a hot air oven at 70°C.

Dried and crushed gulmohar plant leaf powders were kept in plastic storage containers after being size-separated using rotary sieve shakers [1]. In another study, fresh pineapple leaves were first washed. The leaves were then air dried, followed by two days of oven drying at 353 K. The dried leaves were ground into a powder and then passed through a mesh screen to achieve a particle size of approximately 75 micrometres [2]. Another study was performed by Cengiz et al., using Posidonia oceanica leaves. To remove salt and impurities, the collected dead leaves were rinsed with tap water and then washed once more with distilled water. Washed leaves were oven-dried at 70°C for 16 hours. Dried leaves were ground by a crusher machine and sieved into 500 μm particle sizes for experimental use [3]. In this study, Han et al., collected lotus leaves, and to eliminate dust, they were properly rinsed with water multiple times before being washed with distilled water. The freshly cleaned lotus leaf was heated to 65°C for 6 hours. For the adsorption studies, powdered dried materials were passed through a set of sieves with 20–40 mesh particles selected [4]. In another study, Faith et al., used pine tree leaves for bio-adsorbent preparation. A domestic grinder was used to crush the collected pine tree leaves, which were then sieved to produce particles with a size in the 63–500 m range. The leaves were then dried at 70°C for 24 hours [5]. Another study was performed by Deniz et al., for the treatment of dye. The leaves of the princess tree were collected, cleaned with distilled water, dried at 70°C for 24 hours, crushed in a domestic grinder, and sieved to produce particles that ranged in size from 63 to 500 μm.

3. Adsorbent Characterization

After preparation of the bio-adsorbent, it was characterized by various characterization techniques, i.e., surface area, present groups, surface morphology, elemental analysis, point of zero charge, etc. These properties were characterized by various techniques, i.e., FTIR, SEM, TEM, EDX, BET, etc. The properties of these adsorbents and their characteristics are discussed in Table 1. Various elemental groups were found in the bio-adsorbent. In EDX analysis of oak leaf-based bio-adsorbents, Fe, Si, P, S, Ni, and Zn elements were found [6]. Similarly, in cassava leaf powder, C, O, N, and H groups were found [7]. Elements of Platanus leaf bio-adsorbent were found at C, H, N, O, and S. Pineapple leaf elements were found at C, H, and N [8]. Mango leaf elements were found at N, O, and S [9]. The point of zero charge of teak leaf powder was found to be at pH 9.0 [10-12].

4. Process Conditions for the Removal of Dyes

The prepared biosorbent was used for dye removal under various experimental conditions (pH, contact time, temperature, mixing speed, etc.). Various models were used to evaluate the adsorption performance and gain a better understanding of the adsorption mechanism [13]. Detailed experimental conditions and best-fit model behavior are given in Table 2.

Table 2. Summary of experimental conditions for the treatment of dyes using leaf-based bio-adsorbents

S. No.	Bio-Adsorbent	Pollutant (Name of Dye)	Experimental Conditions	Best Fit Model	References
1	Lemon Leaf Grass	Methylene Blue	• Initial dye concentration = 25-500 mg/L • Contact time = 0 to 24 hr • T = 30-60°C • pH = 2-12	• Koble-Corrigan model • Pseudo first-order kinetic model • Endothermic reaction	[12]
2	Platanus SP	Cadmium ion	• pH = 2-8. • Dosage = 25-200 mg • Contact time = 2 hr • Stirring times = 15 to 1,440 min	• Langmuir isotherm adsorption model	[13]
3	Teak Leaf Powder (tectona grandis)	Congo red	• Initial dye concentration = 1000 mg of CR dye in 1 L of double-distilled water • pH = 2-12 • Contact time = 150 min • Room temperature	• Pseudo second order • Langmuir, Freundlich and Temkin isotherm • Endothermic reaction	[10]
4	Oak Leaves (Quercus Peatrea) NZVI	Magenta flexographic dye	• Initial dye concentration (20-180 mgL-1) • pH value (2-10) • Contact time = 60 min	• DSD Model	[6]
5	Cassava Leaf	Methylene Blue	• pH value = 11.0 • Temperature= 45°C • initial dye concentration (0.5 g of CL) • Contact time (60 minutes)	---	[7]
6.	Lotus Leaf	Methylene Blue	• pH: 3.6 - 10.5 • Bio-Adsorbent Dosage (g/L): 0.25 - 10 • Experimental Time: 0-300 min • Dye Concentration (mg/L): 50-150 Temperature (°C): 20-40	**Kinetic model** – Pseudo Second Order Equilibrium Model- Langmuir, Freundlich and Koble-Corrigan isotherm **Process Type-** Endothermic	[4]
7.	Pine Tree Leaves	Basic Red 46 dye	• pH: 2-10 • Bio-Adsorbent Dosage (g/L): 1-6 • Experimental Time: 0-120 min • Dye Concentration (mg/L): 20-100 Temperature (°C): 25-45	**Kinetic model** – Pseudo Second Order **Equilibrium Model-** Langmuir isotherm model **Process Type-** Endothermic	[5]
8.	Posidonia Oceanica	Methyl violet	• pH: 3-11 • Bio-Adsorbent Dosage: 0.025-1 g/30 ml • Experimental Time: 0-175 min • Dye Concentration (mg/L): 10-300 Temperature (°C): 20-45	**Kinetic model** – Pseudo first order & Pseudo Second Order **Equilibrium Model-** Langmuir isotherm model **Process Type-** Endothermic	[3]
9.	Princess Tree Leaf	Basic Red 46 Dye	• pH: 2-10 • Bio-Adsorbent Dosage (g/L): 1-5 • Experimental Time: 5-90 min • Dye Concentration (mg/L): 20-100 Temperature (°C): 25-45	**Kinetic model** – Pseudo Second Order **Equilibrium Model-** Langmuir isotherm model **Process Type-** Exothermic	[11]
10.	Gul mohar Plant Leaves	Methylene Blue	• pH: 2.4-9 • Bio-Adsorbent Dosage (g/L): 0.5-2.5 • Experimental Time: 0-175 min • Dye Concentration (mg/L): 50-200 Temperature (°C): 30-50	**Kinetic model** – Pseudo Second Order **Equilibrium Model-** Langmuir isotherm model and Quadratic model	[1]
11.	Pineapple Leaf Powder	Methylene Blue	• pH: 3.5-9.5 • Bio-Adsorbent Dosage (g/L): 0.3-0.5 • Experimental Time: 0-80 min • Dye Concentration (PPM): 5-40 Temperature (°C): 4-54	**Kinetic model** – Pseudo Second Order **Equilibrium Model-** Langmuir isotherm model **Process Type-** Exothermic	[2]

Conclusion

This study has made an effort to summarize the research published in the literature on the use of leaf-based bio-adsorbents for the elimination of organic dyes. This is because leaf-based bio-adsorbents are inexpensive, widely accessible, and require only minimal preparation. New perspectives and difficulties have emerged, and these are discussed as follows: (i) Since leaves are abundantly found in nature as bio-adsorbents, it is vital to find new, biodegradable, affordable, and effective leaf-based bio-adsorbents for dye removal. (ii) The majority of studies that use leaf-based bio-adsorbents are lab-scale studies.

References

[1] Ponnusami, V., Gunasekar, V., & Srivastava, S. N. (2009). Kinetics of methylene blue removal from aqueous solution using gulmohar (Delonix regia) plant leaf powder: Multivariate regression analysis. *Journal of Hazardous Materials*, *169*(1–3), 119–127. https://doi.org/10.1016/j.jhazmat.2009.03.066.

[2] Weng, C. H., Lin, Y. T., & Tzeng, T. W. (2009). Removal of methylene blue from aqueous solution by adsorption onto pineapple leaf powder. *Journal of Hazardous Materials*, *170*(1), 417–424. https://doi.org/10.1016/j.jhazmat.2009.04.080.

[3] Cengiz, S., & Cavas, L. (2010). A promising evaluation method for dead leaves of Posidonia oceanica (L.) in the adsorption of methyl violet. *Marine Biotechnology*, *12*(6), 728–736. https://doi.org/10.1007/s10126-010-9260-8.

[4] Han, X., Wang, W., & Ma, X. (2011). Adsorption characteristics of methylene blue onto low cost biomass material lotus leaf. *Chemical Engineering Journal*, *171*(1), 1–8. https://doi.org/10.1016/j.cej.2011.02.067.

[5] Deniz, F., & Karaman, S. (2011). Removal of Basic Red 46 dye from aqueous solution by pine tree leaves. *Chemical Engineering Journal*, *170*(1), 67–74. https://doi.org/10.1016/j.cej.2011.03.029.

[6] Kecić, V., Kerkez, Đ., Prica, M., Lužanin, O., Bečelić-Tomin, M., Pilipović, D. T., & Dalmacija, B. (2018). Optimization of azo printing dye removal with oak leaves-nZVI/H2O2 system using statistically designed experiment. *Journal of Cleaner Production*, *202*, 65–80. https://doi.org/10.1016/j.jclepro.2018.08.117.

[7] Theng, M. L., & See, L. (2020). Adsorption of Methylene Blue and Congo Red Dye from Water onto cassava Leaf Powder. *Progress in Energy and Environment*, *12*(March), 11–21.

[8] Hassan, S. S., El-Shafie, A. S., Zaher, N., & El-Azazy, M. (2020). Application of Pineapple Leaves as Adsorbents for Removal of Rose Bengal from Wastewater: Process Composite Design (FCCCD). *Molecular Medicine Reports*, *25*(16), 1–19.

[9] Ong, P. S., Ong, S. T., & Hung, Y. T. (2013). Utilization of mango leaf as a low-cost adsorbent for the removal of cu(II) ions from aqueous solution. *Asian Journal of Chemistry*, 25(11), 6141–6145. https://doi.org/10.14233/ajchem.2013.14290.

[10] Gedam, V. V., Raut, P., Chahande, A., & Pathak, P. (2019). Kinetic, thermodynamics and equilibrium studies on the removal of Congo red dye using activated teak leaf powder. *Applied Water Science*, *9*(3). https://doi.org/10.1007/s13201-019-0933-9.

[11] Deniz, F., & Saygideger, S. D. (2011). Removal of a hazardous azo dye (Basic Red 46) from aqueous solution by princess tree leaf. *Desalination*, *268*(1–3), 6–11. https://doi.org/10.1016/j.desal.2010.09.043.

[12] Ahmad, M. A., Ahmed, N. B., Adegoke, K. A., & Bello, O. S. (2021). Adsorptive potentials of lemongrass leaf for methylene blue dye removal. *Chemical Data Collections*, *31*, 100578. https://doi.org/10.1016/j.cdc.2020.100578.

[13] Siswoyo, E., Endo, N., Mihara, Y., & Tanaka, S. (2014). Agar-encapsulated adsorbent based on leaf of platanus sp. to adsorb cadmium ion in water. *Water Science and Technology*, *70*(1), 89–94. https://doi.org/10.2166/wst.2014.190.

Chapter 33

The Operation of a Modified Rotating Biological Contactor (RBC) for the Enhanced Reclamation of Dye Wastewater

Khyati K. Modi[1,2,*]
and Kaushik Nath[3]

[1]Department of Environmental Engineering, Government Engineering College, Valsad, Gujarat, India
[2]Research Scholar, Gujarat Technological University, Ahmedabad, Gujarat, India
[3]Department of Chemical Engineering, G H Patel College of Engineering and Technology, Vallabh Vidyanagar, Gujarat, India

Abstract

For a long time, a conventional rotating biological contactor (RBC) has been used as a secondary treatment for domestic and industrial wastewater using microbial activity. The prime objective of the present study was to treat synthetic wastewater containing Auramine-O dye and enhance the biological oxygen demand (BOD) by modifying conventional RBCs. To achieve this objective, conventional RBCs were modified by replacing the bank of discs with a rotating drum with a capacity of 10 L, which was divided into three sections. The middle section of the drum consisted of six stationary paddles, which were covered with scrubber material on both sides. The remaining two end sections of the drum were filled with Pall rings covered with porous media. The use of a rotating drum in RBCs enhanced the biofilm attachment surface area up to 2.29 m^2 ($\approx$ 108.2% higher than conventional RBCs), which enhances the performance of RBCs. To prevent dissolved oxygen (DO) from being a limiting factor, four small aerators were placed as a supplemental aeration system in the wastewater collection tank of the modified RBC. To prevent the influence of atmospheric temperature on the biological process, a blast-proof aquarium glass heater was placed in the tank. The wastewater collection tank and the rotating drum of the modified RBCs were fabricated from lightweight high-density polyethylene (HDPE) sheets. Compared to the conventional RBCs, an enhancement of 6.7% in the BOD removal efficiency from the auramine-o-dye was accomplished using the modified RBCs.

Keywords: aerobic process, biofilm, wastewater, dissolved oxygen, rotating drum, aerators

[*] Corresponding Author's Email: kkmgecv.phd@gmail.com.

In: Sustainability in Chemical Processes through Digitalization and Green Chemistry Approaches
Editors: Susarla V. A. R. Sastry, Shravan Kumar and Ashish Kapoor
ISBN: 979-8-89113-257-3

1. Introduction

Wastewater from the textile industry is one of the major sources of various pollutants in the environment. Effluents from the textile industry harm humans, animals, and plantations due to the contamination of pollutants such as organic matter, surfactants, additives, etc. The small proportion of dyes imparts color to the water, lowering its aesthetic value. Conventional treatment methods for chemical oxygen demand (COD) removal from dye-contaminated effluents from the textile industry consist of various combinations of biological, physical, and chemical decolorization methods [1]. Improper disposal of wastewater has led to outbreaks of disease arising from wastewater in many parts of the world, raising the need for wastewater management and treatment and driving the demand for wastewater treatment to higher levels [2]. Due to the existence of complicated chemical compounds in dye wastewater, it contains very low biodegradability and dissolved oxygen (DO). Because organic substances are resistant to biodegradation, common biological treatment technologies for the treatment of dye wastewater do not meet the discharge requirement. As a result, advanced oxidation processes (AOPs) have been used as a pretreatment for industrial wastewater to increase biodegradability or reduce toxicity [3]. Because of its ease of operation and high oxidation potential, Fenton's oxidation process is widely recognized as one of the most promising technologies due to its stable and efficient removal rate from dye wastewater. In the present study, Fenton's oxidation process is used as a pretreatment of synthetic wastewater containing auramine-o-dye.

Biological wastewater treatment is a universal and effective method for the purification of the majority of liquid waste [4]. Biological wastewater treatments are further classified based on the attached growth and suspended growth processes. One of the most efficient fixed-film wastewater treatment technologies is the rotating biological contact (RBC) process [5], which offers the specific advantages of a biofilm system in the treatment of wastewater for the removal of soluble organic substances. The RBC process is a unique adaptation of the moving-medium biofilm system, which facilitates easy and effective oxygen transfer [2]. RBCs are one of the most efficient fixed-film wastewater treatment technologies, with applications including municipal wastewater treatment, food and beverage wastewater treatment, landfill treatment, refinery and petrochemical wastewater treatment, and paper and pulp effluent wastewater treatment [6].

The RBC unit consists of a series of closely spaced large flat or corrugated discs mounted on a common horizontal shaft and partially or completely immersed in wastewater, as shown in Figure 1 [4]. A mechanical motor or compressed air drive continuously rotates the shaft, and a biofilm is formed on the entire surface area of the media, metabolizing the organic materials in the wastewater [7]. RBCs have a distinct advantage in providing effective treatment for biodegradable industrial waste streams with high COD [8].

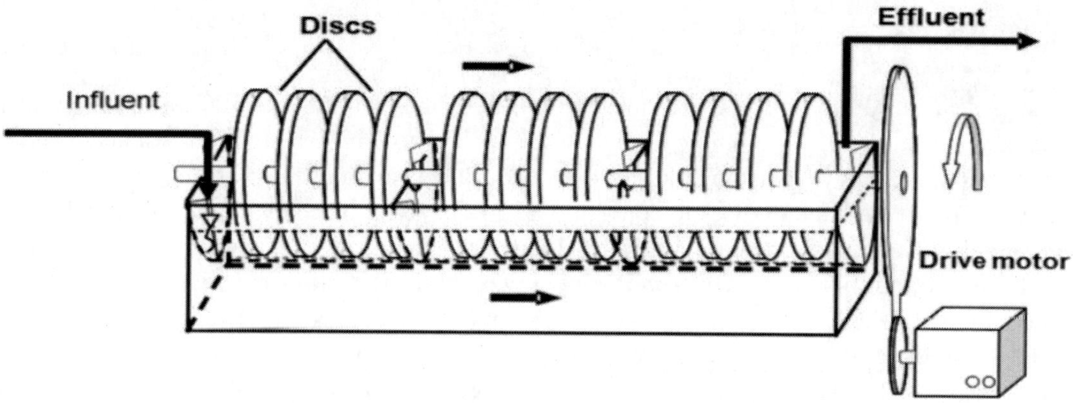

Figure 1. Schematic diagram of conventional RBCs [4].

2. Literature Review

The performance of RBCs depends upon several parameters, such as rotational speed, organic and hydraulic loading rates, hydraulic retention time (HRT), RBC media, staging, temperature, wastewater, biofilm characteristics, dissolved oxygen (DO) levels, effluent and solid recirculation, step-feeding, and submergence of the medium [5]. From the literature review mentioned below, very little work has been carried out to improve the performance of conventional RBCs. Different types of modifications, such as the volume of the reactor, the material of construction of the reactor, the replacement of the disc with a drum, and the use of different types of packing media, are used to enhance the reactor's performance [2].

RBC's two stages were used for treating gray water. The experiments were conducted at different influent COD values (248, 294, 347, 395, and 480 mg/L) and rotational speeds (3, 4.5, and 6 rpm) [9]. The maximum 95.7% removal of COD was achieved at 3 rpm, whereas 95.04% and 94.96% were achieved at 4.5 and 6 rpm, respectively. The authors stated that a high amount of COD removal can be achieved at a rpm lower than 3 in the full-fledged field-level model of RBC. The effluent from the RBC can be used for irrigation and gardening without any risk. A three-stage laboratory-scale RBC was used to treat municipal sewage. The experimental setup was prepared with an acrylic sheet and PVC pipe, and different sizes of discs were used [10]. Based on the experiments, the authors reported that the amount of overall oxygen transfer varies with the surface area of the discs and the rotating speed. In the first stage, the reduction achieved in biological oxygen demand (BOD) was 131.72 mg/L to 58.39 mg/L, COD was 268.82 mg/L to 132.81 mg/L, and TSS removal was up to 80%. Gulhane and Aswin [11] modified the conventional RBC by replacing discs with plastic scrubber-filled rotating drums. The space between the two baffles was filled with media to enable additional biofilm growth, which resulted in better performance of the model. The quality parameters of the effluent were studied after conducting a series of experiments for the analysis of BOD and COD. The model was operated at a rotational speed of 3 rpm for a detention time of 24 h, and a reduction in BOD and COD of 69% and 67% was observed in the fed wastewater, respectively.

3. Objectives

The present study aims to achieve effluent with BOD removal of 60% to 70% from a pretreated (using Fenton's AOP) Auramine-O dye solution and to improve the performance of the conventional RBC by enhancing the surface area for the growth of the biofilm. To achieve this aim, the use of the bank of discs in conventional RBCs was replaced with the rotating drum in modified RBCs, which increases the surface area for the growth of biofilms, resulting in a performance improvement. A lab-scale model was fabricated using lightweight high-density polyethylene (HDPE). A pretreated synthetic solution of Auramine-O dye as wastewater was used for the experimental work. To prevent dissolved oxygen (DO) from being a limiting factor, four small aerators were placed as a supplemental aeration system in the wastewater collection tank of the modified RBC. To prevent the influence of atmospheric temperature on the biological process, a blast-proof aquarium glass heater was placed in the wastewater tank. The experiments were carried out on a lab-scale modified RBC system in the laboratory. It is expected that the modified RBC working on the principle of attached growth will provide a better result compared to normal RBCs. A theoretical comparison was also carried out for conventional and modified RBCs.

4. Material and Methods

In the modified RBC, the bank of discs was replaced by the rotating drum. The reactor had a volume capacity of 10 L and was fabricated from an HDPE sheet. The rotating drum of the modified RBC was mounted on the horizontal shaft and connected to the gear motor. The drum was rotated at 3 rpm in the experiments. The middle section of the drum had six stationary paddles, and scrubber media on both sides of each paddle were attached [12]. The two end sections of the drum were filled with Pall rings covered with porous media [13, 14]. Pall rings are made up of polypropylene material, size: 25 mm, Nos./CU. M.: 51000, surface area m^2/m^3: 210, void adage: 89%, packing factor: 52. The use of a rotating drum in modified RBCs enhanced the biofilm attachment surface area up to 2.29 m^2 ($\approx$ 108.2% higher than conventional RBCs), which enhances the performance of modified RBCs. The inlet and outlet arrangement pipes were provided at the top of the wastewater collection tank and the bottom of the tank. The surface of the modified rotating drum was submerged at 40% for the growth of microorganisms. To prevent DO from being a limiting factor, four small aerators were placed as a supplemental aeration system in the wastewater collection tank of the modified RBCs. To prevent the influence of atmospheric temperature on the biological process, a blast-proof aquarium glass heater was placed in the wastewater collection tank of the modified RBCs. Table 1 depicts the theoretical comparison of conventional RBCs and modified RBCs, which was carried out for similar geometrical parameters and the same capacity of the wastewater collection tank [15–17].

Table 1. Salient features of modified RBCs compared to conventional RBCs

Parameter	Conventional RBC	Modified RBC
Total surface area for microbial growth (m^2)	1.1	2.29
OLR (gm BOD/m^2·d)	1.82	0.873
HLR (m^3/m^2·d)	0.0091	0.0044
Effluent BOD (mg/L)	18.42	3.132
BOD removal efficiency (%)	90.79	96.87

5. Results and Discussion

In the present study, the pretreated synthetic solution of Auramine-O dye wastewater was used as an influent. The industrial aerobic secondary recycling of sludge as biomass for nutrients was used. After the development of the biofilm on the drum surface, a pretreated synthetic solution of Auramine-O dye/wastewater after the adjustment of pH = 7, COD = 308 mg/L, and BOD = 92.40 mg/L was supplied into the wastewater collection tank and kept for a detention time duration of 24 h [18]. The percentage of BOD removal was tested at regular intervals of 1 day and is depicted in Table 2. Figure 2 presents the percentage BOD removal with the day of the progression of experiments. Standard methods were adopted for analyzing the influent and effluent quality of wastewater [19]. The results indicate that the desired level of effluent quality was obtained from the modified RBC.

Table 2. BOD removal efficiency for a detention time of 24 hours at 3 rpm

Day	BOD (mg/L)	% BOD reduction
1	92.40	0.00
2	65.20	29.44
3	55.25	40.21
4	48.30	47.73
5	44.20	52.16
6	39.24	57.53
7	30.45	67.05
8	25.60	72.29
9	20.48	77.84
10	17.02	81.58
11	12.30	86.69
12	12.30	86.69
13	12.30	86.69
14	12.30	86.69
15	12.30	86.69

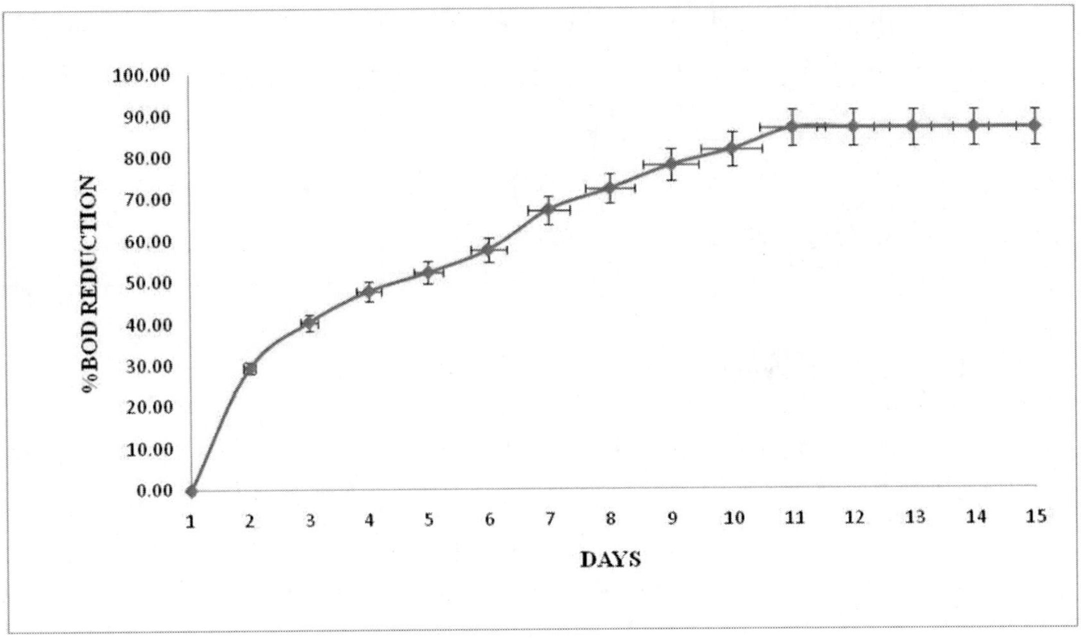

Figure 2. % BOD reduction for a detention time of 24 hours at 3 rpm.

Conclusion

The study revealed that the modified RBC with rotating paddles and drums filled with pall rings showed better performance for wastewater treatment. Using the modified RBC, an 86.69% BOD reduction from pretreated Auramine-O dye was achieved in the experimental work. Using the modified RBC, a pretreated synthetic solution of Auramine-O dye was converted into a biodegradable solution that was harmless and environmentally friendly to dispose of in the atmosphere. The theoretical comparison between the conventional and modified RBCs indicates an enhancement of 6.7% in BOD removal efficiency. The results of the experimental work indicate that modified RBCs are an effective system for the treatment of high organic compound wastewater and can replace conventional RBCs.

References

[1] Patel D. C., Rotliwala Y. C., The Modified Fenton Process for the Treatment of Textile Wastewater, *J. Environ. Res. Dev.* 8 (2014) 983–986.

[2] Gulhane M L, Sahare S V, Modified Rotating Biological Contactor, *Int. J. Mech. Prod. Eng*. (2014) 2320–2092.

[3] Zou H., W. Ma, Y. Wang, A novel process of dye wastewater treatment by linking advanced chemical oxidation with biological oxidation, *Arch. Environ. Prot.* 41 (2015) 33–39. https://doi.org/10.1515/aep-2015-0037.

[4] Ziembińska-Buczyńska A., S. Ciesielski, S. Żabczyński, G. Cema, Bacterial community structure in rotating biological contactor treating coke wastewater in relation to medium composition, *Environ. Sci.* Pollut. Res. 26 (2019) 19171–19179. https://doi.org/10.1007/s11356-019-05087-0.

[5] Cortez S., P. Teixeira, R. Oliveira, M. Mota, Rotating biological contactors: A review on main factors affecting performance, *Rev. Environ. Sci. Biotechnol.* 7 (2008) 155–172. https://doi.org/10.1007/s11157-008-9127-x.

[6] Cortez S., P. Teixeira, M. Mota, *Rotating biological contactors : A review on main factors affecting performance Rotating biological contactors : a review on main factors affecting performance*, (2008). https://doi.org/10.1007/s11157-008-9127-x.

[7] Tonde M. R., S. B. Patil, J. R. Mali, Review Paper on Study of Rotating Biological Contactor for Wastewater Treatment Process, *Int. J. Curr. Eng. Technol.* 5 (2015) 1539–1541.

[8] Selvakumar D., V. Nehru Kumar, Effect of rotational speed of disc for treating sugar and dairy effluent streams in rotating biological contactors, *J. Ind. Pollut. Control.* 23 (2007) 247–250.

[9] Kumar V. N., S.S. Enayathali, Performance of rotating biological contactor for treating segregated gray water for reuse, *Int. J. Recent Technol. Eng.* 1 (2012) 40–41.

[10] Kadu P. A., A. A. Badge, Y. R. M. Rao, *Treatment of Municipal Wastewater by using Rotating Biological Contractors* (Rbc ' s), (2013) 127–132.

[11] Gulhane M., P. Ashwin, Modified rotating biological contactor fitted perforated baffles, *J. Environ. Res.* Dev. 9 (2015) 1003–1005.

[12] Li N., W. Zeng, Y. Yang, B. Wang, Z. Li, Y. Peng, Oxygen mass transfer and postdenitrification in a modified rotating drum biological contactor, *Biochem. Eng. J.* 144 (2019) 48–56. https://doi.org/10.1016/j.bej.2019.01.008.

[13] Ibrahim H. T., H. Qiang, W.S. Al-Rekabi, Y. Qiqi, Improvements in biofilm processes for wastewater treatment, *Pakistan J. Nutr.* 11 (2012) 610–636. https://doi.org/10.3923/pjn.2012.708.734.

[14] Chen S., D. Sun, J. S. Chung, Treatment of pesticide wastewater by moving-bed biofilm reactor combined with Fenton-coagulation pretreatment, *J. Hazard. Mater.* 144 (2007) 577–584. https://doi.org/10.1016/j.jhazmat.2006.10.075.

[15] Abdelkader A. M., Comparative study between membrane bioreactor MBR and rotating biological contactors RBC for graywater treatment, *Int. J. Environ. Sci. Dev.* 12 (2021) 107–111. https://doi.org/10.18178/IJESD.2021.12.4.1326.

[16] Maurício R., R. Dias, V. Ribeiro, S. Fernandes, A.C. Vicente, M.I. Pinto, J.P. Noronha, L. Amaral, P. Coelho, A.P. Mano, 17α-Ethinylestradiol and 17β-estradiol removal from a secondary urban wastewater using an RBC treatment system, *Environ. Monit. Assess.* 190 (2018). https://doi.org/10.1007/s10661-018-6701-8.

[17] Ravi R., L. Philip, T. Swaminathan, Comparison of biological reactors (biofilter, biotrickling filter andmodified RBC) for treating dichloromethane vapors, *J. Chem. Technol. Biotechnol.* 85 (2010) 634–639. https://doi.org/10.1002/jctb.2344.

[18] Ravi R., L. Philip, T. Swaminathan, vapouModifiedrsrotating biological contactor for removal of dichloromethane, *Environ. Technol. (United Kingdom).* 36 (2015) 566–572. https://doi.org/10.1080/09593330.2014.953553.

[19] Chen Z., Q. Wen, J. Wang, F. Li, Simultaneous removal of carbon and nitrogen from municipal-type synthetic wastewater using net-like rotating biological contactor (NRBC), *Process Biochem.* 41 (2006) 2468–2472. https://doi.org/10.1016/j.procbio.2006.06.003.

Chapter 34

Carbonized MOF-5 with Graphitic Carbon Nitride for an Efficient Visible Light-Driven Photocatalytic Performance

Jeny Rachel Biju[1]
Baskaralingam Palanichamy[2]
and Sivanesan Subramanian[1,*]
[1]Environmental Management Laboratory, Department of Applied Science and Technology, Alagappa College of Technology, Anna University, Chennai, Tamil Nadu, India
[2]Department of Chemistry, College of Engineering Guindy, Anna University, Chennai, Tamil Nadu, India

Abstract

The photocatalytic activity of a carbonized MOF-5/g-C_3N_4 heterostructured photocatalyst was investigated in this study. Herein, a metal–organic framework (MOF-5) was carbonized at 800 °C and utilized as a co-catalyst of graphitic carbon nitride (g-C_3N_4). A metal–organic framework (MOF)-templated strategy has been developed to prepare heterostructured nanocatalysts with superior photocatalytic activity. After carbonization, metal organic framework (MOF)-derived ZnO forms heterostructures with g-C_3N_4, significantly narrowing the band gap. All of the nanomaterials generated were effectively analyzed using Fourier transform infrared (FTIR) spectroscopy, scanning electron microscopy (SEM), and X-ray diffraction (XRD), which showed that CMOF800-g-C_3N_4 was better in the presence of visible light than all other synthesized composites. The carbonized MOF's proper conduction band location allows photogenerated electrons to be transferred from g-C_3N_4 to the carbonized MOF, subsequently increasing the charge carrier separation, which shows considerable potential in environmental remediation. This work might open up new possibilities for creating environmentally stable, highly effective photocatalysts for the treatment of wastewater.

Keywords: photocatalysts, metal organic frameworks, graphitic carbon nitride, nanoparticles, wastewater

In: Sustainability in Chemical Processes through Digitalization and Green Chemistry Approaches
Editors: Susarla V. A. R. Sastry, Shravan Kumar and Ashish Kapoor
ISBN: 979-8-89113-257-3

Introduction

Wastewater effluents from the textile, pharmaceutical, paper, rubber, and leather industries are produced in enormous quantities every day around the world. Most of the effluents produced by these companies, especially those that produce dyes, are dumped into the environment without appropriate treatment. These are carcinogenic, mutagenic, and toxicologically harmful to people [1]. For the preservation of human health and the safety of the water supply, harmful dyes must be removed from wastewater. Often, coagulation, flocculation, electrochemical treatment, membrane filtration, and adsorption techniques are used to treat pollutants in wastewater [2]. Adsorption has significant potential for sustainable treatment of industrial wastewater with a number of advantages, such as energy savings, operational simplicity, widespread availability, and high efficiency [3]. However, using only the adsorption approach to deactivate microbial pathogens in wastewater systems is not ideal. Comparatively, the photocatalytic method, as a sophisticated oxidation technique, has the benefits of being environmentally friendly and highly effective under mild conditions and has been extensively utilized in the energy industry [4]. Additionally, owing to the high solubility and toxicity of discharged organic dyes, the ongoing contamination of natural ecosystems with such contaminants negatively affects the accessibility of pure freshwater sources for safe drinking. As a result, it is crucial to develop a sustainable, environmentally friendly method for effectively treating industrial effluents to reduce pressure on the availability of worldwide clean water resources [4]. Researchers have focused on low-cost adsorbents based on solid and industrial pollutants. Metals and dyes may both be successfully removed using the majority of these wastes, as has been proven. The strength, highly active sites, and low bulk of well-known nanoparticles are often favored [4].

Although nanoparticles have comparatively high surface areas, certain nanomaterials have two key benefits: they are simple to synthesize and less expensive to buy, and fewer amounts are needed for effective removal of contaminants. As a result, it is anticipated that nanoparticles will eventually be more cost-effective for adsorption applications. The development of metal-organic frameworks (MOFs) as a class of highly promising crystalline microporous materials supported by the application of a number of proven coordination chemistry principles has attracted much attention [5]. The majority of scientific efforts over the past years have been focused on developing new MOF structures and investigating their many applications [6]. However, MOFs have a few drawbacks, such as low chemical stability. It is essential to add new functions and further improve the qualities of MOFs to fulfil their realistic applications. Materials made of one or more different constituent materials in an MOF are known as MOF composites or hybrids. As a result, a variety of applications can benefit from the exceptional qualities of composites that come from the synergistic interaction of MOFs and other active components [7].

The fascinating visible light photocatalyst graphitic carbon nitride (g-C_3N_4) has an excellent two-dimensional structure, remarkable chemical stability, and a variable electronic structure. Nanocomposites based on g-C_3N_4 have been extensively used in both energy and environmental applications due to their improved performance. Metallic impurities typically provide active adsorption functions, which decrease the band gap and increase the absorption of visible light, giving them special photocatalytic capabilities. Recently, graphitic carbon nitride has drawn much attention in the search for stable and reliable VLD semiconductor photocatalysts. Metal-free g-C_3N_4 has various benefits due to its heptazine ring structure and high condensation degree, including good physicochemical stability and an interesting electronic structure with a medium band gap (2.7 eV) [8].

A heterojunction of g-C_3N_4 and metal organic frameworks has been shown in several investigations to demonstrate the deterioration of complicated dyes and other industrial effluents. After combining the MOF with g-C_3N_4, the analysis revealed that the photoexcited carriers were effectively separated. Furthermore, there is a paucity of knowledge in the current state regarding how the application of carbonized MOF5 and g-C_3N_4 will affect the charge transfer of photoexcited electron-hole pairs and the site of band margins for the treatment of wastewater [9, 10]. For the manufacturing of carbonized products, porous coordination polymers (PCPs) or metal-organic frameworks (MOFs), which have crystalline properties and enormous pores, have recently been used. Moreover, the rapid carbonization of MOFs to produce porous carbon has attracted increasing study interest for a wide range of possible applications [11, 12]. To modify graphitic carbon nitride, MOF5 carbonized at 800^{0}C is shown here. Additionally, a thorough assessment of the effects of the structural, optical, and electrical properties on the treatment of wastewater is provided.

Materials: Urea, zinc nitrate hexahydrate [Zn(NO)6H_2O], diethylformamide, and cis-cyclobutane-1,2-dicarboxylic acid.

Experimentation

Preparation of Graphitic Carbon Nitride

Using melamine as a precursor, g-C_3N_4 was produced through calcination [13]. Thirty grams of melamine was directly calcined in a 60 mL closed crucible at 550 °C for 4 hours at a heating rate of 5 °C/min to produce graphitic carbon nitride. The yellow-colored product thus produced shows the presence of graphitic carbon nitride.

Preparation of MOF-5

According to the reviewed studies, microwave-assisted synthesis was used in the preparation of MOF 5 [14, 15]. In Figure 1, Zn(NO) 6H_2O zinc nitrate hexahydrate was precisely dissolved in ten milliliters of diethylformamide (DEF) along with cis-cyclobutane-1,2-dicarboxylic acid (0.0545 g, 0.378 mol). After that, the solution was sealed in a Pyrex sample vial and heated in a microwave oven. The crystals were found to be square-plate-shaped.

Preparation of MOF-5/g-C_3N_4 and Carbonized MOF-5/g-C_3N_4:

As shown in Figure 2, a MOF5-containing ceramic boat was positioned in the middle of a tubular furnace. To produce carbonized MOF5, the furnace was heated from ambient temperature to 800 °C and sustained there for 0.5 hours.

Nine grams of pure graphitic carbon nitride, 25 mL of water, and 1 g of MOF5 were combined and then subjected to 4 hours of sonication. The powder product was ground in a mortar to produce MOF-5 gC_3N_4.

Likewise, 9 g of pure graphitic carbon nitride, 25 mL of water, and 1 g of carbonized MOF5 were combined. The combination was then subjected to 4 hours of sonication. Subsequently, the powder product was ground to create carbonized MOF-5 gC_3N_4.

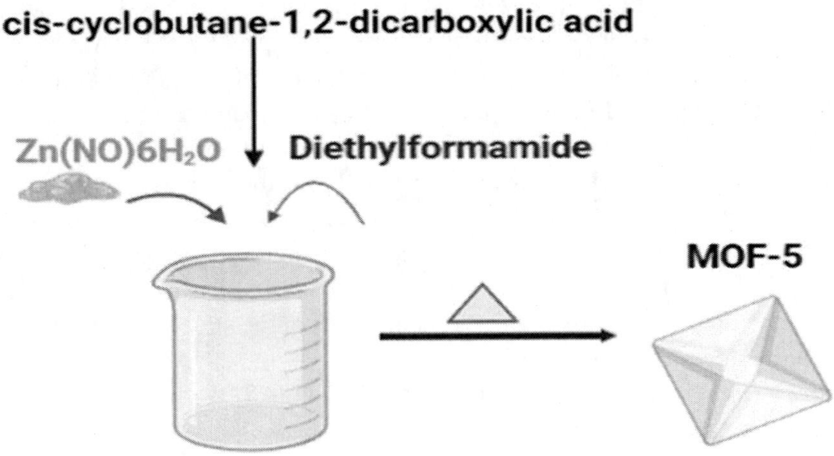

Figure 1. Preparation of MOF-5.

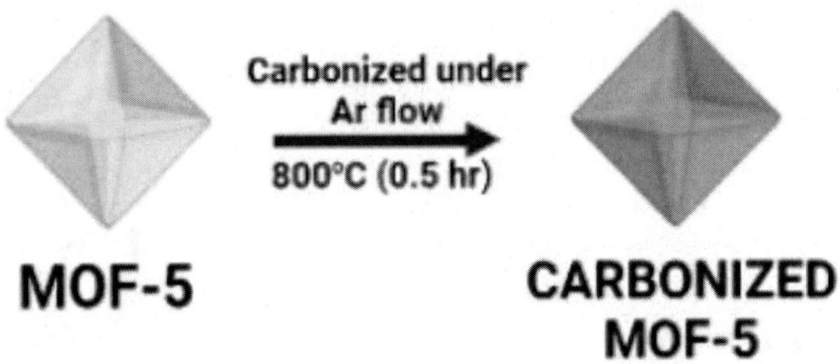

Figure 2. Preparation of carbonized MOF-5.

Results and Discussions

Due to the wide range of natural MOFs, there are several alternatives for derived carbons. MOFs can be pyrolyzed at different temperatures and converted into metal or nanoporous carbon compounds with diverse morphologies because the organic ligands in MOFs have a high carbon content. By carefully controlling the carbonization conditions, which also increase the surface area, pore volume, and porous structure, MOFs can maintain their exceptional structure and performance. The synthetic method used to produce the resulting carbon has the enormous advantage of being easy to set up. In comparison to commercially available carbonized products, those made using this approach have better organized pore structures.

The ambient conditions affected the colors that the MOF-5 derivatives displayed. After pyrolysis, MOF-5 produced two products that both indicated the presence of carbonaceous species: a white product under air and a black product under argon. Interestingly, it was found that MOF derivatives made with argon gas were black in color and nevertheless had good crystallinity. Additionally, broader views showed a number of morphological elements that were slightly different.

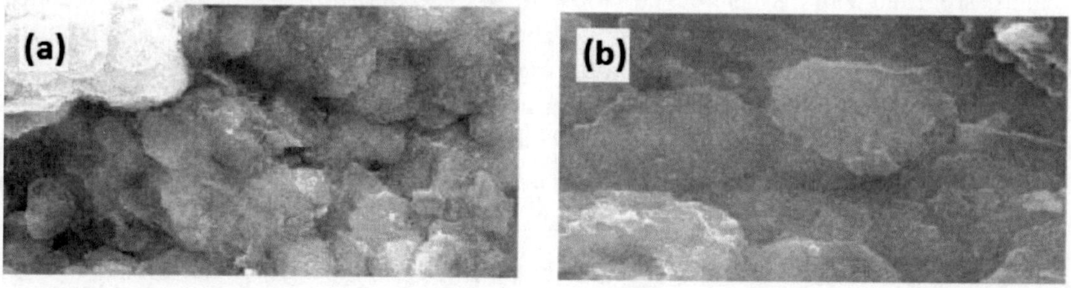

Figure 3. SEM images of MOF5 (a) and CMOF-5 gC_3N_4 (b).

Pictures of MOF5 and its composites with graphitic carbon nitride, such as carbonized MOF5 gC_3N_4, are shown in SEM images. In Figure 3 (a), the SEM image of MOF5 is typical: a small spherical object with a smooth surface and a consistent round and squared shape. After calcination, MOF5 was shown to have decomposed into considerably more homogeneous, tiny particles. ZnO nanoparticles were obtained on the surface of the porous graphitic carbon cubes as a result of their carbonization at 800 °C, as shown in Figure 3 (b). Many irregularities can be seen on the boundaries of the thin layers in the SEM image of gC_3N_4 [16].

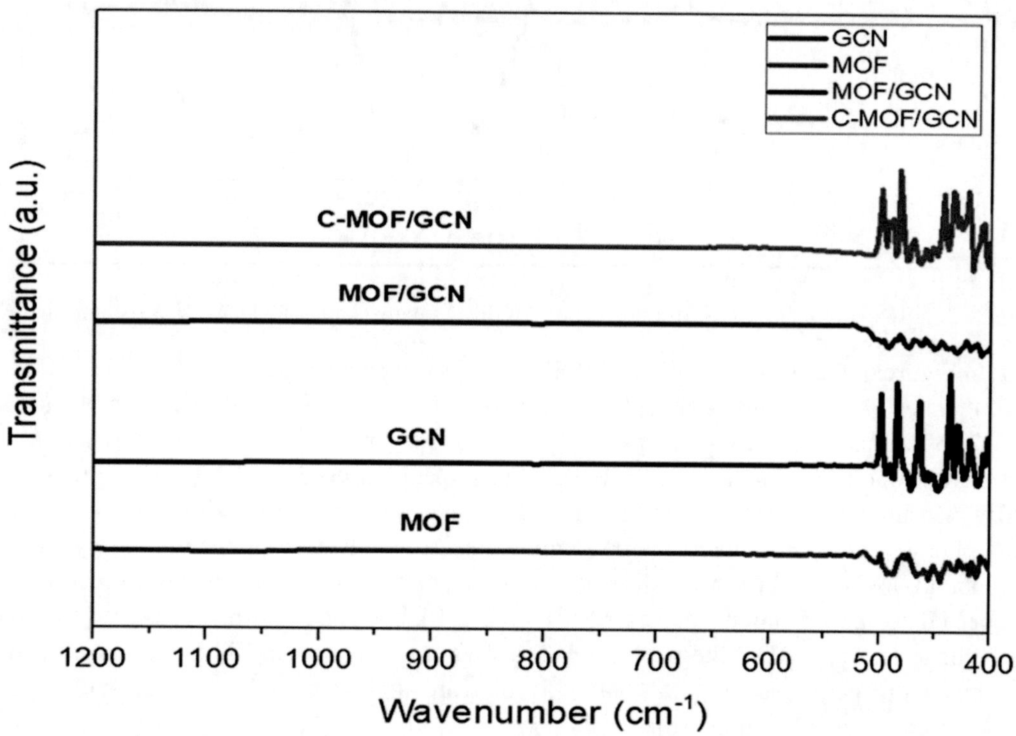

Figure 4. FTIR spectra of gC_3N_4, MOF-5, MOF5– gC_3N_4 and CMOF800– gC_3N_4.

FTIR spectroscopy showed the surface functional groups and production of the pure Zn-derived MOF-5 as well as their composites. The FTIR spectra of MOF-5 and the composites are shown in Figure 4. The FTIR spectrum reveals that gC_3N_4 has a small number of distinctive bands associated with CN-heterocycles (809 cm^{-1}). Characteristic bands at 532, 745, and 1104 cm^{-1} are visible in the spectrum of MOF5. The ZnO cluster of MOF-5 exhibits a symmetric stretching vibration, which results in a weak transmittance peak at 642 cm^{-1}. Most of the peaks fade after MOF5 carbonization due to the thermal degradation of the organic ligand molecules, and only weak peaks are seen. When gC_3N_4 and CMOF800 are combined, all of the peaks associated with gC_3N_4 can be seen.

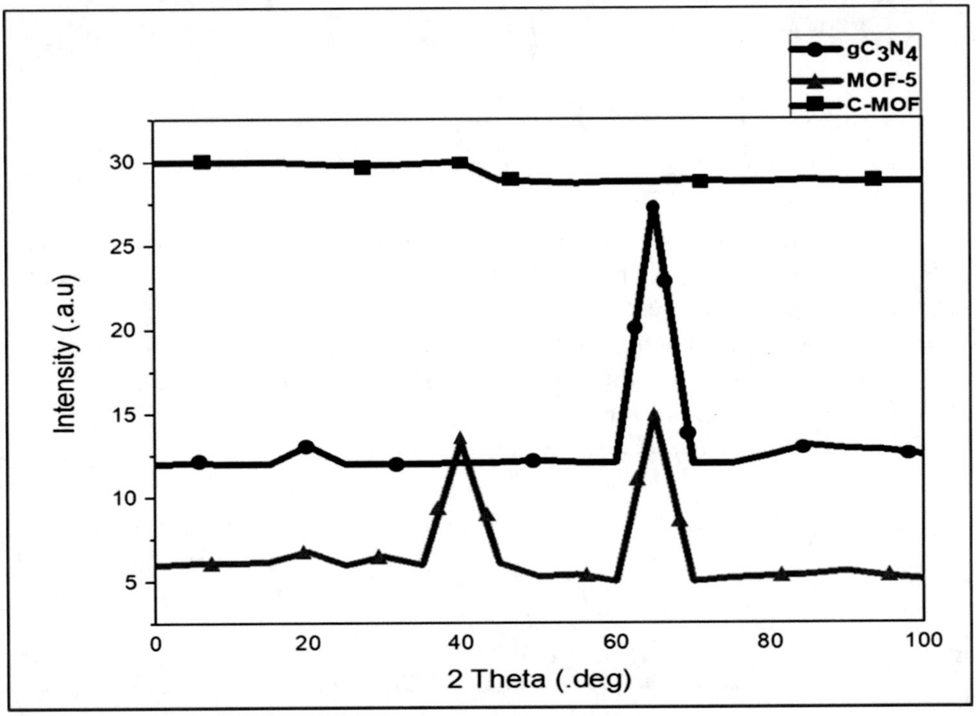

Figure 5. XRD patterns of gC_3N_4 (round shape), MOF5 (triangle shape), and CMOF-5 (square shape).

In Figure 5, the XRD patterns of MOF-5 and its composites are discussed. The (100) and (002) planes of g-C_3N_4 are assigned to two distinct diffraction peaks that are visible at 13.1° and 27.5°. These two peaks are probably due to the conjugated aromatic system and the structure of the tri-s-triazine unit with interplanar spacing, respectively. The outcomes of the XRD measurements revealed that the MOF-5 sample was efficiently fabricated. The crystals thus obtained will be nanocubes. To demonstrate the high crystallinity of the synthesized compound, MOF-5 shows peak sharpness at four peaks, 6.8, 9.6, 13.7, and 15.3, which are connected to the crystalline planes (2 0 0), (2 2 0), and (4 0 0), respectively [17]. After carbonization, the XRD pattern of MOF-5 shows no major peaks. The carbonized sample exhibits large peaks at 20°, 25°, and 45° that are associated with low crystallinity and disorderedly oriented carbon [18].

Field emission scanning electron microscopy (FESEM) can be used to examine the surface morphologies of the samples; high-resolution transmission electron microscopy can be used to observe the lattice spacings; and energy-dispersive spectroscopy (EDS) can be used to analyze the surface elemental distributions. From prior studies, the morphologies of the samples have been confirmed by FESEM analysis. This clearly demonstrates the creation of bulk sheet-like structures for gC_3N_4 [19]. The crystalline structure of MOF-5 was observed, but its morphological structure was not specified [20].

Using ASAP equipment, the sample nitrogen adsorption-desorption isotherms can be determined, and the adsorption-desorption isotherms of the precursors can be seen. The pore diameters and specific surface areas can be determined using the Brunauer–Emmett–Teller technique. These samples have considerable specific surface areas [21].

Photocatalytic Activity

The greater adsorption of pollutants, improved light absorption, and efficient charge transfer and separation are the three key factors that influence the enhanced photocatalytic activity. The photocatalytic mechanism scheme of the carbonized MOF-5/g-C_3N_4 hybrid nanocomposite under visible-light irradiation is depicted in Figure 6. The physical characteristics of photocatalysts, such as the band gap, specific surface area, and crystalline structure, affect the aforementioned parameters. The most crucial variables in the rate and effectiveness of the photocatalysts are, hence, surface area, separation, and transfer of e/h+ pairs [22]. To demonstrate the good photocatalytic activity, pure g-C_3N_4, MOF-5, MOF-5/g-C_3N_4, and C-MOF/g-C_3N_4 were compared. The findings suggest that compared to pure g-C_3N_4 and other composites, carbonized MOF-5/g-C_3N_4 has a higher photocatalytic efficiency. While the holes from the VB of CMOF-5 can transfer to the VB of g-C_3N_4, the photoexcited electrons from the conduction band (CB) of g-C_3N_4 are transported to the CB of CMOF-5, which can reduce the shift in charge recombination and generate more free electrons in the CB of CMOF-5. As a result, charge migration between CMOF-5 and g-C_3N_4 with matching band potentials takes place, which effectively separates photogenerated electron and hole pairs between CMOF-5 and g-C_3N_4 and improves photocatalytic performance under visible light irradiation [23]. Some of the possible reaction equations are listed below:

$$\text{g-}C_3N_4 + h\nu \rightarrow h^+ (\text{g-}C_3N_4) + e^- (\text{g-}C_3N_4) \quad (1)$$

$$\text{CMOF-5} + h\nu \rightarrow h^+ (\text{CMOF-5}) + e^- (\text{CMOF-5}) \quad (2)$$

$$e^- (\text{g-}C_3N_4) \rightarrow e^- (\text{CMOF-5}) \quad (3)$$

Conclusion

Carbonized MOF-5/g-C_3N_4 appears to be more efficient for photocatalytic activity with greater performance when compared to g-C_3N_4/MOF-5, pure g-C_3N_4 and MOF-5. The maximum visible-light-driven photocatalytic activity was achieved by synthesizing pure g-C_3N_4, pure MOF-5, combinations of MOF-5/g-C_3N_4, and MOF-5/g-C_3N_4 carbonized at 800 °C. The lowest band gap and charge carrier recombination rate were expressed by an excellent photocatalyst,

MOF-5/g-C_3N_4, that was synthesized. To further increase MOF-5/g-C_3N_4 photocatalytic activity, it was carbonized at 800 °C with argon gas. Metal organic frameworks (MOFs) derived from ZnO, when combined with g-C_3N_4, considerably narrow the bandgap. Upon further carbonization of MOF-5/g-C_3N_4 at 800 °C, the conduction band position enables the transfer of photogenerated electrons from the carbonized MOF to g-C_3N_4, hence improving the charge carrier separation. In conclusion, MOF-derived nanomaterials containing carbonaceous materials are widely emphasized and investigated for photocatalysis. The characteristics of the products formed from MOF-5 during carbonization were examined for photocatalytic activity, incorporating the benefits of the MOF-based structure. We demonstrate that a rise in carbonization temperature of 800^{0}C alters the shape and structure of the product and its constituent parts, which in turn has an effect on the photocatalytic performance. Therefore, of all manufactured photocatalysts, this one features the highest surface area, the lowest charge carrier recombination rate, the most favorable band gap value, and good photocatalytic activity.

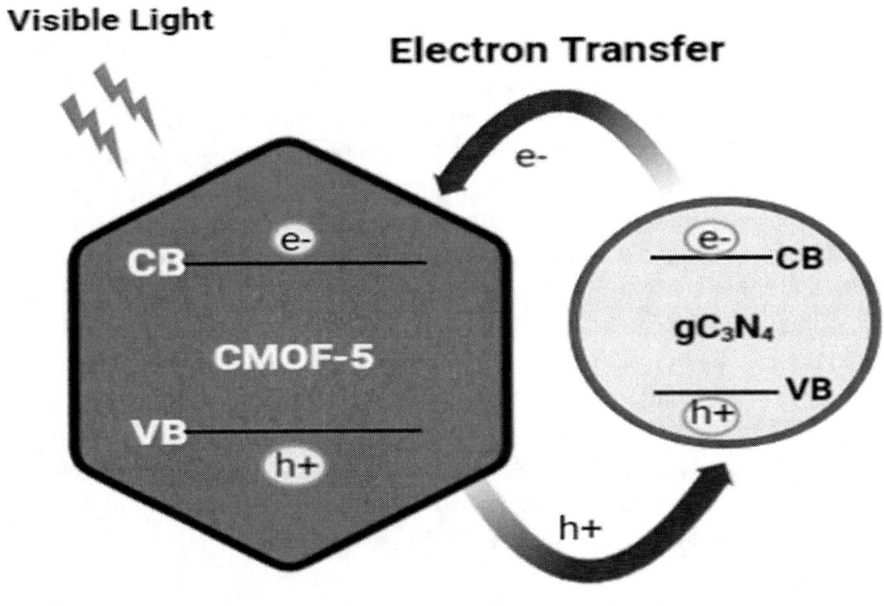

Figure 6. Photocatalytic mechanism scheme of the carbonized MOF-5/g-C_3N_4 hybrid nanocomposite under visible-light irradiation.

References

[1] Younis, S. A., Serp, P., & Nassar, H. N. (2021). Photocatalytic and biocidal activities of ZnTiO 2 oxynitride heterojunction with MOF-5 and g-C 3 N 4 : A case study for textile wastewater treatment under direct sunlight. *Journal of Hazardous Materials*, 410(October 2020), 124562. https://doi.org/10.1016/j.jhazmat.2020.124562.

[2] Tian, S., Xu, S., Liu, J., He, C., & Xiong, Y. (2019). Highly efficient removal of both cationic and anionic dyes from wastewater with a water-stable and eco-friendly Fe-MOF via host- guest encapsulation. *Journal of Cleaner Production,* 239, 117767. https://doi.org/10.1016/j.jclepro. 2019.117767.

[3] Bing, K., Vakili, M., Amini, B., Eong, P., Zuhairi, A., & Salamatinia, B. (2015). Adsorption of dyes by nanomaterials: Recent developments and adsorption mechanisms. *Separation and Purification Technology,* 150, 229–242. https://doi.org/10.1016/j.seppur.2015.07.009.

[4] Dong, S., Cui, L., Zhang, W., Xia, L., Zhou, S., Russell, C. K., Fan, M., Feng, J., & Sun, J. (2020). Double-shelled ZnSnO 3 hollow cubes for efficient photocatalytic degradation of antibiotic wastewater. *Chemical Engineering Journal*, 384(July 2019), 123279. https://doi.org/10.1016/j.cej.2019.123279.

[5] Zhu, Q., & Xu, Q. (2014). Chem Soc Rev. 43(16). https://doi.org/10.1039/c3cs60472a.

[6] Gautam, S., Agrawal, H., Thakur, M., Akbari, A., Sharda, H., Kaur, R., & Amini, M. (2020). Metal oxides and metal organic frameworks for the photocatalytic degradation: A review. *Journal of Environmental Chemical Engineering*, 8(3), 103726. https://doi.org/10.1016/j.jece.2020.103726.

[7] Ma, L., Ai, X., Jiang, W., Liu, P., Chen, Y., & Lu, K. (2022). Zn/Ce metal-organic framework-derived ZnO @ CeO 2 nanoheterojunction for enhanced photocatalytic activity. *Colloid and Interface Science Communications*, 49(May), 100636. https://doi.org/10.1016/j.colcom.2022.100636.

[8] Zhao, Z., Sun, Y., Dong, F., & Dong, F. (2015). *Graphitic carbon nitride based nanocomposites : a review*. 15–37. https://doi.org/10.1039/c4nr03008g.

[9] Titirici, M., & Antonietti, M. (2010). *Chemistry and materials options of sustainable carbon materials made by hydrothermal carbonization*. 103–116. https://doi.org/10.1039/b819318p.

[10] Gong, J., Chen, X., & Tang, T. (2019). Progress in Polymer Science Recent progress in controlled carbonization of (waste) polymers. *Progress in Polymer Science*, 94, 1–32. https://doi.org/10.1016/j.progpolymsci.2019.04.001.

[11] Lim, S., Suh, K., Kim, Y., Yoon, M., Park, H., Dybtsev, D. N., & Kim, K. (2012). Porous carbon materials with a controllable surface area synthesized from metal-organic frameworks. *Chemical Communications*, 48(60), 7447–7449. https://doi.org/10.1039/c2cc33439a.

[12] Xi, K., Cao, S., Peng, X., Ducati, C., Kumar, R. V., & Cheetham, A. K. (2013). Carbon with hierarchical pores from carbonized metal–organic frameworks for lithium sulphur batteries. *Chemical Communications*, 49(22), 2192–2194. https://doi.org/10.1039/c3cc38009b.

[13] Huang, Q., Yu, J., Cao, S., Cui, C., & Cheng, B. (2015). Efficient photocatalytic reduction of CO 2 by amine-functionalized g-C 3 N 4. *Applied Surface Science*, 358, 350–355. https://doi.org/10.1016/j.apsusc.2015.07.082.

[14] Jae, Y. C., Kim, J., Sung, H. J., Kim, H. K., Chang, J. S., & Chae, H. K. (2006). Microwave synthesis of a porous metal-organic framework, zinc terephthalate MOF-5. *Bulletin of the Korean Chemical Society*, 27(10), 1523–1524. https://doi.org/10.5012/bkcs.2006.27.10.1523.

[15] Yang, S. J., Kim, T., Im, J. H., Kim, Y. S., Lee, K., Jung, H., & Park, C. R. (2012). MOF-derived hierarchically porous carbon with exceptional porosity and hydrogen storage capacity. *Chemistry of Materials,* 24(3), 464–470. https://doi.org/10.1021/cm202554j.

[16] Abbasi, M., Noor, T., Iqbal, N., & Zaman, N. (2022). Electrocatalytic study of cu/Ni MOF and its g-C3N4 composites for methanol oxidation reaction. *International Journal of Energy Research*, 46(10), 13915–13930. https://doi.org/10.1002/er.8109.

[17] Cao, J., Qin, C., Wang, Y., Zhang, H., Sun, G., & Zhang, Z. (2017). *Solid-state method synthesis of SnO2-decorated g-C3N4 nanocomposites with enhanced gas-sensing property to ethanol. Materials*, 10(6). https://doi.org/10.3390/ma10060604.

[18] Kukulka, W., Cendrowski, K., Michalkiewicz, B., & Mijowska, E. (2019). MOF-5 derived carbon as material for CO2 absorption. *RSC Advances*, 9(32), 18527–18537. https://doi.org/10.1039/c9ra01786k.

[19] Das, D., Banerjee, D., Das, B., Das, N. S., & Chattopadhyay, K. K. (2017). Effect of cobalt doping into graphitic carbon nitride on photo induced removal of dye from water. *Materials Research Bulletin*, 89, 170–179. https://doi.org/10.1016/j.materresbull.2017.01.034.

[20] Arjmandi, M., Altaee, A., Arjmandi, A., Pourafshari Chenar, M., Peyravi, M., & Jahanshahi, M. (2020). A facile and efficient approach to increase the magnetic property of MOF-5. *Solid State Sciences*, 106(April), 106292. https://doi.org/10.1016/j.solidstatesciences.2020.106292.

[21] Aleksandrzak, M., Sielicki, K., & Mijowska, E. (2020). Enhancement of photocatalytic hydrogen evolution with catalysts based on carbonized mof-5 and g-C3N4. *RSC Advances,* 10(7), 4032–4039. https://doi.org/10.1039/c9ra08388j.

[22] Huang, W., Liu, N., Zhang, X., Wu, M., & Tang, L. (2017). Metal organic framework g-C 3 N 4/MIL-53(Fe) heterojunctions with enhanced photocatalytic activity for Cr(VI) reduction under visible light. *Applied Surface Science*, 425, 107–116. https://doi.org/10.1016/j.apsusc.2017.07.050.

[23] Mohaghegh, N., Tasviri, M., Rahimi, E., & Gholami, M. R. (2015). Comparative studies on Ag 3 PO 4/BiPO 4 -metal-organic framework-graphene-based nanocomposites for photocatalysis application. *Applied Surface Science*, 351, 216–224. https://doi.org/10.1016/j.apsusc.2015.05.135.

Chapter 35

A Performance Analysis of Novel Energy Efficient Electrodes for Capacitive Deionization

V. Dharnidhar Bhardwaj*
and Jyoti Sahu
Department of Chemical Engineering, National Institute of Technology,
Tiruchirappalli, Tamil Nadu, India

Abstract

The global water shortage is a problem that has yet to be solved. Although water covers 71% of the surface of the world, the bulk of it is not deemed suitable for drinking or other uses. Numerous methods for purifying water have been developed, including RO (reverse osmosis), ED (electrodialysis), MSF (multistage flash distillation), and others.

In addition to the methods already stated, capacitive deionization (CDI) is becoming more popular for purifying water. In a CDI cell, which has two porous electrodes (either static or flow electrodes) between which our feed travels, an open channel or porous dielectric material acts as the separator. The ions in the wastewater will travel to their corresponding oppositely charged electrode as a result of electrosorption, and they will be trapped in the electrical double layer that is created due to the ions adhering to the electrode. A CDI model was simulated using COMSOL6 software, wherein the variation in concentration with time for different values of V (electric potential) and U (flow velocity) by deploying standard carbon electrodes and carbon (95% impregnated with silver) for water desalination was studied. The salt adsorption capacity and salt adsorption rate for AC (activated carbon) and silver-doped activated carbon electrodes were calculated. COMSOL Multiphysics software interfaced with MATLAB applications was used to model a novel electrode in different rigorous conditions for CDI processes and to check its performance capacity and efficiency.

Keywords: capacitive deionization (CDI) electrosorption, electrical double layer

* Corresponding Author's Email: dharnilegend@gmail.com.

In: Sustainability in Chemical Processes through Digitalization and Green Chemistry Approaches
Editors: Susarla V. A. R. Sastry, Shravan Kumar and Ashish Kapoor
ISBN: 979-8-89113-257-3

1. Introduction

Water scarcity has been tormenting humans for centuries. It has reached a level wherein it will affect food production, ecosystem function, and developmental activities for years to come [1]. For many of our day-to-day activities, we are dependent on groundwater, which is becoming contaminated and exhausted. There is no plan for accommodating the population increase.

The world's population is growing at 80 million per year. India is the second-most populous country on the globe. India has a population of approximately 140 crores, out of which 163 million lack access to safe drinking water. Water pollution is another lynchpin that has contributed to the cause of scarcity. It affects the entire biosphere. It can be due to discharges from industries, oils, carcasses, feces, etc. Ninety percent of the total water resources are utilized for agricultural activities. Of this, 80% goes to irrigation-related work. The need for food production will rise in the future, which will directly affect agricultural water use, as increased agricultural activities will lead to greater water consumption [2].

Climate change has also been one of the factors contributing to water scarcity, as changes in weather and water patterns have consequently led to the rise of catastrophic occurrences such as floods and droughts [3]. These factors make water less available for use and increase water stress throughout the globe. The lack of water affects a human's wellbeing and deters holistic growth. A lack of safe drinking water can cause severe health impacts [4]. The education of the students will be affected as they are forced to get water from faraway places due to the nonavailability of water nearby. Sanitation and other issues can also arise. According to some reports, in nations with high infant mortality rates, pneumonia, malaria, HIV, and AIDS combined kill fewer infants than a lack of access to good water and sanitation.

A few solutions to tackle water scarcity are being deployed in our day-to-day lives to counter this global phenomenon. Among these are small-scale water harvesting, groundwater aquifers, and water storage in dams. [5] However, the environmental and financial costs of dams have increased. Economic solutions such as increases in pricing often receive critical attention in demand management. Rationing is a good way to manage water demand and can be an equitable way to meet basic needs, but it is unpopular and difficult to administer. New technologies can be implemented to alleviate this issue. Technologies such as water desalination and waste water treatment can reduce water demand and increase productivity through water-saving processes.

Water desalination is a sustainable and intelligent option for countering water scarcity [6]. Between 500 ppm and 35000 ppm of salt are present in both freshwater and saltwater, respectively [7]. Humans have experimented with several purifying methods in an effort to improve the amount of fresh water available. Instead of removing the low amount of salt that is present in the solution, desalination is often carried out by isolating pure water from the solution. but exceptions are always there. The process of desalination means separating minerals from sea water and making it fit for domestic and industrial use. The first desalination plant was set up in the 1950s [8]. Previously, desalination was performed solely by utilizing thermal energy due to the low cost of fossil fuels. As the cost of energy has increased exponentially, we have had to explore other options. Most of the research on desalination has been done in these two categories. 1.) Increase the energy efficiency of commercial technology. 2.) Investigation and proposal of new solutions The essential requirements for a desalination unit are [9] 3.) technological reliability; and 4.) Low energy consumption 5.) No Impact on the Environment 6.) Simplicity of use

The majority of techniques are used for water desalination. Some are popular at a commercial scale, and some are still at a laboratory scale. Evaporation and membrane techniques have been used largely in the commercial sector and have survived a long evolution process spanning over 50 years. Equipment, energy, human resources, chemicals, and other auxiliary resources are all required for any desalination process. Energy consumption per unit of output, the necessity for transfer equipment, and related equipment investment make up the largest portions of costs.

2. Common Techniques Used for Water Desalination

2.1. Reverse Osmosis

Water passes through the semipermeable membrane used in RO desalination, but dissolved particles are rejected. In a pressure vessel, it is contained. To obtain the required flow, the membrane's inlet side is pressurized to a pressure that is higher than the osmotic pressure. A pump powers the RO process [10]. Feed intake, feed water pretreatment, desalination and separation, product water posttreatment, and concentrate management are the phases involved in this process. Pretreatment is essential because the RO feed must have fewer organics, bacteria, and colloidal solids to keep the membranes from becoming fouled. This process might be negatively impacted by mineral scaling, which occurs when the concentrations of poorly soluble minerals (SiO_2) and mineral salts in the RO input increase when the product water is extracted above their solubility limit.

2.2. Electrodialysis

In an electrodialysis cell arrangement, cells are created by arranging a number of anion and cation exchange membranes in an alternating pattern between an anode and a cathode. A potential difference between the electrodes is established, and pumping of the salt solution between these cells is initiated [11]. Positively charged cations travel toward the cathode, while negatively charged ions flow in the opposite direction. Cation exchange membranes that are negatively charged permit cations to pass through but reject negative ions. Overall, this will result in the depletion of some ions while concurrently increasing the concentration of others in specific compartments. While the concentrated solution is referred to as brine, the depleted solution is known as dilute. The electric potential that is applied between the two electrodes powers electrodialysis. High permselectivity, low electrical resistance, good mechanical stability, and excellent chemical stability should all be present in the ideal ion exchange membrane. The desalination of saltwater by electrodialysis has a number of drawbacks, including the inability of electrodialysis to eliminate neutral, harmful components such as viruses and bacteria.

2.3. MSF Distillation

MSF distillation units are used where much cheap energy or waste energy is available [12].

It is a type of thermal desalination process. It can be operated in two modes. once through, wherein no brine recirculation will be there, and the other mode, wherein brine recirculation will occur. The brine heater portion and the heat recovery section are the two parts of the once-through type. The heat rejection section is an additional part of brine recirculation. To use the latent heat for preheating the feed, the consecutive evaporation of brine water into flash steam is followed by condensation inside the stage.

To optimize water and energy recovery, a vacuum is established inside the stages. A steam jet ejector can be used to create a vacuum. Some of the brine is recycled back and combined with incoming sea water to boost the process efficiency. The disadvantages of this include the fact that now more pumps and valves will be used in recirculation. A part of the water is obtained as distillate, which will not have salt content, and the rest will be rejected as brine solution. Some negative effects of multistage flash distillation include the discharge of brine, which pollutes the environment due to the ions present in brine, such as chlorine.

2.4. Multi-Effect Distillation

By heating seawater and letting it evaporate until fresh water is created as our desired product, the multieffect distillation desalinates sea water. In this case, less energy is used than in multistage flash distillation. Numerous multieffect models have been introduced over time. A dual-effect plant with a variety of effects, a train of flashing boxes, and condensers often makes up the traditional style of the model [13]. A turbine, a specific boiler, or flashing steam from a waste energy source are always used to extract the motive steam. salt concentration, saturation pressure, and boiling point elevation. Considering the abovementioned factors, the plant's effects are run at vacuum pressure. The effects of the multiple-effect distillation process are constrained by the temperature difference between the condensing temperatures at the first hot plant effect and the final condenser. The two major problems with these plants are scaling and corrosion. Scale formation can occur in condenser and preheater tubes due to high operating temperatures.

All the techniques mentioned above suffer from one or another drawback when put to use. Some are energy intensive, while others have low efficiencies. [14]. MSF is a procedure that uses a large amount of energy and necessitates both heat and mechanical energy. Operating the plant can improve efficiency but also increases scaling problems. In membrane processes such as reverse osmosis, fouling of membranes can occur. With a dissolved solids content of less than 6 g/L, ED is acceptable for brackish water. Considering all the abovementioned points, we have to look for other cost-effective alternatives. One such alternative is capacitive deionization (CDI). CDI is also an electricity-based desalination process similar to ED. In a CDI system, a potential difference is applied between the two electrodes. The feed or saline water flows through the spacer, which is placed between the electrodes. When dissolved in water, salt molecules take the form of ions. The electrostatic force draws positive ions (cations) to the negative electrode while drawing negative ions (anions) to the positive electrode. This will allow the ions to be removed from the solution. [15] After some time, the electrodes will become saturated with ions. The potential is reversed, due to which the ions move back into the solution, producing a concentrated brine discharge. This step is known as regeneration. The removal of ions from saline water is known as purification. Both of these processes will work in an alternating manner. The applied voltage should not exceed 1.23 V since after that, potential hydrolysis of water starts taking place. The most commonly used electrode material for capacitive deionization is activated carbon. In this work, we tried to incorporate silver into activated carbon at various mass fractions of silver and studied its effect on desalination capacity.

3. Numerical Model

3.1. Assumptions

In capacitive deionization, the ions move under the influence of electrostatic force.

- Laminar flow conditions were assumed for the feed flow between our electrodes.
- The whole ion population that is adsorbed in the electrical double layer that was created by the charges on the electrode surface
- The ion concentration is low. Therefore, the migration of ions is low, which will not affect the macroflow for the feed.
- Activated carbon electrodes were used. There will be no reaction between the ions and the electrode.
- The temperature and pressure remain constant throughout our process.
- For ions in an electric field, the mobility of ions follows the Nernst-Einstein equation.

3.2. Theoretical Basis

3.2.1. Main Control Equations

3.2.2. Navier–Stokes Equations

These are the universal laws of physics. They can model any fluid that we encounter in our day-to-day lives. Equation (1) tells us that mass is conserved, and the second equation refers to Newton's second law, which states that force equal to mass equals acceleration. For 2D, the equations are very accurate, and we will obtain a solution every time we use them [Eq. (2)]. However, for 3D, these equations do not have a solution that is consistent every time. The velocity is low between our electrodes, so we assume we are in the laminar regime, and we will use equations according to laminar flow.

$$\nabla . u = 0 \tag{1}$$

$$\rho \frac{\partial u}{\partial t} = -\nabla P + \mu \nabla^2 u + \rho F \tag{2}$$

where ρ is the fluid density, u = u (u, v, w) is the velocity vector, P is the pressure, and μ is the viscosity. Our dependent variables here are velocity U and pressure P.

Electrostatic Equations

Faraday's laws in differential form tell us that the electric field is irrotational, and their integral form establishes the fact that the electric field is conservative in nature.

The static electric field that surrounds our two electrodes acts similarly to the electric field that is established between the two plates of the parallel plate capacitor. It is an irrotational electric field. For an irrotational electric field, the potential is defined by Eqs. (3) and (4)

$$-\nabla V = E \tag{3}$$

$$\nabla \times \nabla V = 0 \tag{4}$$

E = Electric Field

V = Electric Potential

This is true for sufficiently smooth electric fields. The negative sign is an old convention.

In the case of dielectrics, they have bound charges. These bound charges can be displaced by an external electric field, resulting in an induced dipole. These induced dipoles align themselves with the existing electric field. Due to this, the electric field inside a dielectric will be different from that in free space.

Transport Equations

In the space between the electrodes. The movement of ions is due to convection, diffusion, and electromigration. The Nernst–Planck equation [Eq. (5)] was used in this model to describe the movement of ions in the CDI cell.

$$\frac{\partial c_i}{\partial t} + \nabla \cdot \left(-D_i \nabla c_i - z_i u_{m,i} F c_i \nabla V\right) + u \cdot \nabla c_i = 0 \tag{5}$$

where c_i is the concentration of ion i, D_i is the diffusion coefficient of ion i, V is the electric potential, z_i is the charge number of ion i, and F is Faraday's constant.

96,485 C/mol, and um, i is the ionic mobility of ion i. In this model, the ionic mobility is calculated by the Nernst-Einstein relation.

Nernst Einstein Equation

Table 1 represents the basic data about CDI cells for desalination using activated carbon electrodes, and Table 2 depicts the basic data about CDI cells for desalination using silver-impregnated activated carbon electrodes (5% silver + 95% activated carbon).

Table 1. Basic data about CDI cells for desalination using activated carbon electrodes

Property	Values	unit
Density (ρ)	751.8	kg/m^3
Heat Capacity at constant Pressure (C_p)	0.84	J/(kg.K)
Thermal Conductivity (k)	0.63	W/m^2K
Porosity	0.90	1
Relative Permittivity	8	1
Diffusion Coefficient (Component 1)	1.33*10^-09	m^2/s
Diffusion Coefficient (Component 2)	2.03*10^-09	m^2/s

Table 2. Basic data for CDI cells for desalination using silver-impregnated activated carbon electrodes (5% silver + 95% activated carbon)

Property	Values	Unit
Density (ρ)	796	kg/m^3
Heat Capacity at constant Pressure (C_p)	235	J/(kg.K)
Thermal Conductivity (k)	209.5	W/m^{2}K
Porosity	0.70	1
Relative Permittivity	90	1
Diffusion Coefficient (Component 1)	1.33*10^-09	m^2/s
Diffusion Coefficient (Component 2)	2.03*10^-09	m^2/s

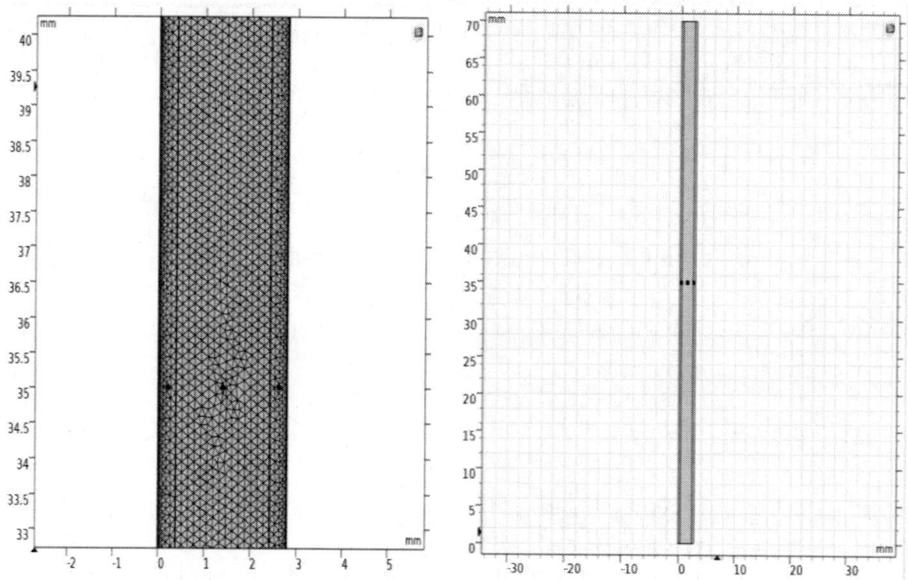

Figure 1. Part of our model showing the 2D tetrahedral mesh that we used in our simulation. b.) The model geometry as used in the simulation.

4. Simulation Conditions

In the capacitive deionization model, activated carbon electrodes with dimensions of 70 mm x 70 mm x 0.38 mm were used. The spacing between the electrodes is 2 mm. The porosity of the carbon electrode is 0.90. A total of 0.49 cm^2 of contact space exists between the activated carbon electrodes, and the salt water diffusion coefficient values were taken [16]. A 2D simulation model was made wherein first symmetrical electrodes were used and then silver-impregnated activated carbon electrodes were studied. The inlet and outlet given to the model were symmetrical. The inlet condition in the laminar flow domain was given as the velocity condition, while the outlet condition was given as the pressure condition. The initial solution ion concentration is equal to the initial inflow concentration. The mesh size was chosen as normal, as a coarser mesh will no doubt reduce the calculation time, but the value might not be that precise. On the other hand, with a finer mesh, the values will be more precise, but it will consume much time, so a trade-off was made between the coarse size and the time taken for operation by choosing a mesh of normal size, as shown in Figure 1.

5. Results and Discussion

5.1. Model Validation

Table 3 shows the comparison of ion absorption capacities (IAC) between the capacitive deionization work [17] and the simulated model created by us using COMSOL Multiphysics. Our model validates itself against the results that were produced in the experimental work with less than 6% deviation. The same initial conditions were followed as in the experimental model, such as the initial concentration of the NaCl solution being kept at 4 mM, the potential given to the two electrodes being between 0.4 V and 1.0 V, and the electrodes being completely free of ions during the start of absorption.

$$\text{IAC (Ion Absorption Capacity)} = \frac{(C_e - C_0).V}{m}$$

where C_0 and C_e are the initial and equilibrium concentrations (M), respectively. V is the volume of the solution (L). m is the mass of the electrode (g). Figure 2 represents the ion absorption vs. applied voltage. Table 3 shows the comparison of ion absorption capacities.

5.2. Velocity Distribution

The velocity does not change much. The surface of the electrode on which fluid sticks is known as the no slip condition. Here, the velocity will be zero, which is shown by the dark blue-colored region on the surface of the electrode. Our solution cannot go inside the electrodes and flow. Therefore, we will have zero velocity for both electrodes. At the start of our simulation, the flow will consist of two zones: the irrotational core region and the boundary layer region, where the viscous effects due to the fluid layers sliding over will be felt. After that, the boundary layer grows with distance and converges on the center line. After that, the velocity profile throughout will remain constant. The center line velocity is the maximum shown by the red-colored region. Decreasing in the horizontal direction from zero at the electrode surface to maximum velocity at the center and then again going to zero at the surface of the electrode due to a no slip condition.

Table 3. Comparison of ion absorption capacities

Applied Voltage(V)	IAC reported in Literature(μmol/gm)	IAC of our Model(μmol/gm)	Percent Deviation
0.4	40	38	5.26%
0.6	64	64	0.00%
0.8	85	90	5.88%
1.0	105	107	1.90%

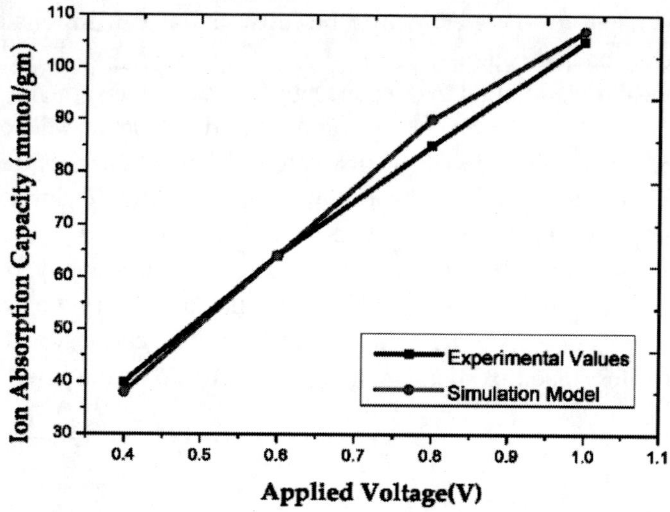

Figure 2. Ion absorption vs applied voltage.

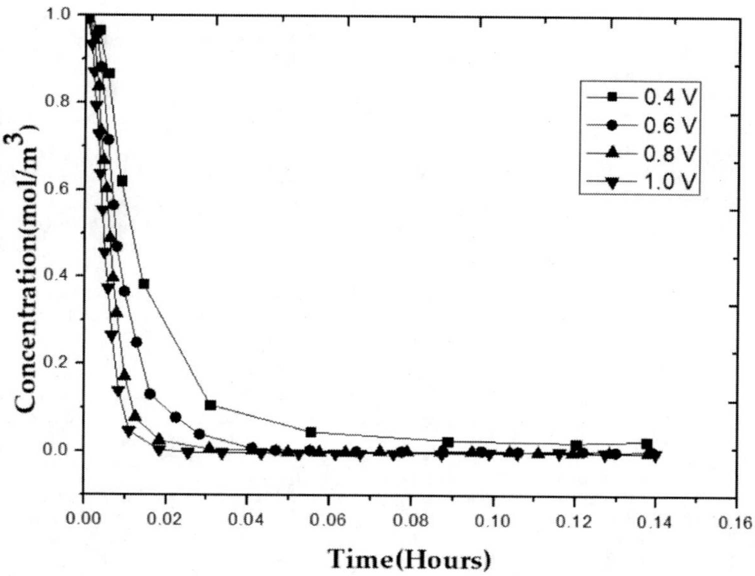

Figure 3. Shows how the voltage variation affects.

5.3. Change in Ion Concentration with Voltage

The inlet concentration and flow rate were kept constant at 1 mol/m3 and 10 ml/min, respectively. When our solution flows between the two electrodes, we vary the potential, ranging from 0.4 V to 1.0 V. As the voltage increases, the electrostatic forces of attraction between our electrode surface and the ions increase, which increases the adsorption rate. The voltage cannot be increased beyond 1.2 V, as above that, hydrolysis of water takes place. Water starts splitting into hydrogen and oxygen. Figure 3 shows how the voltage variation affects the NaCl concentration in our saline solution, and it emphasizes the fact that as the voltage increases, the time taken for the electrode to be fully saturated with ions decreases, thereby improving the rate of ion absorption.

In Figures 4 and 5, the variation in the concentration of ions absorbed with voltage for two different electrodes is shown. One electrode was activated carbon, and the other was activated carbon (95%) impregnated with 5% silver. It was found that for this particular composition of activated carbon and silver, the ion absorption capacity of this electrode decreases when compared with the pure activated carbon electrode. This is because as we add silver to our activated carbon material, the thermal conductivity increases considerably, but at the same time, the porosity and surface area decrease, and since the ion absorption capacity is directly proportional to the porosity and surface area, our ion absorption capacity in the case of activated carbon with silver is less than that of a pure activated carbon electrode.

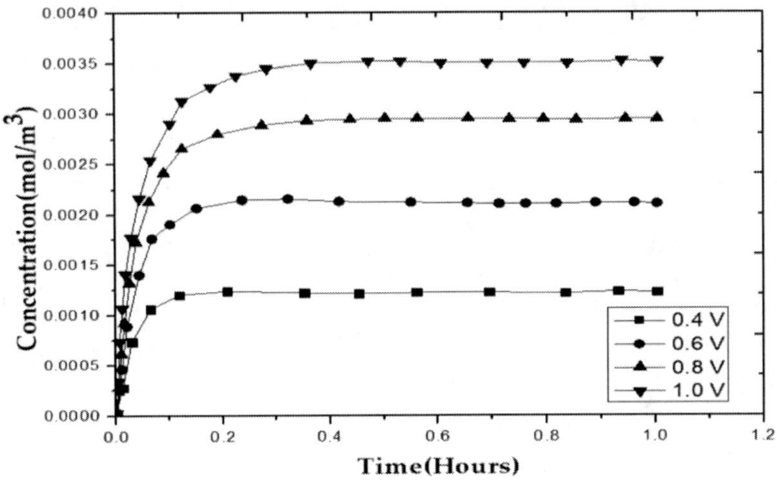

Figure 4. Variation in the concentration of ions absorbed with voltage.

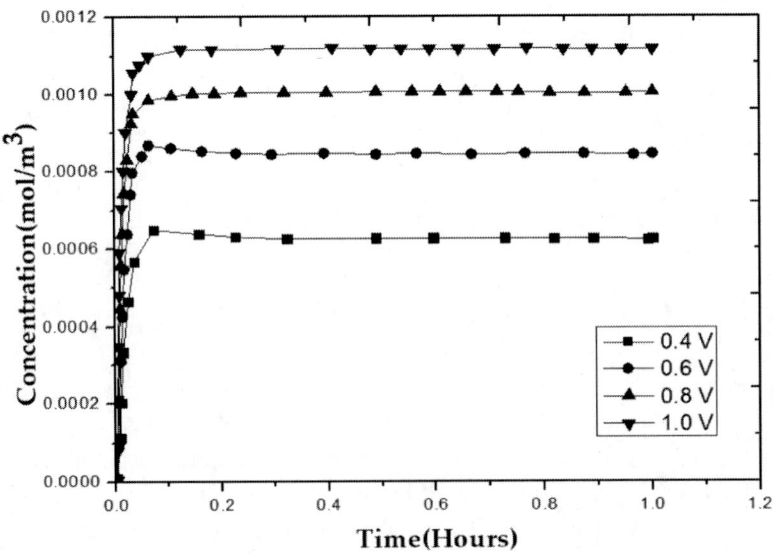

Figure 5. Variation in the concentration of ions absorbed with voltage.

At the same time, the conductivity has increased considerably compared to activated carbon, due to which the rate of ion absorption increases rapidly even though the ion absorption capacity is low.

Figure 5 supports the argument presented in the later half, while Figure 4 shows the variation in the concentration of ions at different voltages for a pure activated carbon electrode.

Figure 6 presents the ion absorption capacity v/s voltage for both electrodes, i.e., activated carbon and activated carbon with silver.

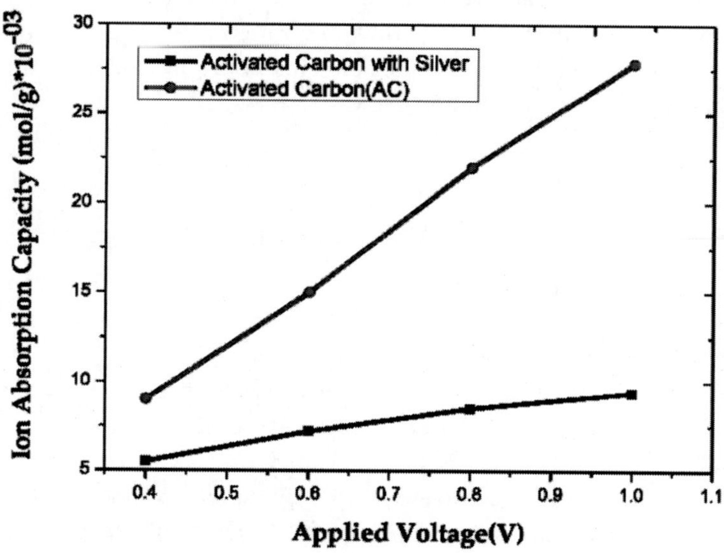

Figure 6. presents us with the ion absorption capacity v/s voltage.

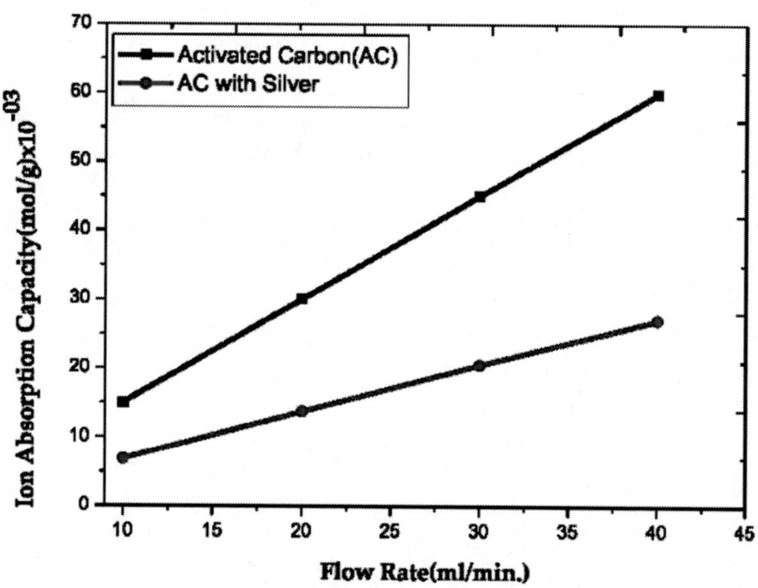

Figure 7. Variation in ion absorption capacity.

5.4. Change in Ion Absorption Capacity with Flow Rate

While keeping the concentration of our NaCl solution at 1 mol/m^3 and the potential constant at 0.6 V, the flow rate varied. As the flow rate increased, it was observed that the ion absorption capacity also increased. This can be attributed to the fact that at low velocity, coion expulsion will be greater. Coion expulsion is the repulsion that takes place between the adsorbed ions on the electrode and the ions in the solution having the same polarity. Therefore, at low velocity due to Coion expulsion, the ion absorption capacity is lower, and as the flow rate increases, the ion absorption capacity also increases. The flow rate varied from 10 ml/m^3 to 40 ml/m^3. Figure 7 shows the variation in ion absorption capacity for both electrodes (activated carbon and activated carbon with silver) as a function of flow rate.

6. Conclusion

1. Ag and Cl in solution interact, causing the Ag-Ag bond to dissolve and create AgCl. High theoretical specific capacities for electrode materials are made possible via conversion reactions.
2. It should be mentioned that although an increase in current density speeds up salt removal, using Faradaic electrodes typically results in a severe reduction in salt removal capacity.
3. When the conductivity of activated carbon with silver added increases, the porosity and surface area for salt adsorption decrease, which results in a significantly lower salt absorption capacity.

For electrode materials, good electronic conductivity is crucial. Low electron conductivity in Faradaic electrode materials is a significant downside since it accelerates the polarization of electrochemical processes and slows down electron transfer.

Electrode materials must exhibit good electronic conductivity. Low electron conductivity, which speeds up the polarization of electrochemical reactions and lowers the rate of electron transfer, is a major drawback of faradic electrode materials and results in sluggish charge-transfer kinetics. It is helpful to add a conductive agent while preparing the electrode and combine carbon and Faradaic materials to create hybrid electrode materials to maximize charge diffusion throughout the electrode.

Future studies will involve including conductive materials such as carbon black and silver with activated carbon and testing the results. Silver is very expensive and dissolves in seawater, which drastically limits its usefulness.

References

[1] Jury, W. A., and Vaux Jr, H. J. (2007). The emerging global water crisis: managing scarcity and conflict between water users. *Advances in agronomy*, 95, 1-76.

[2] Chakkaravarthy, D. N., and Balakrishnan, T. (2019). Water scarcity-challenging the future. *International Journal of Agriculture, Environment and Biotechnology*, 12(3), 187-193.

[3] Du Plessis, A. (2019). Current and future water scarcity and stress. In *Water as an Inescapable Risk* (pp. 13-25). Springer, Cham.

[4] Rijsberman, F. R. (2006). Water scarcity: fact or fiction?. *Agricultural water management*, 80(1-3), 5-22.

[5] Meinzen-Dick, R. S., and Rosegrant, M. W. (2001). Overcoming water scarcity and quality constraints (No. 571-2016-39085).

[6] Manju, S., and Sagar, N. (2017). Renewable energy integrated desalination: A sustainable solution to overcome future fresh-water scarcity in India. *Renewable and Sustainable Energy Reviews*, 73, 594-609.

[7] World Health Organization. (2003). Selenium in drinking-water: Background document for development of WHO guidelines for drinking-water quality (No. WHO/SDE/WSH/03.04/13). World Health Organization.

[8] Curto, D., Franzitta, V., and Guercio, A. (2021). A review of the water desalination technologies. *Applied Sciences*, 11(2), 670.

[9] Semiat, R., and Hasson, D. (2012). Water desalination. *Reviews in Chemical Engineering*, 28(1), 43-60.

[10] Jia, B., and Zhang, W. (2016). Preparation and application of electrodes in capacitive deionization (CDI): a state-of-art review. *Nanoscale research letters*, 11(1), 1-25.

[11] Strathmann, H. (2004, December). Assessment of electrodialysis water desalination process costs. In *Proceedings of the International Conference on Desalination Costing*, Limassol, Cyprus (pp. 32-54).

[12] El-Ghonemy, A. M. K. (2018). Performance test of a sea water multistage flash distillation plant: Case study. *Alexandria engineering journal*, 57(4), 2401-2413.

[13] Al-Shammiri, M., and Safar, M. (1999). Multieffect distillation plants: state of the art. *Desalination*, 126(1-3), 45-59.

[14] Shatat, M., and Riffat, S. B. (2014). Water desalination technologies utilizing conventional and renewable energy sources. *International Journal of Low-Carbon Technologies*, 9(1), 1-19.

[15] AlMarzooqi, F. A., Al Ghaferi, A. A., Saadat, I., and Hilal, N. (2014). Application of capacitive deionization in water desalination: a review. *Desalination*, 342, 3-15.

[16] Zhao, Y., Hu, X. M., Jiang, B. H., and Li, L. (2014). Optimization of the operational parameters for desalination with response surface methodology during a capacitive deionization process. *Desalination*, 336, 64-71.

[17] Hou, C. H., and Huang, C. Y. (2013). A comparative study of electrosorption selectivity of ions by activated carbon electrodes in capacitive deionization. *Desalination*, 314, 124-129.

Chapter 36

The Removal of Lead (Pb) Using Mallet Flower Leaf Powder as an Adsorbent

D. Venkata Padma[1,*]
S. V. A. R. Sastry[2]
D. Krishna[1]
and P. Karthikeya Nikhil[1]
[1]Department of Chemical Engineering, MVGRCE(A), Vizianagaram, A.P., India
[2]School of Chemical Technology, HBTU, Kanpur, U.P., India

Abstract

The toxic metals discharged into the environment from industrial effluents accumulate at an exponential rate. Even with an increase in PPM levels, these metals show adverse effects on human health and all naive animals that depend on water for survival. In the age of modernization, technologies have come up to effectively remove pollutants from industrial effluent, but when pollutants are at a lower concentration, they may face an economic lag in operation. The objective of this paper is to exploit the use of available Mallet flower leaf powder (MFLP) for the elimination of lead from wastewater. Experiments are performed to determine the effects of stirring time, pH of the solution, initial concentration of the adsorbate, temperature of the system, solid adsorbent dosage, and particle size. While keeping the rpm constant, the optimum pH of the solution was 5 at 100 ppm and 1.0 gm of biosorbent with a metal uptake of 8.9 mg/g and a lead metal removal of 91.1% at room temperature. The optimum concentration of the solution was 80 ppm of 1.0 g of biosorbent at room temperature, with the removal of lead metal at 87.37%. The lead metal adsorption isotherms for MFLP fit well with the Freundlich isotherm and Temkin isotherm equations. Atomic absorption spectroscopy was used for the analysis of lead. The final concentrations of the metal ions after adsorption were analyzed through an AAS analysis instrument. The surface of the MFLP was examined by scanning electron microscopy to understand the various functional groups present at the surface sites by FTIR analysis before and after adsorption.

Keywords: lead, AAS, heavy metals, MFLP, adsorption

[*] Corresponding Author's Email: padmakishore9@gmail.com.

In: Sustainability in Chemical Processes through Digitalization and Green Chemistry Approaches
Editors: Susarla V. A. R. Sastry, Shravan Kumar and Ashish Kapoor
ISBN: 979-8-89113-257-3

1. Introduction

The key phrase toxic metal implies any element that is toxic at very low concentrations and is denser compared to salts and others [1]. These are mostly transition metals. As a first step in this direction, a literature survey is being conducted in the domain of adsorption and biosorption of heavy metals for a period of over [2, 3] three decades. The analysis consists of four groups: adsorbates, adsorbents, methods used, and adsorption systems. The study comprises the method of analysis and various parameters affecting the biosorption performance [4–6], such as pH, dosage, time of contact, and batch or continuous operation. Various isotherm models and kinetic models were considered for fitting their [7–9] experimental data to obtain the equilibrium isotherm and kinetic parameters.

Industrial zones that included all levels of production capacity of metals, refineries, and treatment units were built with wastewater treatment plants where contaminants disperse into the groundwater and soil. Modern industrialization is considered a significant source of contamination of the environment [10] with heavy metals. At the cellular level, the organism tends to accumulate this metal and sluggishly develop toxic compounds that become lethal to life sustainability, even at relatively minor levels of exposure.

2. Materials and Methods

2.1. Adsorbent Preparation and Characterization

Mallet leaves were collected from the MVGR College and used as absorbents. The leaves are washed and reduced to a size of [11, 12] 74 microns to 149 microns for future study. On the other hand, lead stock solutions were prepared at different concentrations using lead nitrate as a source for divalent lead. A morphological study attempt on the surface of the absorbent was performed by FTIR.

Experimental Procedure

Laboratory batch adsorption experiments were performed to determine the effect of the initial concentration of metal ions, pH of the solution, size of the adsorbent dosage, [13, 14] temperature of the solution, and running time on the removal of lead metal ions at different sets. The final concentration was analyzed by an AAS analysis instrument. Finally, we obtained the results from various parameters by changing one parameter while keeping others constant. Various studies on kinetics, equilibrium, and thermodynamics calculated and plotted the outcomes. In the present study, it is envisaged (intended) to remove lead (II) from wastewater using a low-cost adsorbent such as Mallet Flower Leaf Powder (MLFP). The parameters covered are agitation time (t), adsorption dosage (l_a), adsorbent particle size (d_p), pH, temperature (T), and initial concentration of lead. The effect of these parameters on percent removal (R_m) or recovery (R_c) and on metal uptake (q_e) was studied by conducting batch experiments. The parameters covered in the case of lead are compiled in Table 1.

Table 1. Various parameters studied in the present process

S.NO	Parameter	Range
1.	Time, min	0-120
2.	Initial concentration of the adsorbate, PPM	10-100
3.	pH of the solution	1-12
4.	Temperature of the system, °C	10-50
5.	Adsorbent size, microns	74-149
6.	Adsorbent dosage, gm/50 ml	0.2-1

At the end of the experiment, the final concentration was analyzed by AAS. The percentage removal and metal uptake [15, 16] are calculated by the following formulas:

Metal uptake = [(initial concentration - final concentration) × 50]/1000

3. Results and Discussion

3.1. Characterization of the MFLP Surface

The FTIR spectra (Figures 1 and 2) show the various functional groups present on the surface before and after adsorption [17]. SEM images reveal that the morphological studies on the surface of the MFLP are represented in Figures 3 and 4.

The percentage removal decreased from 68.5% to 66% for 20 ppm and from 68.5% to 66% by increasing the temperature from 283 K to 323 K (Figure 5). At high temperatures, adsorption was found to be lower. From that, we can presume the temperature. The above graph shows that at pH 5, we obtained the maximum percentage recovery of lead using Schefflera pucckleri as a biosorbent of 91.1% for a concentration of 100 ppm at room temperature (Figure 6). Therefore, the optimum pH is considered to be 5.

The effect of adsorbent dosage on the percentage removal of lead is shown through a plot of data drawn between percentage removal and different adsorbent dosages as removal increased from 48.7% to 83.7% by increasing the adsorbent dosage from 0.2 to 1.0 g (Figure 7). At a high adsorbent dosage of 1 g, the percentage removal achieved a maximum of 79.8%. The experimental data on the percentage removal obtained in the batch adsorption runs The plot is drawn between the size of the adsorbent and the percentage removal of lead from the solution by using two different concentrations.

The temperature, pH, agitation time, and adsorption dosage were kept constant for both cases. Initially, there is an increase in the percentage removal of lead with the decrease in [12, 13] size of the adsorbent from 100 mesh to 200 mesh from 49 to 79.5 and from 51.4 to 85.1 for the concentrations of 60 and 100 ppm, respectively (Figure 8). The percentage removal of lead is observed to be a maximum of 85.1% when the adsorbent size is 200 mesh and the [14, 15] concentration is 100 ppm. The percentage removal is slightly reduced from 44 to 81.7 by raising the initial Pb(II) concentration (C) from 20 ppm to 100 ppm (Figure 9). The negative values of ΔG indicate the spontaneous nature of the adsorption process.

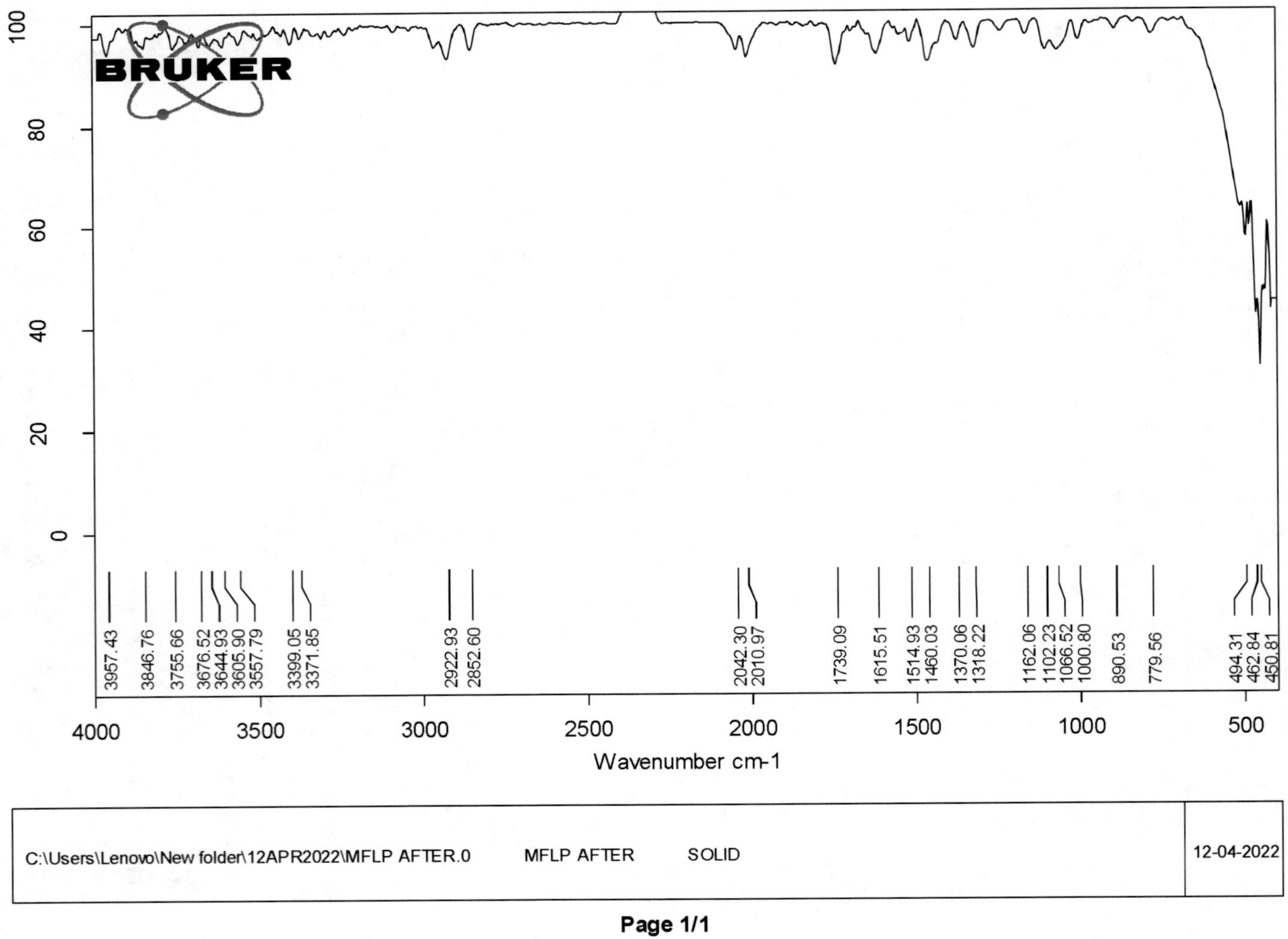

Figure 1. FTIR image of the MFLP adsorbent before lead adsorption.

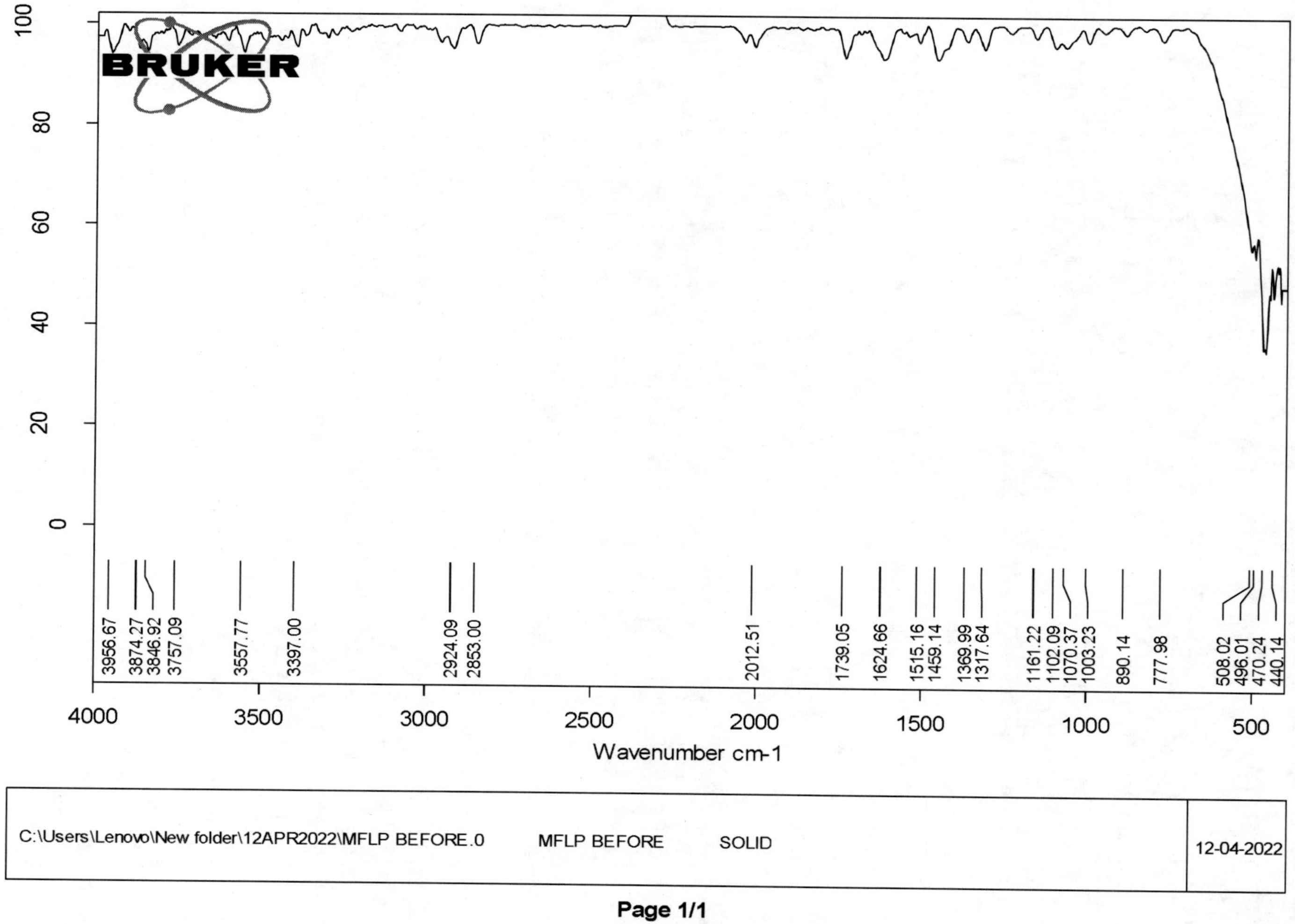

Figure 2. FTIR image of the MFLP adsorbent after lead adsorption.

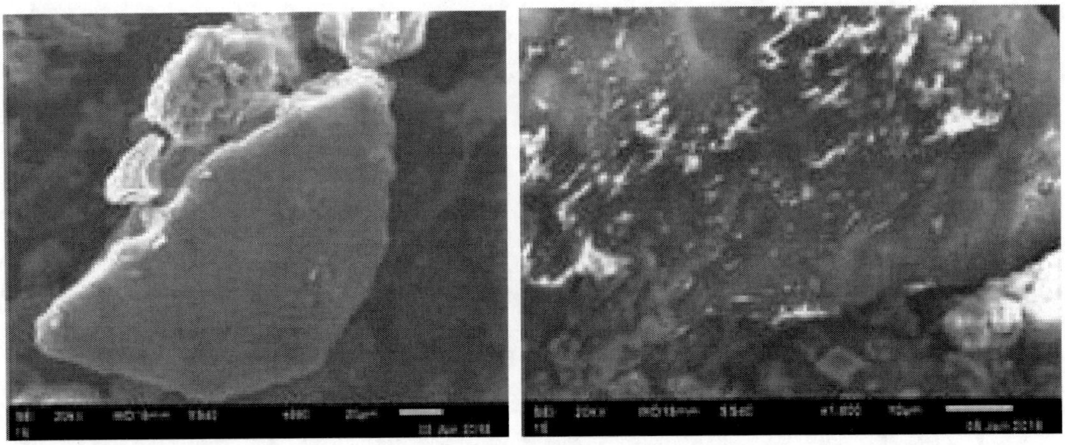

Figure 3. SEM image of the MFLP adsorbent before lead adsorption.

Figure 4. SEM image of the MFLP adsorbent before lead adsorption.

The ΔG values varied from 1486 J/mol to 6275 J/molK as the temperature varied from 283 K to 323 K (Table 2). Adsorption favors 303 K Entropy ΔS varied from 6 to 57.02 J/molK. The time required to reach equilibrium is 40 minutes. The optimum pH of [16, 17] the solution is 5 at 100 ppm at 1.0 gm of biosorbent with a metal uptake of 8.9 mg/g and a removal of lead metal of 91.1% at room temperature.

The optimum concentration of the solution was 80 ppm of 1.0 g of biosorbent at room temperature, with the removal of lead metal at 87.375%. The lead metal adsorption isotherms for the powder fit well with the Freundlich isotherm and Temkin isotherm equations. The coefficient of determination for the isotherms was 0.9998 for the Langmuir isotherm (Table 3), 0.9347 for the Freundlich isotherm (Table 4), and 0.9836 for the Temkin isotherm (Table 5). Thermodynamic studies for the removal of Pb(II) show a value of 0.94.

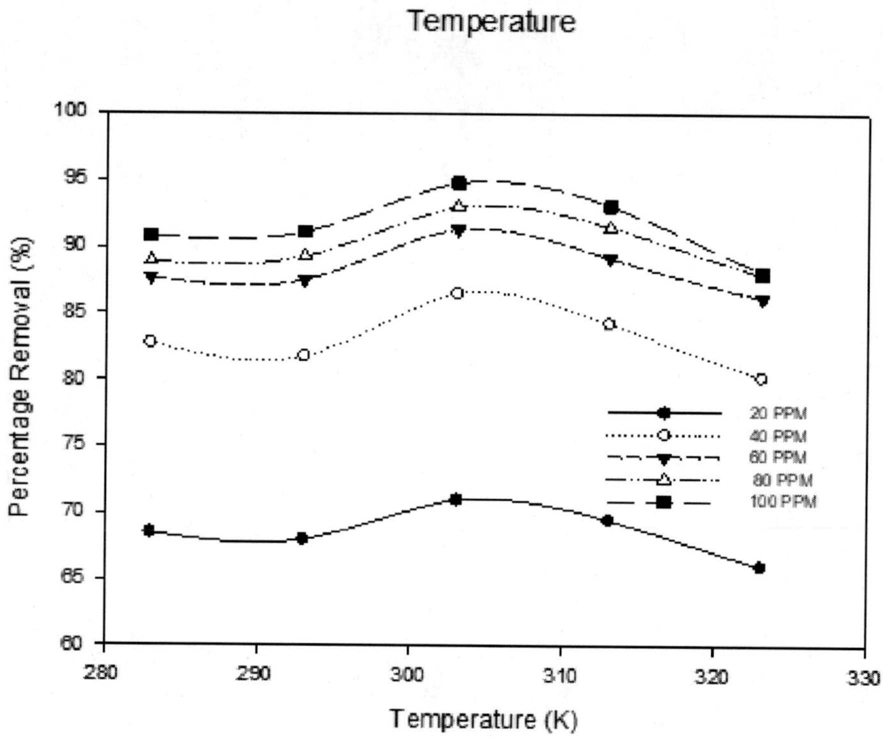

Figure 5. Effect of temperature on the adsorption of Pb(II).

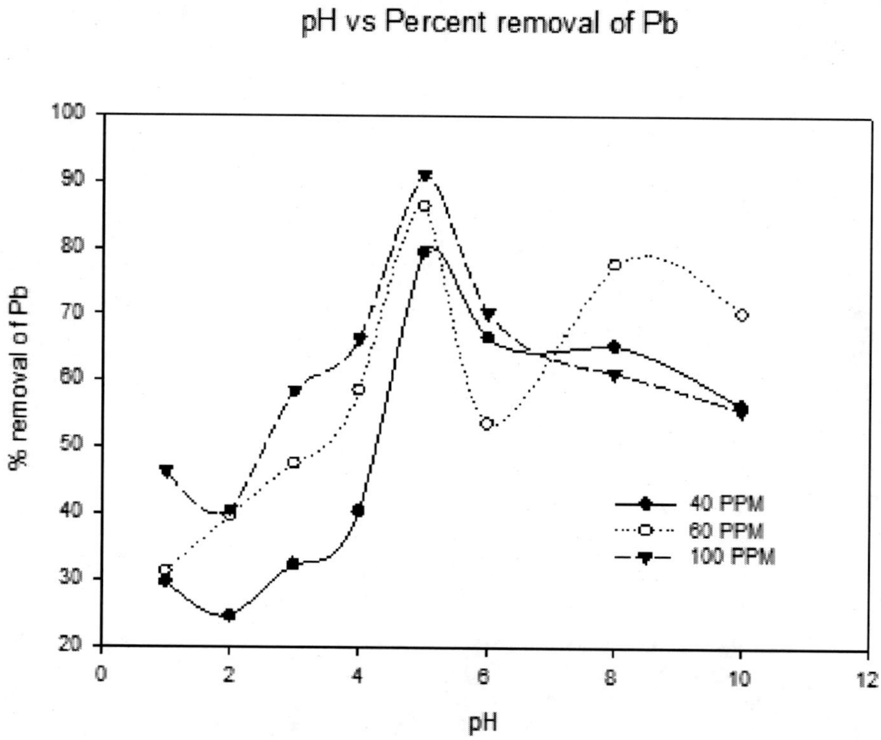

Figure 6. Effect of pH on the percentage recovery of Pb(II).

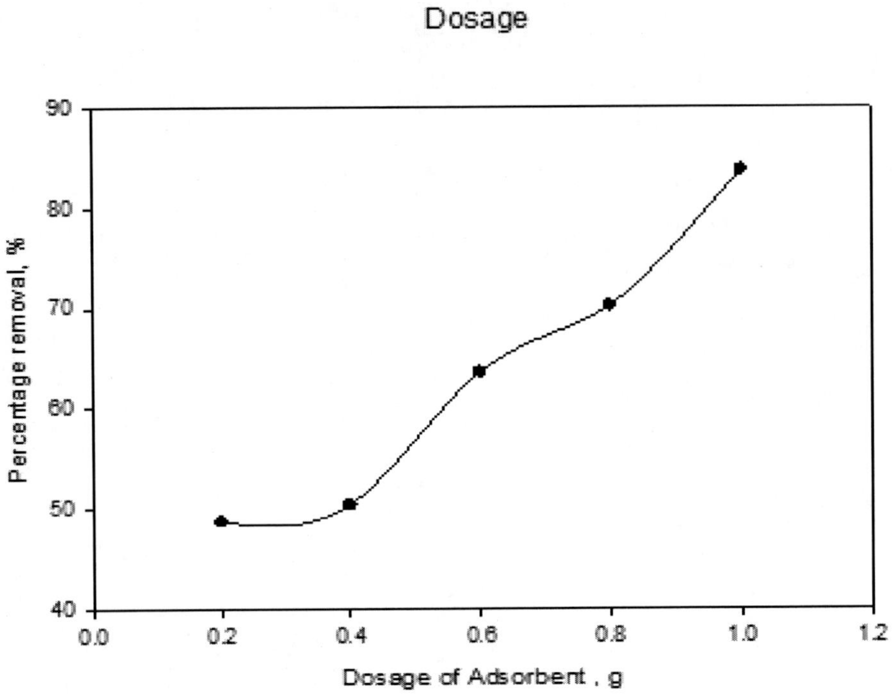

Figure 7. Effect of adsorbent dosage on percentage removal of Pb(II).

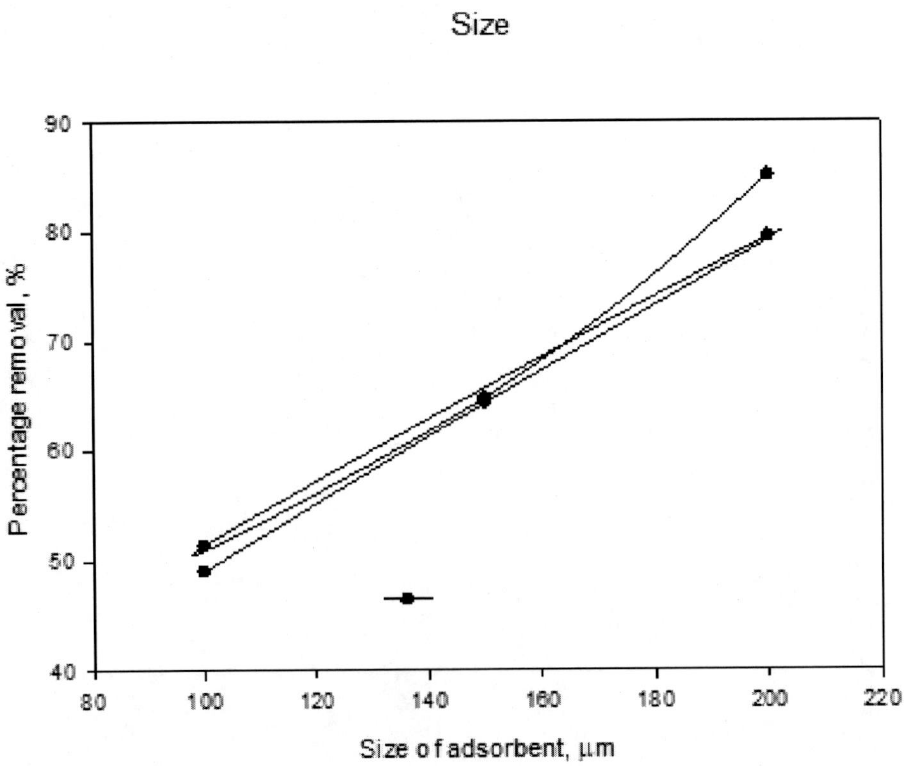

Figure 8. Effect of the size of the biosorbent on the percentage removal of Pb(II).

Table 2. Thermodynamic data of Pb(II)

# PPM	Equation	ΔH (J/mole)	ΔS (J/mol.K)	ΔG (KJ/mol)				
				283K	293K	303K	313K	323K
20	$\ln K_d = 1508 \frac{1}{T} - 4.5216$	12537.512	37.59	-1827.82	-1561.69	-1340.73	-574.818	-430.58
40	$\ln K_d = 113.12 \frac{1}{T} - 1.2266$	-940.47	10.1979	-3609.4	-3652.78	-4679.18	-4363.89	-3764.934
60	$\ln K_d = 58.113 \frac{1}{T} - 1.8476$	-483.15	15.362	-4614.527	-4740.24	-5932.654	-5485.29	-4912.173
80	$\ln K_d = -105.31 \frac{1}{T} + 2.5759$	875.54	21.416	-4889.341	-5155.89	-6516.239	-6183.724	-5318.881
100	$\ln K_d = 208.99 \frac{1}{T} + 1.7333$	-1737.5428	14.4106	-5386.783	-5665.921	-7313.463	-6771.528	-5376.026

Initial concentration.

Table 3. Langmuir isotherm data of Pb(II)

S.No	#T,(K)	Isotherm Equation	q_{max}(mg/g)	b	R_L	R^2
1	283	$\frac{C_e}{q_e} = 1.5529\ C_e - 0.7172$	0.6439	-2.1654	-0.0236	0.6311
2	293	$\frac{C_e}{q_e} = 0.7005\ C_e - 0.6775$	1.4275	-1.03398	-0.0508	0.992
3	303	$\frac{C_e}{q_e} = 0.4226\ C_e - 0.3076$	2.3663	-1.3738	-0.0122	0.9997
4	313	$\frac{C_e}{q_e} = 0.3058\ C_e - 0.2143$	0.3058	-0.7007	-0.024	0.9998
5	323	$\frac{C_e}{q_e} = 0.2394\ C_e - 0.163$	4.17710	-1.468	-0.006858	0.9994

Temperature.

Table 4. Freundlich isotherm data of Pb(II)

S.No	#T,(K)	IsothermEquation	k_f	N	R^2
1	283	$\log q_e = 4.2699\ log C_e - 3.4375$	0.03654	0.23419	0.8746
2	293	$\log q_e = 5.4584\ log C_e - 4.4922$	0.00316	0.1832	0.9347
3	313	$\log q_e = 14.083\ log C_e - 11.115$	7.673×10^{-12}	0.07100	0.9305
4	323	$\log q_e = 3.2378\ log C_e - 2.7255$	1.88×10^{-3}	0.30885	0.8408

[#] Temperature.

Table 5. Temkin isotherm data of Pb(II)

S.No	#T,(K)	IsothermEquation	bT	AT(L/mg)	R^2
1	283	$\log q_e = 20.926\ log C_e - 15.883$	112.4372	0.4618	0.9582
2	293	$\log q_e = 25.858\ log C_e - 20.289$	94.2068	0.4563	0.9534
3	313	$\log q_e = 14.083\ log C_e - 11.115$	37.411	0.4606	0.9836
4	323	$\log q_e = 3.2378\ log C_e - 2.7255$	171.0024	0.4585	0.9577

[#] Temperature.

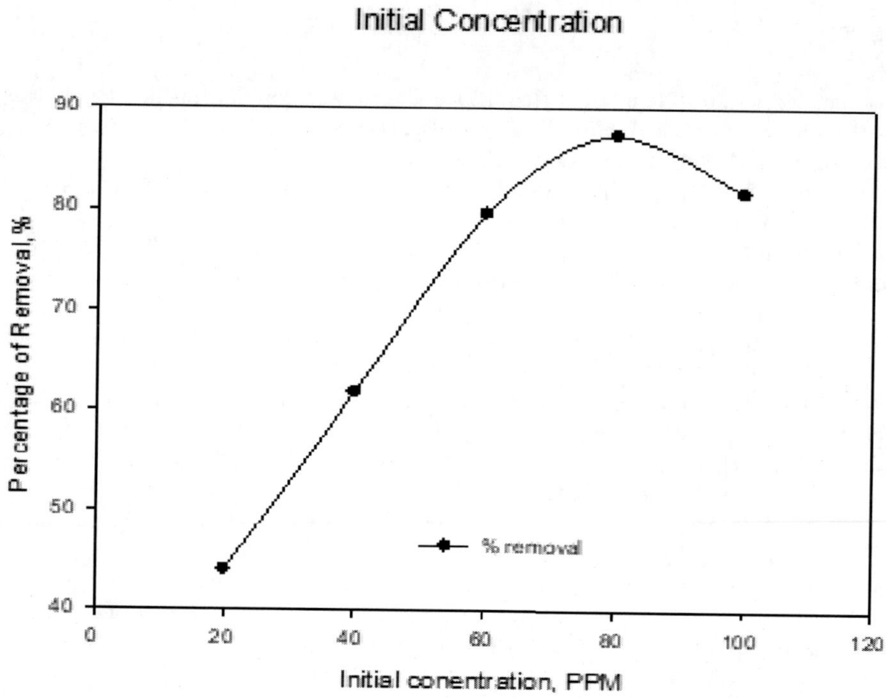

Figure 9. Effect of initial batch concentration in ppm on the percentage recovery of Pb(II).

Conclusion

MFLP is a favorable adsorbent for purifying wastewater from lead pollutants. The operating conditions of the MFLP interaction with Pb(II) solution governed the metal deposit on the active site. The most important variable that generated the optimal time required to attain an equilibrium adsorption rate was approximately 40 minutes. The parameters were the time required for contact, Pb(II) content in solution and its pH, particle mesh size, operating temperature, and weight of MFLP used for each set of experiments. The optimum pH of the solution is 5 at 100 ppm at 1.0 gm of biosorbent with a metal uptake of 8.9 mg/g, and the removal of lead metal is 91.1% at room temperature. The optimum concentration of the solution was 80 ppm of 1.0 g of biosorbent at room temperature, with the removal of lead metal at 87.375%. The lead metal adsorption isotherms for MFLP fit well with the Freundlich isotherm and Temkin isotherm equations. Atomic absorption spectroscopy was used for the analysis of lead. The various tools used for the characterization of MFLP are SEM for understanding the morphology of MFLP before and after adsorption, while FTIR is used for identifying the functional groups. The thermodynamic data, represented as Gibb's free energy (ΔG), change in enthalpy (ΔH) and change in entropy (ΔS), indicate the feasibility, spontaneity, endothermic nature, and enhancement of randomness at the surface of the MFLP.

References

[1] Yarkandi, N. H. (2014) Removal of lead (II) from waste water by adsorption. *International Journal of Current Microbiology and Applied Sciences, 3*(4):207-228.

[2] Ramalingam, S. J. Hidayathulla Khan, T. Pugazhlenthi, M. Thirumurugan, V. (2013) Removal of Pb (II) and Cd (II) ions from industrial waste water using calotropis Procera roots. *International Journal of Engineering Science Invention, 2* (4): 01-06.

[3] Mahammed, A. K. Ali, S. A. Najem, A. M. and Ghaima, K. K. (2013) Effect of some factors on biosorption of lead by dried leaves of water hyacinth (EichhorniaCrassipes). *International Journal of Pure and Applied Sciences and Technology, 17*(2):72-78.

[4] Langmuir, I. Adsorption of gases on glass, Mica, and Platinum. (1918). *Journal of the American Chemical Society, 40*(9): 1361-1403.

[5] Freundlich H. Veberdie adsorption in loesungen (adsorption in solution). (1907). *Z. Phys. Chem., 57*: 385-385.

[6] Baral, S. S. Das, N. Roy Choudary, G. Das, S. N. (2009). A preliminary study on the adsorptive removal of Chromium (VI) using seaweed, HydrillaVerticillata. *Journal of Hazardous Materials, 171*: 358-369.

[7] Weber, W. J., Jr. Morris, J. C. (1963). Kinetics of adsorption on carbon from solution. *Journal of Sanitary Engineering Division AMSE, 89*: 31-59.

[8] Hamdaoui, O. and Naffrechoux, E. (2007). Modeling of adsorption isotherms of phenol and chlorophenols onto granular activated carbon. Part I. Two parameter models and equations allowing determination of thermodynamic parameters. *Journal of Hazardous Materials, 147*: 381-394.

[9] Zouboulis, A. L. Matis, K. A.and Hancock, I. C. (1997). Biosorption of metals from dilute aqueous solutions. *Separation and Purification Methods, 26*(2): 255-295.

[10] Ames Crosta Mills & Company Ltd. and J. R. Sanderson (1973) *Apparatus for the biological treatment of wastewater by the biosorption process.* Great Britain Patent GB1324358.

[11] Kapoor, A. and Vijayraghavan, T. (1998). Biosorption of heavy metals on Aspergillus Niger: effect of pretreatment. *Bioresource Technology 63*(2): 109-113.

[12] Brady, D. Stoll, A. D. Starke, L. and Duncan, J. R. (1994). Chemical and enzymatic extraction of heavy metal binding polymers from isolated cell walls of Saccharomyces cerevisiae. *Biotechnology and Bioengineering, 44*(3): 297-302.

[13] Padma, D. V. and Venkata Ananta Rama Sastry, S. (2020). Kinetic studies on simultaneous biosorption of divalent copper and hexavalent chromium using mallet flower leaf powder. *Journal of Water and Environmental Nanotechnology, 5*(3): 204-217.

[14] Padma, D. V. and Sastry, S. V. A. R. (2022). Biosorption of divalent copper from aqueous solution using mallet flower leaves powder as adsorbent. *International Journal of Environment and Waste Management 29*(4): 444-458.

[15] Padma, D. V. and Sastry, S. V. A. R. (2021). Kinetics, Isotherms and Thermodynamic Studies on Removal of Divalent Copper using Mallet Flower Leaf Powder as Bio-Adsorbent. *Trends in Sciences 18*(24): 1431-1431.

[16] Sastry, S. V. A. R. Rao, B. S. Padma, D. V. (2020). Studies on Selective Batch Adsorption of Cu(II) and Cr(VI) from aqueous solution.*TEST Engineering and Management 83*: 6794 – 6797.

[17] Padma, D. V. and Sastry, S. V. A. R. (2020). Biosorption of Hexavalent Chromium using Mallet Flower Leaves Powder as adsorbent, *TEST Engineering and Management 83*:15714- 15729.

Chapter 37

The Effect of Clay Idols and Plaster of Paris (Pop) Immersion on Water Quality

Hritik Sachan[1]
Prateek Mishra[2]
Vartika Nishad[3]
Shravan Kumar[3,*]
and S. V. A. R. Sastry[4]

[1]Department of Biotechnology and Biochemical Engineering, Indian Institute of Technology, Kharagpur, West Bengal, India
[2,3]Department of Biochemical Engineering, Harcourt Butler Technical University, Kanpur, Uttar Pradesh, India
[4]Department of Chemical Engineering, Harcourt Butler Technical University, Kanpur, Uttar Pradesh, India

Abstract

India is one of the many countries where God is worshipped via sculptures and idols. Plaster of Paris, gypsum, clay, textiles, iron fillings, bamboo, and paints for embellishment are the most commonly used raw materials in the creation of these sculptures and idols. Broken or damaged idols should not be dumped in the open but rather submerged in water bodies, according to Hindu ceremonies and customs. This immersion includes not only damaged or shattered idols but also several festivals, such as Durga Pooja and Ganesh Chaturthi. Immersion of these plaster of Paris idols is capable of causing significant alterations in the water quality at the immersion site, including eutrophication, acidification, and heavy metal toxicity (including mercury, cadmium, arsenic, zinc, and lead). Although the government is attempting to prohibit the production of POP idols, this is impossible because idol production is a source of income for thousands of individuals.

Keywords: plaster of Paris, Clay Idols, heavy metals, paints

---

* Corresponding Author Email: hbti.shravan@gmail.com.

In: Sustainability in Chemical Processes through Digitalization and Green Chemistry Approaches
Editors: Susarla V. A. R. Sastry, Shravan Kumar and Ashish Kapoor
ISBN: 979-8-89113-257-3

1. Introduction

All living creatures, including humans, as well as the environment, are harmed by water pollution, which has turned into a threat to aquatic ecosystems. Pollution affects the physical, chemical, and biological properties of water [1]. The river provides a significant source of water for daily human activity. River water is treasured in India as a precious gift from nature in addition to being used for daily activities, and several religious ceremonies are held along its banks. Water is necessary for all forms of life, and it sustains plants, animals, and aquatic organisms to support human needs such as agriculture and industry [2].

India is a place where several festivals of various cultures and religions are celebrated at the same time. Durga Puja, Ganesh Puja, Tazia, and Laxmi Puja are just a few of India's important holidays. Idols are images that are used to worship God or Goddess. These festivities might last one day, three days, five days, seven days, or 10 days [3]. The idols are submerged in the Lentic and Lotic bodies after they have worshipped for a period of time. Lentic refers to a static body of water, such as ponds, lakes, or marshes, whereas Lotic refers to a moving body of water, such as rivers, streams, or brooks. Freshwater bodies Lentic and Lotic are both impacted by the weather and climate [4].

Ganesh Chaturthi is one of Hinduism's most sacred celebrations. This celebration honors the birth of Lord Ganesh and is mostly commemorated in Gujarat, Maharashtra, and other regions of India. Every year, numerous Ganesh statues of various sizes are submerged in the city's numerous water bodies as part of Surat's extravagant Ganesh Chaturthi celebration [5].

The festival's biggest environmental impact comes from the disposal of plasters of Paris idols in lakes, rivers, and the ocean [6]. When a plaster of Paris idol is submerged in water, it turns into gypsum, which slowly adds a significant amount of decaying material to the water and raises its hardness. These two changes reduce the water's capacity to support life and its quality, which has detrimental environmental effects on any water body's coastal ecology or ecosystem [7].

Plaster of Paris (POP), a less expensive and lighter substance, has become the preferred medium for making Ganesh idols [8]. After the idol is submerged, POP is coupled with the water and settles down after a few minutes due to expansion, smoothing out the water's bottom [9]. POP contains a variety of chemicals, such as phosphorus, gypsum, sulfur, and magnesium. Plastic and thermocol are used to embellish these statues. Thermocol and plastic, out of all the components used to make the idol, are nonbiodegradable and hence harmful [10].

Above all, harmful substances, including lead, mercury, cadmium, arsenic, zinc, and chromium, are included in the pigments used to paint these statues. Strong carcinogens found in the red, orange, blue, and green hues include zinc oxide, mercury, lead, and chromium [11]. The toxic exposure of the general public to these lethal chemicals and heavy metals used in idol production is now being scrutinized more than ever by authorities and civil society groups [12]. Paints are a higher source of hazard in terms of health effects, and the majority of those used for idol decoration are chemical-based [13]. They are high in neurotoxic and nephrotoxic heavy metals, including mercury, cadmium, and lead [14]. These metals are bio accumulative, which means that once they enter marine life forms such as fish, they go up the food chain and eventually wind up in the food we eat. In addition, the brighter the color, the more poisonous it is [15]. Mercury, zinc oxide, chromium, and lead are all known to be higher in red, blue, orange, and green hues. A single drop of mercury on a person's skin has the potential to be lethal [16].

2. Diverse Cultural and Religious Festivals are Organized in India

India is a land of rituals, and its people believe in and adhere to them religiously. The majority of the rites take place along the banks of rivers or other bodies of water. One of the most important Hindu festivals is "Ganesh Chaturthi. Durga Puja, Ganesh Puja, Tazia, and Laxmi Puja are just a few of India's important holidays. Idols are images that are used to worship God or Goddess. These festivities might last one day, three days, five days, seven days, or 10 days [17]. Ganesh Chaturthi is one of Hinduism's most sacred celebrations. This festival commemorates Lord Ganesh's birth and is mostly observed in Maharashtra, Karnataka, Gujarat, and other parts of India [18].

Immersion of idols and puja offerings in rivers has little impact on water quality because these pujas are often performed during the monsoon or postmonsoon seasons, when river flows are at their maximum [19]. The risk of water contamination after immersion of idols in water bodies has increased due to the increase in the quantity and size of idols, the use of materials other than clay to create idols, the use of synthetic paints and varnishes rather than natural dyes to decorate idols in recent years, and the decrease in river flow [20].

In addition, the scattering of leftovers after puja, such as vastras on flowers and ornamental items made of paper and plastic, into water bodies during the immersion procedure has raised the potential for pollution. As a result, the aquatic environment at the immersion locations, as well as downstream, may be impacted [21]. The colors, chemicals, and other materials used to make the idols dissolve when they are submerged drastically alter the water's quality [21].

3. Composition and Ingredients Used in Making Idols Such as Paints

When the idols are submerged, the dyes, chemicals, and other materials used to make the idols dissolve significantly alter the water's purity [22]. These idols are now most often made of plaster of Paris (POP), a less expensive and lighter material. POP contains a variety of chemicals, such as gypsum, magnesium, phosphorus, and sulfur. Plaster of Paris is made up of calcium and sulfate and is derived from gypsum rock [23]. After idol immersion, POP is coupled with water and settles down after a few minutes due to expansion, smoothing the water's bottom. POP takes several years to fully dissolve in water because of the presence of calcium sulfate hemihydrate, which releases poisonous components and decreases oxygen levels in the water body, killing fish and other aquatic organisms. Heavy metals such as lead, zinc, cadmium, and mercury are present in the chemical paints used to embellish the idol, which contribute to the water's acidic character and heavy metal concentration. Other decorations and worship items tossed in the river, such as thermocol, plastic flowers, fabric, incense, and camphor, add stress to the already polluted water. The settling of biodegradable and nonbiodegradable waste is affected by the immersion of idols in water bodies after the rainy season [24].

4. Adverse Effects of Using Heavy Metals in Paints

Chemical colors, such as copper, nickel, cadmium, lead, chromium, and mercury, are commonly employed. If deposited in large amounts, POP renders water alkaline. Water hardness rises when calcium and magnesium concentrations rise [25]. Varnish, watercolors, and other paints can cause considerable changes in water quality following immersion. Heavy elements such as mercury, cadmium, arsenic, chromium, lead, and zinc are included in the paints used to color these idols. Mercury, zinc oxide, chromium, and lead, in particular, are strong carcinogens found in the color's red, blue, orange, and green. Through a process known as bioaccumulation and biomagnification, lead and chromium become extremely harmful to humans, even in very small amounts.

4.1. Mercury (Hg)

Significant health risks are associated with mercury contamination, especially for children and pregnant women [26]. When humans eat infected fish, they risk absorbing deadly levels of mercury. Mercury is difficult to detect and cannot be prevented by clipping off the skin or other sections of the fish because it is odorless and undetectable and accumulates in the meat. Mercury can influence a child's development even at modest levels, putting off walking and talking, becoming less attentive, and getting into academic trouble. High amounts of mercury exposure during pregnancy and infancy can result in mental retardation, cerebral palsy, deafness, and blindness [27]. Mercury poisoning in adults can affect fertility and blood pressure control and cause memory loss, loss of vision, tremors, and numbness in the fingers and toes.

4.2. Cadmium (Cd)

Cadmium is a byproduct of zinc manufacturing. According to Godte et al. (2006), cadmium accumulates most dangerously in the proximal tubular cells of the kidney, where it causes the most damage. Cadmium can also cause bone demineralization, either directly or indirectly, as a result of renal failure. The proximal tubular cells of the kidney, where cadmium is most dangerous, are particularly vulnerable [28]. High cadmium water consumption irritates the stomach, resulting in vomiting, diarrhea, and, in some cases, death. Consuming lower quantities of cadmium over an extended period of time might cause cadmium build-up in the kidneys [29]. Cadmium in the kidney causes renal disease as well as bone damage if the levels rise to a sufficient level [30].

4.3. Arsenic (As)

One of the most toxic substances on earth is arsenic. According to Abdul et al. (2015), inorganic arsenic exposure can cause inflammation of the stomach and intestines, a reduction in the production of white and red blood cells, alterations to the skin, and irritation of the lungs. Large ingestions of inorganic arsenic are thought to enhance the risk of cancer, especially skin, lung, liver, and lymphatic cancers [31]. Women who are poisoned with inorganic arsenic may experience infertility and miscarriages, as well as skin rashes, lowered infection resistance, heart issues, and brain impairment. DNA can be damaged by inorganic arsenic [32].

4.4. Zinc (Zn)

Heavy metal poisoning can be caused by excessive levels of zinc and lead. According to the National Institutes of Health, the most prevalent side effects of zinc poisoning are gastrointestinal problems [33]. Vomiting, diarrhea, and an upset stomach are the most typical signs. After taking considerable doses of zinc, these symptoms, especially nausea and vomiting, can start to manifest as early as 30 minutes later [34]. The main reason for hospitalization for zinc poisoning is decreased urine output, which has also been associated with high zinc dosages [35]. Along with vomiting and nausea, this can also result in sleepiness, a metallic taste in the tongue, convulsions, shortness of breath, low blood pressure, and even shock [36]. Fluids such as water or milk are used to flush out heavy metal toxicity in an emergency.

4.5. Lead (Pb)

Lead is a poisonous chemical that can impair people's health in a variety of ways. Lead harms the kidneys, the central and peripheral neurological systems, and the body's ability to regulate vitamin D levels. Lead inhibits the development of red blood cells. Seizures, comas, and deaths have all been linked to high levels of lead. A child may demonstrate developmental delays, an IQ drop, hyperactivity, learning impairments, behavioral issues, poor hearing, and stunted growth at lower levels of exposure.

5. Effects of Idol Immersion on Water Quality

Because it is integrated into the very fabric of life on earth, water sustains life. Due to uncontrolled urbanization, industry, and countless religious activities, water contamination has become a serious problem in recent years, and the world's water supply is now at risk of catastrophe. Major causes of pollution in religious rituals include the immersion of idols and their accessories, such as adorning materials, plastic flowers, cosmetic items, thermocol garlands, and other nonbiodegradable synthetic colors [37].

5.1. pH

The pH of all fluids is one of the most important chemical variables that explains the ecological aspects of aquatic systems. The addition of organic waste and Ganesh idol components to the water body throughout the study may have caused pH variations. The pH of water (lead, copper, cadmium, etc.) affects the solubility (amount that can be dissolved in water) and biological availability (amount that can be utilized by aquatic life) of chemical components such as nutrients (phosphorus, nitrogen, and carbon) and heavy metals. For instance, pH influences both the amount and kind of phosphorus available in the water as well as whether aquatic life can utilize it. The toxicity of heavy metals is dependent on how soluble they are. Metals are more hazardous because they are more soluble at lower pH levels [25].

5.2. Temperature

When discharged water is much warmer than natural water, it can have a negative impact on water quality. Thermal pollution has a number of serious implications, including decreased dissolved oxygen levels, fish mortality, and alien species influxes. The temperature of the water changes from 25.7 degrees Celsius before immersion to 28.4 degrees Celsius after immersion. Throughout the research period, there was a steady rise in temperature. Increased chemical and biological reactions in water, as well as reduced gas solubility, are caused by rising water temperatures [37].

5.3. Total Hardness

Total hardness is a crucial criterion for determining whether water is suitable for home, industrial, and other uses, even though it is not a pollution indicator. When submerged, it was extremely high [37].

5.4. BOD

BOD is the quantity of oxygen necessary for aerobic bacteria to biochemically oxidize the organic matter in water, lowering the concentration of dissolved oxygen in the water and preventing fish and other aquatic animals from surviving. The BOD results show a significant level of immersion [34].

5.5. Heavy Metal Concentration

The quantity of heavy metals in the water after 48 hours of immersion in lab conditions reveals an increase in lead and cadmium, both of which are hazardous and damaging to the aquatic ecosystem as well as humans. Chromium, zinc, and copper concentrations in the lab water were reduced from control water concentrations, possibly due to metal absorption in PoP or clay after continuous immersion in the water for 48 hours. The water does not show any toxicity in a bioassay test after 48 hours of immersion, indicating that the influence of heavy metals used in color manufacture does not reveal its effect in a short period of time but may be damaging to aquatic life and humans over time [33].

5.6. COD

COD is a pollution indicator, and a high value indicates an inflow of organic and inorganic components into water, such as chemical paints, oil, and grease. During immersion, there was an increase in the level. During the current research, it was discovered that there was an addition of idol ornaments and worship components, which are examples of pollutants [34].

6. Aquatic Life Is Impacted by Idol Immersion

The idols are painted with heavy metal-containing oil paints in a variety of colors that are nonbiodegradable, bioaccumulate and biomagnify along the food chain and are occasionally carcinogenic, neurotoxic, and nephrotoxic. Naturally, the production of synthetic paints in a variety of colors involves the use of heavy metals such as copper, iron, calcium, molybdenum, silicon, magnesium, arsenic, iron, lead, manganese, chromium, zinc, nickel, cadmium, and mercury. Following the immersion of idols, the concentration of these metals (mg/l) increases by a factor of ten. These metals accumulate in the food chain after being absorbed by aquatic flora and wildlife [35].

Conclusion

It has been proven that POP idols do not alter when submerged in river water and continue to function as contaminants over time. The current study indicates that throughout the idol immersion period, the pollution burden on river water increased significantly, which can be detrimental to the aquatic ecosystem. While it is impossible to completely stop this religious practice, there are steps that can be taken to minimize the harm that idol immersion causes to the environment and protect water resources. These steps include the strict application of federal and state laws and regulations, guidelines, and public awareness campaigns that support eco-friendly religious practices.

In the current experiment, the immersion of Ganapati idols had a detrimental effect on the water quality. Before and after Ganapati immersion, the result indicates incremental modifications. The amount of organic materials and pollutants in the water is increasing, which has an impact on the flora and wildlife. The main cause of water pollution is religious practices, with particular blame placed on plaster of Paris, clothing, iron rods, chemical colors, varnish, and paints used in idol creation, as well as the Nirmalya and other items immersed with the idol.

Despite extensive attempts to raise awareness and adopt preventive measures, the results have been inconclusive. The number of idols immersed in water steadily increases year after year. To identify the best eco-friendly solution, it is essential that we comprehend how Ganesh Chaturthi affects the environment. Since the idol immersion ritual is the source of the majority of water pollution, some people are now recommending a small alteration of this practice to avoid water pollution and environmental concerns.

References

[1] Watkar, A. M. and M. P. Barbate (2017) 'Effect of Idol Immersion on Water Quality of Chandrabhaga River in Nagpur,' *International Journal of Researches in Biosciences and Agriculture Technology*, V(1), pp. 19–21. doi:10.29369/ijrbat.2017.05.i.0008.

[2] Abdul, K. S. M. (2015) 'Arsenic and human health effects: A review.,' *Environmental toxicology and pharmacology*, 40 3, pp. 828–846.

[3] Awasthi, G. (2021) *Arsenic Toxicity: Effect on Human Health*.

[4] Bahadure, S. and Sangeetha, P. (2014) '*Sustainable Approach for Conserving and Managing for Urban water bodies*,' (July), pp. 1–15.

[5] Bernard, A. (2008) 'Cadmium & its adverse effects on human health.,' *The Indian journal of medical research*, 128(4), pp. 557–564.

[6] Bhat, N. (2015) '*Pollution Status of Bhoj Wetland before and after Immersion of Idols Pollution Status of Bhoj Wetland before and*,' 5(January 2012).

[7] Bhattacharya, S. (2014a) 'Effects of Idol Immersion on the Water Quality Parameters of Indian Water Bodies: Environmental Health Perspectives,' *International Letters of Chemistry, Physics and Astronomy*, 39(June 2016), pp. 234–263. doi:10.18052/www.scipress.com/ilcpa.39.234.

[8] Bhattacharya, S. (2014b) 'Effects of Idol Immersion on the Water Quality Parameters of Indian Water Bodies: Environmental Health Perspectives,' *International Letters of Chemistry, Physics and Astronomy*, 39, pp. 234–263. doi:10.18052/www.scipress.com/ilcpa.39.234.

[9] Bhattacharya, S. (2014c) '*Hydrology & Meteorology*,' 5(4), p. 7587.

[10] Billore, D. K. and Dandawate, M. (2015) 'Environmental Impact of Idol Immersion on Kakerpura Lake Mhow,' *International Journal of Research -GRANTHAALAYAH*, 3(9SE), pp. 1–4. doi:10.29121/granthaalayah.v3.i9se.2015.3133.

[11] Bose-O'Reilly, S. (2010) 'Mercury exposure and children's health.,' *Current problems in pediatric and adolescent health care*, 40(8), pp. 186–215. doi:10.1016/j.cppeds.2010.07.002.

[12] Burile, A. N. (2018) '*Effect of Idol Immersion on Water Quality of Gandhisagar Lake At Nagpur*.'

[13] Chasapis, C. (2011) 'Zinc and human health: An update,' *Archives of toxicology*, 86, pp. 521–534. doi:10.1007/s00204-011-0775-1.

[14] Dwivedi, S., Mishra, S. and Tripathi, R. D. (2018) 'Ganga water pollution: A potential health threat to inhabitants of Ganga basin,' *Environment International*, 117, pp. 327–338. doi: https://doi.org/10.1016/j.envint.2018.05.015.

[15] Godt, J. (2006) 'The toxicity of cadmium and resulting hazards for human health,' *Journal of Occupational Medicine and Toxicology*, 1(1), pp. 1–6. doi:10.1186/1745-6673-1-22.

[16] Gupta, A. and Chakravarty, A. (2020) *Arsenic Effect On Human Health : A Review*.

[17] Gupta, R., Jana, R. and Kumar, S. (2020) '*Water Pollution Due To Idol Immersion In Water Bodies Of Kolkata*,' Pp. 5423–5427.

[18] Hayet, S. (2021) 'Hemato-Biochemical Profile Of Turkey birds selected from Sherpur district of Bangladesh,' *Int. J. Adv. Res. Biol. Sci*, 8(6), pp. 1–5. doi:10.22192/ijarbs.

[19] Hong, Y. S., Song, K. H. and Chung, J. Y. (2014) 'Health effects of chronic arsenic exposure.,' *Journal of preventive medicine and public health = Yebang Uihakhoe chi*, 47(5), pp. 245–252. doi:10.3961/jpmph.14.035.

[20] Iqbal, K. and Asmat, M. (2012) 'Uses and effects of mercury in medicine and dentistry.,' *Journal of Ayub Medical College, Abbottabad : JAMC*, 24(3–4), pp. 204–207.

[21] Iyer, K. (2019) '*Plaster of Paris*,' in, pp. 527–533. doi:10.1007/978-3-030-15089-1_22.

[22] Jaishankar, M. (2014) 'Toxicity, mechanism and health effects of some heavy metals.,' *Interdisciplinary toxicology*, 7(2), pp. 60–72. doi:10.2478/intox-2014-0009.

[23] Järup, L. (2003) 'Hazards of heavy metal contamination.,' *British medical bulletin*, 68, pp. 167–182. doi:10.1093/bmb/ldg032.

[24] Lokhande, D. P. A. (2019) 'The Effect of Ganesh Idol Immersion on the Water Quality of Gorai Jetty, Mumbai - the Environmental Health Perspective,' *International Journal of Trend in Scientific Research and Development*, Volume-3(Issue-3), pp. 398–402. doi:10.31142/ijtsrd22806.

[25] Mahurpawar, M. (2015) 'Effects of Heavy Metals on Human Healtheffects of Heavy Metals on Human Health,' *International Journal of Research -GRANTHAALAYAH*, 3(9SE), pp. 1–7. doi:10.29121/granthaalayah.v3.i9se.2015.3282.

[26] Plum, L. M., Rink, L. and Haase, H. (2010) 'The essential toxin: impact of zinc on human health.,' *International journal of environmental research and public health*, 7(4), pp. 1342–1365. doi:10.3390/ijerph7041342.

[27] Pollution, S. and Board, C. (2019) '*Impact Of Idol Immersion On Water Quality Of Aquatic Bodies - 2019.*'

[28] Prasad, A. S. (2008) 'Zinc in Human Health: Effect of Zinc on Immune Cells,' *Molecular Medicine*, 14, pp. 353–357.

[29] Rafati Rahimzadeh, Mehrdad (2017) 'Cadmium toxicity and treatment: An update.,' *Caspian journal of internal medicine*, 8(3), pp. 135–145. doi:10.22088/cjim.8.3.135.

[30] Saha, J. C. (1999) 'A review of arsenic poisoning and its effects on human health,' *Critical Reviews in Environmental Science and Technology*, 29(3), pp. 281–313. doi:10.1080/10643389991259227.

[31] Sankhla, M. S., Kumar, R. and Prasad, L. (2019) 'Zinc Impurity in Drinking Water and Its Toxic Effect on Human Health,' *Indian Internet Journal of Forensic Medicine & Toxicology*, 17(4), p. 84. doi:10.5958/0974-4487.2019.00015.4.

[32] Sarbari, N. and Pande, P. K. (2015). Effect of idol immersion on water quality of Yamuna River in Delhi and its potential influence on ground water quality. *Ind. J. Geo. Mar. Sci*. 44: 1545- 1553.

[33] Shukla, N. P. (2014a) 'Study of the Impact of Plaster Of Paris (Pop) and Clay Idols Immersion in Water,' *International Journal of Scientific Engineering and Technology*, 864(7), pp. 2277–1581.

[34] Shukla, N. P. (2014b) 'Study of the Impact of Plaster Of Paris (Pop) and Clay Idols Immersion in Water,' *International Journal of Scientific Engineering and Technology*, (7), pp. 2277–1581.

[35] Tchounwou, P. B. (2012) 'Heavy metal toxicity and the environment.,' *Experientia supplementum (2012)*, 101, pp. 133–164. doi:10.1007/978-3-7643-8340-4_6.

[36] Ujjania, N. C. and Patel, M. S. (2015) 'Ganesh idol immersion and its impact on water quality of Tapi River, India,' *Journal of Environmental Science and Water Resources*, 1(November), pp. 231–235.

[37] Vaghani, M. M. (2016) 'Study of Idol Immersion in Water Bodies,' (March), pp. 478–482.

Chapter 38

An Assessment of Methane Generation from Municipal Solid Waste Using the LandGEM Model for Kanpur City, India

**Shubham Chandra*
and Rajiv Ganguly**
Department of Civil Engineering, Harcourt Butler Technical University, Kanpur, Uttar Pradesh, India

Abstract

Municipal solid waste disposal is an enormous challenge for a developing country such as India due to its growing population, rapid urbanization, and increase in waste generation. The practice of open dumping of municipal solid waste in India leads to major global warming issues due to the presence of methane and carbon dioxide in landfill gases. This chapter assesses the potential of methane generation from MSW determined in Kanpur city by using the LandGEM 3.02 version model developed by USEPA for the duration of 2015–2020. It was calculated that the average and total methane generation for the study period were 228.89 x 10^4 m^3/year and 1373.35x10^4 m^3, respectively. The default input parameters for the values of k and L_0 were modified to be more representative of the study conditions. The k and L_0 values used for the study were 0.05 and 38 m3/Mg, respectively. It was further observed that the methane generation potential from MSW increased by a factor of 7.72 for 2020 compared to the base year of 2016. This suggests that enormous volumes of methane gases are emitted from the open dumpsite in Kanpur, which should be collected and harnessed for fuel and electricity production in Kanpur city.

Keywords: municipal solid waste, methane emissions, LandGEM 3.02

---

* Corresponding Author's Email: sc862278@gmail.com.

In: Sustainability in Chemical Processes through Digitalization and Green Chemistry Approaches
Editors: Susarla V. A. R. Sastry, Shravan Kumar and Ashish Kapoor
ISBN: 979-8-89113-257-3

1. Introduction

Numerous climate changes are brought on by the substantial rise in municipal solid waste generation in India. Uncontrolled municipal waste disposal at open landfills and dump sites is currently a major challenge for municipal authorities. Methane, carbon dioxide, and a very small amount of other gases are produced as a result of these unmanaged waste disposal methods, which have serious negative impacts on the environment. In tropical nations such as India, a large portion of the generated solid waste contains organic matter, which has a significant potential to emit greenhouse gases [1]. Methane gas produced during organic matter decomposition has a global warming potential approximately 20 times greater than that of carbon dioxide, which is the main greenhouse gas released as a result of open dumping.

2. Site Location

According to previous estimates, Kanpur City produces approximately 1500 metric tonnes of municipal solid waste every day, of which 80% is collected and dumped on site [2]. The only location used to dispose of MSW in the city of Kanpur is the Panki dump site. The dump site is located at 26°46'36.72"N and 80°34'51.46"E latitude and longitude, respectively. Approximately 4.6 km^2 of land is designated as a Panki dump site, which has operated for only 21 years. The recorded annual precipitation and temperature by the Municipal Corporation are 804 mm and 28 °C, respectively. There are no facilities for composting or covering landfills.

3. Methodology

3.1. MSW Generation in Kanpur City

Over the course of the study, the total amount of municipal solid waste produced is calculated as the product of Kanpur city's predicted population in a given year and the solid waste generation rate. Eighty percent of the total MSW produced in the city is dumped at the Panki open dumpsite [2], which is taken into account as the overall collection efficiency. In Table 1, the total MSW generation trends are shown.

3.2. Methane Emissions by LandGEM 3.02

The United States Environmental Protection Agency's (USEPA) LandGEM 3.02 modeling tool is used for quantifying greenhouse emissions and several other air pollutants found in municipal solid waste dumps. LandGEM 3.02 follows first-order decay decomposition and calculates the amount of landfill gas generation. The equation below shows the equation for the LandGEM 3.02 model.

$$QCH4 = \sum_{i=1}^{n}\sum_{j=0.1}^{1} kLo\frac{Mi}{10}e^{-ktij}$$

where Q_{CH4} = yearly methane generation (m^3/year), Lo = methane generation potential (m^3/Mg), I = annual time increase, J = the time increment, K = constant of methane generation, and Mi = mass of solid waste in i^{th} year (Mg).

4. Results and Discussion

4.1. Total Municipal Solid Waste Generated

The total quantity of MSW generated in Gg/year in every year is calculated by using the anticipated population and the annual rate of waste generation, i.e., from 2015 to 2020 [3]. The findings show that 350282 Mg of MSW was produced in Kanpur City in 2015. Furthermore, by taking into account the change in the MSW generation rate from 0.398 to 0.910 kg/capita/day for 2015 and 2020, respectively, as indicated in Table 1, the potential of MSW generation in the city increases by 2.37 times, from 350282 Mg in 2015 to 830160 Mg in 2020. Previous research carried out for Kanpur city (2010–2030) assumed a consistent growth rate of 0.6 kg/capita/day for the production of municipal solid waste [3].

Table 1. Total MSW generation in Kanpur City

S.no	Year of forecasting	Estimated population (persons)**	MSW generation rate kg/capita/day	Total MSW (Mg/year)	Total MSW reaching dumping site (Mg/year)
1.	2015	3015060	0.398	437852	350282
2.	2016	3036069	0.6	664899	531919
3.	2017	3058239	0.679	757939	606351
4.	2018	3080571	0.76	854550	683640
5.	2019	3100285	0.840	950547	760438
6.	2020	3123990	0.910	1037633	830107

** Population source https://worldpopulationreview.com/world-cities/kanpur-population).

4.2. Estimation of CH_4 Emissions

Landfill gases generated from MSW generally consist of four components: approx. 50% methane, 45–50% CO2, and a very small amount of NMOCs. In previous literature, it was found that there was no gas generation in the initial year of MSW deposition [3, 4]. By using LandGEM 3.02, the total volumes of LFG and CH_4 were 2746.70 $x10^4$ m^3 and 1373.35 $x10^4$ $m^{3,}$ respectively. The Lo default value is 170 m^3/Mg, which is site-specific for the USA; for India, the value of Lo is 38 m^3/Mg (for Kanpur) [4]. Annual landfill gas emissions and trends for duration (2015–2020) are shown in Table 2.

Table 2. Methane emission from MSW

S.no	Forecasted year	Estimated population (persons)**	Total waste reaching dumping site (Mg/year)	Total Landfill gas generation x 10^4 (m^3/year)	Methane emission x 10^4 (m^3/year)
1.	2015	3014060	350282	0	0
2.	2016	3036069	531919	130.16	65.08
3.	2017	3058239	606351	321.46	160.73
4.	2018	3080571	683640	531.10	265.55
5.	2019	3100285	760438	759.22	379.61
6.	2020	3123990	830107	1004.76	502.38
Total emission				2746.70	1373.35
Average emission				457.78	228.89

Conclusion

In this study, an attempt was made to quantify the methane generation potential for the Panki open dumping site (Kanpur), India, by using the LandGEM 3.02 model. The total landfill gas and methane emissions estimated for 2015 to 2020 by the LandGEM 3.02 model result in 2746.70 x10^4 m^3 and 1373.35x10^4 $m^{3,}$ respectively. The average rate of methane generation was 228.89 x10^4 m^3/year. Due to uncertainty associated with the value of k (ranging between 0.04 and 0.06), the variation in methane emissions was 1098.68x10^4 m^3 to 1648.02x10^4 m^3. It can be concluded that the collection of landfill gases helps Kanpur Municipal Corporation and other management authorities plan for an energy recovery system and efficient utilization of the gases generated.

References

[1] Mehra A. & Bhargava A. (2020). "Sustainable Municipal Solid Waste Management - A case study of Kanpur, India." *Journal of new Development in Chemistry*, Vol-3 Issue-1, 12-22. doi: 10.14302/issn. 2377-2549.jndc-20-3516.

[2] Kaushal A. and Sharma M. P., (2016). "Methane emission from Panki Open Dump Site of Kanpur, India." *Procedia Environmental Sciences*. 35:337-347. https://doi.org/10.1016/j.proenv.2016.07.014.

[3] Rana, R., Ganguly, R., & Kumar Gupta, A. (2017). Evaluation of solid waste management in 447 satellite towns of Mohali and Panchkula–India. *Journal of Solid Waste Technology and Management*, 43(4), 280-294.https://doi.org/10.5276/JSWTM.2017.280.

[4] Srivastava, A. N., & Chakma, S. (2020). Quantification of landfill gas generation and energy recovery estimation from the municipal solid waste landfill sites of Delhi, India. *Energy Sources*, Part A: Recovery, Utilization, and Environmental Effects, 1-14.

About the Editors

Dr. S. V. A. R. Sastry

Dr. S. V. A. R. Sastry has more than 19 years of teaching and research experience in the area of chemical engineering. He received his B.Tech from NIT Jalandhar, his MTech from IIT Delhi and his PhD from JNTU Kakinada, all were in chemical engineering with a special focus on renewable energy. He has authored 12 patents and 20 international books in the area of Renewable Energy, Unit Operations & Heat Transfer. He has published more than 50 research papers in the area of renewable energy & E-waste management in reputed international journals and presented papers in more than 40 international conferences. He is a member scientist of core group of scientists on tree borne oil seeds for promotion, utilization and socio-economic linkages in southern India. He is a senior professional member of institution of engineers, indian institute of chemical engineers, indian desalination association, international association of computer science and information technology and institute for engineering research and publication. He is a reviewer for international journals and permanent reviewer for three Elsevier journals (Energy Conversion & Management; Journal of Cleaner Production; AIChE). He has been awarded outstanding reviewer status by Elsevier, U.S. and granted full free access to SCOPUS by Elsevier, U.S. He is an editor for five international journals. He is an editorial board member for five international journals. His biography has been published in Marquis Who's Who in the World, 31st, 32nd & 33rd editions consecutively. He received the University Best Teacher Award from JNTU Kakinada in 2013. He is a recipient of international awards from IBC, Cambridge, England. He is selected for inclusion in "100 best educators of 2014", "2000 outstanding intellectuals of the 21st century", "cambridge certificate for outstanding educational achievement", and the prestigious "leading professionals of the world-2014" (published as online directory) by International Biographical Centre (IBC), Cambridge, England. He is giving consultation in the design and implementation of 'high speed industrial vaporizer', 'mixing device for chemical absorption process' and 'exploring disruptions, formulations, and technologies in lubricant industry' to Convetit, U.S. He has given more than 10 keynote/invited lectures in international conferences held at Dubai, China, Spain, Brazil and

Canada. He received the University Research Excellence Award for 2022-2023 at HBTU Kanpur. He received the Pillars of the Nation Award for Excellence and Innovation in Education Sector from the Delhi Vidhan Sabha and received the Outstanding Alumni Faculty Award from IIT Delhi in 2023.

Presently, he is the Associate Professor in the Department of chemical engineering; Associate Dean (Research & Development) and Associate Dean (International Student Affairs) at Harcourt Butler Technical University, Kanpur, Uttar Pradesh, India.

Dr. Shravan Kumar

Dr. Shravan Kumar has more than 5 years of teaching and industrial experience in the areas of chemical engineering and biochemical engineering. He received his B.Tech from HBTU Kanpur in Biochemical Engineering, his MTech and PhD, both in chemical engineering from IIT Guwahati, with a special focus on wastewater treatment and bioenergy. He has authored 4 patents. He has published more than 10 research papers in the area of separation technology in reputed international journals and presented papers in more than 15 international conferences. He is a life member of Indian institution of chemical engineers and, Indian desalination association. He is a reviewer for international journals. Presently, he is the Assistant Professor in the Department of biochemical engineering at Harcourt Butler Technical University, Kanpur, Uttar Pradesh, India.

Dr. Ashish Kapoor

Dr. Ashish Kapoor is currently working as Professor and Head, Department of Chemical Engineering, and Dean, International Student Affairs at Harcourt Butler Technical University, Kanpur, Uttar Pradesh, India. He received his BTech and MTech (dual degree) in Chemical

Engineering from the Indian Institute of Technology (IIT) Madras, Tamil Nadu, India. He further earned his PhD from the University of Illinois at Urbana-Champaign, Illinois, USA. Before undertaking his current roles, Dr. Kapoor gained professional experience as a Product Engineer at Lam Research Corporation, California, USA. Additionally, he served as a faculty member in the Department of Chemical Engineering at SRM Institute of Science and Technology, Kattankulathur, Tamil Nadu, India. Dr. Kapoor has contributed significantly to the scientific community with numerous peer-reviewed international publications, encompassing journal articles, conference proceedings, and book chapters. His research interests include microfluidics, lab-on-a-chip technologies, environmental remediation, biomass valorization, and modeling and simulation.

Index

A

B

C

D

E

F

G

H

I

L

M

N

O

P

R

S

T

V

W

Z